CLINICAL CYTOMETRY

ANNALS OF THE NEW YORK ACADEMY OF SCIENCES
Volume 468

CLINICAL CYTOMETRY

Edited by Michael Andreeff

The New York Academy of Sciences
New York, New York
1986

Library of Congress Cataloging-in-Publication Data

Clinical cytometry.

(Annals of the New York Academy of Sciences, ISSN 0077-8923 ; v. 468)
"The papers in this volume were presented at a conference entitled Clinical cytometry, which was held by the Engineering Foundation and the Society for Analytical Cytology on December 7-12, 1983, at the Cloister, Sea Island, Georgia"—Contents p.
Includes bibliographies and index.
1. Flow cytometry—Diagnostic use—Congresses.
2. Cancer cells—Identification—Congresses. 3. Flow cytometry—Congresses. I. Andreeff, Michael, 1943-
II. Engineering Foundation (U.S.) III. Society for Analytical Cytology (U.S.) IV. Series. [DNLM: 1. Flow Cytometry—congresses. W1 AN626YL /
QH 585.5.F56.C641 1983]
Q11.N5 vol. 468   500 s   86-12844
[RC270.3.F56]   [616.07'582]
ISBN 0-89766-331-4
ISBN 0-89766-332-2 (pbk.)

A/PCP
Printed in the United States of America
ISBN 0-89766-331-4 (cloth)
ISBN 0-89766-332-2 (paper)
ISSN 0077-8923

ANNALS OF THE NEW YORK ACADEMY OF SCIENCES

Volume 468
June 15, 1986

CLINICAL CYTOMETRY[a]

Editor
MICHAEL ANDREEFF

CONTENTS

[a] The papers in this volume were presented at a conference entitled Clinical Cytometry, which was held by the Engineering Foundation and the Society for Analytical Cytology on December 7-12, 1983 at The Cloister, Sea Island, Georgia.

VI. Solid Tumor

VII. Bacteria, Chromosomes, and Sperm

Financial assistance received from:

- BECTON DICKINSON
- COULTER ELECTRONICS
- ENGINEERING FOUNDATION
- ORTHO INSTRUMENTS
- SPECTRA PHYSICS

Cytometry in the Clinical Laboratory: Quo Vadis?[a]

BRIAN H. MAYALL

Program for Analytical Cytology
Department of Laboratory Medicine
University of California, San Francisco
San Francisco, California 94143
and
Biomedical Sciences Division
Lawrence Livermore National Laboratory
Livermore, California 94550

INTRODUCTION

Cytometry is a technical discipline that is playing an increasingly significant role in the practice of the clinical laboratory. Over the past decade, it has evolved from being an esoteric novelty found in a few avant garde clinical laboratories to a mature discipline whose techniques are well understood, whose instrumentation is readily available, whose utility is clearly recognized by the biomedical research community, and whose importance to the clinical laboratory is becoming increasingly apparent. This importance is reflected by the many articles in the research and clinical literature that report studies using cytometry.

The series of conferences sponsored by the Engineering Foundation and the Society for Analytical Cytology devoted to "Cytometry in the Clinical Laboratory" is another measure of interest in the clinical applications of cytometry. These proceedings are from the second conference in this series. The first conference has been reported by Shapiro.[1]

In this paper, I will review some of the underlying principles of cytometry that make it singularly well suited to problems in the clinical laboratory, I will summarize some of the current areas of application, and I will attempt to prognosticate about future applications. This can be only a superficial overview; but as an overview, I hope it may provide a framework for the more detailed presentations of the other contributors to these proceedings.

Correspondence: Brian H. Mayall, M.D., Biomedical Sciences L-452, Lawrence Livermore National Laboratory, P.O. Box 5507, Livermore, CA 94550.

[a]A communication from the Program for Analytical Cytology, University of California. This work was performed under the auspices of the U.S. Department of Energy by the Lawrence Livermore National Laboratory under contract number W-7405-ENG-48.

GENERAL CONSIDERATIONS

Cytometry may be defined as the measurement of individual cells and cellular constituents. The operative word is individual. Each cell or cellular organelle is considered as an individual, distinct, discrete entity that is measured separately from all other cells or organelles in the sample. Cytometry concentrates on the individual cell as the measuring unit; but it allows measurement of multiple properties for each cell and it allows these measurements to be made rapidly on many hundreds or thousands of cells from a given sample. The measurements may range from simple properties (such as size as measured by volume, projected area, or mass) through measurements of biochemical constitution (such as DNA content, expression of surface antigens, etc.) to complex measures of physiologic status and enzymatic activity.

The principles underlying cytometry and its clinical applications are discussed in many comprehensive review articles,[2-8] in the proceedings of the Engineering Foundation Conferences on Automated Cytology,[9-12] and in the volumes *Flow Cytometry and Sorting*[13] and *Flow Cytometry IV*.[14] Current information on cytometry and its clinical applications can be found in articles published in *Cytometry*, in *Analytical and Quantitative Cytology*, and increasingly in hematologic, immunologic, oncologic and cytogenetic journals.

The power of cytometry is that it allows measurement of many hundreds or thousands of cells from a heterogeneous population. When the cells are all homogeneous for the property being investigated then cytometry has no advantage over the conventional techniques of biochemistry, but when the property is distributed heterogeneously among the cells then cytometry must be used.

For example, knowing the average amount of DNA per cell in a dividing tissue is not particularly useful, but knowing how the DNA is distributed among the individual cells of the tissue sample may be invaluable. Analysis of the histogram of DNA measurements answers questions such as the proportion of cells having a normal G_0/G_1 diploid DNA content, the presence or absence of a DNA aneuploid population of cells, or the proportion of cells in the S-phase of the cell cycle.[15]

Another example is estimation in a lymphocyte sample of the relative proportion of cells in the lymphocyte subsets that express one or another of the many surface antigens recognized by monoclonal antibodies. Cytometry can differentiate between global effects that affect all cells equally and heterogeneous effects that cause subpopulations of cells within the population to express significantly more or less antigen than anticipated. Techniques that pool information from all cells are poorly suited to such analyses.

Success in cytometry depends on mastery of four major technologies—sample preparation, use of probes and markers, instrumentation, and data display and analysis. Inattention to any of these four interrelated technologies may introduce artifacts and may lead to inaccurate and misleading results. Cytometry must be practiced with a clear understanding of all sources of potential error, and within a clinical or research environment that allows the validity of the cytometric results to be tested against the yardstick of clinical and epidemiological experience.

Sample Preparation

The first requirement of cytometry is that cells should be presented for measurement one at a time. For flow instrumentation, this means a monodisperse suspension; for image cytometry, it means that the system should be able to isolate the image of each cell from those of its neighbors. Not surprisingly, the major clinical applications of cytometry have been in hematology and hematologic malignancies. The cells of the blood are naturally monodisperse, and so are readily analyzed by flow cytometry or spread on a slide for image cytometry. The dispersion of solid tissues, and solid tumors in particular, pose problems that vary with type of tissue, tightness of cellular adhesion, and amount of noncellular matrix. General strategies can be followed to separate cells[16] and recipes have been developed for the reliable dispersion of cells in a wide range of tissues (see, for example, ref. 17).

For many years, it had been felt that fresh material was the key to successful cytometry. This posed a serious limitation on clinical applications because it meant that a tissue sample had to be designated for cytometric analysis at the time it was received. Recently, it has been shown that old slides can be destained and analyzed cytometrically,[18] that thin sections cut from old histology blocks can be measured to give important prognostic information,[19,20] and, most excitingly, that useful single cell suspensions for either flow or image cytometry can be obtained from thick sections cut from old paraffin-embedded material.[21] These developments are particularly relevant to the application of cytometry in clinical oncology. Cytometry now can be used in both retrospective and prospective studies that evaluate cytometric measurements against clinical outcome. It is no longer necessary to wait years to ascertain the putative prognostic powers of a new test.

Probes and Markers

The clinical utility of a measurement depends on the relevance of the property being measured. A marker is a specific cellular property of relevance. Sometimes it may be possible to measure it directly. More usually, it will have to be labeled with a probe and it is the probe that really is measured. Examples of probes include both absorptive and fluorescent dyes, labeled antibodies, and fluorogenic substrates for enzymes. The probe should be specific so that it reacts only with the marker of interest, the reaction should be stoichiometric so that the amount of probe relates directly to the amount of marker, and the product of the probe-marker reaction should be readily measured on the available cytometer.

Some of the more commonly measured cellular properties are listed in TABLE 1, which also indicates the appropriate cytometric approach. This list is by no means complete, but it serves to illustrate the diversity of cellular properties accessible to analysis by cytometry. Some of these properties already have been shown to be of clinical use and others hold much promise for the future.

Cellular DNA content probably is the most widely measured cellular property. It is a marker for both ploidy and cytokinetic state. The clinical utility of DNA measurements are greatly increased when they are combined with markers of cell physiology or phenotype, such as cellular protein,[22] RNA,[23,24] or cell surface antigens.[25] Other cellular markers can be probed cytometrically, for example, receptor-ligand interactions,[26] lectin binding,[27] and cellular cholesterol.[28] The commercial production of monoclonal antibodies is a most important development for clinical cytometry. Monoclonal antibodies provide pure probes of known and reproducible specificity for a wide variety of markers of cellular phenotype. Their clinical usefulness is clearly established in immunology and is extending to other applications.

TABLE 1. Some Cellular Properties Accessible to Cytometry[a]

Type of Cytometer	Flow	Image
Property measured		
Size (volume, mass)	+	+
Content (e.g., DNA, protein)	+	+
Surface antigens	+	+
Intracellular antigens	+	+
Motility	?	+
Enzymatic activity	+	?
Cytokinetics	+	?
Membrane potential	+	?
Intracellular pH	+	+
Membrane fluidity	+	−
Nuclear-cytoplasmic partitioning	?	+
Shape	?	+
Texture	?	+
Viability	+	+

[a]Legend: + Readily accessible; ? Marginally accessible; − Not accessible.

Cytometric Instrumentation

Flow and image cytometry are the two major technical approaches to cytometry.

In flow cytometry, cells are measured in a monodisperse suspension. The fluidics are arranged so that the cells flow singly past one or more measuring stations. The measurements may be electronic (as in the Coulter counter) or optical (as in most flow cytometers). Cells are analyzed at rates up to several thousand cells per second. Cytochemical or immunocytochemical probes are used as labels for the properties of interest. Typically only two or three properties are measured for each cell, but eight or more are possible with experimental systems. Flow cytometry includes the possibility of flow sorting in which cells that satisfy some predetermined criteria may be separated and collected. The sorted cells may then be examined morphologically or used for further biological or biochemical tests.

In image cytometry, cells are presented on a glass slide or other suitable substrate and are analyzed with an imaging system based on the optical microscope. The cell image can be partitioned between nucleus and cytoplasm, and a few or many properties may be measured on each cell. These properties may mimic the morphological descriptions of shape, size, texture, etc. used by pathologists but may also include measurements of content and function similar to those used in flow cytometry. Measurement rates tend to be much slower than in flow, typically being only a hundred or so cells per hour, but the amount of information obtained per cell may be much greater.

The differences between the two approaches are diminishing. Developmental flow systems image cells and developmental image systems process hundreds of cells per second. The two approaches should be viewed as complementary rather than competing technologies.[29] Flow is much faster, it allows cell sorting, it has stimulated the development of a wide range of quantitative cytochemical procedures, and it is able to measure multiple properties of each cell.[30] Image, however, mimics the pathologist more closely and so it seems more familiar and is easier to relate to prior experience.

Data Display and Analysis

Cytometric data represent distributional functions of cellular properties across a population of cells. Such distributions are most readily visualized as simple histograms for a single property, and as histogram surfaces, scatter plots, or contour plots when two properties are measured on each cell. These different displays are illustrated in the figures of the other contributors to these proceedings. It is easy, from such displays, to estimate the location of the peaks and the spread of the measurements. An unusual subpopulation of cells may appear as a peak or cluster of measurements in an unexpected location. When more than two properties are measured, then more complicated techniques are needed to display the data; these may include projecting pairs of measurements, using some of the measurements to select a subset of the data for display, or displaying new parameters that are derived from combinations of the original measurements.

Many methods are available for the description and formal analysis of cytometric data. These range from the simple relative count of cells whose measurements fall within a defined window, as in the differential leukocyte count or lymphocyte subset analysis, to the complex deconvolution of DNA histograms for ploidy and cytokinetic analysis and of two parameter measurements of chromosomes for flow karyograms.

Clinical Relevance

The first major clinical relevance of cytometry is describing heterogeneous populations of cells by the distributions of measurements. One or more properties may be measured to classify the cells into different phenotypes reflecting either normal differentiation or oncogenesis, while analysis of the distribution of properties may allow assessment of cytokinetic or physiologic status of the cells. Most current clinical use relies on such analyses.

The second major way in which cytometry is of clinical relevance is in the detection and characterization of rare events. Again, this application stems from the power of cytometry in measuring and characterizing cells rapidly. The potential of this ability is only just being appreciated. Possible applications include detection and characterization of fetal cells in maternal blood, of dysplastic and neoplastic cells in cervical cytology, and of blast cells in the peripheral blood or the cerebrospinal fluid of leukemia patients who may be in complete clinical remission.

CLINICAL ACCEPTANCE

Before cytometry or any other technique is adopted in the clinical laboratory, it first must prove its usefulness. To be useful, it should provide a clear benefit either by bringing automation to an existing test or by introducing a new approach that goes beyond existing tests.

In the first case, automation should bring all the advantages of machine analysis to an existing well-established test. The existing test is the standard against which the new test is evaluated. The advantages of automation should provide a real benefit that makes economic sense in the clinical laboratory. For example, automation of the white blood cell differential count uses, as the standard, manual counts by the hematology technician. First it had to be shown that automated systems can do the tests as well as the technician, but with the same, or improved, accuracy and precision. Then the criteria for automation become: "do machines do better, in the long run, than technicians?" Better has many meanings; it includes not only objectivity, reliability, reproducibility, accuracy, precision, sensitivity, and selectivity, but also lower cost in both the economic and health sense. Additional factors, including regulatory considerations, service support, patient load, and throughput, also are important. All these criteria have to be factored into the decision to automate an existing test.

In the second case, the new technique should allow us to go beyond an existing test. In many respects, this is the more exciting reason. It is the reason that applies to most of the emerging applications of cytometry in the clinical laboratory. All clinicians are conscious of ways in which they would like to have additional diagnostic information. Improved diagnosis inevitably leads to improved patient management. For example, knowledge of tumor phenotype can predict potential responders or nonresponders to a particular therapy.

It is exciting to go beyond existing procedures, but it also brings its own challenge. The utility of a new test has to be proven by clinical experience and quantified using the classical techniques of biostatistics and prospective epidemiology. Thus, although we are excited by the potential of new tests, we must assess their potential clinical utility in the light of rigorous criteria.

SOME CLINICAL APPLICATIONS

There are clearly many applications of cytometry in the clinical laboratory, some of which already exist and some of which are easy to foresee in the near future. They can be categorized as being in the fields of hematology, immunology, solid tumor and hemato-oncology, cytogenetics, microbiology, and other miscellaneous uses. Each one takes advantage of the unique capabilities of cytometry. I will review briefly some of these applications and I will try to predict developments in the near and more distant future (see TABLE 2).

Hematology and Immunology

Cytometry already is indispensable in the hematology laboratory, both in bringing all the benefits of automation to the routine tests, and in introducing new diagnostic capabilities that are accessible only through quantitative methods.

The *enumeration of red and white blood cells* was the first, and is to date, the most successful clinical application of cytometry. Blood cells are naturally a monodisperse suspension, and it is easy to flow them through an electronic or optical cytometer to get a simple count.

Flow cytometry is the appropriate and preferred method for counting erythrocytes. It provides counting precision because of the large number of cells counted, and it enables heterogeneities within the erythrocyte population to be

TABLE 2. Applications of Cytometry in the Clinical Laboratory[a]

Present Status Time to Clinical Utilization	Clinical Practice 0–2 years	Clinical Research 2–5 years	Experimental 5–? years
Hematology and Immunology			
RBC	++++	++++	++++
Differential WBC	+++	++++	++++
Reticulocyte	+	+++	++++
Thrombocyte	+	++	+++
Bone marrow analysis	+	++	+++
Immunology	++	+++	++++
Transplantation	+	++	+++
Oncology			
Automated cytology	+	++	+++
DNA ploidy	++	+++	++++
Cytokinetics	+	+++	++++
Multiple markers	+	+++	++++
Other			
Cytogenetics	+	++	++
Microbiology	+	++	++++
Semen analysis	+	++	++

[a]Legend: The score is an approximate measure of clinical acceptance of the test, and reflects both relative utilization and value of cytometry. + Possible utility shown in research laboratory; ++ Utility accepted in a few teaching hospitals; +++ Test generally accepted but not universally used; ++++ Test universally used as the method of choice.

detected and analyzed. It is useful for detecting changes in the distribution of erythrocyte sizes, e.g. anisocytosis, and in analyzing anemias.[31,32]

Most large laboratories also use either flow or image cytometry to do the differential white blood cell count. The image approach mimics the manual procedure, using the colors of the Romanovsky stain to differentiate among the types of leukocyte. However, the flow method uses enzymatic reactions to differentiate among the leukocytes types.[33] The latter approach measures large numbers of cells and so gives counting precision, which is particularly relevant in the analysis of leukopenias.[34] It also gives two-parameter scatterplots as output, thus allowing ready visualization of shifts and heterogeneities within the leukocyte classes.

Cytometry, particularly flow cytometry, can detect rarely occurring cells provided they are labeled specifically. This is important in detecting circulating malignant cells, as is discussed in the section on oncology. It also allows detection of fetal cells in the maternal circulation.[35] When flow cytometry is combined with flow sorting, rarely occurring cells may be both detected and then separated for further studies.[36]

The *reticulocyte count*, as done manually, is tedious, subjective, and because the normal incidence is low and large numbers of cells must be screened, statistically unreliable. New cytometric methods allow reticulocytes to be counted either by flow or image cytometers, thus allowing accurate counts to be obtained rapidly and easily.[37,38] This power should greatly increase the usefulness of the reticulocyte count.

Cytometry is being used increasingly for the analysis of *platelets*. They are easy to count in suspension using existing or specifically designed flow instruments.

New cytochemical probes, such as mepacrine, relate to platelet function. Thus it is probable that a cytometric test for platelet function will soon be available in the clinical laboratory.[39]

Analysis of *bone marrow* samples is an application in which cytometry is starting to be used with great promise. Such analysis includes detection of metastatic malignancies, diagnosis and monitoring of leukemias and related malignancies, and assessment of bone marrow status in a range of clinical conditions. The bone marrow is a complex and heterogeneous tissue. Each path of myeloid differentiation is represented by cells in various phases of the cell cycle and in various stages of maturation. However, simple measures of cellular DNA allow detection of DNA aneuploid stemlines associated with malignancy, and new markers of the cell cycle and of differentiation expand the potential utility of cytometry for the analysis of bone marrow.[40]

Evaluation of *immunologic status* is rapidly becoming a most important clinical application of cytometry, particularly flow cytometry.[41-44] A wide variety of monoclonal antibodies are available to the various differentiation antigens expressed by the cells of the immune system. Such antibodies provide reliable reagents for enumeration of the lymphocyte subsets. Major applications are in diagnosis of genetic and acquired immunodeficiency syndromes[45] and in transplantation medicine.[46-48]

Oncology

Malignant cells show genetic and phenotypic differences from normal cells. Many of these differences may be probed and quantified by cytometry and may be exploited to diagnose malignancy, to assess therapeutic responsiveness, and to estimate prognosis. Cytometry is accepted in clinical oncology, but its widespread use is limited by the relatively low specificity of currently accessible markers and by the difficulty of establishing the clinical relevance of cytometric results.

Cancer detection and diagnosis was an early and still is a major application of cytometry. Much of the initial interest was in automating the screening of samples from asymptomatic persons, such as screening gynecologic smears for abnormal cells. This turned out to be a much more difficult task than was originally anticipated. It is not difficult to detect malignant or premalignant cells, but it is extremely difficult to differentiate such cells from the rare artifact that may mimic such cells and occur at a similar frequency. Nevertheless, there are systems, based on either image or flow cytometry, that can screen gynecologic specimens and show acceptably low error rates.[49-51] The commercial viability of these systems is not clear. Commercially successful systems probably must await discovery of cancer markers that are more specific than those currently available.

A different approach to cancer screening may use cytometry to analyze changes in non-tumor cells present in samples taken for screening. Such changes have been found in the chromatin patterns of intermediate squamous cells in gynecologic samples that show malignant or premalignant changes.[52] Similarly, changes have been found in the nuclear chromatin texture of foam cells in nipple aspirate samples taken from women whose breasts show premalignant or malignant disease.[53]

Cytometric assessment of *DNA ploidy* of tumor stemlines is a major clinical application of cytometry in an increasing number of laboratories.

Oncogenesis is invariably associated with changes to the cell genome. These changes are readily identified by cytogenetic techniques and involve addition of supernumary chromosomes, loss of chromosomes, and creation of new chromosomes from balanced or unbalanced translocations. Occasionally, as in chronic myelogenous leukemia, the chromosomal rearrangement may be so nearly balanced that no change in cellular DNA content is detectable.[54] More usually, the aneuploidy of malignancy results in measurable changes in cellular DNA and can be reported as DNA aneuploidy.[55] DNA aneuploidy, when detected, is unquestionably the most universal marker of malignancy.[56] When DNA aneuploidy is found in clinically benign lesions, it is interpreted as presumptive evidence of premalignant changes.[57]

Caspersson developed the quantitative techniques of cytometry and appreciated their application to oncology. His laboratory became a leader in the measurement of DNA content and other properties of cells from human tumors.[58] Many other studies have reviewed the DNA ploidy status of a wide range of human malignancies and have related tumor DNA aneuploidy to clinical course, responsiveness to therapy, and patient prognosis.[2,3,6,59,60] The working hypothesis is that responsiveness to chemotherapy is an expression of tumor phenotype that can be predicted, in some tumors, from the measured DNA ploidy.[61]

The selectivity of DNA aneuploidy as a tumor marker allows malignant cells to be differentiated readily from nonmalignant cells and so to be detected even when they are present in low numbers. DNA cytometry can detect aneuploid leukemic cells in blood, bone marrow, and spinal fluid of patients in complete clinical remission.[62] Early detection permits early resumption of therapy while the malignancy still is undetectable by other methods.

Tumor cytokinetics is essential to understanding the effects of radiotherapy and chemotherapy and may assist in matching therapy to tumor responsiveness.[63]

DNA cytometry reflects the proliferative status of the tumor as well as the DNA ploidy of the stemline. The formal techniques of cytokinetic analysis may be used on well-behaved tumors to estimate from the DNA histogram the fraction of cells in the different compartments of the cell cycle.[15] Cytokinetic analysis has been useful in leukemias; it correlates with phenotype[64] and can monitor response to chemotherapy.[65,66] Cytokinetics also have been used to analyze lymphomas[67] and have been reported to have predictive value for these malignancies.[68]

Unfortunately, most solid tumors are not well behaved. They show local heterogeneities and multiple compartments of noncycling cells that make them poorly suited to DNA cytokinetic analysis.[23,69,70] Nevertheless, both sophisticated cytokinetic analyses and relatively crude analyses of histogram location and shape have been reported as useful in diagnosis, prognosis, and therapeutic management of many solid tumors, including breast,[71-75] kidney,[20] prostate,[76,77] lung,[78-80] bladder,[81] endometrium,[82] uterine cervix,[83] stomach,[84] and chondrosarcoma.[19] Analysis of location and shape of the DNA histogram has also proven useful in the early diagnosis of mycosis fungoides and Sezary syndrome.[85,86]

Cytokinetics has greatly helped our understanding of host and tumor response to therapy, but so far it has had little effect on patient management. In part, this is because of the difficulties inherent in estimating the cytokinetic parameters of tumors from their DNA histograms. This situation could change radically in the near future. Cytometric measurement of monoclonal antibodies to DNA substituted with bromodeoxyuridine greatly facilitates cytokinetic analysis.[87] These antibodies may be obtained from commercial sources. It is probable that a cytokinetic analysis, based on this approach, will soon become a part of the routine clinical workup of any patient who is a candidate for treatment by chemotherapy. When the cytokinetics of the tumor are known, then it is possible to match patient therapy with tumor biology.

Measurement of *multiple markers* allows simultaneous evaluation of tumor cell phenotype and genotype. The potential clinical utility of cytometry is greatly increased when measurements of DNA content are combined with those of one or more phenotypic markers. Combined DNA and RNA cytometry helps diagnosis, classification, and management of leukemias and is used routinely.[88,89] Leukemic cells also express various surface antigens. Many monoclonal antibodies to these surface antigens are available and can be used to characterize the leukemic phenotype. Cytometric detection and quantification of bound antibodies is frequently important for the diagnosis, prognosis, and treatment of leukemias.[41,90,91] Cytometry of lectin binding also is useful in detecting subpopulations of leukemic cells showing phenotypic differences.[92]

Many other markers of phenotypic status are being evaluated in specific situations. For example, quantification of a probe for estrogen receptors has been combined with DNA measurements in the cytometric analysis of breast cancers.[93] The technology for making multiple measurements is well in hand. Clinical acceptance depends on demostrating that the markers are relevant and that the probes are specific.

Other Applications

Chromosome analysis is a repetitive visual analysis task that appears ideally suited to automation using image cytometry. Various groups have been attempting to reach this goal for almost two decades. Part of the lack of success is because investigators had not appreciated the great range of normal variation in the morphology of human chromosomes and part is because the initial specifications for automated systems were overly ambitious. Image cytometry is well suited to the tasks of locating cells in metaphase and estimating the quality of detected spreads.[94] The techniques of image processing also are able to enhance contrast of faint and fuzzy chromosome images, to separate individual chromosomes, and to rearrange chromosomes into a karyogram. Computer assisted karyotype analysis is a viable proposition that is being introduced commercially. However, complete and fully automated chromosome analysis is not reliable and may never be practical because of the inherent complexities of this apparently simple task.

Image cytometry has been used as an adjunct to karyotype analysis and to measure the DNA content of individual chromosomes.[95,96] The measurements show unexpected differences among individuals. These differences reduce the usefulness of DNA measurements for automatic karyotype analysis. But they also may be considered as individual genotypic markers and may be exploited as such.

Flow cytogenetics introduces a powerful new technique for the analysis and isolation of chromosomes.[97–103] Gray and co-workers (personal communication) have shown that it can be applied on a routine basis for the analysis of fetal cells following amniocentesis and that the flow approach is as rapid as conventional cytogenetic analysis. But, because chromosomes from many cells are pooled together, there is a limit to the sensitivity of flow cytogenetics and to its ability to detect cytogenetic defects of clinical relevance. Thus, flow cytogenetics is a powerful research tool and will continue to be used in specific situations, but it is unlikely to be used routinely in the clinical laboratory.

Cytometry has been applied only recently to detection and characterization of *bacteria, viruses, micro-organisms, and intracellular parasites*.[104–109] This is surprising;

micro-organisms are discrete entities and many immunologic and other markers are available for their characterization. Cytometry requires only several thousand organisms in comparison to the millions or billions that are required for microbiological tests used routinely in the clinical laboratory. Cytometry has the potential to diagnose infection and characterize antibiotic sensitivity of pathogens in a fraction of the time currently required while largely avoiding problems of culture with the danger of overgrowth by irrelevant organisms. Microbiology, above all others, seems to be the area of greatest potential growth in the application of cytometry in the clinical laboratory.

Analysis of semen is a routine procedure in the evaluation of human infertility. Sperm are counted, their motility is assessed, and the fraction of abnormal sperm, based on their morphologic appearance, is determined. Cytometry is already used to count sperm and either flow or image cytometry may be used to assess sperm morphology.[110-112] Additionally, it is probable that cytometry may be used to assess the functional properties of the sperm using techniques to monitor enzymatic and antigenic status and the degree of chromatin compaction.[113] Semen analysis is never going to be a major clinical test, but it illustrates the range of clinical problems that can be addressed by cytometry.

QUO VADIS?

The present clinical applications of cytometry already are much greater than is generally realized and the prospect for the future is even brighter (see TABLE 2). Many additional tests based on new developments in monoclonal antibodies, stain technology, and molecular biology, and on advances in cytometric instrumentation and our understanding of the data will become available. The range of applications is growing daily, but some of the current limitations also should be recognized and addressed.

Cytometry is not yet a user-friendly technique. Samples have to be dispersed and stained manually. Cytometers, both flow and image, tend to be excessively complicated; they demand a skilled operator, they do not align and calibrate themselves automatically, and they generally are unable to analyze multiple samples and identify each one unambiguously. Data analysis and reporting is abysmal, rarely going beyond a computer printout that may be illegible, indecipherable, and impossible to incorporate into hospital records. None of these problems are insurmountable, but cytometry will never realize its full clinical potential until they are given more attention than in the past.

In this presentation, I have reviewed some of the major areas of clinical application. Others would have different lists, but all would agree that cytometry has a major and expanding role in the clinical laboratory and in the management of patients. It provides the clinician with objective and relevant information that otherwise is unobtainable. We all welcome the inevitable growth of cytometry in the clinical laboratory through both increased use of existing applications and introduction of new applications. But growth also involves increasing responsibilities for those who are active in the field. It is our responsibility to ensure that this growth builds on firm foundations, that we understand the scientific basis for our discipline, and that we set standards assuring excellence in practice. This is a challenge that I welcome as also do the other contributors to these proceedings.

The future for cytometry in the clinical laboratory is bright; it is a future whose horizons are expanding rapidly to offer great opportunities and great challenges. I believe that we are ready and eager to meet these challenges and to move into this future.

REFERENCES

1. SHAPIRO, H. M. 1982. Cytometry in the clinical laboratory. Cytometry 3:312–313.
2. BARLOGIE, B., M. N. RABER, J. SCHUMANN, T. S. JOHNSON., B. DREWINKO, D. E. SWARTZENDRUBER, W. GOHDE, M. ANDREEFF & E. J. FREIREICH. 1983. Flow cytometry in clinical research. Cancer Res. 43:3892–3997.
3. BRAYLAN, R. C. 1983. Attributes and applications of flow cytometry. Ann. Clin. Lab. Sci. 13:379–384.
4. GOERTTLER, K. & M. STOHR. 1982. Automated cytology. Arch. Pathol. Lab. Med. 106:657–661.
5. KRUTH, H. S. 1982. Flow cytometry: Rapid biochemical analysis of single cells. Anal. Biochem. 125:225–242.
6. LAERUM, O. D. & FARSUND T. 1981. Clinical application of flow cytometry; a review. Cytometry 2:1–13.
7. LOVETT, E. J., III, B. SCHNITZER, D. F. KEREN, A. FLINT, J. L. HUDSON & K. D. MCCLATCHEY. 1984. Application of flow cytometry to diagnostic pathology. Lab. Invest. 50:115–140.
8. TRAGANOS, F. 1984. Flow cytometry: Principles and applications. I. Cancer Invest. 2:149–163.
9. MAYALL, B. H., Ed. 1974. Automated cytology III. J. Histochem. Cytochem. 22:451–765.
10. MAYALL, B. H., Ed. 1976. Automated cytology IV. J. Histochem. Cytochem. 24:1–414.
11. MAYALL, B. H. & B. L. GLEDHILL, Eds. 1977. Automated cytology V. J. Histochem. Cytochem. 25:479–952.
12. MAYALL, B. H. & B. L. GLEDHILL, Eds. 1979. Automated cytology VI. J. Histochem. Cytochem. 27:1–641.
13. MELAMED, M. R., P. F. MULLANEY & M. L. MENDELSOHN, Eds. 1979. Flow Cytometry and Sorting. John Wiley & Sons, Inc. New York.
14. LAERUM, O. D., T. LINDMO & E. THORUD, Eds. 1980. Flow Cytometry IV. Universitetsforlaget. Bergen.
15. DEAN, P. M., J. W. GRAY & F. A. DOLBEARE. 1982. The analysis and interpretation of DNA distributions measured by flow cytometry. Cytometry 3:188–195.
16. THORNTHWAITE, J. T., E. V. SUGARBAKER & W. D. TEMPLE. 1980 Preparation of tissues for DNA flow cytometric analysis. Cytometry 1:229–237.
17. MCDIVITT, R. W., K. R. STONE & J. S. MEYER. 1984. A method for dissociation of viable human breast cancer cells that produces flow cytometric kinetic information similar to that obtained by thymidine labeling. Cancer Res. 44:2628–2633.
18. NASIELL, K., G. AUER, M. NASIELL & A. ZETTERBERG. 1979. Retrospective DNA analyses in cervical dysplasia as related to neoplastic progression or regression. Anal. Quant. Cytol. 1:103–106.
19. KREICBERGS, A. & A. ZETTERBERG. 1980. Cytophotometric DNA measurements of chondrosarcoma. Anal. Quant. Cytol. 2:84–92.
20. BENNINGTON, J. L. & B. H. MAYALL. 1983. DNA cytometry on four-micrometer sections of paraffin-embedded human renal adenocarcinomas and adenomas. Cytometry 4:31–39.
21. HEDLEY, D. W., M. L. FRIEDLANDER, I. W. TAYLOR, C. A. RUGG & E. A. MUSGROVE. 1983. Method for analysis of cellular DNA content of paraffin-embedded pathological material using flow cytometry. J. Histochem. Cytochem. 31:1333–1335.
22. CRISSMAN, H. A. 1981. Simplified method for DNA and protein staining of human hematopoietic cell samples. Cytometry 2:59–62.
23. DARZYNKIEWICZ, Z. & M. ANDREEFF. 1981. Multiparameter flow cytometry: Application in analysis of the cell cycle. Clin. Bull. 11:47–57.

24. SHAPIRO, H. M. 1981. Flow cytometric estimation of DNA and RNA content in intact cells stained with Hoechst 33342 and pryonin Y. Cytometry **2**: 143–150.
25. NORONHA, A. & D. P. RICHMAN. 1984. Simultaneous cell surface phenotype and cell cycle analysis of lymphocytes by flow cytometry. J. Histochem. Cytochem. **32**:821–826.
26. BOHN, B. 1980. Flow cytometry: A novel approach for the quantitative analysis of receptor-ligand interactions on surfaces of living cells. Mol Cell. Endocrin. **20**:1–15.
27. MALIN-BERDEL, J., G. VALET, E. THIEL, J. A. FORRESTER & L. GURTLER. 1984. Flow cytometric analysis of the binding of eleven lectins to human T- and B-cells and to human T- and B-cell lines. Cytometry **5**:204–209.
28. MULLER, C. P., D. A. STEPHANY, D. F. WINKLER, J. M. HOEG, S. J. DEMOSKY, JR. & J. R. WUNDERLICH. 1984. Filipin as a flow microfluorometry probe for cellular cholesterol. Cytometry **5**:42–53.
29. TANKE, H. J., A. M. J. VAN DRIEL-KULKER, C. J. CORNELISSE & J. S. PLOEM. 1983. Combined flow cytometry and image cytometry of the same cytological sample. J. Microsc. **130**:11–22.
30. SHAPIRO, H. M. 1982. Multistation multiparameter flow cytometry: A critical review and rationale. Cytometry **3**:227–243.
31. BESSMAN, J. D. , E. L. HURLEY & M. R. GROVES. 1982. Nondiscrete heterogeneity of human erythrocytes: Comparison of Coulter-principle flow cytometry and Soret-hemoglobinometry image analysis. Cytometry **3**:292–295.
32. BESSMAN, J. D., P. R. GILMER, JR. & F. H. GARDNER. 1983. Improved classification of anemias by MCV and RDW. Am. J. Clin. Pathol. **80**:322–326.
33. MANSBERG, H. P., A. M. SAUNDERS & W. GRONER. 1974. The hemalog D white cell differential system. J. Histochem. Cytochem. **22**: 711–724.
34. ROSS, D. W. & K. MCMASTER. 1982. Neutropenia: The accuracy and precision of the neutrophil count in leukopenic patients. Cytometry **3**:287–291.
35. CUPP, J. E., J. F. LEARY, E. CERNICHIARI, J. C. S. WOOD & R. A. DOHERTY. 1984. Rare-event analysis methods for detection of fetal red blood cells in maternal blood. Cytometry **5**:138–144.
36. WEIL, G. J., W. M. LEISERSON & T. M. CHUSED. 1983. Isolation of human basophils by flow microfluorometry. J. Immunol. Methods **58**:359–363.
37. TANKE, II. J., I. A. B. NIEUWENHUIS, G. J. M. KOPER, J. C. M. SLATS & J. S. PLOEM. 1981. Flow cytometry of human reticulocytes based on RNA fluorescence. Cytometry **1**:313–320.
38. SAGE, B. H., JR., J. P. O'CONNELL & T. J. MERCOLINO. 1983. A rapid vital staining procedure for flow cytometric analysis of human reticuloctyes. Cytometry **4**:222–227.
39. CORASH, L. & Y. MOK. 1983. Platelet suborganelle measurement using fluorescence activated flow cytometry. Blood **62**:253a.
40. HIDDEMANN, W., B. D. CLARKSON, T. BUCHNER, M. R. MELAMED & M. ANDREEFF. 1982. Bone marrow cell count per cubic millimeter bone marrow: A new parameter for quantitating therapy-induced cytoreduction in acute leukemia. Blood **59**:216–225.
41. AULT, K. A. 1983. Clinical applications of fluorescence-activated cell sorting techniques. Diag. Immunol. **1**:2–10.
42. LANDAY, A., G. L. GARTLAND, T. ABO & M. D. COOPER. 1983. Enumeration of human lymphocyte subpopulations by immunofluorescence: a comparative study using automated flow microfluoremetry and fluorescence microscopy. J. Immunol. Methods **58**:337–347.
43. LOKEN, M. R. & L. L. LANIER. 1984. Three-color immunofluorescence analysis of leu antigens on human peripheral blood using two lasers on a fluorescence-activated cell sorter. Cytometry **5**: 151–158.
44. LOKEN, M. R. & A. M. STALL. 1982. Flow cytometry as an analytical and preparative tool in immunology. J. Immunol. Methods **50**:85–112.
45. GEHA, R. S. & E. REINHERZ. 1983. Identification of circulating maternal T and B lymphocytes in uncomplicated severe combined immunodeficiency by HLA typing of subpopulations of T cells separated by the fluorescence-activated cell sorter and of Epstein Barr virus-derived B cell lines. J. Immunol. **130**:2493–2495.

46. COSIMI, A. B. 1983. Diagnostic and therapeutic applications of monoclonal antibodies to human T-cell subsets in renal transplant recipients. Urol. Clin. N. Am. **10**:289–299.

47. GAROVOY, M. R., M. A. RHEINSCHMIDT, M. BIGOS, H. PERKINS, B. COLOMBE & O. SALVATIERRA. 1983. Flow cytometry analysis: The ultimate crossmatch technique? Transplant. Proc. **15**:1939–1944.

48. WILLIAMS, J. M., R. LOERTSCHER, T. COTNER, M. REDDISH, H. M. SHAPIRO, C. B. CARPENTER, J. L. STROMINGER & T. B. STROM. 1984. Dual parameter flow cytometric analysis of DNA content, activation antigen expression, and T cell subset proliferation in the human mixed lymphocyte reaction. J. Immunol. **132**:2230–2237.

49. TANAKA, N., H. IKEDA, T. VENO, A. MUKAWA & K. KAMITSUMA. 1979. Field test and experimental use of CYBEST MODEL 2 for practical gynecologic mass screening. Anal. Quant. Cytol. **1**:122–126.

50. WHEELESS, L. L., S. F. PATTEN, T. K. BERKAN, C. L. BROOKS, K. M. GORMAN, S. R. LESH, P. A. LOPEZ & J. C. S. WOOD. 1984. Multidimensional slit-scan prescreening system: preliminary results of a single blind clinical study. Cytometry **5**:1–8.

51. ZAHNISER, D. J., P. S. OUD, M. C. T. RAAIJMAKERS, G. P. VOOYS & R. T. VAN DE WALLE. Field test results using the bioPEPR cervical smear prescreening system. Cytometry **1**:200–203.

52. WIED, G. L., P. H. BARTELS, M. BIBBO & J. J. SYCHRA. 1980. Cytomorphometric markers for uterine cancer in intermediate cells. Anal. Quant. Cytol. **2**:257–263.

53. KING, E. B., L. K. KROMHOUT, K. L. CHEW, B. H. MAYALL, N. L. PETRAKIS, R. H. JENSEN & I. T. YOUNG. 1984. Analytic studies of foam cells from breast cancer precursors. Cytometry **5**:124–130.

54. MAYALL, B. H., A. V. CARRANO, D. H. MOORE II & J. D. ROWLEY. 1977. Quantification by DNA-based cytometry of the 9q+/22q− chromosomal translocation associated with chronic myelogenous leukemia. Cancer Res. **37**:3590–3593.

55. HIDDEMANN, W., J. SCHUMANN, M. ANDREEFF, B. BARLOGIE, C. HERMAN, R. LEIF, B. MAYALL, R. MURPHY & A. SANDBERG. 1984. Convention on nomenclature for DNA cytometry. Cytometry **5**:445–446.

56. BARLOGIE, B., B. DREWINKO, J. SCHUMANN, W. GOHDE, G. DOSIK, J. LATREILLE, D. JOHNSTON & E. FREIREICH. 1980. Cellular DNA content as a marker of neoplasia in man. Am. J. Med. **69**:195–203.

57. STENZINGER, W., L. SUTER & J. SCHUMANN. 1984. DNA aneuploidy in congenital melanocytic nevi: suggestive evidence for premalignant changes. J. Invest. Derm. **82**:569–572.

58. CASPERSSON, T. 1979. Quantitative tumor cytochemistry. Cancer Res. **39**:2341–2355.

59. ATKIN, N. B. & R. KAY. 1979. Prognostic significance of model DNA value and other factors in malignant tumours, based on 1465 cases. Br. J. Cancer **40**:210–221.

60. FRANKFURT, O. S., H. K. SOCUM, Y. M. RUSTUM, S. G. ARBUCK, Z. P. PAVELIC, N. PETRELLI, R. P. HUBEN, E. J. PONTES & W. R. GRECO. 1984. Flow cytometric analysis of aneuploidy in primary and metastatic human solid tumors. Cytometry. **5**:71–80.

61. LOOK, A. T., F. A. HAYES, R. NITSCHKE, N. B. MCWILLIAMS & A. A. GREEN. 1984. Cellular DNA content as a predictor of response to chemotherapy in infants with unresectable neuroblastoma. New Engl. J. Med. **311**:231–235.

62. REDNER, A., M. ANDEEFF, D. R. MILLER, P. STEINHERZ & M. MELAMED. 1984. Recognition of central nervous system leukemia by flow cytometry. Cytometry **5**:614–618.

63. GRAY, J. W. 1983. Quantitative cytokinetics: Cellular response to cell cycle specific agents. J. Pharmac. Therapeut. **22**:163–197.

64. LOOK, A. T., S. L. MELVIN, D. L. WILLIAMS, G. M. BRODEUR, G. V. DAHL, D. K. KALWINSKY, S. B. MURPHY & A. M. MAUER. 1982. Aneuploidy and percentage of S-phase cells determined by flow cytometry correlate with cell phenotype in childhood acute leukemia. Blood **60**:959–967.

65. GOHDE, W., J. SCHUMANN, T. BUCHNER, F. OTTO & B. BARLOGIE. 1979. Pulse cytometry: Application in tumor cell biology and clinical oncology. *In* Flow Cytometry and Sorting. M. R. Melamed, P. F. Mullaney, M. L. Mendelsohn, Eds.:599–620. John Wiley & Sons, Inc. New York.

66. RICCARDI, A., G. MAZZINI, C. MONTECUCCO, R. CRESCI, E. TRAVERSI, C. BERZUINI & A. ASCARI. 1982. Sequential vincristine, arabinosylcytosine and adriamycin in acute leukemia: Cytologic and cytokinetic studies. Cytometry 3:104–109.

67. SHACKNEY, S. E., A. M. LEVINE, R. I. FISHER, P. NICHOLS, E. JAFFE, W. H. SCHUETTE, R. SIMON, C. A. SMITH, S. J. OCCHIPINTI, J. W. PARKER, J. COSSMAN, R. C. YOUNG & R. J. LUKES. 1984. The biology of tumor growth in the non-Hodgkin's lymphomas. J. Clin. Invest. 73:1201–1214.

68. BRAYLAN, R. C., L. W. DIAMOND, M. L. POWELL & B. HARTY-GOLDER. 1980. Percentage of cells in the S phase of the cell cycle in human lymphoma determined by flow cytometry. Correlation with labeling index and patient survival. Cytometry 1:171–174.

69. DARZYNKIEWICZ, Z., F. TRAGANOS & M. MELAMED. 1980. New cell cycle compartments identified by flow cytometry. Cytometry 1:98–108.

70. DETHLEFSEN, L. A., K. D. BAUER & R. M. RILEY. 1980. Analytical cytometric approaches to heterogeneous cell populations in sold tumors: A review. Cytometry 1:89–97.

71. AUER, G. U., A. G. FALLENIUS, K. Y. ERHARDT & B. S. B. SUNDELIN. 1984. Progression of mammary adenocarcinomas as reflected by nuclear DNA content. Cytometry 5:420–425.

72. AUER, G. U., T. O. CASPERSSON, & A. S. WALLGREN. 1980. DNA content and survival in mammary carcinoma. Anal. Quant. Cytol. 2:161–165.

73. AUER, G., E. ERIKSSON, E. AZAVEDO, T. CASPERSSON & A. WALLGREN. 1984. Prognostic significance of nuclear DNA content in mammary adenocarcinomas in humans. Cancer Res. 44:394–396.

74. EWERS, S-B., E. LANGSTROM, B. BALDETORP & D. KILLANDER. 1984. Flow cytometric DNA analysis in primary breast carcinomas and clinicopathological correlations. Cytometry 5:408–419.

75. RABER, M. N., B. BARLOGIE, J. LATREILLE, C. BEDROSSIAN, H. FRITSCHE & G. BLUMENSCHEIN. 1982. Ploidy, proliferative activity and estrogen receptor content in human breast cancer. Cytometry 3:36–41.

76. SEPPELT, U. & E. SPRENGER. 1984. Nuclear DNA cytophotometry in prostate carcinoma. Cytometry 5:258–262.

77. TRIBUKAIT, B., L. RONSTROM, & P-L. ESPOSTI. 1983. Quantitative and qualitative aspects of flow DNA measurements related to the cytologic grade in prostatic carcinoma. Anal. Quant. Cytol. 5:107–111.

78. BUNN, P. A., D. N. CARNEY, A. F. GAZDAR, J. WHANG-PENG & M. J. MATTHEWS. 1983. Diagnostic and biological implications of flow cytometric DNA content analysis in lung cancer. Cancer Res. 43:5026–5032.

79. ONO, J. & G. AUER. 1983. The significance of DNA measurements for the early detection of bronchial cell atypia. Cytometry 3:340–344.

80. TEODORI, L., D. TIRINDELLI-DANESI, F. MAURO, R. DE VITA, R. UCCELLI, C. BOTTI, C. MONDINI, C. NERVI & S. STIPA. 1983. Non-small-lung carcinoma: Tumor characterization on the basis of flow cytometrically determined cellular heterogeneity. Cytometry 4:174–183.

81. JAKOBSEN, A., S. MOMMSEN & S. OLSEN. 1983. Characterization of ploidy level in bladder tumors and selected site specimens by flow cytometry. Cytometry 4:170–173.

82. MOBERGER, B., G. AUER, G. FORSSLUND & G. MOBERGER. 1984. The prognostic significance of DNA measurements in endometrial carcinoma. Cytometry 5:430–436.

83. JAKOBSEN, A., P. B. KRISTENSEN & H. K. POULSEN. 1983. Flow cytometric classification of biopsy specimens from cervical intraepithelial neoplasia. Cytometry 4:166–169.

84. TEODORI, L, L. CAPURSO, E. CORDELLI, R. DE VIA, M. KOCH, M. TARQUINI, F. PALLONE & F. MAURO. Cytometrically determined relative DNA content as an indicator of neoplasia in gastric lesions. Cytometry 5:63–70.

85. BUNN, P. A., JR., I WHANG-PENG, D. N. CARNEY, M. L. SCHLAM, T. KNUTSEN & A. F. GAZDAR. 1980. DNA content analysis by flow cytometry and cytogenetic analysis in mycosis fungoides and Sezary syndrome. J. Clin. Invest. 65:1440–1448.

86. VAN VLOTEN, W. A., P. VAN DUIJN & A. SCHABERG. 1974. Cytodiagnostic use of

Feulgen-DNA measurements in cell imprints from the skin of patients with mycosis fungoides. Br. J. Dermatol. **91**:365–371.

87. DOLBEARE, F., H. GRATZNER, M. G. PALLAVICINI & J. W. GRAY. 1983. Flow cytometric measurements of total DNA content and incorporated bromodeoxyuridine. Proc. Natl. Acad. Sci. USA **80**:5573–5577.

88. ANDREEFF, M. 1981. Multiparameter flow cytometry: Application in hematology. Clin. Bull. **11**:120–130.

89. ANDREEFF, M., Z. DARZYNKIEWICZ, T. K. SHARPLESS, B. D. CLARKSON & M. R. MELAMED. 1980. Discrimination of human leukemia subtypes by flow cytometric analysis of cellular DNA and RNA. Blood **55**:282–293.

90. RITZ, J. & S. F. SCHLOSSMAN. 1982. Utilization of monoclonal antibodies in the treatment of leukemia and lymphoma. Blood **59**:1–11.

91. WEINBERG, D. S., G. S. PINKUS & K. A. AULT. 1984. Cytofluorometric detection of B cell clonal excess: A new approach to the diagnosis of B cell lymphoma. Blood **63**:1080–1087.

92. BOLDT, D. H. & M. O. NELSON. 1983. Lymphocyte subpopulations in chronic lymphocytic leukemia detected by lectin binding and flow cytometry. Cancer **51**:2083–2089.

93. KUTE, T. E., C. LINVILLE & G. BARROWS. 1983. Cytofluorometic analysis for estrogen receptors using fluorescent estrogen probes. Cytometry **4**:132–140.

94. VAN DEN BERG, H. T. C. M., H. F. DE FRANCE, J. D. F. HABBEMA & J. W. RAATGEVER. 1981. Automated selection of metaphase cells by quality. Cytometry **1**:363–368.

95. BOSMAN, F. T., M. VAN DER PLOEG, P. VAN DUIJN & A. SCHABERG. 1977. Photometric determination of the DNA distribution in the 24 human chromosomes. Exp. Cell. Res. **105**:301–311.

96. MAYALL, B. H., A. V. CARRANO, D. H. MOORE II, L. K. ASHWORTH, D. E. BENNETT & M. L. MENDELSOHN. 1984. The DNA-based human karyotype. Cytometry **5**:376–385.

97. CARRANO, A. V., J. W. GRAY, R. G. LANGLOIS & L. -C. YU. 1983. Flow cytogenetics: Methodology and applications. *In* Chromosomes and Cancer. J. D. Rowley & J. E. Ultmann, Eds. Academic Press. Orland, FL.

98. COLLARD, J. G., E. PHILIPPUS, A. TULP, R. V. LEBO & J. W. GRAY. 1984. Separation and analysis of human chromosomes by combined velocity sedimentation and flow sorting applying single- and dual-laser flow cytometry. Cytometry **5**:9–19.

99. KAMARK, M. E., J. A. BARBOSA, L. KUHN, P. G. MESSER PETERS, L. SHULMAN & F. H. RUDDLE. 1983. Somatic cell genetics and flow cytometry. Cytometry **4**:99–108.

100. LALANDE, M., L. M. KUNKEL, A. FLINT & S. A. LATT. 1984. Development and use of metaphase chromosome flow-sorting methodology to obtain recombinant phage libraries enriched for parts of the human X chromosome. Cytometry **5**:101–107.

101. LANGLOIS, R., L. -C. YU, J. W. GRAY & A. V. CARRANO. 1982. Quantitative karyotyping of human chromosomes by dual beam flow cytometry. Proc. Natl. Acad. Sci. USA **79**:7876–7880.

102. LEBO, R. V. 1982. Chromosome sorting and DNA sequence localization. Cytometry **3**:145–154.

103. WIRSCHUBSKY, Z., C. PERLMANN, J. LINDSTEN & G. KLEIN. 1983. Flow karyotype analysis and fluorescence-activated sorting of Burkitt-lymphoma associated translocation chromosomes. Int. J. Cancer. **32**:147–153.

104. BETZ, J. W., W. ARETZ & W. HARTEL. 1984. Use of flow cytometry in industrial microbiology for strain improvement programs. Cytometry **5**:145–150.

105. INGRAM, M., T. J. CLEARY, B. J. PRICE III & A. CASTRO. 1982. Rapid detection of legionella pneumophila by flow cytometry. Cytometry **3**:134–137.

106. JACOBBERGER, J. W., P. K. HORAN & J. D. HARE. 1983. Analysis of malaria parasite-infected blood by flow cytometry. Cytometry **4**:228–237.

107. STEEN, H. B., E. BOYE, K. SKARSTAD, B. BLOOM, T. GODAL & S. MUSTAFA. 1982. Applications of flow cytometry on bacteria: Cell cycle kinetics, drug effects, and quantitation of antibody binding. Cytometry **2**:249–257.

108. VAN DILLA, M. A., R. G. LANGLOIS, D. PINKEL, D. YAJKO, & W. K. HADLEY. 1982. Bacterial characterization by flow cytometry. Science **220**:620–622.

109. WHAUN, J. M., C. RITTERHAUS & S. H. C. IP. 1983. Rapid identification and detection of parasitized human red cells by automated flow cytometry. Cytometry **4**:117–122.
110. BENARON, D. A., J. W. GRAY, B. L. GLEDHILL, S. LAKE, A. J. WYROBEK & I. T. YOUNG. 1982. Quantification of mammalian sperm morphology by slit-scan flow cytometry. Cytometry **2**:344–349.
111. YOUNG, I. T., B. L. GLEDHILL, S. LAKE & A. J. WYROBEK. 1982. Quantitative analysis of radiation induced changes in sperm morphology. Anal. Quant. Cytol. **4**:207–216.
112. HALAMKA, J., J. W. GRAY, B. L. GLEDHILL, S. LAKE & A. J. WYROBEK. 1984. Estimation of the frequency of malformed sperm by slit scan flow cytometry. Cytometry **5**:333–338.
113. EVENSON, D. P., Z. DARZYNKIEWICZ & M. R. MELAMED. 1980. Relation of mammalian sperm chromatin heterogeneity to fertility. Science **210**:1131–1133.

The Little Laser That Could: Applications of Low Power Lasers in Clinical Flow Cytometry[a]

HOWARD M. SHAPIRO

The Center for Blood Research
800 Huntington Avenue
Boston, Massachusetts 02115

INTRODUCTION

The first practical flow cytometers and electronic and electrooptical cell counters were firmly established as clinical laboratory instruments within ten years after their introduction in the 1950s. Although tremendous technical progress has been made during the past ten years in the development of flow cytometric apparatus, reagents, and analytical methods, the new generation of flow cytometers has not yet had its predecessors' impact on laboratory medicine.

Until recently, the cost of apparatus would not have impeded the diffusion of the new technology of flow cytometry into clinical laboratories, which continued to operate on a "free market" basis long after cost containment measures were instituted in most other areas of hospitals and clinics. Within the past few years, however, the *modus operandi* has been changed, with the result that we are forced to consider the cost of acquiring information about patients as well as the utility of that information in the diagnosis and treatment of those patients' illnesses, whether the information is obtained using flow cytometers, NMR scanners, or stethoscopes. Under the present rules of the game, those flow cytometric clinical assays which can be implemented using relatively inexpensive apparatus are those which are most likely to find a niche in the clinical laboratory.

Increasing cost and complexity have demonstrably made it more difficult for laboratories to acquire and operate the newest generation of flow cytometers. The multiwatt ion lasers now favored as light sources in these instruments, which typically account for about half the production cost of the apparatus, also increase installation and operating costs because laser power and cooling water requirements generally mandate extensive replumbing and rewiring of a laboratory before a flow cytometer can be placed in service. It is legitimate to ask whether smaller, simpler, less expensive instruments employing low-power light sources might be useful for clinical purposes.

The clinical applications of flow cytometry, like applications of the technology in biomedical research, require the measurement of one or more of a relatively small number of cellular characteristics or parameters, as can be noted from

[a]Portions of this work were supported by Grant P01-CA19589 and Contract N01-AI32687 from the National Institutes of Health (NIH) (Bethesda, MD), and by Contract DAAG-29-82-C-0011 from the Army Research Office (Research Triangle Park, NC). Other portions were refused support by the NIH and by the National Science Foundation (Washington, DC).

TABLE 1. I and my colleagues have previously shown[1-7] that most of these parameters can be measured, using reagents presently available (TABLE 2), by small, relatively inexpensive flow cytometers employing low-power laser sources, with sensitivity and precision equal to and, in some cases, better than those obtainable using larger, more costly instruments. In this paper, I will consider the possible uses of such apparatus for some existing and projected clinical applications.

CLINICAL APPLICATIONS AND MEASUREMENT PARAMETERS

TABLE 1 lists a number of clinical tests in the fields of hematology, immunology, oncology, microbiology, parasitology, and genetics which can or could be done by flow cytometry, and shows the cellular parameter(s) measured for each of these applications. TABLE 2 shows reagents which can be used to make the required measurements using illumination from low-power laser sources. The sources are the helium-cadmium (He-Cd) laser, which emits at 325 nm (ultraviolet (UV)) (typical power 2–10 mW) and 441 nm (deep blue) (typical power 10–40 mW), the air-cooled argon laser, usually configured to emit 10–25 mW at 488 nm (blue-green), and the helium-neon (He-Ne) laser, from which 1–10 mW can be obtained at 633 nm (red).

Blood cell counting and sizing, for which flow cytometry is now the standard method, is based on a cell size measurement made either by the Coulter principle (electronic volume sensing) or by analysis of light scattering or extinction signals. Among cell counting instruments now in clinical use, only those from Ortho (Ortho Diagnostic Systems, Westwood, MA) use laser sources; these are 1mW He-Ne lasers. Other, more elaborate, laboratory-built instruments which make multiangle light scattering measurements of cells[8,9] also employ low-power (5 mW) He-Ne laser sources.

While several flow cytometric techniques for differential leukocyte counting have been described, only one commercial instrument currently provides a five-part differential count (i.e., counts of lymphocytes, monocytes, basophils, eosinophils, and neutrophils) by flow cytometry. This is the Hemalog system[10,11] from Technicon Instruments Corporation (Tarrytown, NY). Two flow cytometers with quartz-halogen lamp illumination are used to measure absorption and scattering of light by cells in separate sample streams, one stained with Astra blue, a basic dye used to identify basophils, and the other with a chromogenic peroxidase substrate, which allows identification of lymphocytes, monocytes, neutrophils, and eosinophils. Differential counts can be obtained in a similar manner from He-Ne laser light scattering and extinction measurements of cells stained with chromogenic enzyme substrates.[12,13]

Various fluorescent dyes[14] can also be used for differential counting. Methods employing the green orthochromatic and red metachromatic fluorescence of acridine orange (AO)[15-18] can be implemented in apparatus using low-power argon ion laser sources. My colleagues and I described an alternative procedure involving staining of cells with a mixture of acidic and basic fluorescent dyes[14,19] and measurement in a multibeam flow cytometer[20] with He-Cd and argon laser sources. Multiangle (forward and orthogonal or 90°) light-scattering measurements, as well as fluorescence measurements, were used for leukocyte classification in this system. The traditional Romanowsky staining procedures used to differentiate leukocytes on stained smears combine the acid dye eosin, which stains protein, with a mixture of basic dyes including methylene blue and the

TABLE 1. Measurement Parameters Used for Existing and Projected Clinical Applications of Flow Cytometry

Field/Clinical Test	Cell Size	90° Scatter	DNA Content	DNA A-T/G-C	Chromatin Structure	RNA Content	Antigens	Total Protein	Enzymes	Receptors	Membrane Potential
Hematology											
Cell counts	X										
Cell size	X										
Differential WBC	X	X	X		?		X	X	X		?
Reticulocytes	X					X					?
Immunology											
Lymphocyte subsets							X				
Cellular immune responses	X		X		?	X	X			X	X
Tissue typing							X				
Oncology											
Cytodiagnosis and classification	X	X	X		?	X	X	X	X	X	
Microbiology											
Bacterial identification/sensitivity testing	X	X	X	X		?	X	X	X		?
Virus identification	X	X	X	X			X				
Parasitology											
Detection of intra-erythrocytic parasites			X								X
Genetics											
Karyotyping	X		X	X			X		X	X	
Carrier state detection							X		X	X	
Prenatal diagnosis							X		X	X	

TABLE 2. Fluorescent Probes Usable for Flow Cytometric Measurements with Low Power Laser Sources[a]

Excitation Wavelength/ Source	Emission Wavelength	Probes Used for Measurement of						
		DNA Content	RNA Content	Chromatin Structure	Antigens/ Receptors	Total Protein	Enzymes	Membrane Potential
325 (He-Cd)	470	Hoechst dyes, DAPI (A-T)			SITS, coumarins	SITS	umbelliferone based substrates	
325 (He-Cd)	600	propidium						
441 (He-Cd)	500	olivomycin (G-C)	thioflavin T					
441 (He-Cd)	530	mithramycin (G-C)				BSF	fluorescein based substrates	oxacarbocyanines $(DiOC_n(3))$
488 (Ar ion)	530	acridine orange		acridine orange	FITC	FITC	fluorescein based substrates	oxacarbocyanines $(DiOC_n(3))$
488 (Ar ion)	580		pyronin Y		phycoerythrin			indocarbocyanines $(DiIC_n(3))$
488 (Ar ion)	620	propidium	acridine orange	acridine orange				
633 (He-Ne)	670		oxazine 1		allophyco-cyanin			indodicarbocyanines $(DiIC_n(5))$

[a]Wavelengths are in nm. Abbreviations: BSF, brilliant sulfaflavine; FITC, fluorescein isothiocyanate; SITS, 4-acetamido-4'-isothiocyanatostilbene-2,2'-disulfonic acid.

related azure dyes, which stain nucleic acids and glycosaminoglycans. Mixtures of acidic and basic fluorochromes, e.g., fluorescein isothiocyanate (FITC) and propidium iodide (PI), or sulphorhodamine 101 and DAPI (4',6-diamidino-2-phenylindole), which have been used for estimation of total protein and DNA content of cells, can also be employed for differential counting; neither application of these dye combinations requires the use of lasers emitting more than a few milliwatts for fluorescence excitation.

Three-part differential counts, i.e., counts of lymphocytes, monocytes, and granulocytes, can be derived from distributions of electronic cell volume measurements[21] or of orthogonal light scattering intensity,[9,14,20,22–24] which is low in lymphocytes, higher in monocytes, and highest in granulocytes. Orthogonal scatter measurements have also been used[22–24] for gating purposes, permitting immunofluorescence analyses of lymphocytes to be done on buffy coat or whole blood samples without the need for a preliminary density gradient separation to remove granulocytes. Low-power lasers are, as mentioned previously, adequate sources for scattering measurements.

In principle, a differential leukocyte count could be done using fluorescent antibodies specific for the various cell types. It is unlikely that this approach to differential counting would be cost-effective, since there exist less expensive reagents, i.e., dyes, which can be used with simpler staining procedures to produce an accurate and precise five-part differential count. Where morphology and classical cytochemistry cannot be used to identify closely related cell types, as in the analysis of lymphocyte subpopulations or the determination of the lineage of leukemic cells, fluorescent antibody techniques, with and without flow cytometry, are finding an increasing range of clinical applications.

It is customary to use an argon ion laser emitting 100 mW or more at 488 nm for excitation of FITC-labeled antibodies. Most people seem to believe that increasing the laser power used increases the sensitivity of immunofluorescence measurements. This is demonstrably not the case when cells stained with FITC-antibodies are excited at 488 nm; under these circumstances, the average intensity of cellular autofluorescence is generally several times the intensity of the fluorescence emission from the few thousand FITC-antibody molecules which could be detected on a cell in the absence of autofluorescence.[3,6,7] I and others (R. Hoffman, personal communication) have been able to make adequate measurements of immunofluorescence in human lymphocytes stained with FITC-labeled anti-subset antibodies using flow cytometers with air-cooled argon laser sources emitting 20 mW or less at 488 nm.

The use of phycobiliproteins such as phycoerythrin (PE) and allophycocyanin (APC) for fluorescent labeling of antibodies[6,7,25] offers some advantages over the classical FITC labeling method. Phycobiliproteins, which are protein components of the photosynthetic apparatus of algae and cyanobacteria, typically contain several chromophores per protein molecule, and have high extinction coefficients and high quantum efficiencies. The absorption (and excitation) and emission maxima of phycobiliproteins lie at longer wavelengths than do those of FITC and of the majority of autofluorescent materials in cells and tissues; this can permit better discrimination between labeled and unlabeled cells. The red fluorescence (emission maximum about 575 nm) of PE can be excited much more effectively by the 488 nm argon ion laser line than can the fluorescence of conventional antibody labels such as tetramethylrhodamine and Texas red. Thus, a flow cytometer with a single low-power argon laser source can be used to make simultaneous measurements of FITC (emission maximum about 520 nm) and PE immunofluorescence. I and my colleagues[6] have shown that fluorescence from cells stained with antibodies labeled with APC (emission maximum about 660 nm) can be detected in a flow cytometer using a 7 mW He-Ne laser source.

Until phycoerythrin-labeled antibodies became available, two-color immuno-fluorescence studies were typically done with a dual-laser instrument in which an argon laser was used for excitation of FITC-labeled antibodies and a krypton laser, emitting light at 568 nm, was used to excite antibodies labeled with Texas red or rhodamine 101 isothiocyanate (XRITC). In addition to being very expensive, and hungry and thirsty for electrical power and cooling water, such an apparatus is more difficult to use than is a single-beam flow cytometer. Substitution of a tunable dye laser, pumped by the argon laser, in place of the krypton laser reduces power requirements and, to some extent, instrument cost; however, the apparatus may become even more difficult to operate owing to the idiosyncrasies of dye lasers. It is much simpler to do two-color immuno-fluorescence using the FITC/PE combination as described above; this should facilitate the development of clinical assays in which cells are labeled with two antibodies.

I do not think it is likely that there will ever be a clinical assay which will require anything more elaborate than three-color immunofluorescence measure-ments. Using FITC-, PE-, and APC-labeled antibodies, it is now possible to make such measurements in a dual-beam flow cytometer using low-power argon and He-Ne laser light sources. Further developments in phycobiliprotein reagents may make the task even simpler. Glazer and Stryer[26] have recently investigated the fluorescence characteristics of a "tandem" conjugate of PE and APC; this material emits at 660 nm when excited at 488 nm, and could presumably be used as a third label, allowing three-color immunofluorescence studies to be done in an instrument using only a single low-power argon laser.

The reticulocyte count is a good example of clinical assay which, while informative and procedurally straightforward, is so labor-intensive that it may be underutilized on that account. The fraction of juvenile erythrocytes in peripheral blood is estimated by examination of a smear after a basic dye such as new methylene blue or brilliant cresyl blue has been used to precipitate the residual RNA in these young cells. Flow cytometric approaches to reticulocyte counting rely on the use of fluorescent basic dyes to demonstrate RNA. Ortho has obtained approval from the U. S. Food and Drug Administration for a procedure using AO as the RNA stain; alternative methods use pyronin Y[27-28] and thioflavin T.[29] AO and pyronin Y, both of which can be excited adequately using a low-power 488 nm argon laser,[2] are structural homologues of methylene blue which, like other oxazine and thiazine dyes, is itself fluorescent when excited with red light. I have used the laser dye oxazine 1 with He-Ne laser excitation for measurements of RNA content in lymphocytes,[2] and I would expect this dye and/or the more classical reticulocyte stains to be usable in the same fashion for reticulocyte counting.

Jacobberger et al.[30,31] have identified reticulocytes by flow cytometry using the cyanine dye dimethyloxadicarbocyanine (DiOC1(3)) with argon laser excitation. Staining of reticulocytes by DiOC1(3) may depend as much on the higher membrane potentials of reticulocytes, as compared to mature erythrocytes, as on the younger cells' RNA content, since partitioning of lipophilic basic dyes such as cyanines into cells is dependent upon the electrical potential difference across the cytoplasmic membrane. Cell membrane potentials may be estimated by flow cytometry using a variety of cyanine dyes;[1] no more than a few milliwatts of laser power are needed for excitation of these compounds.

Among the applications of flow cytometry in clinical immunology, the identification and enumeration of lymphocyte subpopulations based upon surface markers and tissue typing based upon demonstration of various HLA antigens require only immunofluorescence measurements, although light scattering or cell volume measurements are usually also made and used to gate the fluorescence signals. Measurements of a wider range of parameters may be useful in defining functional characteristics of cells of the immune system, for example, in demonstrating activated T-cells among lymphocytes in blood and other body fluids.

While quantitative assessment of T-cell activation and proliferation in response to polyclonal mitogens such as lectins and antibodies might be made to serve as a general test of cellular immune function, the clinical relevance of such a measurement is not well established. At present, there is somewhat more interest in the detection of T-lymphocyte activation in response to specific antigenic stimulation. In autoimmune diseases and in graft rejection, a substantial amount of tissue damage is mediated and/or caused by activated T-cells. One would therefore expect that increases and decreases in the number of circulating activated T-cells would occur in parallel with exacerbations and remissions of the disease process in these conditions, and that monitoring the fraction of activated cells might be useful in assessing disease activity.

T-cell activation may be detected by flow cytometric measurements of several parameters. The DNA synthesis in activated cells which is customarily gauged by cells' uptake of tritiated thymidine can be detected with greater sensitivity by flow cytometry of cells stained with DNA fluorochromes. Increases in RNA content,[2,7,32-33] membrane potential[1,7,34] and cell size, and the display of new surface antigens and receptors,[7,35-36] all of which occur earlier during activation than does DNA synthesis, are also useful in demonstrating the presence of activated cells in samples.

My colleagues and I are currently evaluating several flow cytometric approaches to early detection of T-cell activation in the mixed lymphocyte reaction (MLR),[37] with the objective of making it possible to use MLRs for tissue matching in renal transplantation from cadaver donors. Activation of a prospective recipient's lymphocytes by lymphocytes from a prospective donor is now assessed by measurements of tritiated thymidine uptake at 96-120 hours. The MLR has been found to be a better predictor of histocompatibility than is tissue type, and the test is now used for matching renal transplants from living donors, related to the recipient. In transplantation from cadaveric donors, it is not possible to maintain function of the donor organ for the 4–5 days required to read an MLR by radioisotope techniques. We have found that enumeration of cells bearing the 4F2 or Tac (interleukin-2 receptor) activation antigens allows us to predict reactivity in the MLR at 12–18 hours; this flow cytometric assay should therefore be usable for organ matching. The 4F2 antigen can readily be detected using APC-labeled antibody and low-power He-Ne laser excitation.[6]

While a great deal of time and money has been expended on efforts to develop flow cytometric sytems for cancer cytodiagnosis, none has yet come into clinical use. Instead, flow cytometry has been successfully applied to other problems in clinical oncology.[38-40] The demonstration and characterization of abnormalities in tumor cell DNA and/or RNA content and chromatin structure[32] can aid in establishing a diagnosis, estimating prognosis, and selecting and monitoring therapy. Analyses of tumor antigens and receptors can also provide clinically relevant data. It is likely that strong correlations exist between DNA (and RNA) content of tumor cells, on the one hand, and surface antigen patterns, on the other; whether this is so remains to be established by multiparameter studies. All of the flow cytometric parameters now thought to be relevant for clinical work in oncology can be measured in apparatus using low-power laser sources.

Relatively little work has been done on applications of flow cytometry to the diagnosis of bacterial, viral, and parasitic diseases. The same characteristics can be used to identify bacteria as are used to identify larger cells, e.g., cell size and light scattering properties, nucleic acid and protein content, and the presence and quantity of antigens and enzymes. The ratio of (adenine + thymine) to (guanine + cytosine) (A-T/G-C) varies widely among bacterial species; this can be estimated from flow cytometry of bacteria stained with mixtures of DNA dyes such as Hoechst 33258 (specific for A-T) and chromomycin A3 (specific for G-C).[41] I am now evaluating the He-Cd laser as a light source for flow cytometry of bacteria (and chromosomes) stained with this combination of dyes. I have also recently described the possible uses of flow cytometry of bacterial membrane potentials for detection, identification, and determination of antibiotic sensitivity of bacteria in clinical specimens.[7] Intracellular viruses and parasites may be detected by their contributions to the nucleic acid content and/or composition of infected cells, or by the presence of specific antigens.[7] Red cells containing malaria parasites can also be detected by their increased membrane potentials.[30]

Much of the recent progress in the flow cytometric analysis and sorting of human chromosomes has resulted from the use of double staining techniques to demonstrate different A-T and G-C rich regions in chromosomes of similar size and nucleic acid content.[42–43] It is not clear that flow karyotyping will ever be a practical clinical assay; the present practice of using dual 12-watt ion lasers as light sources in flow cytometry of chromosomes makes it difficult even to install the apparatus in most institutions. It remains to be seen whether He-Cd laser source instruments can provide the sensitivity and precision required for measurements of chromosomes. Other projected applications of flow cytometry in clinical genetics, for example, detection of disease carrier states and prenatal diagnosis, which could be based upon measurements of enzymes, antigens, and/or receptors, would presumably not require instruments with large laser sources.

CONCLUSIONS

The discussion just completed, after considering virtually all of the existing and projected clinical applications of flow cytometry, yields a relatively short list of required measurement parameters, all of which can now be measured effectively in instruments using low-power laser sources. A flow cytometer with four illuminating beams (325, 441, 488, and 633 nm) derived from He-Cd, argon, and He-Ne lasers could be used to make measurements of these parameters and of almost any other parameter which can now be measured by flow cytometry. Further developments in metal-vapor (e.g., He-Cd and helium-selenium) and in solid-state lasers should provide efficient sources of green and yellow excitation for those few fluorescent probes (e.g., the rhodamine dyes and phycocyanin) which cannot be effectively excited at any of the four wavelengths just mentioned.

It is now feasible to produce both the "ultimate" flow cytometer just described and simpler instruments, also using low-power laser sources, suitable for a more restricted range of clinical applications. I am optimistic that such apparatus will be found to be cost-effective for clinical laboratory use in the future.

[NOTE ADDED IN PROOF: Green-(543 nm) and yellow- (594 nm) emitting He-Ne lasers with power output from 1–3 mW became available in 1985 and were successfully used for detection of phycoerythrin (green excitation) and Texas red (yellow excitation) immunofluorescence in flow cytometers.]

REFERENCES

1. SHAPIRO, H. M. 1981. Flow cytometric probes of early events in cell activation. Cytometry **1**:301.
2. SHAPIRO, H. M. 1981. Flow cytometric estimation of DNA and RNA content in intact cells stained with Hoechst 33342 and pyronin Y. Cytometry **2**:143.
3. SHAPIRO, H. M. 1983. Multistation multiparameter flow cytometry: A critical review and rationale. Cytometry **3**:227.
4. SHAPIRO, H. M., D. M. FEINSTEIN, A. S. KIRSCH & L. CHRISTENSON. 1983. Multistation multiparameter flow cytometry: Some influences of instrumental factors on system performance. Cytometry **4**:11.
5. SHAPIRO, H. M. 1983. Building and Using Flow Cytometers: The Cytomutt Breeder's and Trainer's Manual. Howard M. Shapiro, M. D., P. C., West Newton, MA.
6. SHAPIRO, H. M., A. N. GLAZER, L. CHRISTENSON, J. M. WILLIAMS & T. B. STROM. 1983. Immunofluorescence measurement in a flow cytometer using low-power helium-neon laser excitation. Cytometry **4**:276.
7. SHAPIRO, H. M. 1985. Practical Flow Cytometry. Alan R. Liss. New York.
8. SALZMAN, G. C., J. M. CROWELL, C. A. GOAD, K. M. HANSEN, R. D. HIEBERT, P. M. LA BAUVE, J. C. MARTIN, M. L. INGRAM & P. F. MULLANEY. 1975. A flow-system multiangle light-scattering instrument for cell characterization. Clin. Chem. **21**:1297.
9. SALZMAN, G. C., J. M. CROWELL, J. C. MARTIN, T. T. TRUJILLO, A. ROMERO, P. F. MULLANEY & P. M. LA BAUVE. 1975. Cell identification by laser light scattering: identification and separation of unstained leukocytes. Acta Cytol. **19**:374.
10. ORNSTEIN, L. & H. R. ANSLEY. 1974. Spectral matching of classical cytochemistry to automated cytology. J. Histochem. Cytochem. **22**:453.
11. MANSBERG, H. P., A. M. SAUNDERS, & W. GRONER. 1974. The Hemalog D white cell differential system. J. Histochem. Cytochem. **22**:711.
12. KAPLOW, L. S. & E. LERNER. 1977. Computer-assisted monocyte esterase assay by flow cytometry. J. Histochem. Cytochem. **25**:590.
13. KAPLOW, L. S. 1977. The application of cytochemistry to automation. J. Histochem. Cytochem. **25**:990.
14. SHAPIRO, H. M. 1977. Fluorescent dyes for differential counts by flow cytometry: Does histochemistry tell us much more than cell geometry? J. Histochem. Cytochem. **25**:976.
15. HALLERMANN L., R. THOM & H. GERHARTZ. 1964. Elektronische Differentialzahlung von Granulocyten und Lymphocyten nach intravitaler fluorochromierung mit Acridinorange. Verh. Deutsch. Ges. Inn. Med. **70**:217.
16. ADAMS, L. R. & L. A. KAMENTSKY. 1971. Machine characterization of human leukocytes by acridine orange fluorescence. Acta Cytol. **15**:289.
17. STEINKAMP, J. A., A. ROMERO & M. A. VAN DILLA. 1973. Multiparameter cell sorting: Identification of human leukocytes by acridine orange fluorescence. Acta Cytol. **17**:113.
18. ADAMS, L. R. & L. A. KAMENTSKY. 1974. Fluorimetric characterization of six classes of human leukocytes. Acta Cytol. **18**:389.
19. SHAPIRO, H. M., E. R. SCHILDKRAUT, R. CURBELO, C. W. LAIRD, R. B. TURNER & T. HIRSCHFELD. 1976. Combined blood cell counting and classification with fluorochrome stains and flow instrumentation. J. Histochem. Cytochem. **24**:396.
20. SHAPIRO, H. M., E. R. SCHILDKRAUT, R. CURBELO, R. B. TURNER, R. H. WEBB, D. C. BROWN & M. J. BLOCK. 1977. Cytomat-R: A computer-controlled multiple laser source multiparameter flow cytophotometer system. J. Histochem. Cytochem. **25**:836.

21. VAN DILLA, M. A., M. J. FULWYLER & I. U. BOONE. 1967. Volume distribution and separation of normal human leukocytes. Proc. Soc. Exptl. Biol. Med. **125**:367.
22. HOFFMAN, R. A., P. HANSEN, P. C. KUNG & G. GOLDSTEIN. 1980. Simple and rapid measurement of human T lymphocytes and their subclasses in peripheral blood. Proc. Natl. Acad. Sci. USA **77**:4914.
23. HOFFMAN, R. A. & W. P. HANSEN. 1981. Immunofluorescent analysis of blood cells by flow cytometry. Intl. J. Immunopharmacol. **3**:249.
24. RITCHIE, A. W. S., R. A. GRAY & H. S. MICKLEM. 1984. Right angle light scatter: A necessary parameter in flow cytofluorimetric analysis of human peripheral blood mononuclear cells. J. Immunol. Method. **64**:109.
25. OI, V. T., A. N. GLAZER & L. STRYER. 1982. Fluorescent phycobiliprotein conjugates for analyses of cells and molecules. J. Cell. Biol. **93**:981.
26. GLAZER, A. N. & L. STRYER. 1983. Fluorescent tandem phycobiliprotein conjugates. Emission wavelength shifting by energy transfer. Biophys. J. **43**:383.
27. TANKE, H. J., I. A. B. NIEWENHUIS, G. J. M. KOPER, J. C. M. SLATS & J. S. PLOEM. 1981. Flow cytometry of human reticulocytes based on RNA fluorescence. Cytometry **1**:313.
28. TANKE, H. J., P. H. ROTHBARTH, J. M. J. J. VOSSEN, G. J. M. KOPER & J. S. PLOEM. 1983. Flow cytometry of reticulocytes applied to clinical hematology. Blood **61**:1091.
29. SAGE, B. H., JR., J. P. O'CONNELL & T. J. MERCOLINO. 1983. A rapid, vital staining procedure for flow cytometric analysis of human reticulocytes. Cytometry **4**:222.
30. JACOBBERGER, J. W., P. K. HORAN & J. D. HARE. 1983. Analysis of malaria parasite-infected blood by flow cytometry. Cytometry **4**:228.
31. JACOBBERGER, J. W., P. K. HORAN & J. D. HARE. 1984. Flow cytometric analysis of blood cells stained with the cyanine dye $DiOC_1(3)$: reticulocyte quantification. Cytometry **5**:589.
32. DARZYNKIEWICZ, Z., F. TRAGANOS & M. R. MELAMED. 1980. New cell cycle compartments identified by multiparameter flow cytometry. Cytometry **1**:98.
33. NORONHA, A. B. C., D. P. RICHMAN & B. G. W. ARNASON. 1980. Detection of in vivo stimulated cerebrospinal fluid lymphocytes by flow cytometry in patients with multiple sclerosis. N. Engl. J. Med. **303**:713.
34. DARZYNKIEWICZ, Z., L. STAIANO-COICO & M. R. MELAMED. 1981. Increased mitochondrial uptake of rhodamine 123 during lymphocyte stimulation. Proc. Natl. Acad. Sci. USA **77**:6696.
35. COTNER, T., J. M. WILLIAMS, L. CHRISTENSON, H. M. SHAPIRO, T. B. STROM & J. L. STROMINGER. 1983. Simultaneous flow cytometric analysis of human T cell activation antigen expression and DNA content. J. Exp. Med. **157**:461.
36. WILLIAMS, J. M., L. CHRISTENSON, J. L. ARAUJO, C. B. CARPENTER, E. L. MILFORD, H. M. SHAPIRO & T. B. STROM. 1983. A new approach to the monitoring of kidney transplant patients via flow cytometric analysis of T-cell subsets, activation antigens, and DNA content. Transpl. Proc. **15**: 1957.
37. BACH, F. & K. HIRSCHHORN. 1964. Lymphocyte interaction: A potential histocompatibility test in vitro. Science **143**:813.
38. LAERUM, O. D. & T. FARSUND. 1981. Clinical application of flow cytometry: A review. Cytometry **2**:1.
39. BARLOGIE, B., B. DREWINKO, J. SCHUMANN, W. GOHDE, G. DOSIK, D. A. JOHNSTON & E. FREIREICH. 1980. Cellular DNA content as a marker of neoplasia in man. Am. J. Med. **69**:195.
40. BARLOGIE, B., M. N. RABER, J. SCHUMANN, T. S. JOHNSON, B. DREWINKO, D. E. SWARTZENDRUBER, W. GOHDE, M. ANDREEFF & E. J. FREIREICH. 1983. Flow cytometry in clincial research. Cancer Res. **43**:3982.
41. VAN DILLA, M. A., R. G. LANGLOIS, D. PINKEL, D. YAJKO & W. K. HADLEY. 1983. Bacterial characterization by flow cytometry. Science **220**:620.
42. LANGLOIS, R. G., A. V. CARRANO, J. W. GRAY & M. A. VAN DILLA. 1980. Cytochemical studies of metaphase chromosomes by flow cytometry. Chromosoma **77**:229.
43. LANGLOIS, R. G., L.-C. YU, J. W. GRAY & A. V. CARRANO. 1982. Quantitative karyotyping of human chromosomes by dual beam flow cytometry. Proc. Natl. Acad. Sci. USA **79**:7876.

Flow Cytometric Instrumentation for Scientific and Clinical Use

VOLKER KACHEL

*Max-Planck-Institut für Biochemie
bei München
Federal Republic of Germany*

INTRODUCTION

Flow cytometric analysis allows a fast and precise determination of one or more physical and biochemical features of biological cells. Flow cytometry has been a valuable tool for scientific cell research for many years. Flow cytometry, however, can also provide a powerful means for clinical diagnosis of cell samples, e.g. in hematology, immunology, and oncology.[1-5]

Today's flow cytometric instrumentation, consisting either of individually built systems or commercially available flow cytometers and sorters, was primarily developed for scientific application; that is, the systems are characterized by high accuracy and resolution and they can be adapted to the different requirements of scientific experiments with cells.[6] Such systems, which are relatively expensive, are not optimized for clinical applications. They are not simple to operate: they require highly trained personnel, the throughput of samples is not quick enough, and the operation of the data evaluation computers requires persons who are familiar with computer languages. This paper describes flow cytometric instrumentation with improved features for clinical applications.

CELL PARAMETERS TO BE MEASURED

Fluorescence is the most important parameter measured by flow cytometry. Specific stains are bound in or at the surface of cells and in this way a large number of different cell features can be measured: DNA, RNA,[7] immunological surface markers,[3] membrane potential,[8] intracellular pH,[9] etc. An overview on fluorescence methods in flow cytometry is given by Shapiro.[10]

Some of these cell features are interesting not only for their absolute value, but also for the relative value normalized to the cellular volume or surface. That is easily understood when we imagine a large and a small cell containing identical quantities of fluorescent antibodies per surface unit. Both cells are measured in a fluorescence flow cytometer with different signals, which are proportional to the cell surfaces. Assuming, that the primary interest in such an experiment is directed toward the surface density quantitation of the cell surface, another parameter that allows an independent quantitation of the surface area must be simultaneously determined. Such a parameter is the cell volume, from which the cell surface can be derived and which can be measured in the best possible way by the electrical resistance pulse method (Coulter volume[11,12]).

A clinically useful multiparameter flow cytometer should be able to measure at least two fluorescences and a cell size–related parameter.

LIGHT SOURCES FOR EXCITING CELL FLUORESCENCE

Cellular fluorescence must be excited by an exciting light source, which must be able to deliver light of the characteristic exciting wavelength range of the stain used. The light source has an important role in determining features, quality, and price of flow cytometers. There are two main kinds of light sources in use: (1) Ion lasers, more recently also combined with tunable dye lasers and (2) high pressure mercury or xenon arc lamps.

Ion lasers deliver a highly intense beam of monochromatic and coherent light of distinct wavelengths. The wavelength restriction can be a handicap when dyes must be excited in a wavelength range that is not generated by the laser. By combining an ion laser with a dye laser the wavelength can be tuned over a certain wavelength range. Arc lamps, e.g. the high pressure mercury arc lamp HBO 100, deliver intense incoherent light within the whole visible and UV spectrum with several very intense lines especially in the blue and UV range.

Although laser light sources are superior to arc lamps with regard to monochromatic light intensity and minimal divergence of the beams, comparative investigations of Peters[13] have shown that HBO 100 mercury arc lamp illumination yields resolution equivalent to a 5W argon laser at considerably lower cost, when dyes are used that are compatible with the mercury arc emission spectrum.

These points indicate that reasonably priced clinical flow cytometers should be equipped with arc lamp illumination. The HBO 100 mercury arc lamp is used in several proven optical flow cytometers[14,15] and also used by us in the Fluvo-Metricell instruments,[16] which combine multiparameter fluorescence and electrical cell volume analysis.

TABLE 1 compares several features of scientific and clinical flow cytometers. The three last points of the list, fast throughput, data reduction, and automation, are not yet sufficiently realized in today's flow cytometers and new developments will be shown below, which aim toward an improvement of flow cytometers for clinical use.

TABLE 1. Flow Cytometers

	Scientific	Clinical
Accuracy	+ +	+ +
Calibration	+ +	+ +
Flexibility	+ +	+
Fast throughoutput	+	+ +
Data reduction	1	+ +
Automation	+	+ +

REQUIREMENTS AND REALIZATION OF TRANSDUCERS

In today's state-of-the-art flow cytometers, the cell suspension is supplied from a supply vessel through a more-or-less long tube into the point of measurement of the transducer chamber. That is mainly conditioned by the necessity to hydrodynamically focus the cell stream into the point of measurement.[15-18] The tubes, however, delay the supply of the cell samples and require a substantial cleaning process between subsequent samples and the thin ejection tip impedes a fast throughput of cell suspension and cleaning fluid.

FIGURE 1 shows a cell sample supply principle, where the cells are fed into the point of measurement without an interconnecting tube.[19] The vessel itself, which can be plugged easily into the transducer, feeds the cells into the point of

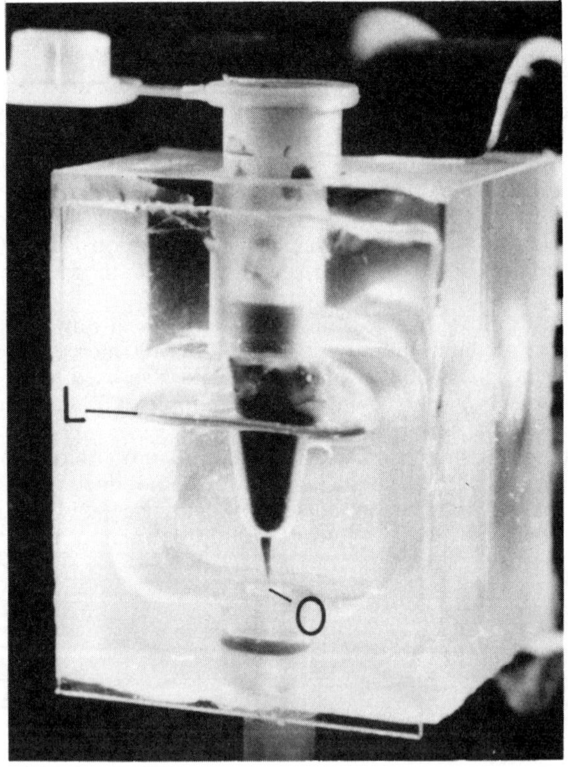

FIGURE 1. Demonstration chamber explaining the principle of the tubeless transducer. The reaction vessel, containing black ink, has a hole of about 100 μm diameter in its bottom. It is plugged into a block of plexi glass, which was excavated and closed on its front by a glass plate. The level of sheath water at L. The ink is hydrodynamically focused directly from the particle container into the point of measurement in the orifice (o). Suction of about 0.2 bar is applied through o.

measurement. This kind of sample supply has the following advantages particularly for a clinical cytometer:

(1) There is no tube that must be passed by the cells: the cell measurement starts immediately after plugging in the vessel. The initial dead time for each sample is minimized.

Figure 2 shows the time course of an experiment with the new principle. The x-axis shows time and the y-axis the particle rate of an experiment with fixed erythrocytes, which were introduced into the transducer in time phase 24 by plugging in their vessel. In time phase 40 the vessel was pulled out from the transducer head. Shortly after inserting the vessel, the particles are measured with the full particle rate and immediately after removing the vessel the particle rate is decreased to negligible values without intermediate cleaning. This kind of particle supply allows a very quick succession of samples.

(2) No sedimentation problems occur with larger or heavier cells, since the up to down injection flow works with and not against the sedimentation.

(3) The principle is suited for automatic exchange of samples.

(4) The cell sample is always accessible for the operator, which can perform kinetic experiments.

Figure 3 shows a schematic diagram and Figure 4 the photoprints of a three-parameter transducer system that is able to measure two fluorescence parameters and cell volume electrically. Details of the device are described in reference 19.

This type of a manually operated transducer is not optimized for routine use

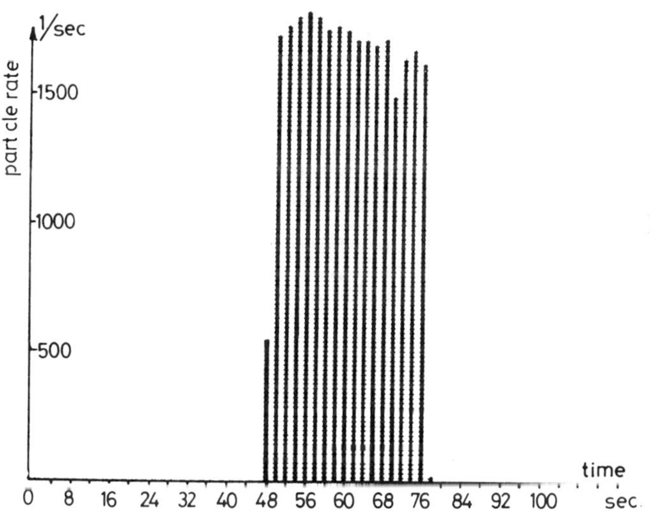

FIGURE 2. Dynamic behavior of the new transducer, demonstrated with a time sequence measurement of fixed erythrocytes, where every 2 seconds a histogram was collected and the total number of cells was determined.[19] Parameter measured: electrical cell volume. Each channel of the x-axis represents 2 seconds. y-axis: particle rate. In the time phase 24 the vessel with the erythrocytes was plugged into the transducer, the full particle rate is immediately recorded after that. In time phase 40 it was removed without cleaning the chamber. The measurement is started and finished with negligible delay.

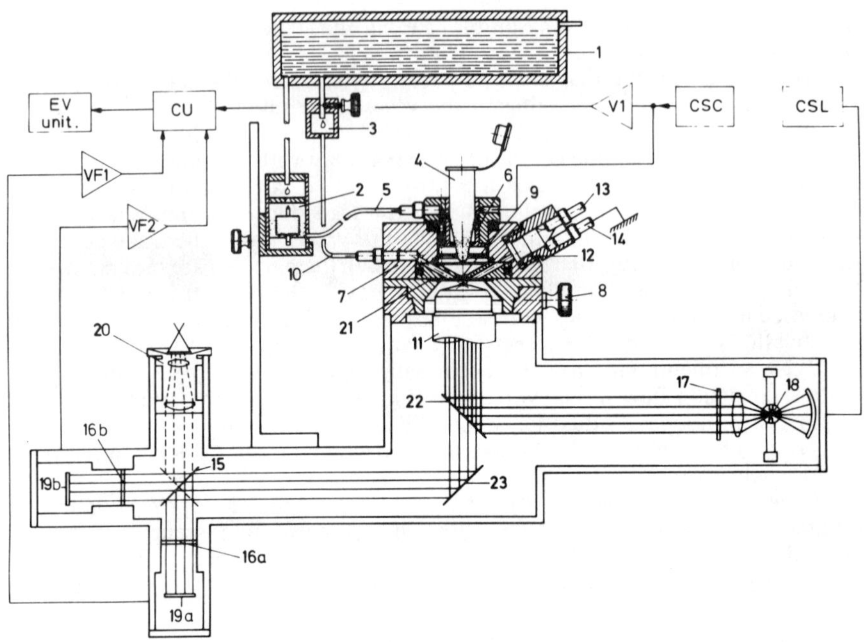

FIGURE 3. Schematic diagram of a 3-parameter Fluvo-Metricell transducer system equipped with a tubeless transducer. 1: particle-free sheath fluid; 2: dropping chamber isolating the transducer and the supply vessel 1 and regulator for keeping a constant fluid level inside the regulating chamber. The particle flow from 4 into the transducer is controlled by the level difference in 2 and 4; 3: isolating and regulating device for cleaning fluid; 4: pluggable vessel supplying the cell suspension; 5: tube supplying the particle-free suspension; 6: active electrode of the sizing circuit; 7: transducer block; 8: adjustment device for the optical system; 9: measuring orifice; 10: supply of cleaning fluid; 11: objective of the epi-illuminating fluorescence system; 12: grounded electrode; 13: suck connection; 14: ground connection; 15: switchable mirror enabling observation of the orifice via ocular lens 20; 16a, b: fluorescence filters for two different wavelengths; 20: ocular device for observation and adjustment of the orifice; 21: cover glass between orifice and objective; 22: dichroic mirror separating exciting and fluorescence ight; 23: mirror; CSL: stabilized current source for the illumination lamp; CSC: current source for the electrical cell-sizing system (0-1 mA); CU: control unit interconnecting the three signals; V1, VF1, VF2: analog amplifiers; EV: digital data evaluation unit.

in clinical cytometry. For such applications cell samples should be supplied and measured automatically. This can be done as mentioned by automatically exchanging the particle vessels. This method is quick and does not require cleaning between samples. Such vessels, however, with a precise bottom orifice are not yet available as mass-produced one-way products.

Alternatively, with a tube transducer an improvement can be achieved, by replacing the single tube injector by a two- or multi-tube injector device. As shown in FIGURE 5 a carrying and leading off tube end in the proper injection tip, which is here a kind of bypass to the tubes. Via both tubes an unthrottled, quickly

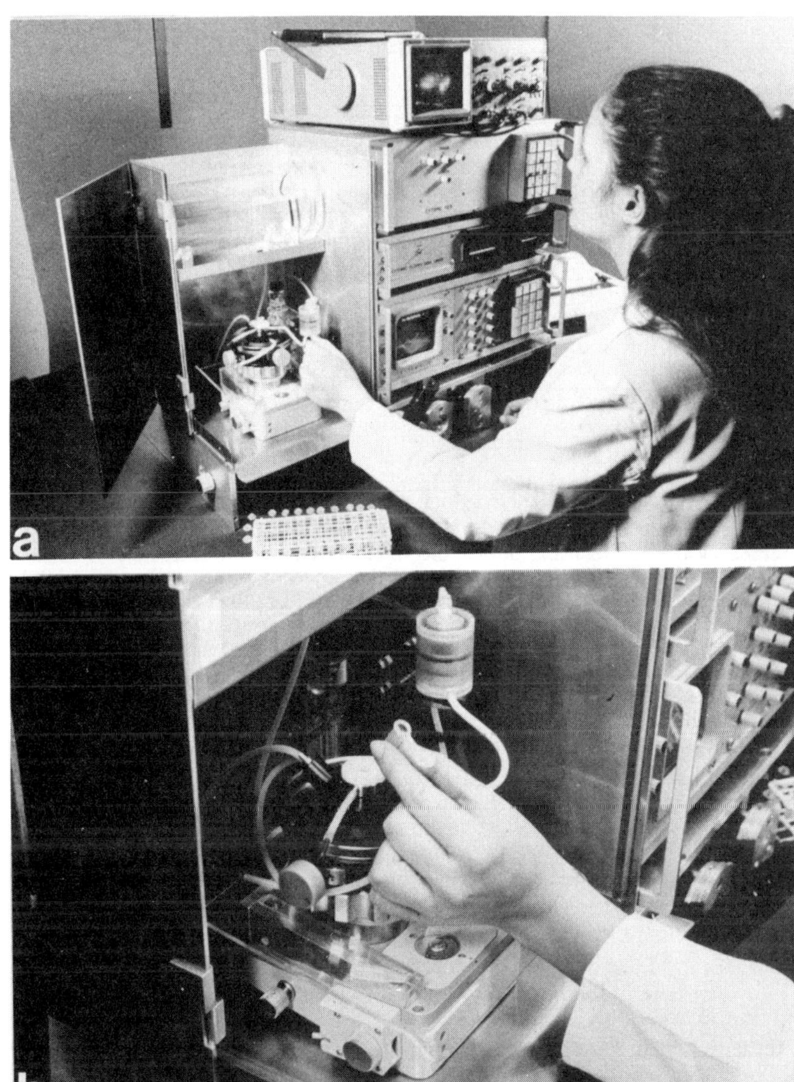

FIGURE 4. (a) Prototype of a tubeless Fluvo-Metricell transducer with a Cytomic 12 and 123 data module. (b) A particle vessel is about to be inserted into the transducer head.

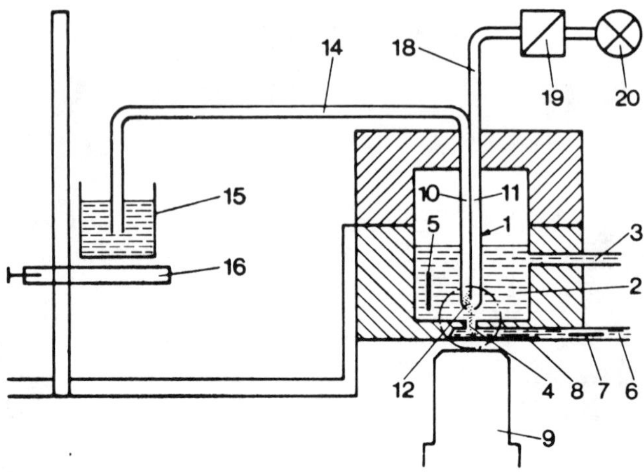

FIGURE 5. Example of a double tube supply device, which enables fast exchange of samples without exchange of vessels in the transducer. By using the double tube principle the bottleneck of each focusing device, the small injector tip, 12, no longer restrains the particle supply and cleaning processes. This particle injection orifice in this sytem is a kind of bypass in a tube system of considerable cross section (14,10,11,18). Via the large cross section tubes the particles can be quickly transported to the orifice 12 and removed via valve 19 and pump 20 when the measurement was finished and the inlet of 14 was connected to cleaning fluid. 1: double tube injector; 2: particle-free fluid, supplied through the tube, 3; 4: measuring orifice; 5: active electrode; 6:suction tube; 7: grounded electrode; 8: cover glass; 9: fluorescence objective; 10: carrying part of the double tube; 11: leading off part; 12: injector tip; 14: particle supply tube; 15: particle supply vessel; 16: regulation platform; 18: waste tube; 19: valve; 20: pump.

cleanable pathway is available conducting the cells to the injector tip and removing the rest of the sample when the measurement is finished.

DATA SYSTEM REQUIREMENTS

The data system in general collects the analog pulse height signals delivered by the transducer, converts them into digital numbers, stores the values in its memory or on disk or tape, and allows processing of the data sets.

The data system of a clinical flow cytometer should be able to concentrate the measured results into few numbers or curves. It must be able to be operated without special knowledge of computer programming. Such features are available in the Cytomic data system modules:[20-22] They are equipped with a set of preprogrammed functions that are specially developed for flow cytometric data evaluation. Each function can be called by pressing a key of a pocket calculator-like keyboard. By pressing a sequence of keys, the user can perform more or less complicated tasks. This manual operation is more suited for use in scientific experiments, where data evaluations according to very different schedules are required.

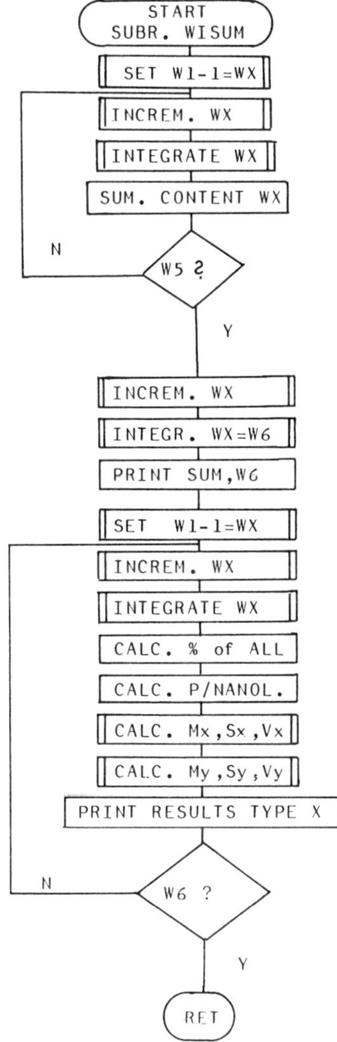

FIGURE 6. Flow diagram of the "Multiple Window Evaluation" Program, which performs the following task: Calculates and prints from 6 predefined two parameter windows (1) the total sum of cells in cyte window, (2) the absolute content of each window, (3) the relative content of each window, related to the total number of cells, (4) the concentration of cells in each window, (5) the mean Mx, standard dev. Sx, coeff. of variation Vx, and (6) mean My, standard dev. Sy, and coeff. of variation Vy. The user is only required to press one key. Execution of the routine including printing on a low cost matrix printer needs 13 seconds. An application of the program is shown in Figure 7.

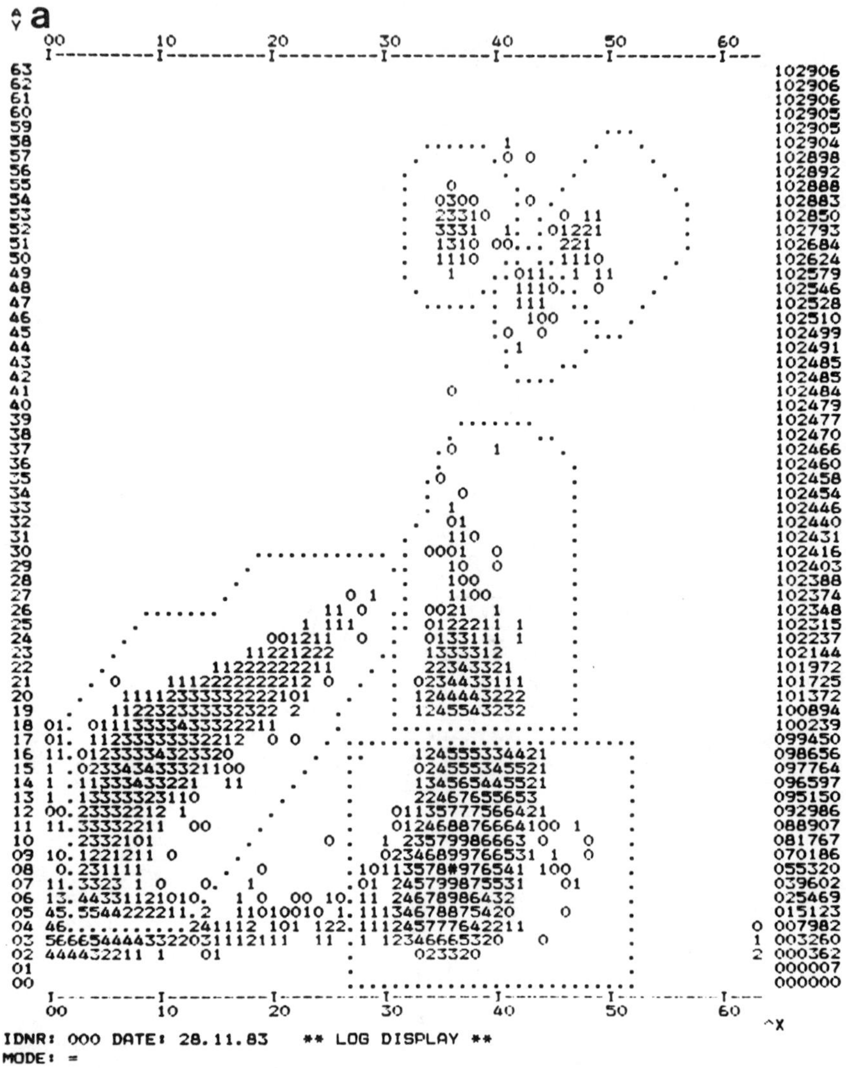

FIGURE 7. (a) Example of evaluation of a blood cell distribution which was stained according to reference 2. Six different cell populations can be distinguished and evaluated by a set of 6 windows: wdw 1=thrombocytes, wdw 2=erythrocytes, wdw 3=reticulocytes, wdw 4=lymphocytes, wdw 5=granulocytes, wdw 6-calibrator particles. The 6 evaluation windows shown by the pointed traces are stored in the instrument and automatically applied by the evaluation program. x-axis = cell volume (log 64 channels resolution), y-axis = fluorescence (log 64 channels resolution). The frequency scale is also log. The number inside the field marks the frequencies of the channels. The numbers at the right side are the integral line frequencies. An example of an evaluation print is shown in TABLE 2.

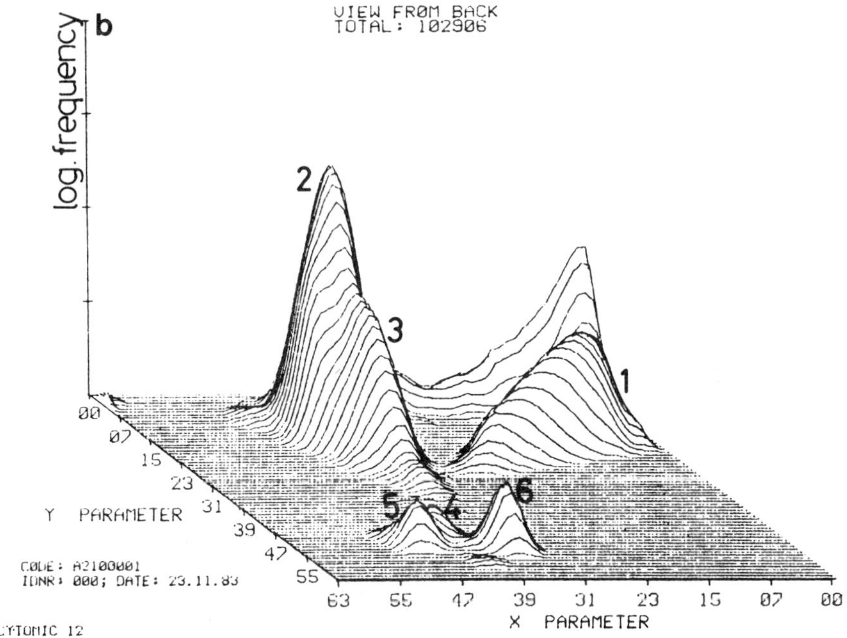

FIGURE 7. (b) Prospective view of the blood cell distribution shown in (a). The curve, plotted with the Cytomic 12 plot program, is seen from the backside, i.e., the coordinate origin is at the opposite right side. Before plotting, the histogram was smoothed by the Cytomic 12 smoothing program. The numbers correspond with the numbers in (a).

The function set of the Cytomic 12 analyzer also contains functions of considerable complexity, which are already suited for routine evaluation of clinical samples. One of those functions "Multiple window evaluation for blood cell differentiation" was specially tailored for numerical integration of a two-parameter field of blood cells.[2]

EXAMPLES OF EVALUATION PROGRAMS

Complex Preprogrammed Functions

FIGURE 6a shows a typical two-parameter field printed in the so-called top view mode. Six different cell populations can be discriminated: (1) thrombocytes, (2) erythrocytes, (3) reticulocytes, (4) lymphocytes, (5) granulocytes, and (6) test particles. The cells have been stained with 3,3-dihexyloxacarbo-cyanine and acridine orange.[2]

The program, whose flow chart is shown in FIGURE 7, determines the total content of cells in the field, the numbers of the different cell types, the absolute number, the percentage of each cell type in relation to total cell number, the cells/microliter of each cell type and calculates the mean, standard deviation, and coefficient of variation of cell size and fluorescence of each cell type and finally a

TABLE 2.

```
CELL  SAMPLE:- - - - - - - - - - - - - - - - - - - - - - -

IDNR: 000 DATE: 28.11.83
CONTENT ALL WINDOWS: 100176
CONTENT WI6 (TESTP): 000218

THE SAMPLE VOLUME IS CALCULATED FROM THE CONTENT OF WI6
REFERENCE PART.CONC.  IN WI6 MUST BE 5.0 P/MICROL.*250
TOTAL BLOOD DILUTION REQUIRED:    1:250

WI  SUM    %ALL  P/MKLX10 MX    SX    VX     MY    SY    VY    TYPE
1   005246 05.23 030080   08.39 05.73 68.32  10.58 06.48 61.19 TYP1
2   092786 92.62 532032   37.36 01.58 04.24  08.39 02.50 29.85 TYP2
3   001960 01.95 011238   38.04 01.84 04.86  20.60 03.17 15.41 TYP3
4   000067 00.06 000384   43.87 01.23 02.81  47.14 01.60 03.40 TYP4
5   000117 00.11 000670   48.14 01.45 03.02  51.21 01.57 03.07 TYP5
6   000218 00.21 001250   37.46 01.35 03.61  52.07 01.57 03.02 TEST
```

table of the results is printed. The flow chart of FIGURE 7 explains the window evaluation program. The integration windows shown by the pointed lines in FIGURE 6a must be defined before calling up the program. The separate definition of the integration boundaries allows the adaption of the program to very different evaluation problems since the body of the function is fixed; the set of integration windows, however, can be varied arbitrarily. The evaluation protocol as shown in TABLE 2 is printed in a few seconds by the Cytomic 12 analyzer.

Programming by Teaching Command Sequences

The function bodies of functions like the window evaluation are programmed in assembler and cannot be changed by the user. He can, however, build himself more or less complex data evaluation programs by teaching the instrument to execute command sequences. Up to 1000 function calls including complex functions like the window evaluation can be included in a command sequence.

A command sequence may be programmed by switching the instrument into the "command learning mode" and then entering the commands desired. The instrument learns its program directly from the keys entered and the user does not need to know a programming language. A large number of command sequences can be stored on disk.

TABLE 3 shows an example of a command sequence program. The middle column of the table shows the list of commands in normal text, which is printed on request. Commands belonging to a loop, i.e. a repetitive sequence of commands, are displayed to the right. The command sequence program of TABLE 3 collects in its first loop five subsequent two-parameter histograms and stores them on disk. The collecting time, selectable between 1 and 127 seconds, and the starting file code number, which identifies the data set on disk, have to be defined before starting the execution of the sequence.

After readjustment of the initial file code to have the results in the same order as the data collection, the histograms collected in the first loop are evaluated: the

TABLE 3.

```
EXECUTION PROTOCOL            COMMAND LIST

                             PRINT   CR - LINEFEED   132
                             PRINT   CR - LINEFEED   132
                             PRINT   CR - LINEFEED   132
FILE-CODE:  A2200000   DATE 27.05.84    SET COMMND LOOP START 134
FILE-CODE:  A2200001   DATE 27.05.84        COLLECT GRX PRIOR.-X- 2 E
FILE-CODE:  A2200002   DATE 27.05.84        DISPLAY GROUP GRX.      2 0      1. LOOP:
FILE-CODE:  A2200003   DATE 27.05.84        INSERT DISPLAY PERIOD 130       COLLECT AND STORE
FILE-CODE:  A2200004   DATE 27.05.84        STORE ON FLOPPY 2 PAR 248       ON FLOPPY 5 2-P
                                            STORE ON FLOPPY 2 PAR 248       HISTOGRAMS
                                            CLEAR CONTENT OF GRX. 2 3
 (1)  GXO.  FILE CODE: A2200000             CLEAR CONTENT OF GRX. 2 3
      GXO.INTEGRAL=S000785. .2K             LOOPEND,  DEF. LOOPS  135
      K=02 001-D63=S000785. .2K           - NUMBER OF LOOPS: 005
      M=024.40 3=07.41 V=30.39%   CALL USERN-DATE-CODEN. 13C
      K=03 001-D63=S000785. .2K   DECREMENT INPUT DIGIT 13E
      M=016.34 3=10.46 V=63.99%   DECREMENT INPUT DIGIT 13E
                                  DECREMENT INPUT DIGIT 13E
                                  DECREMENT INPUT DIGIT 13E
 (2)  GXO.  FILE CODE: A2200001   DECREMENT INPUT DIGIT 13E        READJUST THE
      GXO.INTEGRAL=S000773. .2K   PRINT   CR - LINEFEED   132      FILE CODE
      K=02 001-D63=S000773. .2K   PRINT   CR - LINEFEED   132
      M=024.34 3=07.28 V=29.91%   PRINT   CR - LINEFEED   132
      K=03 001-D63=S000773. .2K   SET COMMND LOOP START 134
      M=016.15 3=10.22 V=63.29%       READ FROM FLOPPY 2PAR 249
                                      READ FROM FLOPPY 2PAR 249
 (3)  GXO.  FILE CODE: A2200002        PRINT ACTUAL TEXTLINE 131
      GXO.INTEGRAL=S000778. .2K        INTEGR. INSIDE TRACE. 237
      K=02 001-D63=S000778. .2K        PRINT ACTUAL TEXTLINE 131
      M=024.34 3=07.30 V=30.00%        DISPLAY GROUP KXX.      1 0
      K=03 001-D63=S000778. .2K        PRINT ACTUAL TEXTLINE 131   2. LOOP:
      M=016.16 3=10.22 V=63.27%        CALCULATE: M - S - CV 1 1   READ FROM DISK THE
                                       PRINT ACTUAL TEXTLINE 131   5 HISTOGR. MEASURED.
 (4)  GXO.  FILE CODE: A2200003        INCR. GROUPNUMBER KXX 1 5   INTEGRATE 1 WINDOW.
      GXO.INTEGRAL=S000784. .2K        PRINT ACTUAL TEXTLINE 131   CALCULATE M. S AND CV
      K=02 001-D63=S000784. .2K        CALCULATE: M - S - CV 1 1   OF THE X AND Y WIN-
      M=024.44 3=07.42 V=30.38%        PRINT ACTUAL TEXTLINE 131   DOW PROJECTIONS.
      K=03 001-D63=S000784. .2K        PRINT   CR - LINEFEED   132 PRINT THE RESULTS
      M=016.38 3=10.52 V=64.26%        DISPLAY GROUP GRX.      2 0
                                       LOOPEND,  DEF. LOOPS  135
 (5)  GXO.  FILE CODE: A2200004      - NUMBER OF LOOPS: 005
      GXO.INTEGRAL=S000784. .2K   PRINT   CR - LINEFEED   132
      K=02 001-D63=S000784. .2K   PRINT   CR - LINEFEED   :32
      M=024.34 3=07.37 V=30.28%   PRINT   CR - LINEFEED   132
      K=03 001-D63=S000784. .2K
      M=016.28 3=10.33 V=63.43%
```

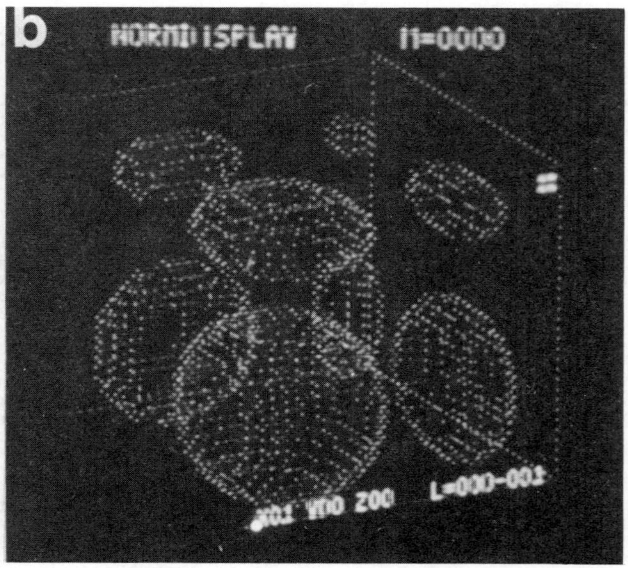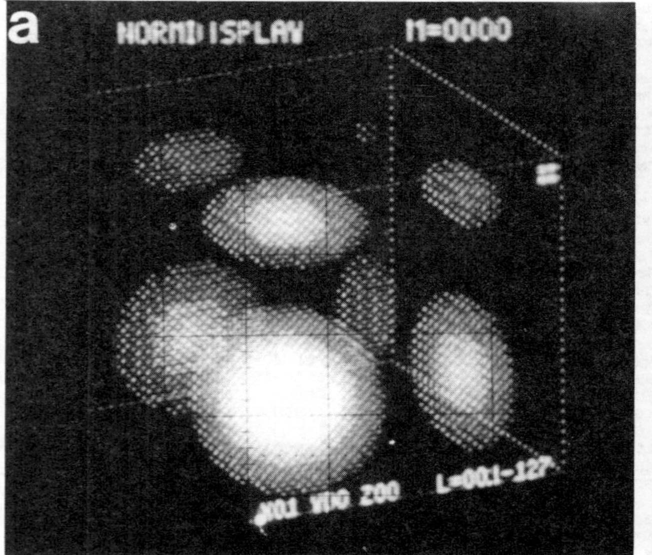

FIGURE 8. (a) Example of a three-parameter cube display taken from the screen of the Cytomic 123 analyzer. The test histogram contains 8 different clusters or clouds of cells of different shape and frequency contents. The frequencies are shown by the brightness of the channel dots. Each axis of the cube, which consists of 32,000 channels, represents a cell feature with a resolution of 32 channels.

(b) Same test histogram as shown in (a). Here, however, only those channels of the clouds are displayed which contain a particle count of 1. The structure looks like the skin of the clouds.

histograms are read from the disk, a predefined integration window is put into the field and evaluated, the projection histograms to X and Y (i.e. the one-parameter histograms of the X- and Y-parameters) inside the window are generated and their means, standard deviations, and coefficients of variation are calculated and printed.

The execution protocol at the left hand side of TABLE 3 shows in its first five lines a print initiated by the disk system during the first loop, which certifies code number and date of each histogram stored.

The printed blocks (1) to (5) contain the evaluation results of the five histograms stored in loop 1.

Three- and Four-Parameter Data Display and Evaluation

With further development of flow cytometry, flow cytometric measurements are performed more frequently with three and more cell parameters. In this case the so-called list mode technique can be applied, where the cell features are stored on disk without classification. When the data uptake is finished the stored cell data are reprocessed from the disk; by gated classification, for example.

The list mode technique is not well suited for clinical application, since the reprocessing time is too long for real-time evaluation of clinical samples.

A suitable approach for online display and processing of multi-parameter data is the Cytomic 123 data module,[21] which is able to display and process in real-time three-parameter, and in gated mode four-parameter, data fields.

The striking advantage of this data device against list mode devices is that 3- and 4-parameter data are immediately available in the form of a classified field, which gives an obvious graphical representation of the data with the ability of interactive definition of spatial evaluation-windows. FIGURE 8 gives an example of a multipeak three-parameter field, which has been taken directly from the screen of the Cytomic 123 prototype. Each axis of the cube represents one of the three parameters with a resolution of 32 steps, e.g. x-axis: cell volume, y-axis: fluorescence 1, and z-axis: fluorescence 2. The whole field consists of $32 \times 32 \times 32 = 32768$ channels. The particle frequencies in the different peaks are emphasized by the brightness of the dots and can be made additionally visible by selecting frequency levels. FIGURE 8a shows the eight-peak three-parameter field including the channels with contents above 1. In the display of FIGURE 8b, however, only channels of the same field are shown with contents of 1. A skin-like structure surrounding the structures of FIGURE 8a is made visible. The structures visible on the screen may be evaluated by an integration function.

The Cytomic 12 and 123 modules can be combined into a data system with two- and three-parameter capabilities, which collects in parallel two- and three-parameter histograms. Data are stored on a disk common to both. The data system shown in FIGURE 4 is such a combined system.

DISCUSSION AND CONCLUSION

Flow cytometers for clinical routine use must have a high degree of automation in order to speed up the sample throughput and to relieve the personnel of tedious routine work. The transducers and the data modules described are the first results of our efforts to fill the gap in this field of flow cytometric instrumentation.

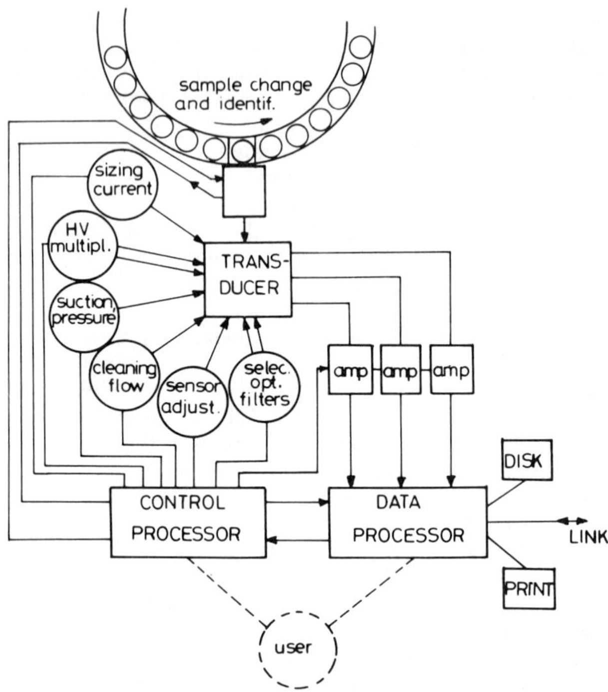

FIGURE 9. Block diagram of a fully automated flow cytometer with automated sample exchange and a control processor, which supervises the transducer settings. The data processor, which can be a Cytomic module, collects and supervises online the data delivered by the transducer and informs the control processor when data are not in the range expected. (HV) = high voltage of the photomultipliers.

The combination of fluorescence measurements with electrical cell volume measurements is certainly the most accurate method for density and concentration measurements. Also, several commercially available flow cytometers use light scatter instead of the electrical method for cell size determination. The use of light scatter for cell volume measurement is problematic since the results are in many ways influenced by the scatter angles, the size range of the cells, and their internal and surface structure, as is pointed out in references 23–25.

Light scatter can be applied when the physical detection of cells is the primary task or cell clusters must be discriminated without exact volume quantification. The electrical Coulter method can be used for the same purpose, but is, however, much better suited for absolute and relative volume quantification and in this way for determination of concentrations and surface densities. A Coulter device on the other hand is more sensitive to electrical distortions and blockages of the relative small sizing orifice. Both of these problems can be overcome by an appropriate set up of the transducer system and the use of carefully prepared and prefiltered fluids and samples.

An automatic flow cytometer needs not only an automatic sample exchange but also several other automatically controlled and adjusted devices. FIGURE 9 shows a schematic diagram of an automatic flow cytometer. The automatic adjustments are performed by a control processor, which regulates and supervises a considerable number of instrumental parameters like the sizing current, high voltage of the multipliers, suction, cleaning, etc. The control processor communicates with the data processor, which collects and evaluates the histograms. Error messages from the control processor may stop the data uptake of the data processor and this on the other hand must be able to supervise the shape and position of the histograms of calibrator particles and inform the control processor that the measuring conditions are in the range expected. This procedure requires a standardized preparation of the cell- and calibrator samples.

The data processor should be able to calculate and print the evaluation results immediately, or in a larger clinic to transfer the data to a central computer during the time the automatic transducer needs to change the cell samples.

The instrumentation described consists as far as possible of standard elements that are available for moderate cost; e.g. 8-bit microprocessor eurocards, and standard objectives, lenses, multipliers, and illumination systems. In this way the instruments can be produced for reasonable cost.

REFERENCES

1. DARZYNKIEWICZ, Z., M. ANDREEFF, F. TRAGANOS & M. R. MELAMED. 1981. Proliferation and differentiation of normal and leukemic lymphocytes as analyzed by flow cytometry. Acta Pathol. S.A.:329–397.
2. VALET, G. 1984. A new method for fast blood cell counting and partial differentiation by flow cytometry. Blut **49**:83–90.
3. LOKEN, M. R. & A. M. STALL. 1982. Flow cytometry as an analytical and preparative tool in immunology. J. Immunol. Methods **50**:85–112.
4. FRANKFURT O.S. & R. P. HUBEN. 1984. Clinical applications of DNA flow cytometry for bladder tumors. Urology **23**:29 34.
5. VALET, G., H. H. WARNECKE & H. KAHLE. 1984. New possibilities of cytostatic drug testing on patient tumor cells by flow cytometry. Blut **49**:37–43.
6. MELAMED, M. R., P. F. MULLANEY & M. L. MENDELSOHN, Eds. 1979. Flow Cytometry and Sorting. John Wiley. New York.
7. TRAGANOS, F., Z. DARZYNKIEWICZ, T. SHARPLESS & M. R. MELAMED. 1977. Simultaneous staining of ribonucleic and desoxyribonucleic acids in unfixed cells using acridine orange in a flow cytofluorometric system. J. Histochem. Cytochem. **25**:46–56.
8. SHAPIRO, H. M., P. J. NATALE & L. A. KAMENTSKY. 1979. Estimation of membrane potentials of individual lymphocytes by flow cytometry. Proc. Natl. Acad. Sci. USA **76**:5728–5730.
9. VALET, G., A. RAFFAEL, L. MORODER, E. WUNSCH & G. RUHENSTROTH-BAUER. 1981. Fast intracellular pH determination in single cells by flow cytometry. Naturwiss. **68**:265–266.
10. SHAPIRO, H. M. 1983. Multistation multiparameter flow cytometry: A critical review and rationale. Cytometry **3**:227–243.
11. KACHEL, V. 1979. Electrical resistance pulse sizing. *In* Flow Cytometry and Sorting. M. R. Melamed, P. F. Mullaney & M. L. Mendelsohn, Eds. John Wiley. New York.
12. KACHEL, V. 1982. Sizing of cells by the electrical resistance pulse technique: Methodology and application in cytometric systems. *In* Cell Analysis, Vol. 1. N. Catsimpoolas, Ed. Plenum. New York.
13. PETERS, D. C. 1979. A comparison of mercury arc lamp and laser illumination for flow cytometers. J. Histochem. Cytochem. **27**:241–245.

14. STEEN, H. B., T. LINDMO & O. SOERENSEN. 1980. A simple high resolution flow cytometer based on a standard fluorescence microscope. *In* Flow Cytometry IV. O. D. Laerum, T. Lindmo & E. Thorud, Eds.:31–33. Universitetsforlaget. Bergen, Oslo.

15. WEGMANN, F. Eds. 1979. Pulse Cytophotometer ICP-11 and ICP-22. *In* Flow Cytometry and Sorting. M. R. Melamed, P. F. Mullaney & M. L. Mendelsohn, Eds.:674–678. John Wiley. New York.

16. KACHEL, V., E. GLOSSNER, E. KORDWIG & G. RUHENSTROTH-BAUER. 1977. Fluvo Metricell, A combined cell volume and cell fluorescence analyzer. J. Histochem. Cytochem. **25**:804–812.

17. FULWYLER, M. J., C. W. McDONALD & J. L. HAYNES. 1979. The Becton Dickinson FACS IV Cell Sorter. *In* Flow Cytometry and Sorting. M. R. Melamed, P. F. Mullaney & M. L. Mendelsohn, Eds. John Wiley. New York.

18. SUPER, B. S. 1979. The Ortho Fluorograf. *In* Flow Cytometry and Sorting. M. R. Melamed, P. F. Mullaney & M. L. Mendelsohn, Eds. John Wiley. New York.

19. KACHEL, V., E. GLOSSNER & H. SCHNEIDER. 1982. A new flow cytometric transducer for fast sample throughput and time resolved kinetic studies of biological cells and other particles. Cytometry **3**:202–212.

20. KACHEL, V., H. SCHNEIDER & K. SCHEDLER. 1980. A new flow cytometric pulse height analyzer offering microprocessor controlled data acquisition and statistical analysis. Cytometry **1**:175–192.

21. KACHEL, V. & H. SCHNEIDER. 1982. Ineractive three-parameter data acquisition and display module. Analyt. Cytol. Cytom. IX. Abstr. p. 105.

22. KACHEL, V., K. SCHEDLER, H. SCHNEIDER & L. HAACK. 1984. "Cytomic" Data system modules—Modern electronic devices for flow cytometric data handling and presentation. Cytometry **5**:299–303.

23. MULLANEY, P. F., J. M. CROWELL, G. C. SALZMAN, J. C. MARTIN, R. D. HIEBERT & C. A. GOAD. 1976. Pulse-height light-scatter distributions using flow-systems instrumentation. J. Histochem. Cytochem. **24**:298–304.

24. SALZMAN, G. C. 1982. Light scattering analysis of single cells. *In* Cell Analysis. N. Catsimpoolas, Ed.:**1**:111–143. Plenum. New York.

25. KERKER, M. 1983. Elastic and inelastic light scattering in flow cytometry. Cytometry **4**:1–10.

Cell and Nuclear Growth During G_1: Kinetic and Clinical Implications[a]

ZBIGNIEW DARZYNKIEWICZ[b] AND FRANK TRAGANOS

Memorial Sloan-Kettering Cancer Center
New York, New York 10021

and

LISA STAIANO-COICO

Cornell University Medical College
New York, New York 10021

INTRODUCTION

Many antineoplastic agents are cell cycle- or phase-specific, affecting only cell populations progressing through the cycle. However, in addition to tumor cells, normal cells from a variety of tissues (e.g., bone marrow, gut) actively proliferate. Therefore, knowledge of both tumor and normal (stem) cell kinetics should be of great value in designing effective drug treatment protocols.

Despite advances in the monitoring of cell cycle kinetics, especially since the introduction of flow cytometry, it is still not possible to adequately implement analysis of cell cycle kinetics in the clinic. If this were possible, one could envisage the routine analysis of the cell cycle as a procedure yielding data that would then be used for the selection of optimal drug combinations, customized to individual patients and/or stage of disease. The limitation of flow cytometry for this application is that it generally provides only a static "snapshot" of the data and little kinetic information. Simple enumeration of cells in $G_0 + G_1$ versus S versus $G_2 + M$ without the information on the kinetics in each of these phases, their proliferative potential, or proportion of stem cells is of little practical value. This is especially true in light of the accumulating evidence on the presence of noncycling or slowly cycling S or G_2 phase (S_Q, G_{2Q}) cells in tumors (see for example refs. 1 and 2).

Two distinct flow cytometric approaches have been used to obtain cell cycle kinetic information. The first approach is direct and is based on the use of sequential "snapshots" in time of the DNA distribution of cell populations. This is done in conjunction with S-phase cell labeling by BrdUrd followed by cytochemical[3-5] or immunochemical[6,7] detection of this precursor, or by ^{3}H-Tdr labeling, cell sorting, and ^{3}H-detection.[8] Limitation of these direct techniques stems from the fact that cell cycle progression is determined *in vitro* after tumor

[a]Supported by USPHS Grants CA23296 and 28704.

[b]Address correspondence to: Dr. Z. Darzynkiewicz, Sloan-Kettering Institute, Walker Laboratory, 145 Boston Post Road, Rye, New York 10580. Telephone: (914) 698-1100, ext. 237.

excision. It is a common observation that cell cycle progression is often altered after transferring isolated tumor cells to tissue culture. Thus, since labeling requires *in vitro* growth, at least for a short period of time, the results of such an analysis may not be representative of the *in vivo* situation.

The second approach is indirect and is based on the observed correlation between various cellular metabolic parameters and the rate of cell progression. Thus, there is a close relationship both between the RNA content of cells and their progression through G_1 and S[9,10] and their cycling versus noncycling status.[1,11,12] Likewise, the sensitivity of DNA *in situ* to denaturation correlates with cell cycle kinetics.[1,13] Therefore, a multivariate analysis of cell populations (e.g., RNA or protein content versus DNA content) can provide an estimate of the kinetic behavior of cells. However, with this indirect approach the correlation may not always be apparent.[14] As a result, this approach (see review in ref. 15) proves useful only when it is applied to carefully characterized cell systems in which the correlation between the desired parameters have already been established.

The emphasis in the present study is on the analysis of exponentially growing cell populations in which a correlation could be observed between RNA or protein content, nuclear nonhistone protein content, and/or resistance of DNA *in situ* to acid denaturation and cell kinetic behavior.

MATERIALS AND METHODS

Cells

Chinese hamster ovary (CHO) cells obtained from Los Alamos National Laboratory were maintained as growing monolayer cultures in F10 medium supplemented with 10% fetal bovine serum and antibiotics.[9,16] A detailed description of these cells, culturing conditions, and experimental procedures is given elsewhere.[16] The murine leukemia L1210 cell line was maintained in Dulbecco-Vogt modified Eagle's medium containing 10% fetal bovine serum, as described.[17] Details of the kinetic experiments are given elsewhere.[16] HL-60 cells were obtained from Dr. Harry Crissman of the Los Alamos National Laboratory and were grown in suspension in RPMI 1640 medium containing 10% fetal calf serum.

Nuclei Isolation

The cells were rinsed in phosphate-buffered saline solution and then transferred to a solution containing 0.32 M sucrose, 10 mM Tris, 5 mM $MgCl_2$, 1 mM $CaCl_2$ and 0.1% Triton X-100 at pH 7.2. Cells were broken using a glass-pestle homogenizer. The nuclei were rinsed twice in the sucrose solution described above minus the Triton X-100 and resuspended in the sucrose buffer. This procedure yielded clean nuclei from L1210 cells. To isolate nuclei from HL60 or CHO cells the method of Prentice and Gurley[18] was used.

Cell Staining, Flow Cytometry

The method of simultaneous staining of DNA versus RNA with acridine orange (AO) is described in detail in earlier publications.[14,20] Staining of total cell protein versus DNA was performed as described by Crissman and Steinkamp[21] using the combination of Hoechst 33342 and rhodamine 640. Partial denaturation of DNA *in situ* was induced in fixed, RNase-treated cells exposed to pH 1.5 and then stained with AO at pH 2.6.[13] Proteins and DNA of isolated nuclei were stained by combination of fluorescein isothiocyanate (FITC) and propidium iodide (PI) according to Crissman and Steinkamp[21,22] in the Pollack *et al.* modification.[23] The fluorescence of individual cells stained with AO, FITC, and PI was measured using the FC200 or H50 flow cytometers (Ortho Diagnostics, Westwood, MA.) as described.[19,20] Cells stained with PI and rhodamine 640 were measured in the dual laser system in Los Alamos National Laboratory as described.[16] The measurements of whole cells (DNA versus RNA, DNA versus protein, partial DNA denaturation) and of the isolated nuclei were done for all these cell lines, with essentially similar results (e.g., the distribution of G_{1A} and G_{1B} compartments). To avoid redundancy, only three examples, one each for DNA, RNA, and protein (CHO cells, Fig. 1); DNA and protein (isolated nuclei HL-60 cells, Fig. 2); and DNA *in situ* measurements (L1210 cells, Fig. 3), are given.

RESULTS

The distributions of individual cells with respect to their DNA versus RNA or DNA versus protein content are shown in Figure 1. These two distributions are similar indicating a close relationship between RNA and protein content in individual cells across the cell cycle. It is not surprising to see this relationship inasmuch as the total RNA detected in the cells is predominantly (~85%) rRNA and the rate of protein synthesis is expected to correlate with the number of ribosomes. The populations, however, are more heterogeneous with respect to cellular protein than RNA content.

A conspicuous feature of these distributions (Fig. 1) is the presence of a threshold RNA or protein content (marked by the arrows) in G_1. It is quite evident that cells with a subthreshold amount of RNA or protein (G_{1A}) do not enter S phase. Thus, in exponentially growing, unperturbed populations, cells have to accumulate a critical RNA or protein content prior to initiation of DNA replication. This restriction is rigorously controlled. We have studied the RNA content of numerous normal and tumor cell types, all of which demonstrated this characteristic.

To discriminate G_{1A} versus G_{1B} cells reproducibly, an objective criterion is needed. Toward this end, a gating window (based on DNA content) can be set in early-S phase, the cells measured within the window, the mean value and standard deviation of cellular RNA or protein content obtained, and the threshold established at the level of the mean value minus three standard deviations (Fig. 1). Thus, by definition, the G_{1A} cells have an RNA or protein content significantly lower than S-phase cells. An important point should be emphasized here, namely, that the detection of the G_{1A} and G_{1B} compartments

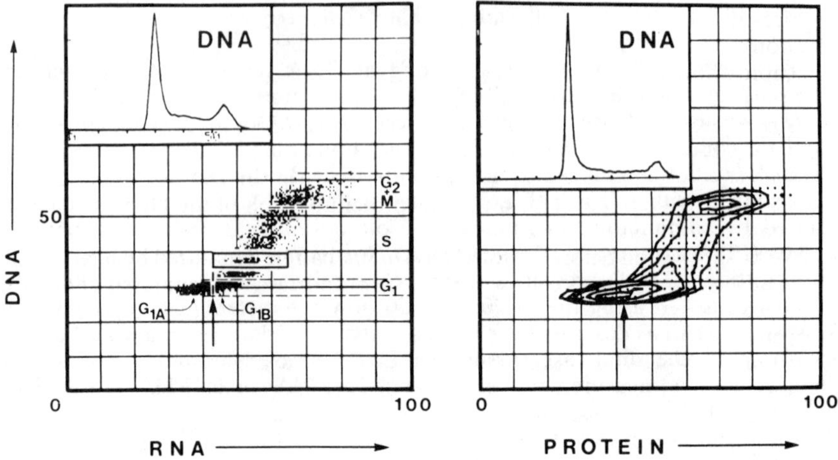

FIGURE 1. The cytogram and contour map representing the distribution of CHO cells from exponentially growing cultures, with respect to cellular DNA versus RNA and DNA versus protein content, respectively.

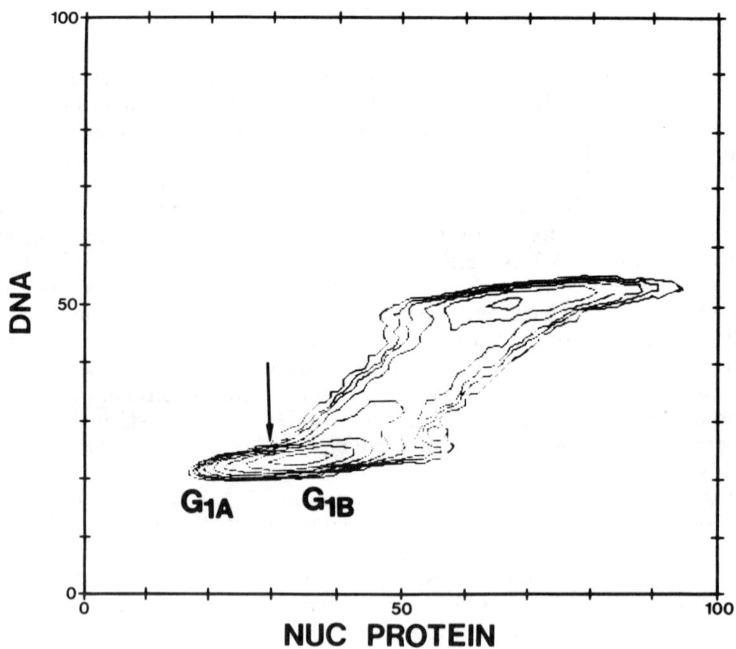

FIGURE 2. Protein and DNA content of nuclei isolated from the exponentially growing HL-60 cells.

requires both high cellular viability and accurate measurements (high resolution). Dead or broken S-phase cells make determination of the appropriate thresholds difficult or impossible.

FIGURE 2 illustrates the distribution of nuclei isolated from exponentially growing HL-60 cells with respect to their protein and DNA content. Here, as in the case of whole-cell RNA or protein content (FIG. 1), it is also evident that cells must attain a critical content of nuclear proteins prior to their entrance to S phase. No cells originating from the low protein portion of the G_1 cluster can be seen entering S phase.

We have reported before[13,24] that DNA *in situ* can be denatured by heat or acid and, after staining of such cells with AO, the proportions of denatured and native DNA can be estimated from the degree of red and green luminescence, respectively. Interestingly, there are significant differences in DNA sensitivity to denaturation between cells in different phases of the cell cycle. This principle has been used to estimate proportions of cells in different compartments of the cell cycle including discrimination between cells in G_0 and G_1 or G_2 and M phases.[25] The application of this staining technique is demonstrated in FIGURE 3 in combination with a stathmokinetic experiment to study cell exit from G_1.

Analysis of exponentially growing cells (FIG. 3a) reveals the presence of the threshold value of α_t for the G_1 population (arrow). Thus, as in the case of the previously described thresholds for total RNA or protein or nuclear protein content, there is a critical value of α_t (resistance of DNA *in situ* to denaturation) in G_1. As we have shown before,[13,24,25] the resistance to denaturation correlates strongly with chromatin condensation. One can deduce, therefore, that following mitosis, the cells' nuclear chromatin must decondense by some critical degree prior to initiation of DNA replication. Here again, the restriction is strictly controlled and no cells entering S with high α_t values can be seen, regardless of the cell type.

In the stathmokinetic experiment as illustrated in FIGURE 3, the kinetics of cell exit from G_1 and transition from G_{1A} to G_{1B} can be measured. By establishing the gating windows on both axes, i.e. DNA content and α_t, the number of cells in G_{1A} and G_{1B} at different times after addition of Colcemid can be calculated and plotted (FIG. 3d). The data show that the exit from G_1 or G_{1A} has an exponential component represented as a straight slope on a semilogarithmic scale. In the case of G_{1A} population, this exponential slope is evident from the onset of stathmokinesis. The first rate kinetics of exit from G_{1A}, reproduced with different cell types[25] demonstrates that cells have very heterogeneous residence times in G_{1A}, with a characteristic exponential "probabilistic-like" component.

DISCUSSION

Do cells attain a critical mass or RNA content before initiating DNA replication? Despite extensive studies there is still a controversy regarding the idea of "critical" mass. There are numerous publications in support of the threshold mass in G_1 (e.g., 1,25,26, and a review in 27). On the other hand, there are a few reports in which the authors were not able to observe such thresholds.[28, 29] The possible causes for this discrepancy are difficult to assess. It is obvious from the bivariate flow cytometric data shown here for CHO cells and observed by us[1,25] and others[30,31] in numerous other cell systems that during nonperturbed exponential growth no cells with a subthreshold RNA content enter S phase. The only

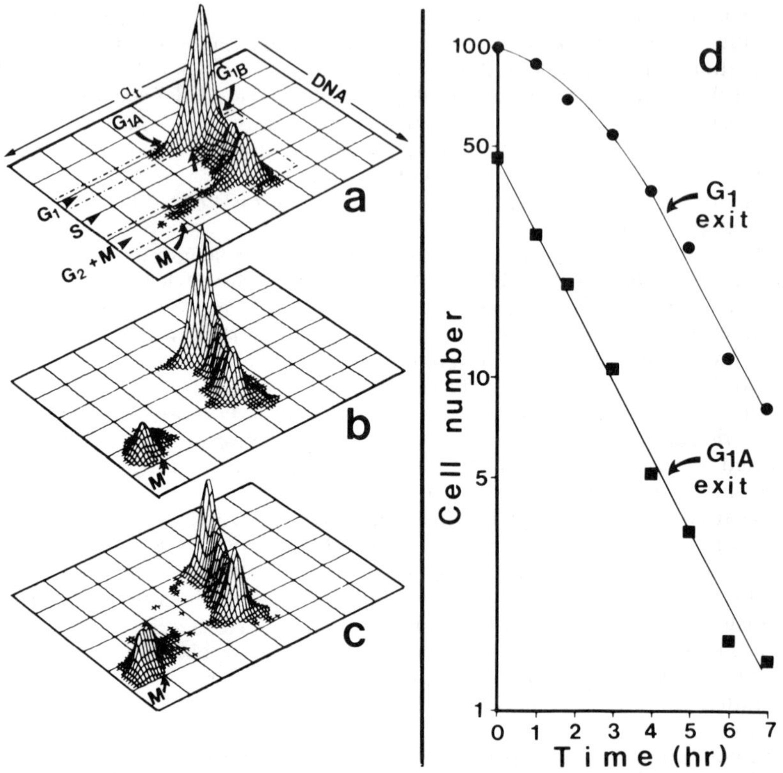

FIGURE 3. Recognition of different cell cycle compartments as shown in this figure, can be obtained after partial denaturation of DNA *in situ* with acid followed by staining with AO. Denatured DNA stains metachromatically in red whereas the native portion fluoresces green. The α_t value represents the proportion of denatured DNA.[24] In this kinetic experiment, exponentially growing CHO cells were treated with colcemid and the cells were sampled at hourly intervals. The histograms at 0 (a), 2 (b) and 4 hr (c) during the stathmokinesis are shown. The number of cells in G_{1A} and G_1 are plotted versus the time in culture after addition of colcemid (d).

situations in which well-defined thresholds could not be observed occurred when a significant number (>5%) of cells in the samples were nonviable, or when cells grew under adverse conditions (e.g. suboptimal pH) resulting in nonexponential growth rates. It is possible, however, that in certain cell types the content of RNA and protein per cell in G_1 cells is not correlated and thus the presence of the RNA thresholds may not coincide with the thresholds based on protein dry mass or volume measurements. The RNA threshold, however, is clearly a universal feature of all exponentially growing, unperturbed cell populations.

It should be emphasized that critical cell size (RNA-protein-content), although observed during exponential growth, is not the prerequisite for cell entrance into S phase. When cell growth is perturbed, e.g., by adenovirus infection,[32] protein-synthesis suppression,[33] or RNA-synthesis suppression (our observation, unpublished) the cells may initiate DNA replication despite their subthreshold size. Thus, cell growth in G_1 and initiation of DNA synthesis can be dissociated as originally pointed out by Mitchison[34] indicating that these two processes are under different regulatory controls and are not causally related. These processes, however, are coordinated in unison during unperturbed exponential growth. This subject was recently reviewed by Baserga.[35]

There is less evidence as to whether there is a critical nuclear size (i.e., nuclear protein content) necessary for entrance into S phase. The data of Auer[36] and Auer and Zetterberg[37] indicate a requirement for growth in nuclear size (dry mass) before initiation of DNA synthesis. There is also compelling evidence from cell fusion experiments that initiation of DNA replication in nuclei of G_0 cells reactivated in heterokaryons is preceded by growth in nuclear size (see review in ref. 38). Likewise, flow cytometric data of Roti Roti *et al.*[22] and Pollack *et al.*[23] indicate that DNA replication is initiated only in those nuclei which attained some critical size. On the other hand, Yen and Pardee[39] and Johnson *et al.*[40] reported that G_1 cells can enter S phase without an increase in nuclear size. Here again the reasons for the contradictory data are unknown.

Because the histone/DNA ratio remains constant during the cell cycle, the observed differences in nuclear mass (protein) in G_1 cells must be due to an altered ratio of nonhistone/histone proteins. The threshold content of nuclear proteins (FIG. 2) thus indicates that following mitosis there is an influx of nonhistone proteins to the nucleus and that cells do not initiate DNA replication until a critical nonhistone/histone ratio is established. The proteins that enter the nucleus may be involved in decondensation of nuclear chromatin, which takes place after mitosis and prior to DNA replication.

Our data on DNA denaturation *in situ* support the above assumption. Namely, the sensitivity of DNA *in situ* to acid-induced denaturation, which was shown to correlate with chromatin condensation,[1,13,41] changes during G_1 (FIG. 3).

As in the case of cellular RNA or protein content, it is possible to discriminate G_{1A} and G_{1B} compartments based on changes in DNA stability to denaturation. In direct comparisons, the proportions of cells in G_{1A} versus G_{1B} were similar regardless of whether determined by RNA content or DNA stability to denaturation, under various growth conditions.[41] This would also suggest that decondensation of chromatin in G_1 is correlated with accumulation of RNA (predominantly rRNA) in the cell. Molecular mechanisms responsible for the observed differences in stability of DNA *in situ* are unknown.

The data discussed above indicates that following mitosis cells grow, increasing their RNA and protein content and nonhistone/histone nuclear protein ratio. There is a simultaneous decondensation of nuclear chromatin, the latter reflected by a decrease in the sensitivity to denaturation of DNA *in situ*. DNA replication is initiated only when a threshold value of the above parameters is reached. Thus, prior to S phase, cells equalize with respect to their growth parameters and chromatin condensation. We proposed to distinguish this growth phase as a distinct metabolic compartment termed G_{1A}.[1,25] Cells leave G_{1A} as a metabolically uniform cohort and enter the prereplicative compartment termed G_{1B}. According to Mitchinson's concept of the cell cycle,[34] G_{1A} is the cell growth phase which is distinct from the DNA division cycle. Whereas only cell growth takes place during G_{1A}, during G_{1B} and other phases both the growth and DNA division cycle proceed.

The immediately postmitotic cell population is very heterogeneous with respect to protein or RNA content.[16] This heterogeneity is the consequence of both the variability of the mother (mitotic) cell population and asymmetric cytokinesis.[16] It was shown before that the duration of the G_1 phase of individual cells correlates inversely with the size[42] or RNA content they inherit.[16] It would be expected, therefore, that the "growth" or "equalization" subphase (G_{1A}) will be heterogeneous, reflecting the metabolic inequality of the postmitotic cells. Indeed the residence times in G_{1A} show a very wide distribution and are characterized by an exponential-like component (refs. 1 and 25; see FIG. 3d). This first-order kinetics of cell equalization is most likely the main cause of the "probabilistic" distribution of cell division times and equivalent to the "A" state in the Smith and Martin transition probability model of the cell cycle.[43] The "probabilistic" event in the metabolic model, however, is a reflection of the heterogeneity of mother cell population plus their unequal cytokinesis. Actually, asymmetric cytokinesis may indeed be a genuine probabilistic event generating the probabilistic-like distributions of the subsequent kinetic events.

There are striking similarities between the "noncompetent" state of the cell cycle as proposed by Pledger[44-46] and the pre-G_{1B} portion of the cycle of our model.[1,16] In the case of exponentially growing cells, the noncompetent state would include the deep quiescent state as established by metabolic parameters (G_{1Q}) plus G_{1A}.[1]

There are two significant clinical implications of the cell cycle model discussed above. The first derives from the fact that between divisions, each cell passes sequentially through two distinct phases, growth and DNA replication/division. It is conceivable that metabolic processes, nutrient requirements, and sensitivity to various antimetabolites or poisons are different for each of these phases. So far, in designing new antineoplastic agents or analyzing their effects, the emphasis has been placed on DNA replication/division while little attention has been given to cell growth. Also, few studies have been done to compare the growth versus DNA replication/division phases of normal and transformed cells. Better understanding of the metabolic differences between these phases and their relationship in the lifetimes of normal versus tumor cells may initiate new strategies for selective treatments of tumor cell populations.

The second implication of the model relates to the wide distribution of the metabolic-equalization times, i.e. times of cell residence in G_{1A}. The distributions show the exponential-like component, such as seen in FIGURE 3d for each of the cell types studied, regardless of whether G_{1A} estimates were based on RNA content or chromatin structure measurements. Because of this heterogeneity, eradication of tumor cell populations even consisting of only cycling cells by the cell cycle–specific drugs, requires much longer treatment than predicted based on the mean estimates of the linearly distributed cell cycle times. Indeed, the sensitivity of cells to most antitumor treatments, when plotted either versus dose or duration of treatment is similar to that of the G_{1A} residence time.

It is possible that there is a causal relationship between the sensitivity of cells to certain treatments and duration of G_{1A}. The following example may illustrate this relationship. When G_{1A} cells are treated with a DNA-damaging agent (e.g., ionizing radiation) the consequences of the DNA damage with respect to cell survival become apparent only after cells enter the DNA replication/division cycle. If cells can repair the damage while in G_{1A}, they should remain viable. Hence, distributions of residence times in G_{1A} and cell survival distributions should show similarities, which appears to be the case (exponential-like distribution).

The degree of heterogeneity of cell generation times or degree of metabolic heterogeneity of cell populations can be estimated from the proportion of cells residing in G_{1A}. With the knowledge of the half-time of cell residence in G_{1A} and the proportion of cells in this compartment, it may be possible to estimate the duration of treatment needed for a given log-cell-kill effect or to select drugs which have preference either towards cells in the growth (G_{1A}) or DNA replication/division phases.

Recognition of the metabolic compartments of the cell cycle may thus be of use in the clinic in designing treatment protocols for individual patients.

ACKNOWLEDGMENT

We thank Miss Robin Nager for help in the preparation of this manuscript.

REFERENCES

1. DARZYNKIEWICZ, Z., F. TRAGANOS & M. R. MELAMED. 1980 Cytometry 1:98–108.
2. ALLISON, D. C., J. M. YUHAS, R. F. RIDOLPHO, S. L. ANDERSON & L. S. JOHNSON. 1983. Cell Tissue Kinet. 16:237–243.
3. LATT, S. A. 1974. J. Cell. Biol. 62:546–550.
4. BÖHMER, R. M. & J. ELLWART. 1981. Cell Tissue Kinet. 14:653–658.
5. DARZYNKIEWICZ, Z., M. ANDREEFF, F. TRAGANOS & M. R. MELAMED. 1978. Exp. Cell Res. 115:31–35.
6. GRATZNER, H. G. & R. C. LEIFF. 1981. Cytometry 1:385–389.
7. DOLBEARE, F., H. GRATZNER, M. G. PALLAVICINI & J. W. GRAY. 1983. Proc. Natl. Acad. Sci. USA 80:5573–5577.
8. GRAY, J. W., J. H. CARVER, Y. S. GEORGE & M. L. MENDELSOHN. 1977. Cell Tissue Kinet. 10:97–107.
9. DARZYNKIEWICZ, Z., D. P. EVENSON, L. STAIANO-COICO, T. K. SHARPLESS & M. R. MELAMED. 1979. J. Cell Physiol. 100:425–438.
10. FUJIKAWA-YAMAMOTO, K. 1982. J. Cell Physiol. 112:60–66.
11. BAUER, K. D. & L. A. DETHLEFSEN. 1981. J. Cell Physiol. 108:99–112.
12. WALLEN, A. C., R. HIGASHIKUBO & L. A. DETHLEFSEN. 1984. Cell Tissue Kinet. 17:65–77.
13. DARZYNKIEWICZ, Z., F. TRAGANOS, T. K. SHARPLESS & M. R. MELAMED. 1977. Cancer Res. 37:4635–4640.
14. BROCK, W. A., D. E. SWARTZENDRUBER & D. J. GRDINA. 1982. Cancer Res. 42:4999–5003.
15. DARZYNKIEWICZ, Z. 1984. *In* Growth, Cancer and the Cell Cycle. P. Skehan & S. J. Friedman, Eds.:249–278. Humana Press. Clinton, NJ.
16. DARZYNKIEWICZ, Z., H. CRISSMAN, F. TRAGANOS & J. STEINKAMP. 1982. J. Cell Physiol. 113:465–474.
17. TRAGANOS, F., L. STAIANO-COICO, Z. DARZYNKIEWICZ & M. R. MELAMED. 1981. Cancer Res. 41:780–789.
18. PRENTICE D. A. & L. R. GURLEY. 1983. Biochim. Biophys. Acta. 740:134–144.
19. DARZYNKIEWICZ, Z., F. TRAGANOS, T. SHARPLESS & M. R. MELAMED. 1976. Proc. Natl. Acad. Sci. USA 76:358–362.
20. TRAGANOS, F., Z. DARZYNKIEWICZ, T. SHARPLESS & M. R. MELAMED. 1977. J. Histochem. Cytochem. 25:46–46.
21. CRISSMAN, H. A. & J. STEINKAMP. 1982. Cytometry 3:84–90.
22. ROTI ROTI, J. L., R. HIGASHIKUBO, O. C. BLAIR & N. UYGUR. 1982. Cytometry 3:91–93.
23. POLLACK, A., H. MOULIS, N. L. BLOCK & G. L. IRWIN III. 1984. Cytometry 5:473–481.

24. DARZYNKIEWICZ, Z., F. TRAGANOS, T. SHARPLESS & M. R. MELAMED. 1975. Exp. Cell Res. **90**:411–428.
25. DARZYNKIEWICZ, Z., T. SHARPLESS, L. STAIANO-COICO & M. R. MELAMED. 1980. Proc. Natl. Acad. Sci. USA **77**:6696–6699.
26. KILLANDER D. & A. ZETTERBERG. 1965. Exp. Cell Res. **38**:272–284.
27. PRESCOTT, D. M. 1976. Reproduction of Eukaryotic Cells. Academic Press, New York.
28. FOX, T. & A. PARDEE. 1970. Science **167**:80–81.
29. GERSHEY, E. L., R. M. D'ALISA & R. M. ZUCKER. 1979. Exp. Cell Res. **122**:9–14.
30. MONROE, J. G. & J. C. CAMBIER. 1983. J. Immunol. **130**:626–631.
31. KRISTENSEN, F., C. WALKER, F. BETTENS, F. JONCOURT & A. L. deWECK. 1982. Cell Immunol. **74**:140–149.
32. POCHRON, S, M. ROSINI, Z. DARZYNKIEWICZ, F. TRAGANOS & R. BASERGA. 1980. J. Biol. Chem. **255**:4411–4414.
33. RØNNING, O. W. & T. LINDMO. 1983. Exp. Cell Res. **144**:171–179.
34. MITCHISON, J. M. 1971. The Biology of the Cell Cycle. Cambridge University Press. London.
35. BASERGA R. 1984. Exp. Cell Res. **151**:1–5.
36. AUER, G. 1972. Exp. Cell Res. **75**:231–236.
37. AUER, G. & Z. ZETTERBERG. 1972. Exp. Cell Res. **75**:245–253.
38. RINGERTZ, N. R. & R. E. SAVAGE. 1976. Cell Hybrids. Academic Press. New York.
39. YEN, A. & A. B. PARDEE. 1979. Science **204**:1315–1317.
40. JOHNSON, T. S., D. E. SWARTZENDRUBER & J. C. MARTIN. 1981. Exp. Cell Res. **134**:201–205.
41. DARZYNKIEWICZ, Z., F. TRAGANOS, S. B. XUE & M. R. MELAMED. 1981. Exp. Cell Res. **136**:279–293.
42. KILLANDER, D. & A. ZETTERBERG. 1965. Exp. Cell Res. **40**:12–20.
43. SMITH, J. A. & L. MARTIN. 1983. Proc. Natl. Acad. Sci. USA **70**:1263–1266.
44. PLEDGER, W. J., C. D. STILES, H. N. ANTONIADES & C. D. SCHER. 1978. Proc. Natl. Acad. Sci. USA **75**:2839–2843.
45. STURANI, E., L. TOSCHI, R. ZIPPEL, E. MORTEGANI & L. ALBERGHINA. 1984. Exp. Cell Res. **153**:135–144.
46. SEUWEN, K., U. STEINER & G. ADAM. 1984. Exp. Cell Res. In press.

Flow Cytometric Analysis of Cell Kinetic Responses Using Measurements of Correlated DNA and Nuclear Protein[a]

A. POLLACK,[b] H. MOULIS, D. L. PRUDHOMME,
N. L. BLOCK, AND G. L. IRVIN III

Research Service
Veterans Administration Medical Center
and
Departments of Surgery, Urology, and Microbiology
University of Miami School of Medicine
Miami, Florida 33125

INTRODUCTION

Single parameter flow cytometric histograms of DNA content can be resolved into three compartments—G_0+G_1, S, and G_2+M. Darzynkiewicz and colleagues[1-4] have shown that, using acridine orange to stain single- and double-stranded nucleic acids, G_1 can be subdivided into early and late compartments (G_{1A} and G_{1B}, respectively), and under certain conditions G_2 can be separated from M. Moreover, these investigators have been able to quantitate resting G_{1Q} (i.e. G_0) cells in certain populations.[3] Their data show that two-parameter flow cytometry represents an important improvement over single parameter DNA histogram analysis—with the potential for providing valuable information regarding recruitment and an added sensitivity to cell kinetic changes.

One of the major drawbacks of most two-parameter procedures is that they are not suitable for the analysis of solid tumor tissue because they require whole cells. Preparation of single cells from solid tumor tissue is often time-consuming and may not be feasible when large numbers of samples are analyzed daily. Since Roti Roti *et al.*[5] described a very promising technique for multiparameter cell kinetic analysis that involves nuclei, we have worked to simplify and adapt their procedure for application to solid tumors.[6] We report here that multiple cell cycle compartments can be identified using a rapid method for the simultaneous staining of DNA and nuclear protein in stripped nuclei. In this procedure,

[a]Supported in part by: Veterans Administration Merit Review (4321-001), National Institutes of Health Biomedical Research Support Grant (S07 RR 05363), and Weeks Endowment Fund, Department of Urology, University of Miami School of Medicine.

[b]Address correspondence to: Alan Pollack, Ph.D., Research Service 151, Veterans Administration Medical Center, 1201 N.W. 16th St., Miami, Florida 33125. Phone: 305-324-3186.

nuclear isolation, DNA staining with propidium iodide (PI),[c] and nuclear protein staining with fluorescein isothiocyanate (FITC) are accomplished without centrifugation. Log phase EL4 murine leukosis cells and mitogen-stimulated human peripheral blood lymphocytes were used to characterize the different compartments in the two parameter histograms. The R3327-G rat prostatic adenocarcinoma model was used to establish the application of the DNA/nuclear protein method to the analysis of solid tumors.

MATERIALS AND METHODS

EL4 Suspension Cultures

EL4 murine leukosis tumor cells[7] were grown in suspension cultures with RPMI 1640 medium supplemented with 1 mM L-glutamine, 20% heat inactivated fetal bovine serum, 0.23% sodium bicarbonate, 100 U/ml penicillin, and 100 µg/ml streptomycin (GIBCO, Grand Island, NY). On day 0 of culture, 1–2×10^6 cells were suspended in 9 ml of supplemented RPMI 1640 in 3025 Costar flasks (Cambridge, MA). Following an overnight incubation at 37°C, in a 100% humidified-5% CO_2 environment, the cells were treated with colcemid (2.0 µg/ml) for various periods of time as described in the figure legends.

Lymphocyte Microcultures

Human peripheral blood lymphocyte (HPBL) microcultures were prepared as described previously.[7,8] Briefly, the lymphocytes were isolated using Lympho-Paque (Nyegaard & Co., Oslo, Norway) and washed in Medium 199 containing 1 mM L-glutamine, 100 U/ml penicillin, 100 µg/ml streptomycin, 20% heat-inactivated fetal bovine serum, and 0.23% sodium bicarbonate. The cells were cultured at a concentration of 1×10^5 per 0.2 ml microculture in the presence or absence of PHA (highly purified, Welcome Reagents Limited, Beckenham, England). Treatment of lymphocytes with hydroxyurea (2mM), colcemid (1.0 µg/ml), or ^{3}H-TdR (5 µCi/ml) was initiated 12–20 h prior to culture termination. The above compounds were obtained from Sigma (St. Louis, MO), except for the ^{3}H-TdR (2 Ci/mmole; methyl-^{3}H) which was obtained from New England Nuclear, Boston, MA.

R3327-G Solid Tumors

The R3327-G rat prostatic adenocarcinoma was grown in Copenhagen $\times$ Fischer F_1 male rats. The procedures for cell dispersion using enzymes and subcutaneous inoculation have been described in detail previously.[9,10] R3327-G

---

[c]Abbreviations: Acridine Orange= AO; diploid= 2C; coefficient of variation= CV; ethylenediaminetetraacetic acid= EDTA; flow cytometry= FCM; fluorescein isothiocyanate= FITC; human peripheral blood lymphocytes= HPBL; nonidet P40= NP40; nuclear isolation buffer= NIB; phytohemagglutinin= PHA; propidium iodide= PI; tritiated thymidine= ^{3}H-TdR.

cells (2×10^7/rat) implanted on day 0 into 100 g male rats form tumors that are 0.5–1 cc in volume at 2–4 weeks. Following treatment by castration (or sham castration), the tumors were excised, weighed, and cut into small pieces for flow cytometric analysis. The tumor pieces were stored dry at -80°C in cryotubes (Nunc, Denmark).

Sample Preparation

The procedure used was a modification of that recently reported by Roti Roti *et al.*[5] Prior to staining, lymphocytes, or EL4 cells were washed once in Dulbecco's phosphate-buffered saline and approximately 5×10^5 pelleted cells were suspended in 1 ml nuclear isolation buffer (NIB; 0.5% NP40, 0.05 M Trizma-base: Trizma-HCl pH=7.4, 0.05 M NaCl, 1.0 mM EDTA). Frozen R3327-G solid tumors were prepared for staining by teasing in 1.0 ml teasing buffer (0.1 M Trizma-base, 5 ml/liter 38% HCl, 0.08 M NaCl, pH=7.4), filtering through 105 micron and again through 62 micron nylon meshes, and adding 3.0 ml NIB for counting. Solid tumor nuclei were diluted to 1.25–2.00×10^5/ml in NIB. Each sample was vortex-mixed (3 sec) prior to staining.

In some of the histograms shown chicken red blood cell (CRBC) nuclei were used as an internal standard. Heparinized chicken blood was washed twice in a large volume of Hank's-balanced-salt solution and 1 drop of concentrated CRBC aliquotted per tube. The CRBC were pelleted and frozen at -70°C. Just prior to staining, CRBC were thawed and diluted in NIB for counting. The CRBC were adjusted to 2×10^6/ml in NIB and 10 μl added per sample before staining. The CRBC were, therefore, stained at the same time as the sample.

Sample Staining

Staining was accomplished by adding to each 1 ml of nuclear suspension 0.9 ml 0.03 M NaCl-0.03 M NaHCO$_3$ (Bicarbonate buffer, pH=8.1), 0.1 ml 3 μg/ml FITC in bicarbonate buffer, and 1.0 ml 70 μg/ml PI in bicarbonate buffer, in rapid succession. The final dye concentrations were 0.1 μg/ml FITC and 23.3 μg/ml PI. Nearly identical histograms were obtained with 3.0 μg/ml FITC, however, the lower FITC concentrations were found to be optimal in terms of coefficients of variation (CVs). The staining conditions were arrived at through trial-and-error using the conditions of Roti Roti *et al.*[5] and Crissman and Steinkamp[11] as guides.

After staining, the suspensions were incubated on ice for 30 min–5 h before being analyzed on an EPICS V flow cytometer (Coulter Electronics, Hialeah, FL) equipped with a 515 nm long pass filter, 560 nm dichroic filter, 630 nm long pass filter (red PMT, DNA), 520 nm narrow band pass filter (green PMT, protein), and an argon-ion laser emitting at 488 nm (400mW). All solutions used were less than 2 weeks old. The FITC was obtained from Miles-Yeda (Elkhart, Ind) and the PI from Calbiochem (La Jolla, CA).

The data in the figures are displayed as 64×64 histograms with DNA content on the X-axis, protein content on the Y-axis, and number of nuclei on the Z-axis. Two types of graphs are show—isometric and contour. Both types of graphs were scaled such that the peak channel in the Z-plane is 1. The isometric plots show all points in the Z-plane between 0.01 and 1.0. The contour plots were sliced at levels 0.8, 0.4, 0.2, 0.1, 0.04, and 0.015 in the Z-plane. The isometric and contour plots show the scaled single parameter DNA and nuclear protein histograms on the X-axis and Y-axis, respectively. In the panels displaying the contour plots, the single parameter DNA histogram in the upper right-hand corner is a peak-plot, or outline, of the two-parameter histogram in the Z-X plane.[8] By showing only the peak channels, the relative slice levels can be accurately displayed along side the single parameter histograms.

RESULTS

Nuclear Isolation

Several different nuclear isolation and staining protocols were investigated. The nuclear isolation buffer that we describe herein was arrived at empirically.[6] Recently, we found that the nuclear isolation medium (NIM) of Thornthwaite *et al.*[12] gives nearly identical results.

All samples were checked for cytoplasmic tags and were found to have less than 1% tags. It should be noted that some cell types, notably epidermal cells, require more rigorous nuclear isolation. The accuracy of the technique is dependent on stripping the cytoplasm off the nuclei. Syringing or vortex-mixing of the cell suspensions was used to ensure nuclear isolation; however, care was taken to avoid the disaggregation of mitotic nuclei.

DNA/Nuclear Protein Analysis of EL4 Cells

The two-parameter DNA/nuclear protein histograms of log phase EL4 cells were divided arbitrarily into G_{1A}, G_{1B}, S, G_{2A}, G_{2B}, and M. In order to compare our results with those of Darzynkiewicz *et al.*[13] G_1 was compartmentalized in accordance with their nomenclature. Cells with G_1 DNA contents and nuclear protein contents below that of S phase cells were termed G_{1A} cells, whereas those G_1 cells with nuclear protein contents equal to that of S phase cells were termed G_{1B} cells. Evidence from communications by Roti Roti *et al.*,[5] and Pollack *et al.*,[6] suggests that G_2 might also be compartmentalized. Cells in G_2 with nuclear protein content equal to that of cells in S phase were, therefore, tentatively termed G_{2A} cells, while G_2 cells with nuclear protein content above that of cells in S phase were termed G_{2B} cells.

FIGURE 1 shows typical DNA/nuclear protein histograms of log phase EL4 cells that have been exposed to colcemid for various periods of time. The area labeled M appears to consist of stripped mitotic nuclei (referred to as nuclei, although no nuclear membrane is present). Sorting of nuclei falling in this high DNA–low protein region shows that they have the appearance of cytoplasm-free particles of condensed chromosomes held together in a matrix.[8] A majority of these intact M nuclei are most likely from prophase.

Further evidence that these high DNA–low protein nuclei are from M cells is shown in FIGURE 1. Treatment of EL4 with colcemid over time resulted in a significant increase in the M area within 2 h from 1.9% to 7.7%. The accumulation of cells in M from colcemid was reflected as a decrease in the proportion of cells in G_{1A} from 12.8% to 5.7%. The M block caused a more rapid depletion of cells from G_{1A} than from G_{1B}; G_{1B} decreased from 43.7% to 35.7%. After 6 h of colcemid

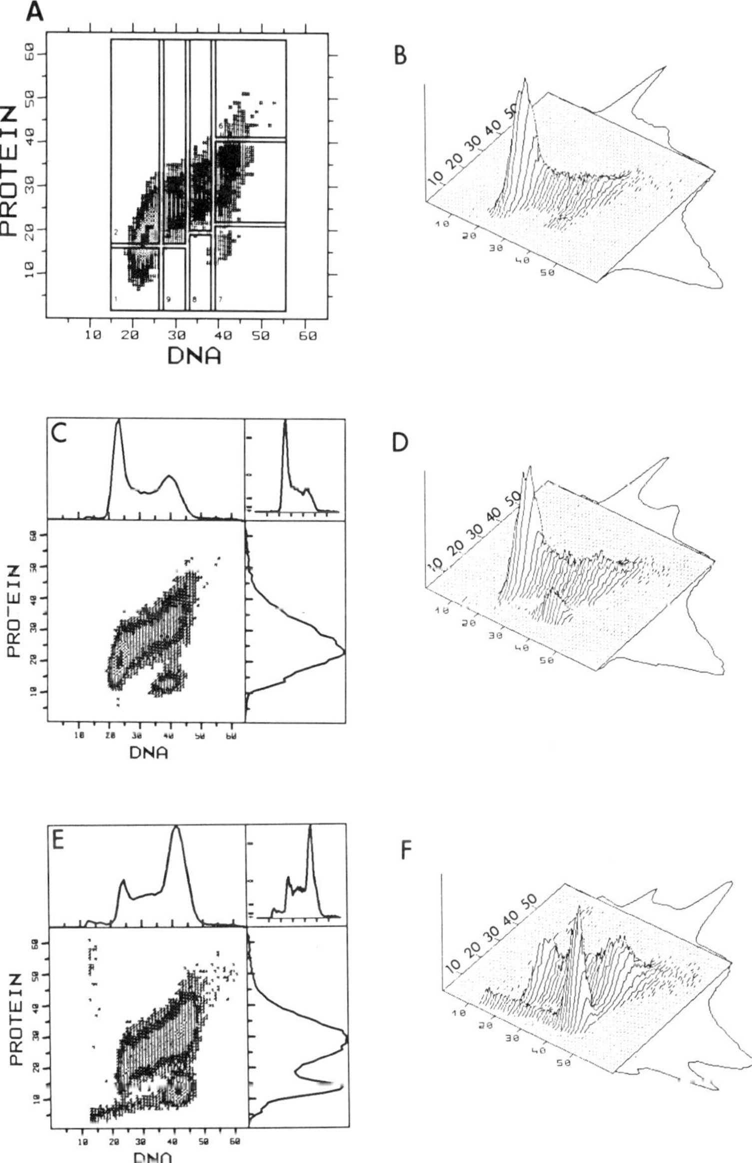

FIGURE 1. DNA/nuclear protein analysis of EL4 cells treated with colcemid. A & B, untreated; C & D, colcemid-treated (2 µg/ml) for 2 h; E & F, colcemid-treated for 6 h. A, C, and E show the contour displays and B, D, and F show the isometric displays. In Panel A, G_{1A} = area 1, G_{1B} = area 2, S = areas 3 & 4, G_{2A} = area 5, G_{2B} = area 6, and M = areas 7 & 8 & 9. See MATERIALS and METHODS for details of scaling.

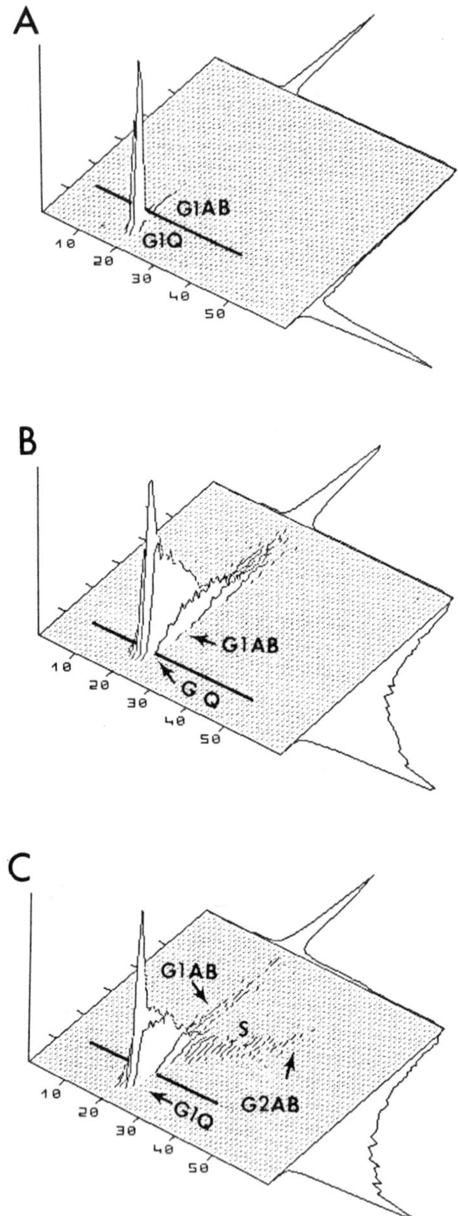

FIGURE 2. DNA/nuclear protein analysis of HPBL cultured for 48 h. A, unstimulated HPBL; B, PHA-stimulated HPBL treated with 2 mM hydroxyurea for 20 h; C, PHA-stimulated HPBL; The 64 × 64 isometric displays show nuclear protein (ordinate) versus DNA (abscissa).

treatment M increased to 22.6%, there were almost no cells in G_{1A}, and G_{1B} was reduced to 10.9%.

Comparisons of the numbers of M nuclei in DNA/nuclear protein histograms with mitotic indices counted manually[8] showed that 60–80% of the M nuclei were retained by our preparation procedure. It is apparent from the histograms in FIGURE 1 that some of the M nuclei disaggregated; a line of partially degraded M nuclei (i.e., loss of chromosomes) extends from the M area.

DNA/Nuclear Protein Analysis of Mitogen-stimulated Human Lymphocytes

Darzynkiewicz *et al.*[3] have used mitogen-stimulated lymphocytes to model the G_0–G_1 transition. Acridine orange staining of DNA/RNA showed that resting G_0 (G_{1Q}) lymphocytes increase in RNA content prior to the onset of DNA synthesis. These investigators were able to distinguish resting G_{1Q} cells from early G_{1A} and late G_{1B} cells.

We have observed that, qualitatively, the DNA/nuclear protein method provides similar results. Exposure of lymphocytes to the mitogen PHA for 24 h resulted in an increase in nuclear protein content in some cells, while DNA content remained constant.[6] Using the DNA/nuclear protein technique, G_{1Q} cells were delineated from activated G_1 cells (G_{1A} and G_{1B}); however, owing to some overlapping, all three compartments (G_{1Q}, G_{1A}, and G_{1B}) were not clearly distinguished. In the case of PHA-stimulated lymphocytes, therefore, G_{1A} and G_{1B} cells were grouped and termed G_{1AB} cells.

FIGURE 2 illustrates the nuclear DNA/protein distributions of PHA-stimulated human peripheral blood lymphocytes after 48 h of cell culture. Nuclear protein levels in stimulated lymphocyte G_1 nuclei were dramatically elevated at this time and this was reflected in the rato of G_{1AB}:G_{1Q} (TABLE 1). The significance of the elevation of nuclear protein content in some G_1 nuclei to levels above those of S phase cells is unclear; this event might occur in all proliferating lymphocytes prior to entry into S phase, or might be representative of some other phenomenon

TABLE 1. Ratio of G_{1AB}:G_{1Q} in Cultured HPBL

Culture[a,b]	24 Hours[c]	48 Hours	72 Hours
No PHA	0.09[d]	0.10	—
PHA Alone	0.21	1.22	1.40
PHA + Hydrox	—[e]	1.62	1.60
PHA + Colcemid	—	—	0.88
PHA + ^{3}H-TdR	—	—	1.04

[a]Table taken from results in ref 6.
[b]Microculture conditions. Hydrox = 2 mM hydroxyurea treatment (20-h treatment at 48 h, 12-h treatment at 72 h).
[c]Time in cell culture.
[d]Estimate of the G_{1AB}:G_{1Q} ratio in 64 × 64 DNA/nuclear protein histograms using integration between boundaries 15–25 DNA and 3–13 protein for G_{1Q}, and 15–25 DNA and 14–63 protein for G_{1AB}. G_{1Q} was positioned approximately on channel 19 DNA and 5 protein. Representative experiment of two.
[e]Not done.

(e.g. terminal differentiation). Cytoplasmic tags were not detected on these high protein activated lymphocytes.

FIGURE 2 also shows that 48-h cultures of PHA-stimulated HPBL have small numbers of cells in S phase. Treatment of activated lymphocytes with the DNA synthesis inhibitor hydroxyurea prevented entry into S phase and promoted a build-up of cells in G_{1AB}.

Histograms of PHA-stimulated HPBL after 92 h of cell culture show a considerable increase in the number of cells in G_{1AB}, S, G_{2A}, G_{2B}, and M (FIG. 3). Moreover, a considerable number of G_{1Q} cells are evident at this time. Colcemid treatment of 72 h (ref 6; TABLE 1), or 92 h HPBL (not shown), for 12 h results in two obvious changes in DNA/nuclear protein histograms. First, there is a distinct increase in the proportion of nuclei in the low protein–high DNA M area, and second, the ratio of G_{1AB}:G_{1Q} decreases. These data suggest that more cells enter G_{1AB} from M than cells entering G_{1Q} from M.

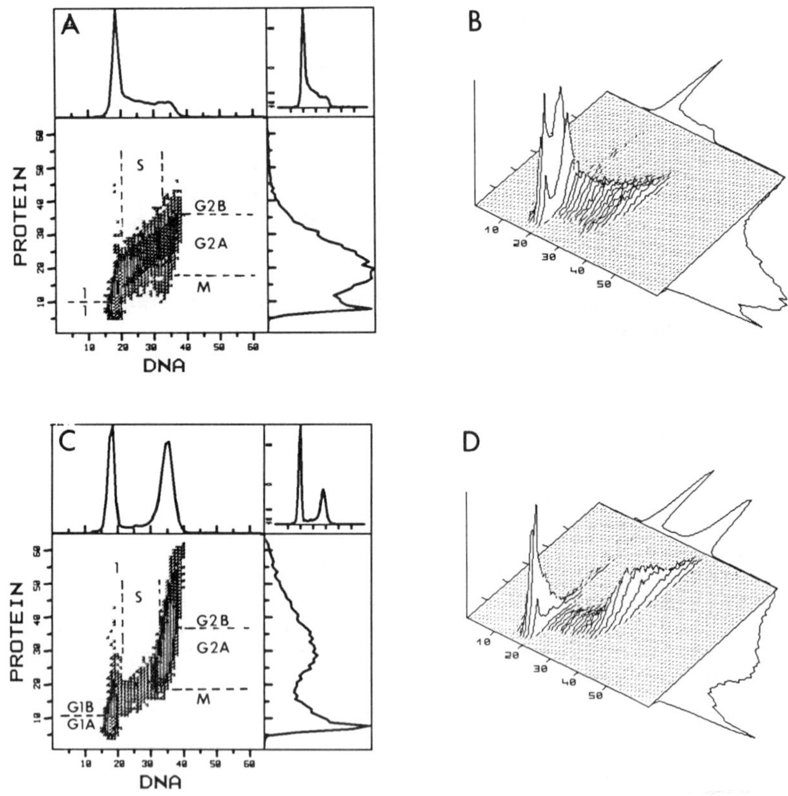

FIGURE 3. DNA/nuclear protein histograms of HPBL cultured for 92 h and exposed to radiotoxic levels of ^{3}H-TdR. A & B, PHA-stimulated HPBL; C & D, PHA-stimulated HPBL exposed to 5 µCi/ml ^{3}H-TdR for the 18 h prior to culture harvest.

Evidence for the Subdivision of G2

Previously, we reported[14] that exposure of proliferating human lymphocytes to radiotoxic levels of ^{3}H-thymidine (^{3}H-TdR) caused an increase in the $G_2 + M$ area in single parameter DNA histograms. Concomitant with this increase in the number of cells with 4C DNA, the number of cells entering mitosis (estimated by counting mitotic indices) was reduced. The advantage of using the DNA/nuclear protein technique for examining radiation effects is that information on both the accumulation of cells in G_2 and the decrease in M can be acquired simultaneously.[8]

FIGURE 4 shows the two parameter display of 92-h lymphocytes treated with a radiotoxic level of ^{3}H-TdR for 18 hours. As expected, the proportion of cells in M decreased from 1.7% to 0.5%, while the proportion of cells in G_2 increased from 2.4% to 21.1%. The most striking feature of the two parameter analysis was the unexpected finding that approximately 45% of the nuclei in G_2 had nuclear protein levels above that of S phase nuclei (G_{2B}). Unperturbed PHA-stimulated HPBL had only 13% of the G_2 nuclei in G_{2B}. The obvious difference between the proportions of G_{2B} cells in untreated and ^{3}H-TdR perturbed cultures suggests that this compartment may provide a highly sensitive measure of cell cycle changes in response to radiation exposure.

Another characteristic of the response of HPBL to ^{3}H-TdR is that there was a significant reduction in the proportion of G_{1AB} cells. This effect was similar to that observed as a consequence of colcemid treatment.

FCM-DNA/Protein Analysis of R3327-G Solid Tumors

R3327-G rat prostatic adenocarcinomas are made up of two distinct cell populations as evidenced by DNA content[10] and karyotype.[15] One population has 2C DNA in G_1 and the other has 3.2C DNA in G_1. The growth rates of tumors resulting from the implantation of R3327-G cells are sensitive to androgen.[16] These poorly differentiated tumors reflect their human tumor counterparts in almost all ways and provide an ideal system for the development of new methods for analyzing human solid tumors.

The CVs of DNA/nuclear protein prepared tumor samples ranged from 3.5–7.0 and were slightly more variable and higher than the CVs observed when tumor samples were analyzed for DNA content alone.[10] The best results were obtained when low concentrations of tumor nuclei were used (1.25–2.00×10^5/ml NIB).

FIGURE 4 shows typical DNA/nuclear histograms of tumors grown in intact control and castrated rats. In this example, castration (or sham castration for the intact controls) was performed 48 h prior to tumor excision. The two parameter analysis shows that tumors grown in castrate rats have proportionally fewer cells in aneuploid mid-S phase.[6] These data are similar to those reported previously using single parameter DNA histogram analysis of propidium iodide-stained nuclei.[17] The results also show that the high protein aneuploid G_{1B} compartment increased from 55% of the total aneuploid population in the intact control to 70% in the castrate. Thus, the two-parameter analysis indicates that castration effected a G_{1B} block that resembles in appearance hydroxyurea treatment of proliferating HPBL (FIG. 2).

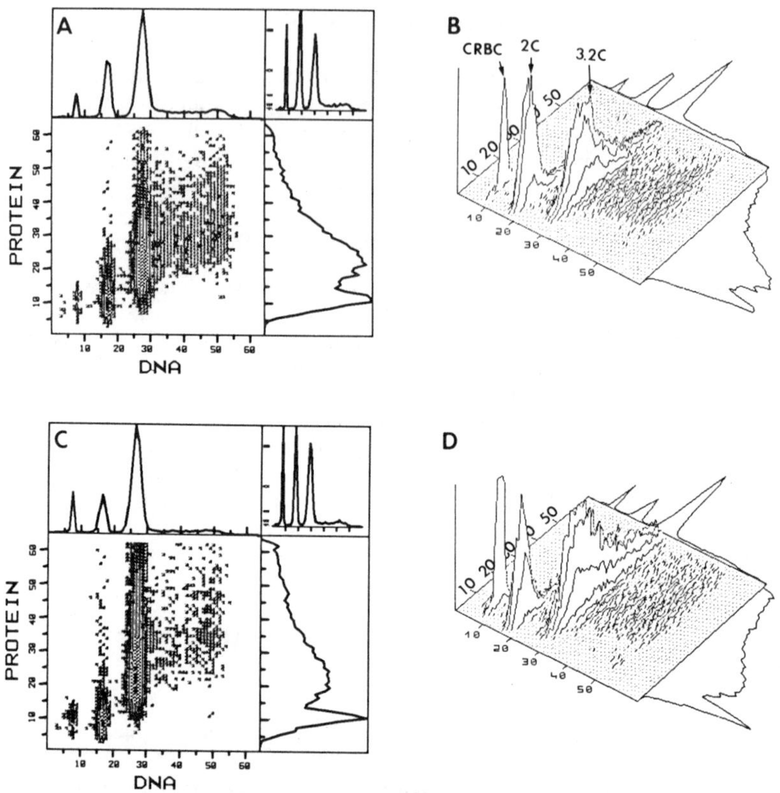

FIGURE 4. DNA/nuclear protein analysis of R3327-G rat prostatic solid tumor specimens. A & B, tumor grown in an intact male rat that was sham castrated 48 h prior to tumor excision; C & D, tumor grown in a castrate rat that was castrated 48 h prior tumor excision. The scaling routines in these figures excluded the CRBC peak, because the CRBC peak was the highest peak. The two-parameter histograms were scaled to the highest channel after channel 12 (DNA). Those channels having a height greater than 1 (i.e., CRBC peak) after scaling were set equal to 1.

DISCUSSION

The results described show that the DNA/nuclear protein technique is a rapid multiparameter cell kinetic method that has potential for the analysis of solid tumors. Tentative subdivisions of G_1 in DNA/nuclear protein histograms of log phase EL4 cells into early G_1 (G_{1A}) and late G_1 (G_{1B}) provided results analogous to those of Darzynkiewicz *et al.*[13] using acridine orange. Five different cell culture lines have been tested thus far and in all cases two classes of G_1 cells were observed—G_1 cells with less nuclear protein than S phase cells (G_{1A}) and G_1 cells with the nuclear protein contents equal to that of S phase cells (G_{1B}). Despite the fact that G_{1A} is depleted prior to G_{1B} when cells are blocked from entering G_1 (i.e.,

G_2 or M block), it is unclear whether these compartments truly contain early and late G_1 cells.

Since the G_{1A} compartment is below that of S phase in terms of nuclear protein, there is no overlapping between G_{1A} and S. In contrast, the G_{1B} compartment does overlap with S. If the treatment of log phase EL4 cells with colcemid prevented the entry of cells into G_{1A} and G_{1B} at the same rate, then the proportion of cells in G_{1A} would be reduced faster than that of G_{1B}. Thus, it can not be assumed from the data presented that the G_{1A} and G_{1B} compartments in DNA/nuclear protein histograms represent early and late G_1, respectively. It should be noted, however, that regardless of whether the G_{1A} compartment represents early G_1, it is apparent that this subdivision of G_1 provides useful information and an added sensitivity to cell kinetic changes because the fractional depletion of cells from G_1 can be recognized easily by monitoring the G_{1A} compartment. In the case of acridine orange staining, evidence using synchronized cells shows that G_{1A} does in fact represent an earlier part of G_1 than G_{1B}.[13,18]

The increase in the nuclear protein levels in resting G_{1Q} lymphocytes in the early states of mitogenic activation (24–40 h) has been attributed to the nuclear acidic protein component.[19,20] Roti Roti et al.[5] have suggested that the FITC fluorescence from the staining of nuclear protein by FITC in the presence of PI reflects the nonhistone chromatin proteins more than the histones.

The G_{1Q} compartment as visualized in DNA/nuclear protein histograms appears to be unique to certain cell populations such as HPBL. We have analyzed several different types of solid tumors and have not observed a distinct G_{1Q} population. The distributions of the R3327-G rat prostatic adenocarcinoma shown in FIGURE 4 are typical of many human solid tumors; G_1 can only be subdivided into G_{1A} and G_{1B}. The subdivision of G_1 in the histograms of these tumors showed that castration caused an increase in the proportion of G_{1B} cells and a decrease in the proportion of S cells. The two-parameter analysis, therefore, provided additional cell kinetic information that could not be obtained by measuring DNA content alone. Further investigations are needed to determine whether this added information has any relation to the extent of clinical response.

Roti Roti et al.[5] divided G_2 into early and late compartments on the basis of the transit of radiolabeled cells. Our results concerning the effects of radiotoxic levels of incorporated ^{3}H-TdR (FIG. 3) support their conclusions; radiation from intranuclear ^{3}H appeared to cause an initial accumulation of cells in early G_2 (G_{2A}) followed by the movement of these cells into G_{2B}. Preliminary experiments, using cell lines grown *in vitro* and solid tumors grown *in vivo*, show that externally administered ionizing radiation also causes an accumulation of cells in G_{2B}. Since the DNA/nuclear protein technique afforded the separation of G_2 from M in two parameter histograms, information regarding G_2 arrest and M depletion were obtained simultaneously.

ACKNOWLEDGMENTS

The authors are grateful to Dr. Howard Gratzner of the Institute for Cell Analysis, University of Miami School of Medicine, Miami, Florida, for his generosity in permitting our use of the EPICS V flow cytometer at his facility. The authors also thank Mercedes Fuentes and Rigoberto Ledesma for their excellent technical assistance, Scott Morgan for typing the manuscript, and Richard O. Ward, Leroy Ivey, and S. Diane Amos of Medical Media Service, Veterans Administration Medical Center, Miami, Florida, for the photography.

REFERENCES

1. DARZYNKIEWICZ, Z., T. SHARPLESS, L. STAIANO-COICO & M. R. MELAMED. 1980. Proc. Natl. Acad. Sci. USA **77**:6696–6699.
2. DARZYNKIEWICZ, Z., F. TRAGANOS & M. R. MELAMED. 1980. Cytometry **1**:98–108.
3. DARZYNKIEWICZ, Z., F. TRAGANOS, T. SHARPLESS, M. R. MELAMED. 1976. Proc. Natl. Acad. Sci. USA **73**:2881–2884.
4. DARZYNKIEWICZ, Z., F. TRAGANOS, T. SHARPLESS & M. R. MELAMED. 1977. J. Histochem. Cytochem. **25**:875–880.
5. ROTI ROTI, J. L., R. HIGASHIKUBO, O. C. BLAIR & N. UYGUR. 1982. Cytometry **3**:91–96.
6. POLLACK, A., H. MOULIS, N. L. BLOCK & G. L. IRVIN III. 1984. Cytometry. **5**:473–481.
7. POLLACK, A., D. L. PRUDHOMME, D. B. GREENSTEIN, G. L. IRVIN III, A. J. CLAFLIN & N. L. BLOCK. 1982. Cytometry **301**:28–35.
8. POLLACK, A., H. MOULIS, D. B. GREENSTEIN, N. L. BLOCK & G. L. IRVIN III. 1985. Cytometry **6**:428–436.
9. CLAFLIN, A., E. C. MCKINNEY & M. A. FLETCHER. 1977. Oncology **34**:106–110.
10. POLLACK, A., C. B. BAGWELL, N. L. BLOCK, G. L. IRVIN III. A. J. CLAFLIN & B. J. STOVER. 1981. J. Surg. Oncol. **18**:389–398.
11. CRISSMAN, H. A. & J. A. STEINKAMP. 1982. Cytometry. **3**:84–90.
12. THORNTHWAITE, J. T., E. V. SUGARBAKER & W. J. TEMPLE. 1980. Cytometry **1**:229–237.
13. DARZYNKIEWICZ, Z., H. CRISSMAN, F. TRAGANOS & J. STEINKAMP. 1982. J. Cell. Physiol. **113**:465–474.
14. POLLACK, A., C. B. BAGWELL & G. L. IRVIN III. 1979. Science **203**:1025–1027.
15. CLAFLIN, A. J., A. POLLACK, T. MALININ, N. L. BLOCK & G. L. IRVIN III. 1982. J. Nat. Cancer Inst. **69**:79–87.
16. POLLACK, A., N. L. BLOCK, B. J. STOVER, G. L. IRVIN III, M. P. FUENTES, A. J. CLAFLIN & T. I. MALININ. 1983. J. Natl. Cancer Inst. **70**:907–914.
17. POLLACK, A., N. L. BLOCK & G. L. IRVIN III. 1984. J. Urol. **131**:123A.
18. ASHIHARA, T., F. TRAGANOS, R. BASERGA & Z. DARZYNKIEWICZ. 1978. Cancer Res. **38**:2514–2518.
19. JOHNSON, E. M., J. KARN & V. G. ALLFREY. 1974. J. Biol. Chem. **249**:4990–4999.
20. POLET, H. & H. SPIEKER-POLET. 1980. Exp. Cell Res. **128**:419–430.

The Assessment of the Chemotherapeutic Effects of Vinca Alkaloids by Immunofluorescence and Flow Cytometry[a]

H. G. GRATZNER, G. FOUNTZILAS,
J. H STEIN, AND A. A. YUNIS

*Institute for Cell Analysis and
Department of Medicine
University of Miami School of Medicine
Miami, Florida 33101*

Flow cytometry (FCM) has been shown to be an invaluable technique for the assessment of various parameters related to cell growth and identification. Among the most useful of these methods are those which can be applied to the measurement of cell proliferation.[1] The conventional techniques for assessing proliferation employ the DNA histograms produced by the quantitative staining of DNA with fluorescent dyes such as propidium iodide (PI).[2]

Problems arise with the application of the DNA histogram technique to cells which are perturbed by drugs or other modalities or in which cells of different DNA content are present. This is because the segment of the population undergoing DNA replication must be extracted from the DNA histogram, an operation performed by means of computer-assisted mathematical "stripping" of the S phase fraction.[2,3] Part of the S phase fraction underlies the G_1 and G_2+M peaks, and particularly with perturbation of cell cycle traverse, it is difficult to distinguish those cells of a population which are replicating their DNA and hence in S phase, from those cells which are blocked in cycle or which underlie part of G_1 or G_2+M. The problem is compounded where cells of more than one DNA content are present, for example, in tumors where there may exist a diploid, "normal" line together with one or more aneuploid or polyploid tumor cells.[4-6]

We have attempted to develop techniques that could alleviate these problems and also be used to measure cell kinetic parameters by FCM in a way not previously possible. The approach we have used is to produce antibodies that are highly specific for the base analogues, 5-bromodeoxyuridine (BrdUrd) and 5-iododeoxyuridine, which are incorporated into DNA in the place of thymidine.[7]

In this report, we will present data from some of our recent efforts to develop this immunofluorescence FCM technique for clinically oriented applications, specifically the screening of chemotherapeutic drugs against pancreatic carcinoma, using cell lines derived from human pancreatic tumors. The methods permit the analysis of DNA synthesis across S phase in addition to following the

[a]This study was supported by March of Dimes Grant 15-37, NIH Grant ES 03407 and Training Grant AM07114, and a gift from Mr. Issam Fares.

fate of the drug-treated cells subsequent to a brief pulse of BrdUrd.[8] The studies with vinca alkaloids reported here suggest that the dual-parameter, DNA/BrdUrd technique is effective for measuring DNA synthesis rate and monitoring cohorts of drug-perturbed cells as they pass through the cycle and provide a means to distinguish those cells which are blocked in G_2+M from those which only appear to be blocked but which are still undergoing DNA synthesis.

MATERIALS AND METHODS

Tumor Cell Line

The MIA Pa Ca-2 cell line, which was established in one of our laboratories,[9] was maintained in Dulbecco's modified Eagle's medium (320-1965, GIBCO, Grand Island, New York, NY) supplemented with 5% heat inactivated fetal bovine serum (200-6140, GIBCO), 3% horse serum (#230-6050, GIBCO), and antibodies in 100×20 mm tissue culture dishes (#25020, Corning Glass Works, Corning, NY). The *in vitro* doubling time of this cell line under conditions of routine passage is about 24 hours.

Chemicals

The chemicals and concentrations are as follows: Dihydroxyanthracenedione (NSC 301739, Lederle) (DHAD); vincristine (VCR); and vindesine (NSC 245467, Eli Lilly) (VDS). These are dissolved in phosphate buffered saline (PBS) and aliquots were kept at 70°C until use.

Cell Treatment and BrdUrd Labeling

Survival Studies

The MIA Pa Ca-2 cells were exposed to either a high dose (LD_{80}) or a low dose (LD_{20}) that is, the concentration of drug required to kill 80% and 20% of cells, respectively, as measured by trypan blue exclusion. These concentrations were determined from dose response curves which were published previously.[10]

The doses of the drugs used and their cytotoxic effect expressed as LD_{80} or LD_{20} were as follows:

$$VCR, 1.2 \times 10^{-8} \text{ M, } LD_{20}; 1 \times 10^{-7} \text{ M, } LD_{80}$$

$$VDS, 1.2 \times 10^{-8} \text{ M, } LD_{20}; 1 \times 10^{-7} \text{ M, } LD_{80}.$$

Tissue culture dishes (100×20 mm) with exponentially growing cells (approximately 1×10^6 cells/dish) were exposed to drugs for 1 or 24 hours in a 5% CO_2 incubator. After the end of the exposure period, the cells were washed free of drug.

BrdUrd at 10 μM in 10 ml medium/dish was added for an additional 30 minutes. The cells were then washed free of BrdUrd, re-fed with fresh medium, and incubated. At selected intervals (0, 8, 12, 18, 24, 36 hours), subsequent to the BrdUrd exposure the cells were treated with trypsin-EDTA (#610-5300, GIBCO). The detached cells were fixed in 70% ethanol and maintained at 4°C until use.

Staining of Cells

The fixed cells were stained for BrdUrd incorporation and DNA content by the method previously described (Dolbeare *et al.*[10]). Briefly, the cells were treated with 1.5 M HCl, washed with PBS, and then stained with monoclonal anti-BrdUrd antibody. The antibody was produced as described (Gratzner[7]). The preparation used was diluted 1:20 in PBS containing 0.05% NP 40 (Shell). 300 µl were incubated with about 2×10^6 cells at room temperature for 60 minutes. The cells were then washed twice with PBS and then incubated for 30 minutes in fluorescein-labeled goat anti-mouse Ig (Cappel), which had been diluted 1:60 in PBS. The cells were then washed twice in PBS, and 1 ml 15 µg/ml propidium iodide in PBS added one hour before FCM analysis.

FCM Analysis

The cells were then analyzed with the EPICS V Cell Sorter (Coulter Electronics). Fluorescence was excited by the 488 nm line of the argon laser set at 400 mW. Green fluorescence was analyzed through a 520 ± nm interference filter; red fluorescence (propidium iodide) was analyzed at greater than 630 nm with a long pass interference filter. Two-parameter histograms were generated by means of programs resident in the Coulter MDADS Computer. The programs enabled the formation of 64 channel BrdUrd histograms at any point in the DNA histogram by a projection from the three-dimensional histograms generated from the BrdUrd versus PI data (FIG. 1). These 64 channel histograms were transferred via serial line to a PDP 11/34 computer (Digital Equipment Corp.), where a program derived a value for the weighted average BrdUrd incorporation per cell for each histogram.

RESULTS

FCM Studies of DNA Synthesis Rate across S Phase

The rate of DNA synthesis rate across S phase is plotted in FIGURE 2. This parameter in effect measures the average BrdUrd incorporation rates at selected intervals in S phase, employing 64 channel histograms, which were projected from the bivariate plots as described in MATERIALS AND METHODS. In FIGURE 2 are plotted data from untreated and treated exponentially growing cell populations. Little difference was observed in the shape of the curves for untreated cells from six different experiments, demonstrating the reproducibility of the method. The rate of synthesis accelerates in early S phase, peaks in mid-S, and decreases in the final third segment of S phase. In FIGURE 2 are also shown the DNA synthesis rates derived from an experiment in which cells were treated with high or low doses of VCR and VDS. The treatment with these drugs was compared to cells

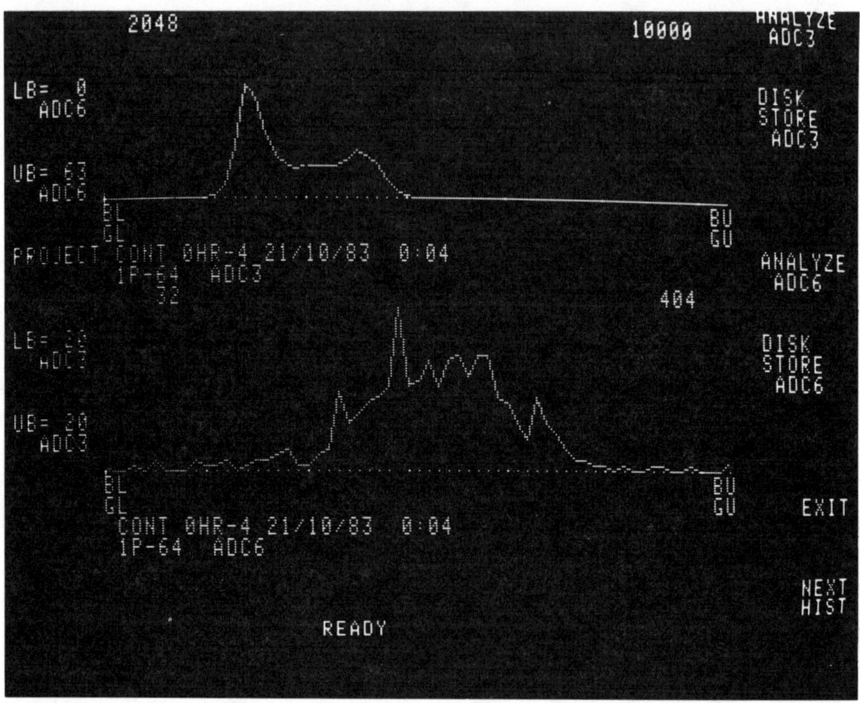

FIGURE 1. Computer display showing the method for calculation of BrdUrd incorporation across S phase. Lower histogram represents the BrdUrd incorporation at mid-S phase (with vertical line on upper histogram at channel 20). The average BrdUrd incorporation was calculated from the (*lower*) BrdUrd histogram to derive the curves in FIGURE 1.

that were treated with DHAD, which has been shown to block DNA replication in MIA Pa Ca-2 cells at the concentrations employed in these experiments, 10^{-7} M.[10] The differences in the effects of the treatment with the two drugs are apparent: DHAD causes alterations in the BrdUrd incorporation rates across S phase which are not found with the vinca alkaloids.

Traverse of Untreated Cells

Three-dimensional histograms of cells which were pulsed with BrdUrd, but not treated with drug are shown in FIGURE 3. The propidium iodide staining for DNA content produces the typical histogram expected, showing G_1, S and G_2+M stages of the cycle (FIG. 3a). The G_1 peak is wider in FIGURE 3a and has a higher coefficient of variation than the G_1 peak in FIGURE 3b. Presumably, this is due to the fact that the DNA-synthesizing cells in the non-BrdUrd–labeled cells of FIGURE 3a are close to the G_1/S border and are not displaced along the y axis as they are in FIGURE 3b, where the cells have incorporated BrdUrd for 30 minutes

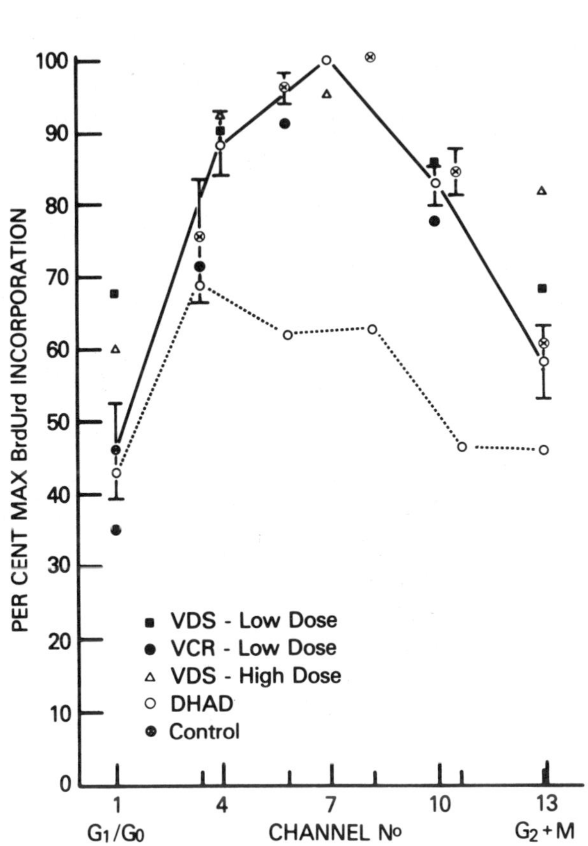

FIGURE 2. DNA synthesis rate, as measured by incorporation across S phase in control and drug-treated cells. Error bars are the SE of control cells from 6 separate experiments.

and are thus fluorescently stained with the anti-BrdUrd antibody. No cells remain in the region of unlabeled S phase cells (between the G_1 and G_2+M peaks) in FIGURE 3b, probably because all the cells in the culture are actively growing (a growth fraction of one), with no "S-blocked" cells. In FIGURE 3c, at 8 hours after the BrdUrd pulse, the BrdUrd-labeled cells can be seen to pass into the S phase region, and the cells which were formerly in G_2+M at 30 minutes post-BrdUrd pulse have now traversed mitosis, where the DNA content has been halved and are at 8 hours in the G_1 phase of the subsequent cell generation. The BrdUrd-labeled cells in this next G_1 phase have half the fluorescence intensity of the cells of the previous G_2+M, as is expected because of the previous mitotic division. At 12 hours after the BrdUrd pulse (FIG. 3d), unlabeled cells are passing through S phase and the cells which were labeled during the first generation are

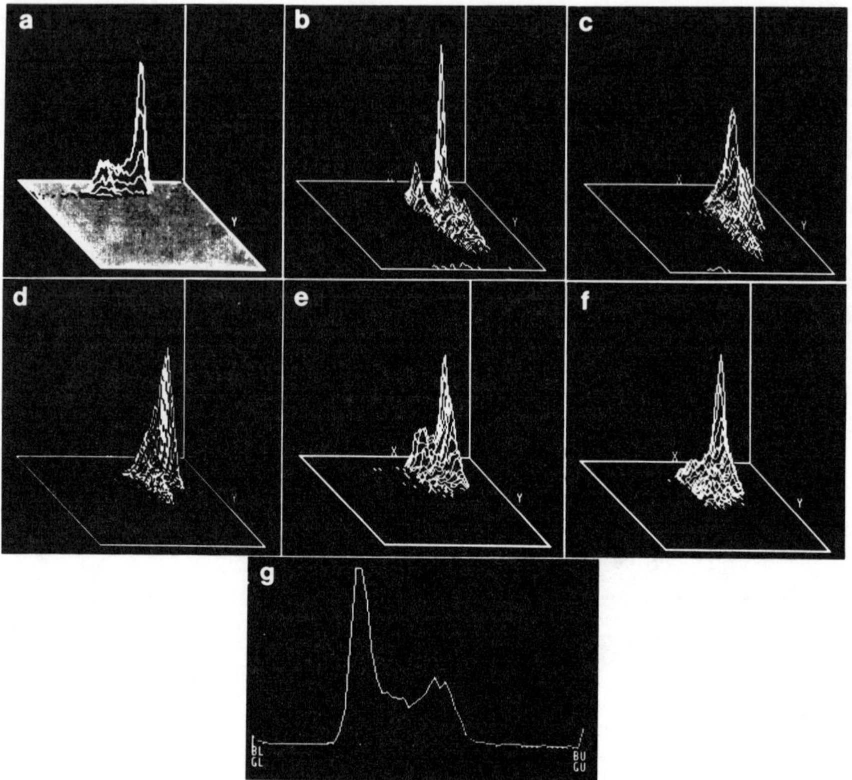

FIGURE 3. Traverse of untreated cells through the cell cycle. The cells were pulsed with 10 µM BrdUrd for 30 minutes, chased with warm medium, and samples taken at the designated time intervals subsequent to the chase, fixed and stained by the BrdUrd/DNA technique.

Times post-BrdUrd pulse: a, No BrdUrd; b, 0 time; c, 8 hours; d, 12 hours; e, 18 hours; f, 24 hours; g, representative DNA histogram (1 parameter).

G_1/G_0 peaks are at the right along their x axis, G_2+M to the left, since the histograms are rotated 150°; y axis, relative DNA content.

being damped to produce diffuse fluorescence with continuing passage of these cells around the cell cycle in FIGURE 3e and f, 18 and 24 hours respectively. In FIGURE 3g is shown the typical DNA histogram representative of the untreated cells at the various stages of the traverse of the cells depicted in FIGURE 3.

One-Hour Treatment with Vincristine

The effects of a one hour treatment with 10^{-7} M VCR are shown in FIGURE 4. Pulsing the cells with BrdUrd results in DNA synthesis, but with some perturbation, since the G_1 peak is wider than the G_1 peak of FIGURE 3b, and this could be interpreted as a situation analogous to that shown in FIGURE 3a, where no BrdUrd was added to the medium. There appears to be more cells in the S phase region of the histogram, between the G_1 and G_2+M peaks shown in FIGURE 4a that are not incorporating BrdUrd.

Within 8 hours after treatment, a dramatic accumulation of cells occurs in S and G_2+M (FIG. 4b). The relatively low resolution of the DNA histogram, caused by the HC1 treatment required for antibody binding, precludes assignment of the block to late S or early G_2 (FIG. 4d). A portion of the population appears to escape the block, however, and traverse the cycle into the subsequent G_1 as shown by the appearance of a new G_1 peak at 24 hours (FIG. 4h). Additionally, at 18 hours after the BrdUrd pulse, (FIG. 4f), cells appear that are greater than the tetraploid DNA content, and these polyploid cells increase in number by 24 hours after the BrdUrd pulse (FIG. 4h).

24 Hour Vincristine Treatment

When the cultures are treated for 24 hours with the high dose (10^{-7} M), VCR (FIG. 5a), the cells are blocked in G_2+M. These G_2+M cells, which appear to have arrested in this cell cycle stage, are actually synthesizing DNA, as demonstrated by a large number of BrdUrd-labeled cells along the y axis in FIGURE 5a through 5d (compare with FIG. 1a). As shown in FIGURE 6, the rate of DNA replication increases with increasing DNA content, which suggests that these cells have entered a polyploid S phase. Cohorts of cells can be followed through windows in specific phases of the cell cycle, and these plots confirm that a fraction of the surviving cells divide and pass into the subsequent diploid G_1 (data not shown). Large amounts of debris can be observed which results from cells killed by the drug. Microscopic examination of the immunofluorescently stained cells confirmed the existence of large, multinucleate cells. Some of these cells, observed after the anti-BrdUrd staining, but prior to PI staining, display extremely bright fluorescence, which is indicative of very high (off scale) DNA synthesis (FIG. 6).

DISCUSSION

In this study, we demonstrate the effects of vinca alkaloids on the cell cycle and the DNA replication rate monitored with a new method, the BrdUrd/DNA technique. This method permits the assessment of DNA *content* concurrently with DNA *synthesis*. Thus, the procedure enables one to simultaneously monitor *de novo* DNA synthesis and the segment of S phase in which synthesis occurs. In addition, cohorts of cells can be followed through the cell cycle.

Previous studies have shown that VCR and VDS inhibit MIA Pa Ca-2 cells in late S or G_2+M.[10,11] The drugs, in most cases, do not appear to directly inhibit DNA synthesis, although VDS has been shown to inhibit ^{3}HdTHd incorporation in Ehrlich ascites tumor cells.[12] In the experiments reported herein, no consistent effects on BrdUrd incorporation were observed.

The BrdUrd/DNA technique has advantages for studying the kinetics of perturbed cells: (1) The fraction of S phase cells can be measured directly, eliminating the need for the usual computer stripping programs employed to analyze S phase populations in DNA histograms; (2) In drug-perturbed

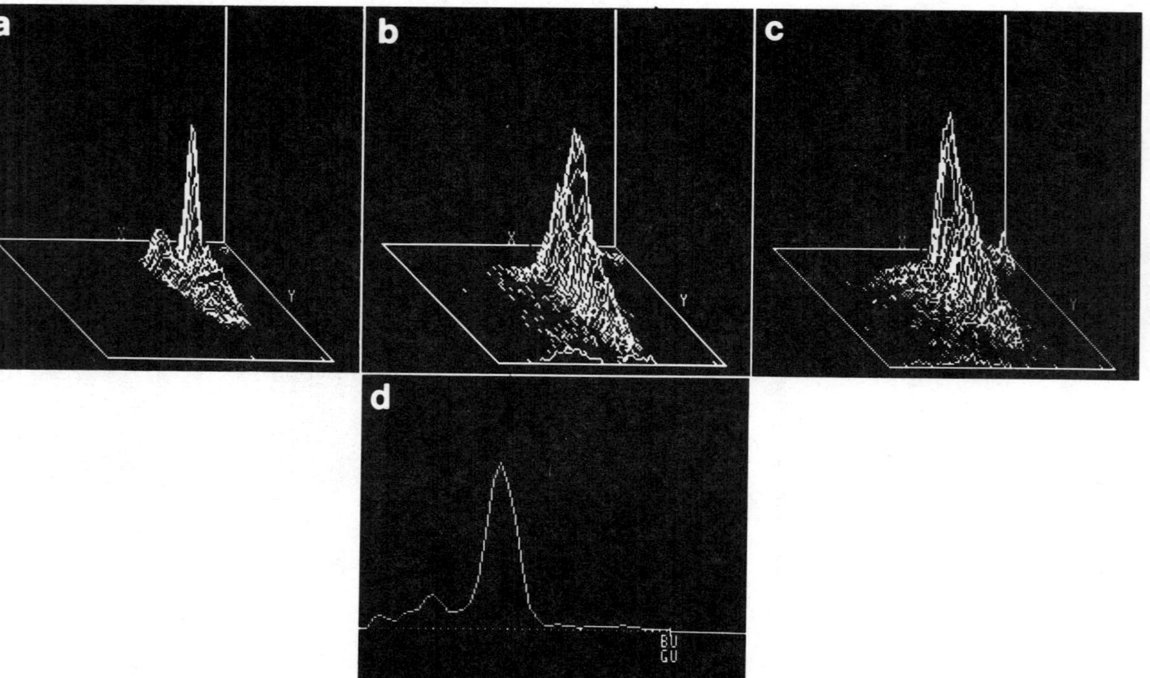

FIGURE 4. Cell cycle traverse of MIA Pa Ca cells treated for one hour with 1×10^{-7}M VCR. Drug treatment was followed by a 30 minute pulse of BrdUrd. The cells were then stained by the BrdUrd/DNA technique.

Time post pulse: a, 0 hours; b, 8 hours; c, 12 hours; d, DNA histogram projected from 4f; e, DNA histogram projected from 4c; f, 18 hours; g, 24 hours; h, DNA histogram projected from 4g.

The three dimensional histograms were rotated 150°.

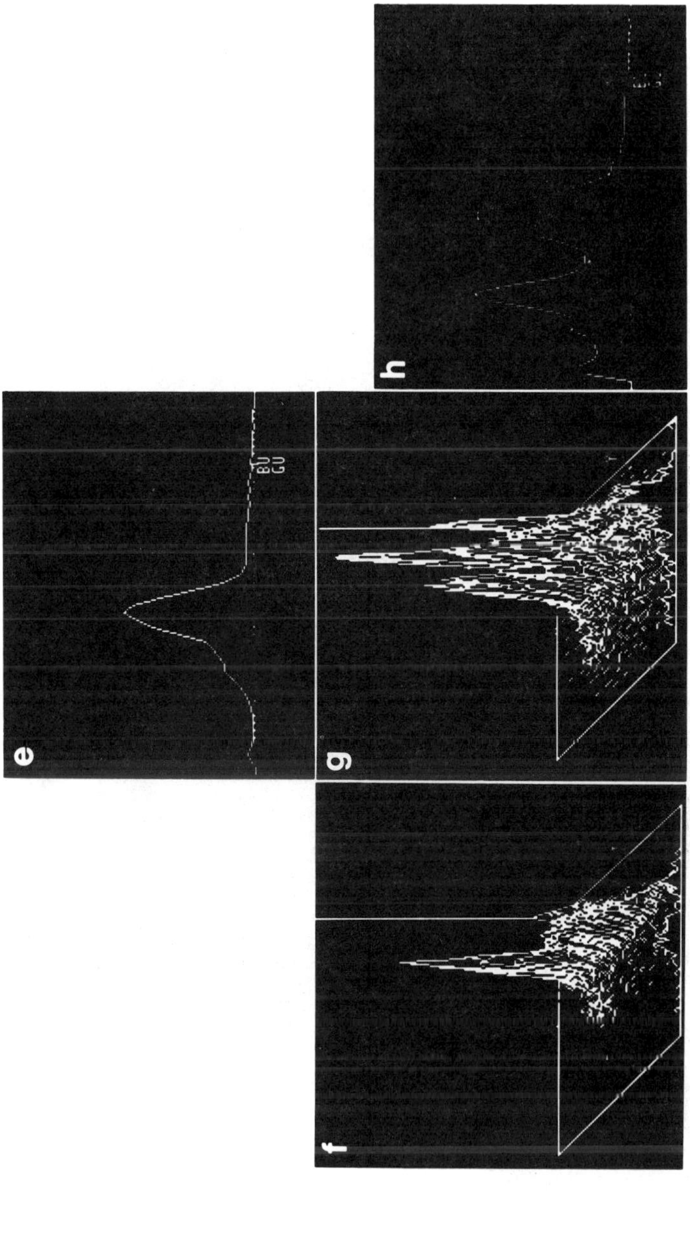

FIGURE 4 *(continued)*

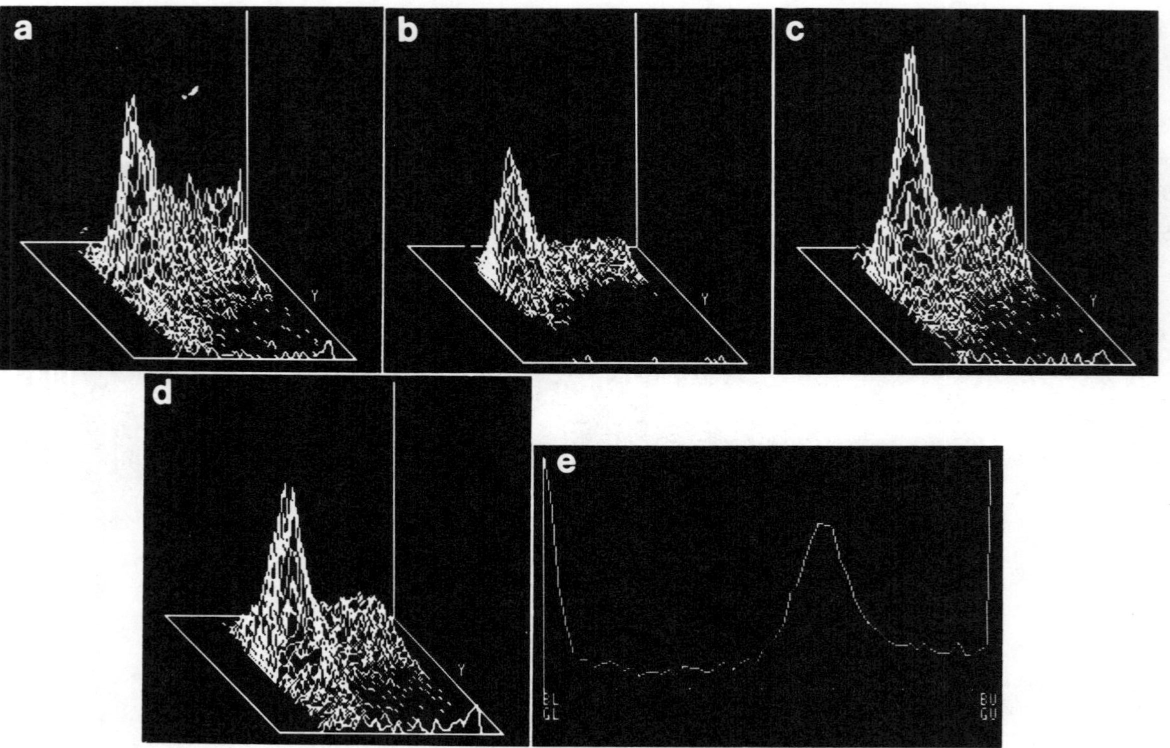

FIGURE 5. Effect of 24-hour treatment with VCR. The cells were exposed to VCR (1×10^{-7} M) for 24 hours, washed and then pulsed with BrdUrd for 30 minutes, fixed and stained by the BrdUrd/DNA method.

Times post pulse: a, 0 time; b, 8 hours; c, 18 hours; d, 24 hours; e, DNA histogram projected from 5a (0 time) culture.

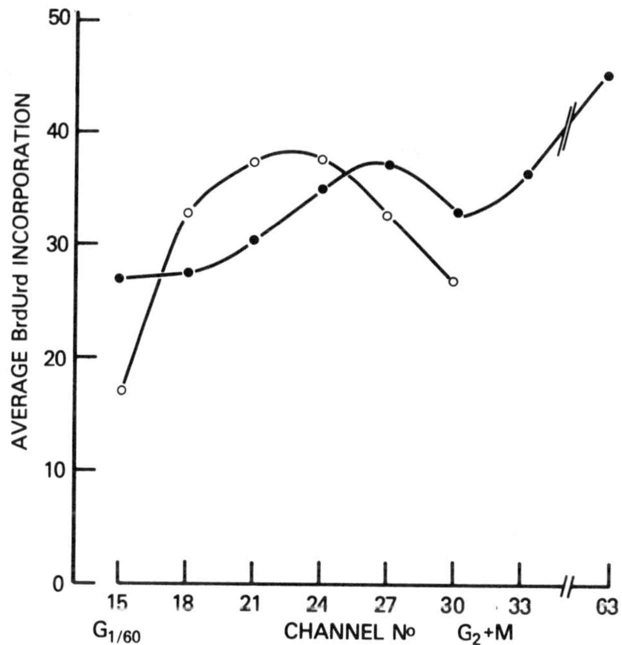

FIGURE 6. DNA synthesis rate across S phase as measured by BrdUrd incorporation in control (-O-) and cells treated for 24 hours with 1×10^{-7} M vincristine (-●-). Conditions for drug exposure and BrdUrd incorporation were as described for FIGURE 5.

populations, an increased proportion of cells detected in S phase can be due to a metabolic block rather than to cells actively synthesizing DNA. The dual parameter technique permits analysis of such perturbed populations.

By measuring incorporation of cells as they traverse designated "windows," such as G_2+M, it is possible to follow the fate of untreated cells. When cells have been treated with drugs such as VCR for 24 hours, the cells appear to be blocked in G_2+M when analyzed by conventional single-parameter DNA histogram analysis. However, if these cells are pulsed with BrdUrd and then chased though the cycle, a large percentage of the cells incorporate this base analogue into DNA. The nuclear fluorescence produced by immunocytochemical staining is extremely intense, in fact some of the cells' fluorescence is off-scale at the instrument settings employed with the flow cytometer. This is probably because many of the cells in G_2+M and greater are actually polyploid and are entering a new, polyploid S phase. The fate of these polyploid cells can be followed by direct examination of the DNA histograms comparing histograms flowing through G_2+M and hence into the subsequent G_1, since a fraction of these cells divide and enter G_1.

Direct examination of the brightly fluorescent cells confirms their aberrant condition, since many were observed to have fragmented or extremely large nuclei.

The stoichiometric nature of the immunofluorescent staining of BrdUrd provides for the quantitation of DNA synthesis at the single cell level. Thus, the determination of DNA synthesis rates across S phase is possible. The efficacy of the method is demonstrated by the observation that the replication rate, as measured by the BrdUrd/DNA technique, is inhibited by DHAD, an agent which has been shown to inhibit DNA synthesis directly, whereas little effect on synthesis was produced by the vinca alkaloids, which usually have no direct effect on DNA synthesis.

In addition to the studies discussed in this report, the clinical applicability of the anti-BrdUrd, immunocytochemical technique for the measurement of blastogenic response (in preparation) and *in vivo* assessment of tumor kinetics in patients[13] has also been demonstrated. With the recent introduction of the commercial anti-BrdUrd reagent, we can look forward to many imaginative uses of this reagent.

SUMMARY

The effects of the vinca alkaloids on the rates of DNA synthesis in the human pancreatic carcinoma line, MIA Pa Ca-2 have been studied by a new technique for measuring cell kinetics and DNA synthesis by flow cytometry and immuno-fluorescence. The method employs a monoclonal antibody that is highly specific for bromodeoxyuridine or iododeoxyuridine. The drugs vincristine and vindesine do not appear to have a direct effect on DNA synthesis rate across S phase, whereas DHAD, a compound that has been found previously to affect DNA synthesis, does appear by this technique to inhibit DNA synthesis at specific segments of S phase. That vincristine does not block cells in S phase by inhibiting DNA synthesis is borne out of the observation that cells blocked in S or G_2+M can still incorporate BrdUrd at a high rate in these phases of the cell cycle.

REFERENCES

1. GRAY, J. W., P. H. DEAN & M. L. MENDELSOHN. 1979. Quantitative cell cycle analysis. *In* Flow Cytometry and Sorting. M. R. Melamed, P. F. Mullany & M. L. Mendelsohn, Eds. John Wiley & Sons. New York.
2. DEAN, P. N. 1974. Mathematical analysis of DNA distributions derived from flow microfluorometry. J. Cell Biol. **60**:523.
3. FRIED, J. 1976. Method for the quantitative evaluation of data from flow cytometry. Comp. Biomed. Res. **9**:263.
4. MANN, R. C., R. E. HAND, JR. & G. R. BRASLAWSKY. 1983. Parametric analysis of histograms as measured in flow cytometry. **4**:75.
5. RITCH, P. S., S. E. SHACKNEY, W. H. SCHUETTE, T. H. TALBOT & C. H. SMITH. 1983. A practical graphical method for eliminating the fraction of cells in S in DNA histograms from clinical tumor specimens aneuploid cell populations. **4**:66.
6. BARLOGIE, B., M. W. RABER, J. SCHUMANN, T. J. JOHNSON, B. DREWINKO, D. E. SWARTZENDRUBER, W. GOHDE, M. ANDREEFF & J. FREIRIECH. 1983. Flow cytometry in clinical cancer research. Cancer Res. **43**:398.
7. GRATZNER, H. G. 1982. Monoclonal antibody to 5-bromo and 5-iododeoxyuridine: A new reagent for detection of DNA replication. Science **218**:474–475.
8. DOLBEARE, F., H. G. GRATZNER, M. G. PALLAVICINI & J. W. GRAY. 1983. Flow cytometric measurement of total DNA content and incorporated bromodeoxyuridine. Proc. Natl. Acad. Sci. USA **80**:5573:5577.

9. YUNIS A. A., G. K. ARIMURA & D. J. RUSSIN. 1977. Human pancreatic carcinoma (MIA Pa Ca-2) in continuous culture: Sensitivity to asparaginase. Int. J. Cancer **19**:128–135.
10. FOUNTZILAS, G., H. G. GRATZNER, L. O. LIM & A. A. YUNIS. 1984. Sensitivity of human pancreatic carcinoma cells to dihydroxyanthracenedione. Int. J. Cancer **33**:347–353.
11. HILL, B. & R. D. H. WHELAN. 1981. Comparative cell killing and kinetic effects of vincristine or vindesine in mammalian cell lines. J. Natl. Cancer Inst. **67**:437.
12. CREASEY, W. H. 1981. Biochemical effects of vindesine. Br. J. Cancer **44**:921.
13. MORTSYN, G., S. M. HSU, T. KINSELLA, H. G. GRATZNER, A. RUSSO & J. B. MITCHELL. 1983. Bromodeoxyuridine in tumors and chromosomes detected with a monoclonal antibody. J. Clin. Invest. **72**:1844–1850.

Flow Cytometric Monitoring of Anthracycline Transport in Tumor Cells[a]

AWTAR KRISHAN

*Department of Oncology and
Comprehensive Cancer Center for the State of Florida
University of Miami Medical School
Miami, Florida 33136*

INTRODUCTION

Anthracyclines such as adriamycin and daunomycin are important cancer chemotherapeutic agents.[1] Major differences in cellular transport and binding characteristics of various anthracyclines have been described.[2,3] Most of the anthracyclines are fluorescent and can be easily excited with the 488 nm laser emission from an argon laser.[4,5] Some of the well-known anthracyclines such as adriamycin are slowly transported across cellular membranes while others (e.g., daunomycin) are rapidly transported.[6,7] Whereas adriamycin and daunomycin bind to chromatin,[8] anthracyclines such as AD 32 do not bind to chromatin.[7] Most of the anthracylines are highly fluorescent and thus are ideally suited for analysis by laser excitation and flow cytometry. This technique offers the advantage of identifying subpopulations based on their cellular fluorescence,[4,5] rapid monitoring of drug uptake and the added advantage that selected cells can be sorted for further analysis by other means. In earlier reports we have used this method for monitoring anthracycline transport in tissue culture cells, in drug resistant cells, and in tissues isolated from animals injected with adriamycin.[4,5]

Cellular resistance to anthracyclines has been suggested to be due to a mechanism for rapid drug efflux from resistant cells.[9] Thus, drugs that will affect drug efflux and enhance drug retention can modify cellular resistance to anthracyclines. Several recent studies have shown that in fact calcium channel blockers and phenothiazines will inhibit drug efflux from resistant cells and thereby render them drug sensitive.[10-16]

It would be useful if before the administration of anthracyclines and agents that affect their cellular efflux *in vivo*, we could screen tumor cells *in vitro* for their anthracycline transport characteristics. As shown in the following sections, we have used laser flow cytometry (for excitation of intracellular anthracycline content) to monitor drug fluorescence in tumor cells and to monitor the effect of drug efflux blocking agents on drug retention.

[a]These studies were supported by NIH-CA 29360.

MATERIALS AND METHODS

Murine leukemic lymphoblasts of P388 cell line and its adriamycin-resistant subline P388/R84 were grown in suspension cultures and nourished with Eagle's minimal essential medium supplemented with 10% fetal bovine serum and antibiotics, penicillin, and streptomycin. Tumor cells from a human cancer of cervix xenograft grown in athymic nude mice and a suspension cell line derived from the xenograft and maintained by serial subcultivation was used for comparative studies. Cells recovered from pleural fluid of solid tumor patients (received in the laboratory for chemosensitivity testing) were used for drug incubation studies. Cells (approximately 10^6/ml) from the cell lines, ascites, or collagenase-DNase digests made from solid tumor biopsies were incubated for 2 h in media containing 2 µg/ml of adriamycin or daunomycin. A parallel set of tubes was incubated with media containing both adriamycin or daunomycin and 5–10 µM chlorpromazine (CpZ).

After incubation at 37° for 2 h, samples were analyzed in a Coulter Electronics Epics V cell sorter. Excitation from a spectraphysics 5 watt laser (488 nm laser line, at 500 mW) was used to excite cellular drug fluorescence. Data were collected, scaled, and analyzed in a Coulter Electronics MDADS system. Details of our methods and instrumentation have been published earlier.[4,5]

RESULTS AND DISCUSSION

Histograms A and B in FIGURE 1 are of P388 leukemic cells incubated with daunomycin for 0–10 min and adriamycin 15–25 min, respectively. The abcissa in these histograms records intracellular drug fluorescence; the ordinate shows time (length of incubation) covering a span of 10 minutes. The height of the bars (maximum of 1000) indicates the number of cells recorded in each channel. As seen in histogram A, cells exposed to daunomycin rapidly become fluorescent and the maximum number of fluorescent cells recorded is reached in less than 10 min of drug exposure. In contrast, adriamycin fluorescence is slower to appear in cells and between 15 to 25 min of incubation the number of fluorescent cells is still increasing.

Thus we can use this analytical mode to compare the kinetics of drug uptake in different tumor cell populations.

Instead of using time as a parameter, light scatter, which approximates cell size, is often used as a second parameter to compare and correlate cellular drug fluorescence with cell size. Histograms in FIGURE 2 show the intracellular drug fluorescence of P388 S and P388/R84 cells incubated with THP-adriamycin, 2 µg/ml, for 30 minutes. Note the increased fluorescence of the P388/S cells in A.

As described earlier in drug resistant cells, a rapid drug efflux mechanism has been shown to be responsible for the reduced drug retention. Several agents such as phenothiazines or the calcium channel blocker, verapamil will block drug efflux in resistant cells and increase drug retention. This is clearly demonstrated by comparison of histograms 2 A and B with 2 C and D. Histograms B and D in this figure show the effect of co-incubation with 10 µM verapamil (Vpl) on the intracellular drug fluorescence. In P388/S cells, co-incubation with Vpl did not make any significant difference in drug retention. In contrast, Vpl increased the

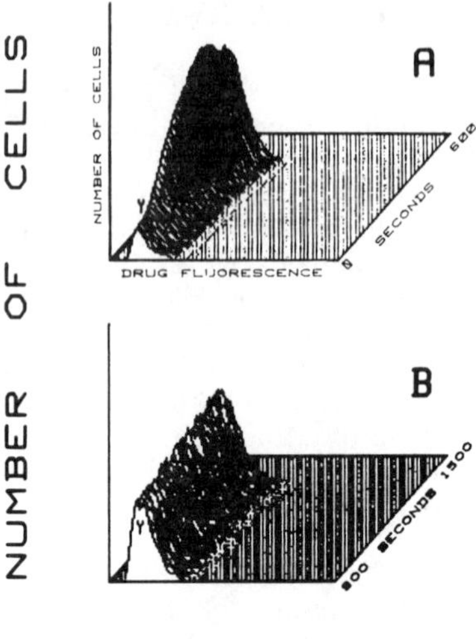

FIGURE 1. Multiparameter histograms compare uptake of daunomycin (A) and adria-mycin (B) in P388 drug-sensitive cells. Ordinate records length of incubation, while on the abcissa drug fluorescence is recorded.

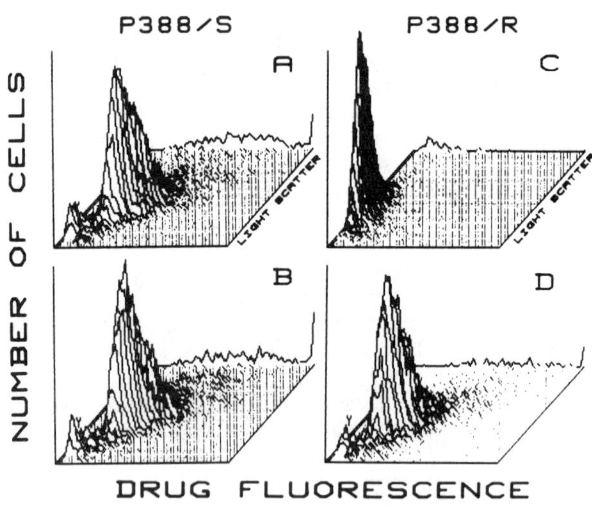

FIGURE 2. Histograms show the effect of Verapamil on cellular THP-AdR fluorescence in P388/S and P388/R84 cells.

Histograms A and C are of P388/S and P388/R84 cells incubated and 2 µg/ml of THP-AdR, respectively.

Histograms B and D are of P388 S and P388/R84 cells so-incubated with THP-AdR (2 µg/ml) and Verapamil (10 µg/ml), respectively.

THP-AdR retention in P388/R cells, presumably by decreasing drug efflux. In soft agar assays, this effect is confirmed by enhanced sensitivity of P388/R cells to AdR or other anthracyclines in the presence of CpZ or Vpl. Similarly, enhancement of intracellular anthracycline retention and chemosensitivity by several phenothiazines and calcium channel blockers has been reported in drug-resistant cells.[10-16]

Some of our recent flow cytometric observations have shown that anthracycline transport in a cell population is not uniform and neither is the response to agents that block drug efflux. Similarly, cells from different cell lines differ in their response to phenothiazines and other drug efflux blocking agents. For example in cells from a human squamous cell carcinoma xenograft and its *in vitro* cell line, we have noted a pronounced effect of CpZ on drug efflux but not that of the calcium channel blocker verapamil on adriamycin retention (unpublished data).

In some preliminary clinical studies, we have used the laser excitation method to monitor adriamycin uptake and the effect of phenothiazine, chlorpromazine, on the drug retention in tumor cells from pleural fluid and ascites of solid tumor patients. On the basis of a preliminary analysis of data from 25 patient specimens, we have so far recognized the following subsets:

1. Tumor cells or subpopulations that do not transport adriamycin or daunomycin and are not affected by phenothiazines (histograms/3A and B). These cells presumably are resistant owing to either inability to transport or rapid drug efflux.

2. Tumor cells that transport adriamycin or daunomycin but efflux it rapidly. This efflux is blocked by co-incubation of cells in the presence of CpZ or Vpl. Thus, a certain subpopulation shows enhanced drug retention in the presence of efflux modifying agent chlorpromazine (histograms 3C and D). These cells may

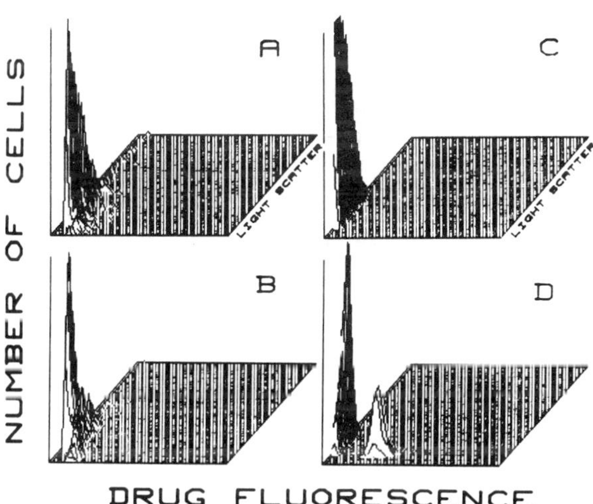

FIGURE 3. Histograms A and B are of patient specimen incubated with daunomycin (A) and incubated with daunomycin plus chlorpromazine (B).

Histograms C and D are of different patient specimen incubated with daunomycin (C) and daunomycin plus chlorpromazine (D).

have an efficient drug efflux mechanism which can be blocked and the resistant cells rendered sensitive to anthracyclines.

3. Tumor cells that transport the drug but do not efflux it rapidly and are not sensitive to phenothiazines. These cells may represent cells that are intrinsically sensitive to anthracyclines and lack an influx blocking or enhanced efflux mechanism.

It is important to note that laser-excited flow cytometry can rapidly identify tumor cells or subpopulations and their drug transport characteristics which may be exploited for enhancing clinical response. Current studies in our laboratory are focused on seeking correlation between these easily monitored drug transport characteristics and drug cytotoxicity in tumor cells.

REFERENCES

1. CROOK, S. T., & S. D. REICH, Eds. 1980. Anthracyclines: Current Status and New Developments. Academic Press Inc. New York.
2. SKOVSGAARD, T. & N. NISSEN. 1975. Adriamycin an antitumour antibiotic: A review with special reference to Daunomycin. Dan. Med. Bull. **22**:62–73.
3. DANO, K. 1976. Experimentally developed cellular resistance to Daunomycin. Acta Pathol.Microbiol. Scandinavia **256** (Section A, Suppl):1–78.
4. KRISHAN, A. & R. GANAPATHI. 1980. Laser flow cytometric studies on intracellular fluorescence of anthracyclines. Cancer Res. **40**:3895–3900.
5. KRISHAN, A. & R. GANAPATHI. 1979. Laser flow cytometry and cancer chemotherapy: Detection of intracellular anthracyclines by flow cytometry. J. Histochem. Cytochem. **27**:1655–1656.
6. EGORIN, M. J., R. E. CLASWON, J. L COHEN, L. A. ROSS & N. R. BACHUR. 1980. Cytofluorescence localization of anthracycline antibiotics. Cancer Res. **40**:4669–4676.
7. KRISHAN, A., M. ISRAEL, E. J. MODEST & E. FREI III. 1976. Differences in cellular uptake and cytofluorescence of Adriamycin and N-Trifluoracetyladriamycin-14-valerate. Cancer Res, **36**:2114–2116.
8. DIMARCO, A. 1975. Adriamycin (NSC-123127): Mode and mechanism of action. Cancer Chemother. Rep. **6**:91–106.
9. INABA, M. & R. K. JOHNSON. 1978. Uptake and retention of Adriamycin and Daunomycin by sensitive and anthracycline resistant sublines of P388 leukemia. Biochem. Pharmacol. **27**:2123–2130.
10. TSURUO, T., H. LIDA, M. NOJINE, S. TSUKAGOSHI & Y. SAKURAI. 1983. Circumvention of Vincristine and Adriamycin resistance *in vitro* and *in vivo* by calcium influx blockers. Cancer Res. **43**:2905–2910.
11. GANAPATHI, R. & D. GRABOWSKI. 1983. Enhancement of sensitivity to Adriamycin in resistant P388 leukemia by calmodulin inhibitor Trifluoperazine. Cancer Res. **43**:3696–3699.
12. GANAPATHI, R., D. GRABOWSKI, R. TURINIC & R. VALENZUELA. 1984. Correlation between potency of calmodulin inhibitors and effects on cellular levels and cytotoxic activity of Doxorubicin (Adriamycin). Eur. J. Can. Clin. Oncol. **20**:799–806.
13. KRISHAN, A, & L. Y. W. BOURGUIGNON. 1984. Cell cycle related phenothiazine effects on adriamycin transport. Cell Biol. Int. Rep. **8**:449–456.
14. TSURUO, T., H. LIDA, S. TSUKAGOSHI & Y. SAKURAI. 1982. Increased accumulation of Vincristine and Adriamycin in drug resistant P388 tumor cells following incubation with calcium antagonist and calmodulin inhibitors. Cancer Res. **42**:4730–4733.
15. TSURUO, T., H. LIDA, M. YAMASHIRO, S. TSUKAGOSHI & Y. SAKURAI. 1982. Enhancement of Vincristine and Adriamycin-induced cytotoxicity by Verapamil in P388 leukemia and its sublines resistant to Vincristine and Adriamycin. Biochem. Pharmacol. **31**:3138–3140.
16. KRISHAN, A., A. SAUERTEIG & L. L. WELLHAM. 1985. Flow cytometric studies on modulation of anthracycline transport by phenothiazines. Cancer Res. **45**:1046–1051.

Assessment of Leukocyte Alkaline Phosphatase by Image Analysis

LEONARD S. KAPLOW[a]

Laboratory Service
Veterans Administration Medical Center
West Haven, Connecticut 06516

Departments of Pathology and Laboratory Medicine
Yale University School of Medicine
New Haven, Connecticut 06510

JILL Y. CROUCH

Veterans Administration Medical Center
West Haven, Connecticut 06516

JERRY ANN MEYERS, HANS KUNZ,
AND GERARDO L. GARCIA[b]

Bioengineering Department
Coulter Biomedical Research Corp.
Coulter Drive
Concord, Massachusetts 01742

Almost twenty years ago, at a symposium sponsored by the New York Academy of Sciences, one of us (LSK) described methods for cytochemically assessing alkaline phosphatase activity in peripheral blood neutrophils and reviewed the clinical applications of such assays.[1] It is thus with a sense of *deja vu* that two decades later, the same author addresses the same topic again published by the Academy. The thrust of this presentation, however, is directed at and illustrative of the technological changes and advances that have occurred in medicine and in science in general during that period.

Alkaline phosphatase remains an ubiquitous enzyme found in many human tissues. Its presence in neutrophils has been known for almost three decades. The enzyme occurs almost exclusively in mature neutrophils. The phrase "leukocyte alkaline phosphatase activity" (LAPA) is well established and will be used in this discussion rather than the more descriptive but less commonly used designation "neutrophil alkaline phosphatase" (NAP). The function of LAPA is unknown and activity is completely unrelated to serum levels. LAPA is known, however, to be influenced by hormones of the pituitary-adrenal axis.

[a]Address all correspondence to: L. S. Kaplow, M.D., Chief, Laboratory Service (113), Veterans Administration Medical Center, West Haven, CT 06516.

[b]Present address: New England Medical Center, 171 Harrison Ave, Box 246, Boston, MA 02111.

TABLE 1. Manual and Instrument Precision

81 Patients	r	SE
Manual precision	0.95	22.3
Tech variability	0.95	24.2
diff3.50 Precision	0.95	20.7

Quantification of this enzyme has many important and underutilized[2] clinical applications, but is primarily used as an aid in confirming the diagnosis of chronic myelogenous leukemia.[3] It may be accurately assayed by biochemical analysis of separated and washed leukocytes, a procedure that is time consuming and rarely used. Most clinical laboratories measure activity by a cytochemical staining and scoring technique that is dependent upon visual, subjective interpretation of the amount of precipitated reaction product in band and mature neutrophils. One-hundred stained neutrophils are rated from 0 to 4+, based on the observer's estimation of the intensity of stain deposited within the cytoplasm. The sum of the ratings is considered the score for a given blood sample. Thus, scores can vary from zero to a maximum of 400.[4] Each laboratory establishes its own normal range. Despite the subjectivity of this approach, intralaboratory reproducibility by trained technologists can be excellent as demonstrated in this report (TABLE 1).

The present study was carried out to determine whether a commercial automated differential white cell counter, the diff3.50 (Coulter Biomedical Research Corp. Hialeah, FL 33010), using image recognition and analysis techniques, could be adapted to automatically assay for this enzyme.

Peripheral blood smears are prepared using the Coulter Model 401 Blood Spinner to insure an even monocellular distribution of cells and then fixed and cytochemically stained manually. Counterstaining is carried out on the Hematek II, Model 4414 stainer (Ames, Elkhart, IN, 46514), using the the lowest setting on the stain intensity dial. The staining procedure used for the diff3.50 is based on Sigma reagents[5] except for the counterstain, which is the Coulter Romanowski, pure dye Stain-Right preparation.[6] Differences between the manual and automated staining procedures are shown in TABLE 2. The substrate used for instrumental analysis is naphthol AS-MX phosphate, whereas in this

TABLE 2. Comparison of Manual and Machine Stains

	Manual Method	Automated Method
Fixative	60% Buffered acetone	60% buffered acetone
Fixation time	10 Seconds	30 seconds
Substrate	Naphthol AS-BI phosphate	Naphthol AS-MX phosphate
Coupler	Fast violet B salt	Fast violet B salt
Buffer pH	9.4–9.6	8.6
Incubation time	10 minutes	60 minutes
Counterstain	Mayer's hematoxylin	Coulter Stain-Right[a]

[a]Pure dye Romanowsky stain.

laboratory (West Haven VA Medical Center), the manual procedure calls for naphthol AS-BI phosphate. When coupled to Fast Violet B Salt, both yield a reddish azo dye. The fixation time for instrumental analysis is 30 seconds as compared to 10 seconds for manual staining and the incubation time is increased from 10 to 60 minutes. The pH of the incubation mixture (8.6) for the diff3.50 is lower than that used for the manual method (9.5). A polychrome counterstain is required for use with the diff3.50, rather than a hematoxylin nuclear stain, primarily to impart a red color to red cells. This is necessary to enable the instrument to establish focus and calibration. The longer incubation period is needed to compensate for inhibition of the enzymatic reaction due to the extended fixation time and to impart a more intense stain than that obtained with the manual technique.

An algorithm to assess the amount of reaction product in each of 100 neutrophils, for use with the diff3.50 instrument, was developed by the Coulter Biomedical Research group and is illustrated in schematic form (FIG. 1).

The stained slide is automatically scanned by the analyzer. The slide is stopped whenever the instrument encounters what it interprets as a dark black

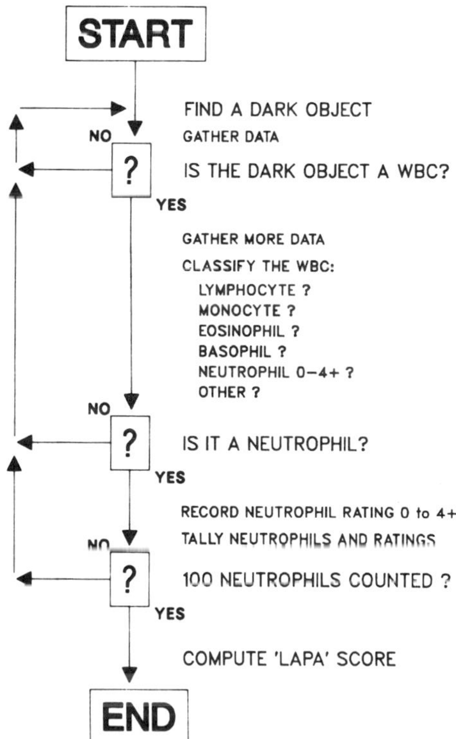

FIGURE 1. Schema illustrating the algorithm used by the diff3.50 automated differential leukocyte counter in assessing leukocyte alkaline phosphatase activity.

object. Forty-five features are measured and, on the basis of such feature data and on pre-set thresholds, the instrument attempts to identify the dark object as a leukocyte or as an artifact (debris, overlapping red cells or other). If the object is not identified as a leukocyte then the search continues until the next black object is found. If identified as a cell other than a neutrophil, the cell will be ignored and the search continued. Upon finding and classifying a cell as a neutrophil, the instrument assigns a rating of 0 to 4+ to that cell based on selected feature data. This approach closely mimics the manual method. Such data are banked and after rating 100 consecutive neutrophils, the instrument calculates the total score. The time required for machine assay varies with the total white blood cell count and averages 2 to 3 minutes per sample.

The anticoagulant EDTA chelates zinc and is not recommended for LAPA assays. Heparin is the recommended anticoagulant for this purpose. However, EDTA is routinely used as an anticoagulant for white cell differential counts and consequently, numerous blood samples anticoagulated with this compound are available daily in clinical laboratories. Because of this convenience, all the investigations to be described were carried out on specimens anticoagulated with EDTA, but were limited to a comparison of manual to machine performance only. Heparin remains the anticoagulant of choice for clinical studies of LAPA.

A cell-by-cell analysis of staining intensity (ratings) of 1600 cells on duplicate slides from eight patients, comparing ratings made by the instrument to subjective ratings made by a highly trained technologist is illustrated in FIGURE 2. The data indicate that 93.2% of all cells were similarly rated by both technologist and instrument. There were 39 cells (2.4%) which were identified as debris or overlapping red cells by the technologist and rated as either 3+ or 4+ by the diff3.50. Despite these differences, precision and accuracy for the instrument were excellent (TABLE 1 and FIGURE 3). As would be expected, the instrument was highly precise (r=0.95) based on one technologist scoring duplicate slides. A similar result was obtained when two highly skilled technicians were compared to each other.

(8 samples — 1600 cells)

MANUAL

		0	1+	2+	3+	4+	*
diff3.50	0	712	18	0	0	0	2
	1+	16	551	12	0	0	1
	2+	0	8	179	6	0	0
	3+	0	0	1	27	0	21
	4+	0	0	0	2	27	18

*DEBRIS, OVERLAPPING RBC'S, NRBC'S

FIGURE 2. Confusion matrix comparing ratings for 1600 neutrophils determined automatically by instrument to subjective ratings assigned by a trained technologist.

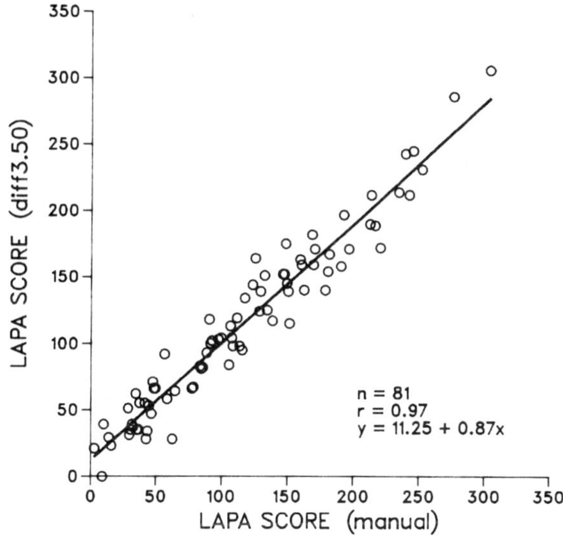

FIGURE 3. Comparison of instrumental LAPA assays to manual assays based on two technicians scoring two slides each.

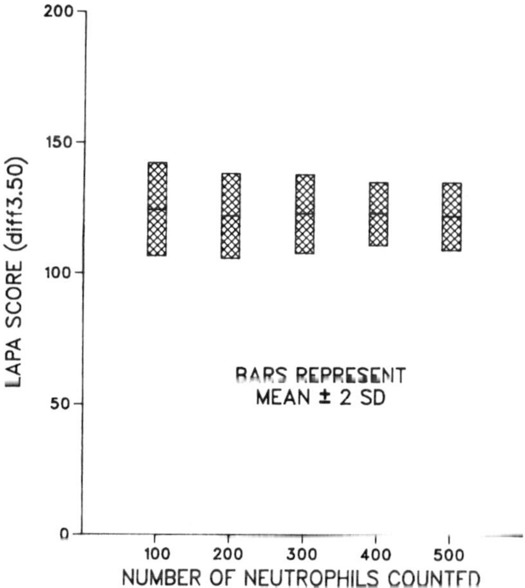

FIGURE 4. Mean and standard deviation of scores derived by the instrument dependent on the number of cells rated.

We were also interested in determining how many cells had to be counted to achieve reasonable precision. An experiment was carried out in which instrument scores were obtained based on counting 100, 200, 300, 400, and 500 cells on 10 duplicate smears from one patient. The results are given in FIGURE 4. Differences were minimal and statistically and clinically insignificant, indicating that 100 cell instrument assays could provide reliable and reproducible data.

Instrumental accuracy (FIGURE 3) was assessed by comparing the results of 100 cell analyses by the machine to scores obtained on the same stained slides derived from the results of two technicians each rating 100 cells on two different smears. Thus, for this purpose, "truth" was defined as the average of a 400 manual score. Agreement was excellent (r=0.97, SE = 17.2).

One of the problems associated with cytochemical assays has been the instability of precipitated azo dyes. The final reaction products obtained in staining for alkaline phosphatase, and for other enzymes by this technique, are dissolved by oil and by organic solvents. As a consequence, it has heretofore been impossible to preserve stained smears once they have been examined microscopically under oil or mounted. This has been a serious handicap since it does not permit re-examining stained preparations or allow review or use of such slides for teaching purposes. Much to our surprise, the counterstained azo dye product obtained by the automated method appears to be insoluble in oil and in Polymount (Polysciences, Inc., Warrington, PA 18976). The stability of stained slides after constant exposure to immersion oil for 3 months is shown in FIGURE 5

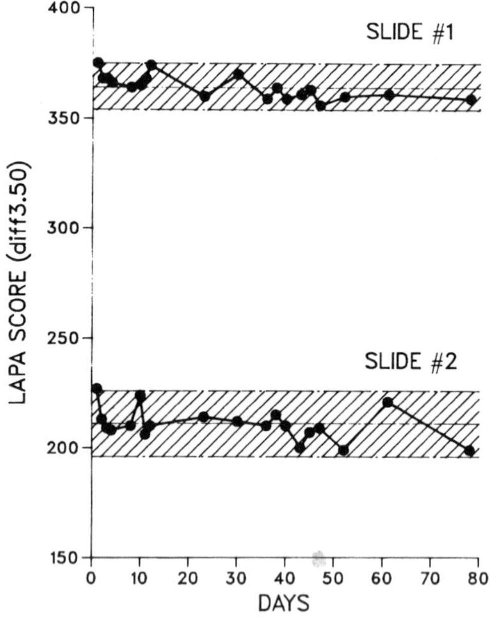

FIGURE 5. Analysis of stability of stained slides after prolonged exposure to immersion oil based on instrumental assessment of leukocyte alkaline phosphatase activity.

TABLE 3. Stability in Polymount

Sample Number	Baseline (tech A)	12 Months (diff3.50)	18 Months (diff3.50)	18 Months (tech B)
110	12	52	29	7
104	14	14	1	4
125	27	30	36	33
123	31	35	53	40
101	32	45	37	33
105	48	28	32	41
102	52	66	77	46
129	53	75	91	61
111	63	50	27	62
108	75	80	77	94
126	76	111	98	110
117	86	88	86	86
115	88	130	127	147
127	88	118	99	126
130	89	89	97	89
121	99	146	144	161
131	112	155	139	139
109	118	96	70	114
114	118	174	174	177
124	122	162	158	199
118	123	132	143	142
106	124	127	130	146
100	136	141	137	166
119	139	164	161	181
113	141	186	183	196
107	149	132	137	150
120	151	153	137	189
116	155	195	209	231
122	186	222	213	283
128	213	234	233	278
112	253	284	280	300
Mean	104	121	117	129
SE	10.2	12.0	11.9	14.6

and after mounting in Polymount for prolonged periods in TABLE 3. No significant changes in activity were observed. Preliminary studies from this laboratory suggest that the second fixation (methanol) which occurs during counterstaining, stabilizes the precipitated azo dye so that it is resistant to oil and some other solvents. Possibly other ingredients in the Coulter Stain-Right reagent contribute to this phenomenon.

SUMMARY

We have shown that it is possible to automate the assessment of leukocyte alkaline phosphatase by using an azo dye cytochemical staining procedure and a commercial, highly sophisticated image analysis instrument originally designed specifically as a differential white cell counter. The data to date indicate that values obtained by this approach are at least as precise and accurate as current manual techniques. Instrumental analysis avoids the subjectivity associated with manual interpretation of staining intensity and should permit meaningful interlaboratory comparisons. The stability of the stained smears upon exposure to immersion oil or Polymount mounting medium proved to be an unexpected bonus. In addition to such functional data on leukocytes as illustrated by this report, these instruments, with appropriate staining methods and software, can also provide clinically useful quantitative data on red cells, as have been described for reticulocytes.[7] We hope to see more clinical applications in the future for these expensive and target-oriented image analysis instruments. They are capable of automatically providing objective quantitative information on a cell by cell basis—providing feature data that cannot be obtained by other means.

ACKNOWLEDGMENT

The support and encouragement provided by the Coulter Biomedical Research Corp. are gratefully appreciated.

REFERENCES

1. KAPLOW, L. S. 1980. Leukocyte alkaline phosphatase cytochemistry: Applications and methods. Ann. N.Y. Acad. Sci. **155**:911–947.
2. KAPLOW, L. S. 1974. Letter to Editor. N. Engl. J. Med. **290**(20):1144.
3. KAPLOW, L. S. 1971. Leukocyte alkaline phosphatase in disease. Crit. Lab. Sci. **2**:243–278.
4. WILLIAMS (Beutler). Text of Hematology. 3rd. edition.
5. Sigma Tech. Bull. No. 85, Sigma Chem. Co., P.O. Box 14508, St. Louis, MO 63178.
6. Coulter Stain-Right Pack Instructions, 7507–436A (rev 8–81), Coulter Electronics, Inc., Hialeah, FL 33010.
7. KAPLOW, L. S. & J. Y. CROUCH. 1982. Automated reticulocyte counts. Lab. Med. **13**:167–169.

Controls for Flow Cytometers in Hematology and Cellular Immunology

MICHAEL C. BROWN, ROBERT A. HOFFMAN,
AND STEFAN J. KIRCHANSKI

Ortho Diagnostic Systems Inc.
410 University Avenue
Westwood, Massachusetts 02090

INTRODUCTION

The availability of controls and standardized procedures will have significant impact on the widespread acceptance of new methods of analysis. Additionally, controls should have applicability to both manual and automated methodologies to allow for a smooth transition between the two. Controls for assays involving intact viable cells, notably those involving flow cytometry, represent a major challenge to the modern clinical laboratory involved with hematology or immunology. The current generation of analyzers and reagents, including monoclonal antibodies, offers a wealth of information hitherto unavailable. The newness of much of this technology makes problems of calibration and control even more acute. While fresh biological samples can be used with very careful (and time consuming) preparation and calibration, they do not represent a viable method for the great majority of laboratories. Controls must be designed to be simple and convenient to use. Ideally such a control replaces the sample to be analyzed and checks both the procedure and the instrument, insuring that both are functional and have not changed from previous assays. Additionally such a control should have a shelf life of several months to alleviate problems associated with frequent recalibration and changing control values.

CONTROLS FOR HEMATOLOGY

For a number of years stabilized preparations of human or animal blood cells have been used for daily calibration and control of hematology counters, including both impedance and optical flow cytometers. Unfortunately these controls do not necessarily duplicate many of the important aspects of fresh whole blood and as a consequence may give different results on differing types of instrumentation.[1-4] Of course, the more closely a control is matched to fresh whole blood the smaller this discrepancy should be. This problem is accentuated with the latest generation of hematology analyzers which augment the more traditional seven or eight parameter measurements (RBC count, WBC count, platelet count, hemoglobin, hematocrit, mean cell volume, mean cell hemoglobin, and mean cell hemoglobin concentration) with information about leukocyte classes and red cell morphology. While it may be relatively simple to control for any single parameter, the demands of the busy hematology laboratory require that a single control fulfill all requirements, and this may necessitate some

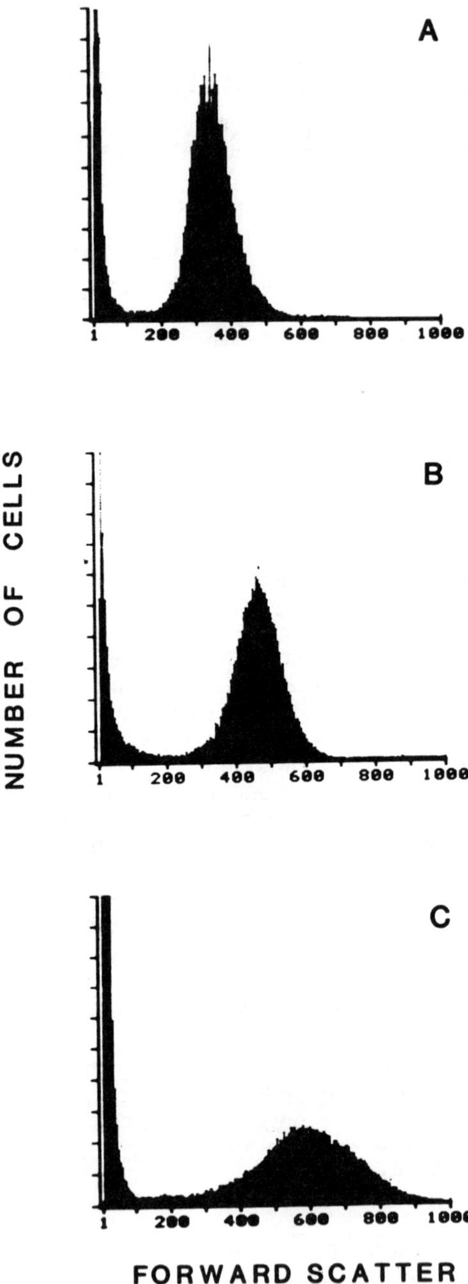

FIGURE 1. Foward scatter histograms obtained during the white cell counting cycle on Ortho ELT-8 Hematology analyzer and displayed on Ortho 2150 Computer System. Only those cells above channel 200 are counted as leukocytes. (a) fresh whole blood; (b) commercial control using fixed human erythrocytes as leukocyte analogue; (c) commercial control using avian erythrocytes as leukocyte analogue.

94

compromise in control performance. The typical eight-parameter control consists of four parts: erythrocytes, platelets or an analogue, leukocytes or an analogue, and a stabilizing diluent to serve as an analogue of serum. These components are generally processed separately then blended into the final product.

Maintaining the desired characteristics of the erythrocytes is frequently the determining factor in the shelf life of the blended control. These stabilized red cells must have a stable mean cell volume by the appropriate sizing method and a stable hemoglobin concentration throughout the control's life. In the ideal case they should interact with the instrument's saline diluent in the same manner as a fresh erythrocyte. Finally the red cells must be readily lysable with the WBC lysing reagent, which may be either osmotic or detergent based. It is these interactions with the saline and the lysing reagent that mandate that the red cells not be fixed but rather stabilized by the diluent in the control.

The desired characteristics of a leukocyte control are that the cells are sensed as a white blood cell based on the size (light scatter or impedance) in the WBC lysing reagent and that they have a shelf life at least equal to that of the red cell component. A number of manufacturers use fixed red cells, either human or avian, as the leukocyte analogue. Both types of fixed cells have served adequately in optical flow cytometers, although neither shows the desired light characteristics of fresh whole blood (FIG. 1). The addition of a three part screening differential to the Ortho ELT-8 however makes these controls somewhat obsolete since they control only for total leukocyte count and not for the subclass percentages (FIG. 2). This is typical of the types of problems facing manufacturers of instruments and controls, and customers as the technology and the scope of the assays rapidly change.

Platelets in a control must be stabilized in such a manner as to avoid clumping over the life of the control. This problem is not present if latex particles are used in place of platelets, but their size distribution is not equivalent to that of platelets, frequently resulting in erroneous calibration values. Similar counting problems have been observed with animal platelets. Thus it must be concluded that stabilized human platelets are the ideal component to use.

The diluent used in the control replaces serum and must serve as an energy source for the viable cells in the preparation and adequately buffer against the catabolites produced by these cells. Other than this, the function of the diluent is to interact as little as possible with the cells to allow them to maintain characteristics similar to whole blood.

CONTROLS FOR CELLULAR IMMUNOFLUORESCENCE

Cellular immunology represents a relatively new area for automated cell analysis in the clinical and clinically oriented research laboratories. Our approach towards the problem of controls in this area has been to consider instrument performance and calibration as an issue separate from the performance characteristics of the antibodies. These two issues then may be dealt with by two separate controls.

The most crucial aspect of instrument evaluation in an immunofluorescent analysis is the ability of the instrument to adequately resolve two differing populations of cells, such as unstained cells versus those cells which have reacted with the antibody. In order to test this criterion a control must consist of two or more differing populations of varying fluorescent intensity. It is highly desirable

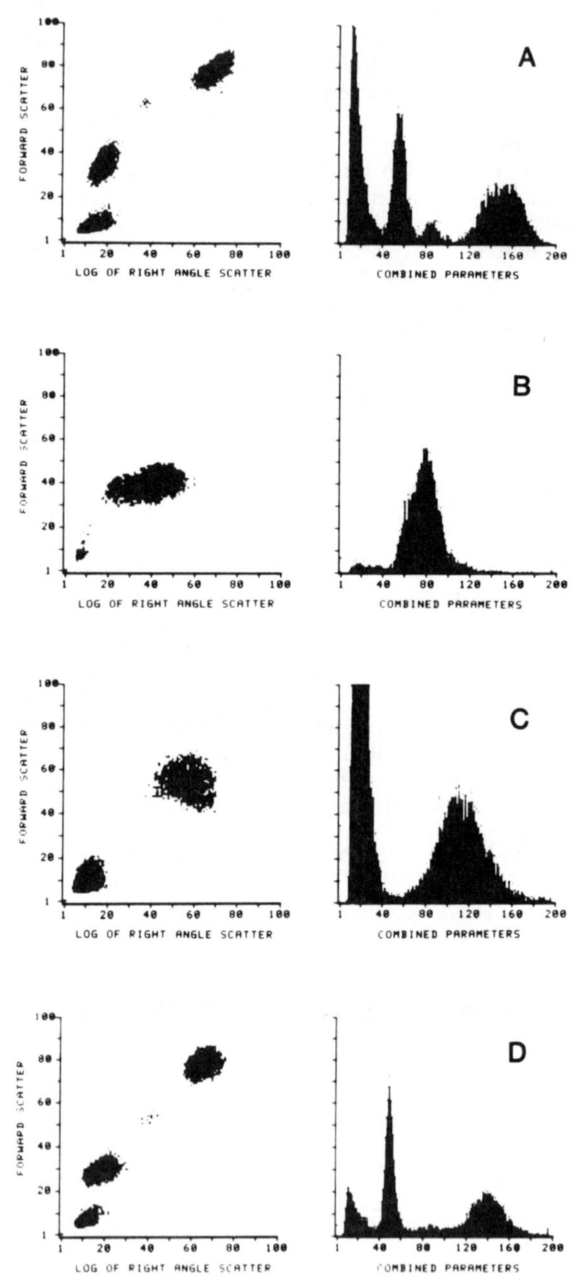
FORWARD SCATTER
LOG OF RIGHT ANGLE SCATTER
COMBINED PARAMETERS
A
FORWARD SCATTER
LOG OF RIGHT ANGLE SCATTER
COMBINED PARAMETERS
B
FORWARD SCATTER
LOG OF RIGHT ANGLE SCATTER
COMBINED PARAMETERS
C
FORWARD SCATTER
LOG OF RIGHT ANGLE SCATTER
COMBINED PARAMETERS
D

that these intensity levels approximate those levels commonly observed with the antibodies of interest. If in fact the actual number of fluorophores attached to each particle is known, then, with some caution, the actual number of sites occupied per cell may be estimated. The spectral properties of the control particles should be carefully matched to those of the fluorophore-labeled antibody to ensure that results obtained on one instrument will be comparable to those obtained on another instrument, which may have a different optical configuration. Finally the particles should have a shelf life of several months or more.

Our strategy in the production of the base particle for such a control has been to utilize nuclei obtained from calf thymocytes. The nuclei may be easily prepared from readily available material and are highly uniform in size. When fixed, the nuclei have a shelf life at 4°C of over one year. Using a biological particle offers significant advantages over synthetic materials (e.g., polystyrene beads) in that the fluorescent labeling methodology is essentially the same as the conjugation of the fluorophore to antibodies. Thus quenching of the fluorophore and the spectral shifts when bound to the fixed nuclei can be anticipated to be identical to those observed with soluble proteins (FIG. 3).

Fluorescence is introduced into the particles by the treatment with known quantities of "activated" derivatives of the dye (i.e. FITC, TRITC, XRITC) followed by extensive washing to remove free chromophore. Extent of the nonspecific (noncovalent) binding can be assessed by blockage of the activated dye with free amines.

Fluorescence of the particles is readily quantitated using conventional fluorometric methods in either or both of two complementary techniques. In the first method dye bound per particle is calculated based on measuring residual dye in the supernatant of the labeling. Essentially (dye added − free dye)/number of particles = dye per particle. In the second method known concentration of labeled particles is measured in a spectrofluorometer relative to standards of free dye so that the number of dye molecules per particle are directly measured.

The most appropriate standard for these measurements is the free non-derivatized parent dye compound. Slight spectral variations relative to the dye conjugated to protein can be readily compensated for by using a large bandpass on the fluorometer, ensuring of course that scatter characteristics of the particle do not influence the measurement. Using free dye avoids introducing the serious problem of quenching into the standard. This quenching occurs to various extents depending upon the chromophore, the degree of substitution of the protein with the chromophore, and to a lesser degree the nature of the protein, all of which are issues best avoided in selection of an absolute standard. Thus for fluorescein isothiocyanate (FITC) or dichlorotriazinyl-aminofluorescein (DTAF) a standard of sodium fluorescein in alkaline-buffered solution would be used. For tetramethylrhodamine isothiocyanate (TRITC) a standard of rhodamine B would

---

FIGURE 2. Two-dimensional histograms and combined parameter histogram obtained during the white cell counting cycle on the Ortho ELT-8/DS Hematology Analyzer with three part differential leukocyte screen as displayed on Ortho 2150 Computer System. (a) fresh whole blood. Peaks from left to right on combined scatter parameter represent debris, lymphocytes, monocytes, and granulocytes. Only those cells above the debris are counted as white cells; (b) commercial control using fixed human erythrocytes as leukocyte analogue; (c) commercial control using avian erythrocytes as leukocyte analogue; (d) control using fixed human leukocytes.

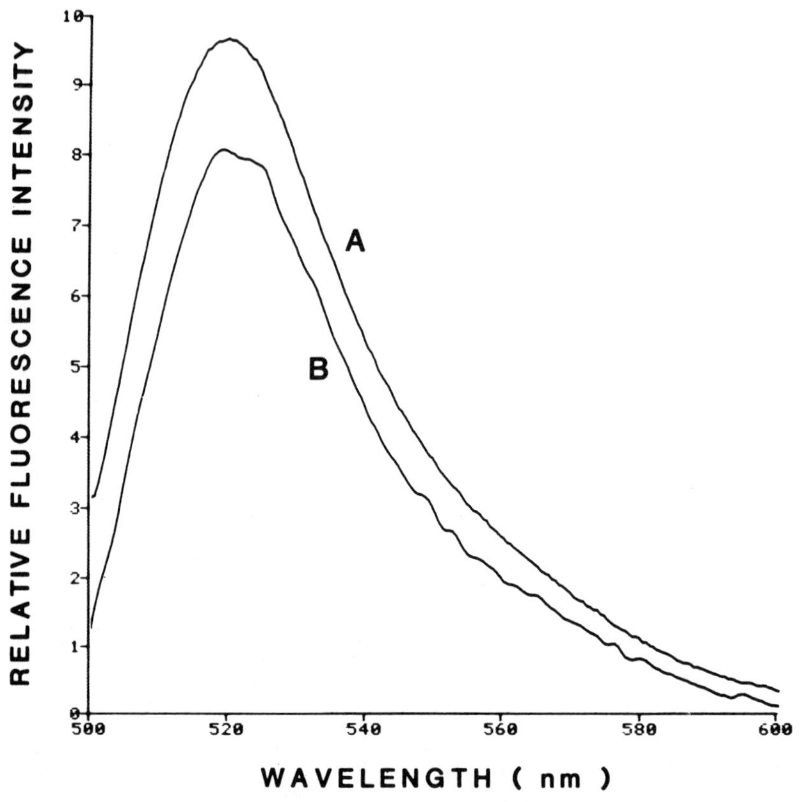

FIGURE 3. Comparison of fluorescence spectra obtained from (a) FITC-labeled IgG antibody and (b) FITC-labeled calf thymocyte nuclei. Samples were analyzed in an Aminco SPF-500 Spectrofluorometer using 490 nm excitation.

be used. Of course a correction factor must be applied to the value obtained for the nuclei relative to such a standard. For example the fluorescence yield of FITC-labeled IgG (fluorescein/protein ratio of 1:1) is 35% of the yield of an equivalent amount of free fluorescein. Therefore if a single particle has the fluorescence equivalent to 10,000 free fluoresceins it would actually have 10,000/ 0.35 or 28,600 fluoresceins conjugated to the protein in the nuclei, assuming that the quenching of the dye bound to protein of the nuclei is equivalent to that observed for soluble proteins. As shown in FIGURE 4, this appears to be a valid assumption since the values obtained by the subtractive method of measurement and the bulk method of measurement are in good agreement. It should be noted that unstained nuclei do have autofluorescence in the region of interest and this must be carefully accounted for in assigning values to low levels of fluorescence. As anticipated, bulk measurements are in good agreement with those obtained by flow cytometry (FIG. 5).

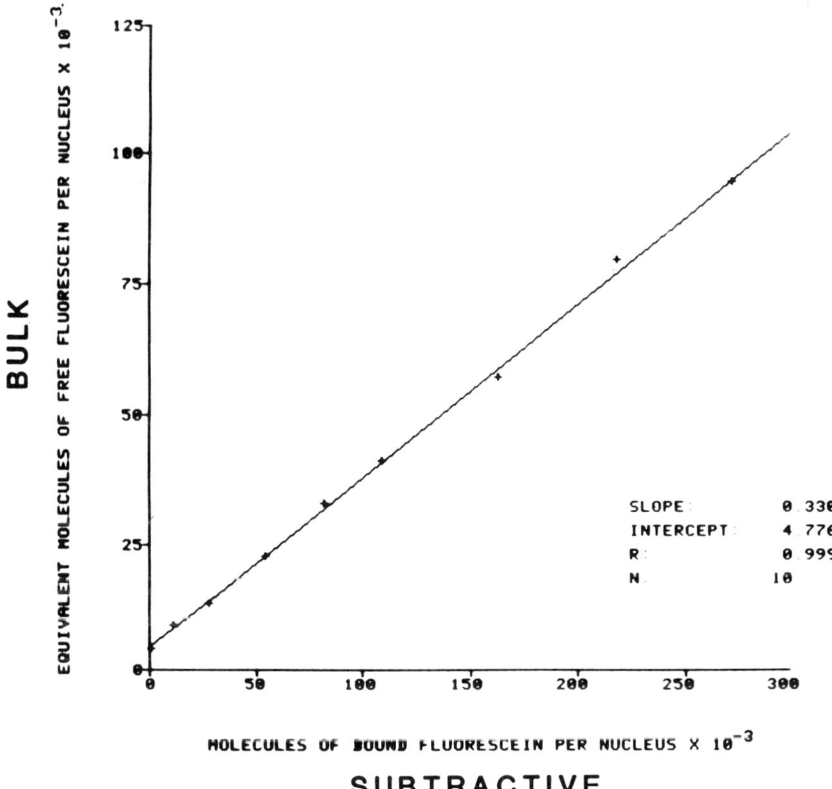

SUBTRACTIVE

FIGURE 4. Comparison of subtractive and bulk methods of measurement on labeled thymocyte nuclei. Subtractive measurements were made by measuring residual dye in the supernatant after labeling. Bulk measurements were made relative to standards of free sodium fluorescein. The slope of the line (0.33) represents the fluorescence yield of fluorescein conjugated to nuclei relative to free dye. The y intercept (4.77×10^3) represents the autofluorescence of unlabeled nuclei.

A control for instrument calibration, which we have termed "FluoroTrol," consists of three fluorescent levels of these particles. (1) an unstained population; (2) a population with signal equivalent to 20,000 free fluorescein molecules; and (3) a population with signal equivalent to 72,000 free fluorescein molecules. Use of FluoroTrol as a control consists of ensuring that the brightest population falls in a given mean channel and that adequate resolution exists between the low and unstained populations. Such conditions assure the ability of the instrument to adequately resolve the fluorescence of positively and negatively stained leukocytes of the commonly used monoclonal antibodies. FIGURE 6 shows the distribution of FluoroTrol relative to those obtained with several Ortho-mune OKT antibodies.

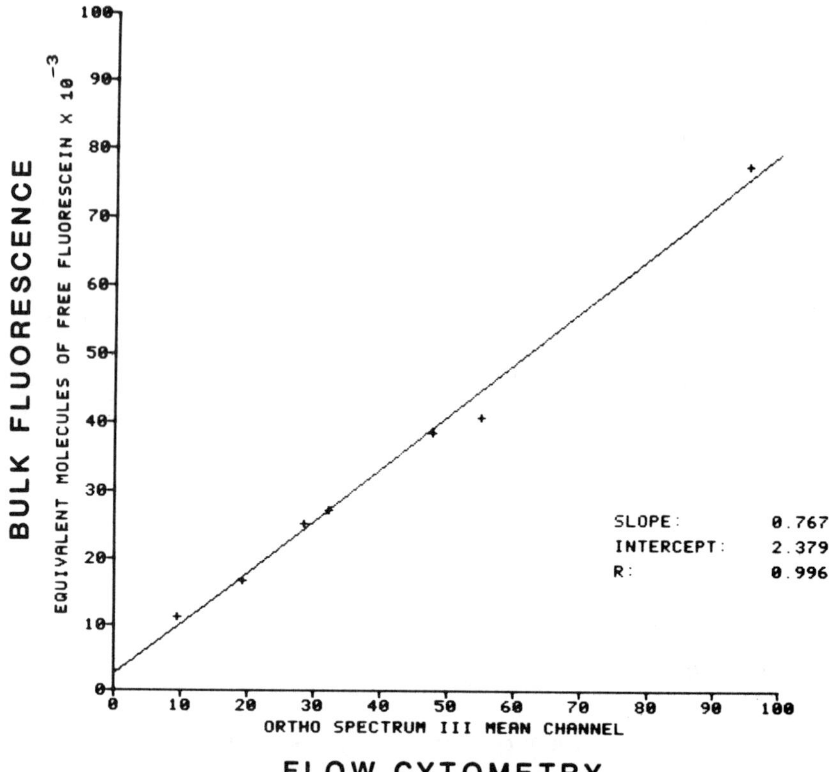

FIGURE 5. Comparison of results obtained by bulk fluorometric analysis with results obtained by flow cytometry. Bulk analysis was performed on Aminco SPF-500 spectrofluorometer using 490 nm excitation. FITC labeled nuclei were suspended at 5×10^6 nuclei per ml in 0.1 M NaCl, 0.05 M borate and measured relative to standards of free sodium fluorescein in the same buffer. Results are expressed per nuclei. Flow cytometric analysis was performed using an Ortho Spectrum III using an argon ion laser (488 nm) as the excitation source.

A more complicated issue than that of instrument control and calibration is that of controlling for the antibody and associated staining procedures. Since the most crucial aspect is whether the antibody has the desired immunologic activity (titer, intensity, and specificity) such a control is required to possess the antigenic determinants for the antibody in question. In order to replicate all aspects of the procedure the control should have similar physical characteristics, such as shape, size, and volume, of the cells being probed. In addition to staining in a specific manner with the antibody, the control should also be sensitive to any contaminants in the sample that might produce nonspecific staining effects. These combined criteria effectively eliminate controls derived from any material other than of human origin.

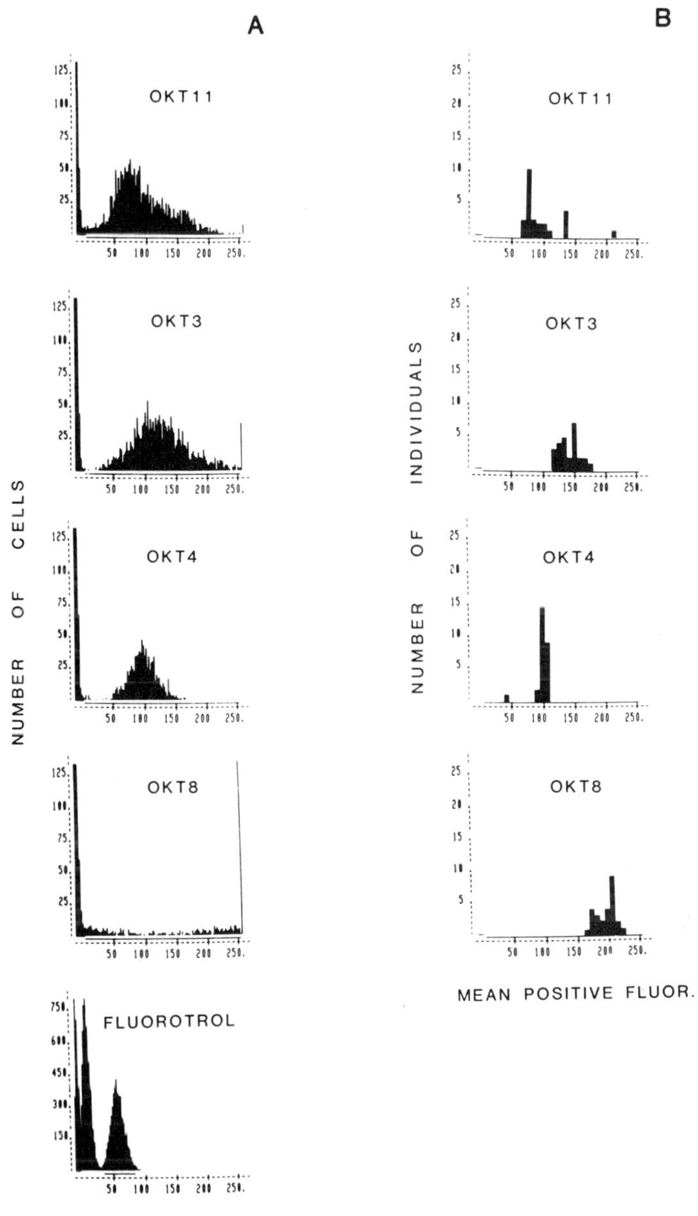

FIGURE 6. (a) Comparison of fluorescence histograms obtained with Ortho-mune OKT antibodies and normal donor leukocytes with fluorescence histogram of FluoroTrol. (b) Distribution of mean positive channel of OKT-positive cells among normal individuals. OKT4 shows relatively consistent results among donors and may serve as an additional source of control.

Since a minimum stability of several months is desired special steps must be taken to preserve these relatively fragile cells. The ideal situation would be to preserve the viability of the cells by freezing the cells in a solution of glycerol or dimethyl sulfoxide in liquid nitrogen. An aliquot could then be thawed and washed prior to use. Unfortunately, while applicable to individual research laboratories, this is not a viable alternative for the majority of clinical laboratories or for commercial production of such a control.

A more practical approach involves the stabilization or fixation of the cells, keeping in mind that fixation will most likely induce subtle changes in the cells' physical properties. Fixatives must be employed that will not react with crucial determinants at the antigenic sites or must be used at sufficiently low concentrations to minimize the effect. Unfortunately the lower the concentration of fixative used the less the anticipated stability will be. Fixatives, such as glutaraldehyde, that induce autofluoresecence of the cells must be excluded as well as those which may result in nonspecific staining due to free functional groups unless such groups are blocked.

Commercial production of such controls involves pooling cells from a wide number of donors. Since antigenic density, with the exception of OKT 4, varies widely among individuals, the pool will have broader fluorescent distributions than a typical single donor. It may be desirable to prescreen all donors prior to adding them to the pool to maximize uniformity.

As an alternative to fixed or stabilized reagent controls a normal fresh whole blood method may be used, much the same as it is in hematology. Studies have shown that normal blood may be kept at room temperature for at least 48 hours before significant changes in immunofluorescent analysis of lymphocyte subsets occur.[5] Thus, normal whole blood samples can serve adequately as staining controls until stabilized control material becomes available on a widespread basis.

SUMMARY

The necessity of using freshly prepared biological samples to control and calibrate flow cytometers has impeded the utilization of flow cytometry in the clinical laboratory where regulations demand careful control calibration and yet workload and technician training militate against the preparation of specialized control samples. For a number of years, cell preparations with stabilized properties have been used routinely in hematology laboratories to evaluate the daily performance of automated blood cell counters including optical flow cytometers. Unfortunately, the stabilized preparations do not necessarily duplicate many important aspects of whole blood, resulting in variable performance of a given control preparation on different types of instrumentation. The new generation of hematology analyzers measures a host of new parameters and will require more sophisticated controls for these enhanced parameters. Some practical aspects of present and future hematology controls will be discussed. Immunology, a new area for automated cell analysis, can benefit greatly from controls both for instrument performance and for cell staining. Up to this time there has been little availability of control materials suitable for clinical immunology. We have been evaluating fluorochrome-stained cell nuclei for use as an instrument control and for calibration of immunofluorescence analyses. Our development and use of this type of control will be described. We will also discuss approaches and experiments aimed at producing cells with stabilized antigenic properties which can be stained with monoclonal antibodies and can thus act as actual reagent controls.

REFERENCES

1. ALLISON, F. S. 1983. A historical review of quality control in hematology. Am. J. Med. Tech. **49**:633–642.
2. DREWINKO, B., P. BOLLINGER, M. ROUNTREE, D. JOHNSTON, G. CORRIGAN, W. DALTON & J. TRUJILLO.1982. Eight parameter automated hematology analyzers: Comparison of two flow cytometric systems. Am. J. Clin. Pathol. **78**:738–747.
3. GILMER, P. R. & L. J. WILLIAMS. 1980. The status of methods of calibration in hematology. Am. J. Clin. Pathol. **74**:600–605.
4. MAYER, K., B. CHIN, J. MAGNES, H. THALER, C. LOTSPEICH & A. BAISLEY. 1980. Automated platelet counters. Am. J. Clin. Pathol. **74**:135–150.
5. SHIELD, C. F. III, P. MANLETT, A. SMITH, L. GUNTER & G. GOLDSTEIN. 1983. Stability of human leukocyte differentiation antigens when stored at room temperature. J. Immunol. Methods **62**:347–352.

Extended Leukocyte Parameters
for Hematology Analyzers

DEBORAH PEEBLES, STEFAN KIRCHANSKI,
MICHAEL BROWN, AND ROBERT HOFFMAN

Ortho Diagnostic Systems Inc.
410 University Avenue
Westwood, Massachusetts 02090

INTRODUCTION

ORTHO[™a] ELT Hematology Analyzers are laser-based flow cytometers that automatically produce a complete blood count (CBC) consisting of red cell, white blood cell (WBC), and platelet counts as well as hematocrit, hemoglobin concentration, mean corpuscular volume (MCV), and two calculated indices (mean corpuscular hematocrit and mean corpuscular hemoglobin). In these instruments a fluidics system first performs automatic dilutions of whole blood, and the cellular components of the diluted blood are then counted by means of the narrow-angle foward light scatter of a helium-neon laser focused on the sample flowing in a quartz cuvette. White blood cells are distinguished from red cells by diluting part of the blood sample with a detergent-containing reagent that lyses erythrocytes but leaves the nuclei of the leukocytes to be counted. Data calculation, storage, and display are performed automatically by an integral microcomputer.

Narrow-angle forward light scatter is used in the analyzers to count and to size cells, as it has been demonstrated that such light scatter is proportional to cell size.[1] However, light scattered at a wide angle (90° ± 37°) has been found to be sensitive to variations in the index of refraction of the scattering cells.[1] Thus, cellular inclusions such as granules having a different index of refraction from the ground cytoplasm can be detected by wide-angle light scatter. In fact, leukocytes can be separated into three populations (lymphocytes, monocytes, and granulocytes) on the basis of wide-angle and foward light scatter.[2] The enhanced analyzers (ELT-8/WS) are able to produce leukocyte counts of the three subpopulations through the addition of a new optical bench that contains an optical train at right angles to the flow cell to detect the wide-angle light scatter (FIG. 1).

When the new optics are used with an improved lysing reagent which lyses the red cells without significantly altering the light scatter of the white cells, the optical detectors produce signals that can be processed into a four-peaked histogram (FIG. 2). In this histogram the ordinate represents cell number while the abscissa represents a combination of forward light scatter (cell size) and wide-angle light scatter (cell granularity). Thus, the extreme left-hand peak represents very small objects with little or no granularity; it contains red cell ghosts and some platelets. The extreme right-hand peak represents large cells with pro-

---

[a]Trademark - Ortho Diagnostic Systems Inc.

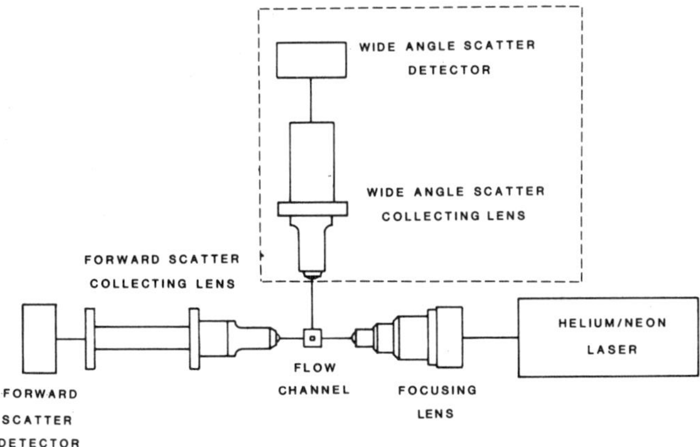

FIGURE 1. Diagram of optical bench of ORTHO ELT-8/WS Hematology Analyzer. Components surrounded by dashed line represent the newly added wide-angle scatter optics.

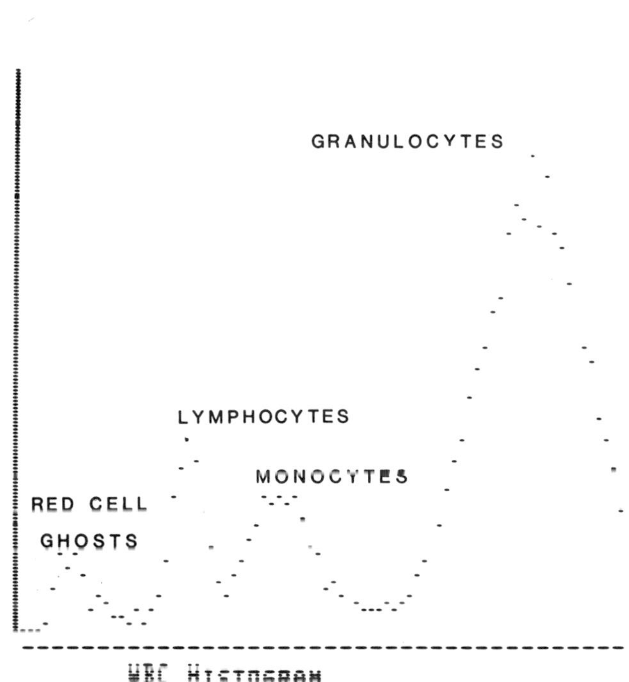

FIGURE 2. WBC histogram photographed from the screen of an ORTHO ELT-8/WS before the computer analyzed the peaks and determined the thresholds.

nounced granularity (a maximum of both scatter signals), the granulocytes, while the peak next to the red cell ghosts contains small, relatively nongranular cells, the lymphocytes. The peak between the lymphocytes and the granulocytes is made up of large cells with an intermediate degree of granularity, the monocytes. This separation between the leukocyte subpopulations is achieved solely on the basis of the light scatter characteristics of the cell types.

Blood from a normal individual produces a white cell histogram with the three leukocyte subpopulations clearly resolved and the proportion of cells in each subpopulation within certain normal ranges. However, the exact size and granularity of the leukocyte subpopulations does vary from individual to individual. Consequently, fixed thresholds cannot be used to accurately resolve the histogram peaks. Instead, the computer analyzes the histogram looking for maxima and minima, and then sets calculated thresholds to exactly define the subpopulations of each different blood sample (FIG. 3). This process allows the sytem to correctly differentiate leukocytes in spite of normal individual variations in cell characteristics. What is more, this analytical process allows the instrument to flag abnormal blood samples that lack one or more subpopulations, have inadequate separation between subpopulations, or have subpopulations whose light scatter characteristics fall outside the limits of normal variation.

In theory, the extended leukocyte counts of the ELT/WS instruments can have an important impact on the work flow and overall efficiency of the clinical

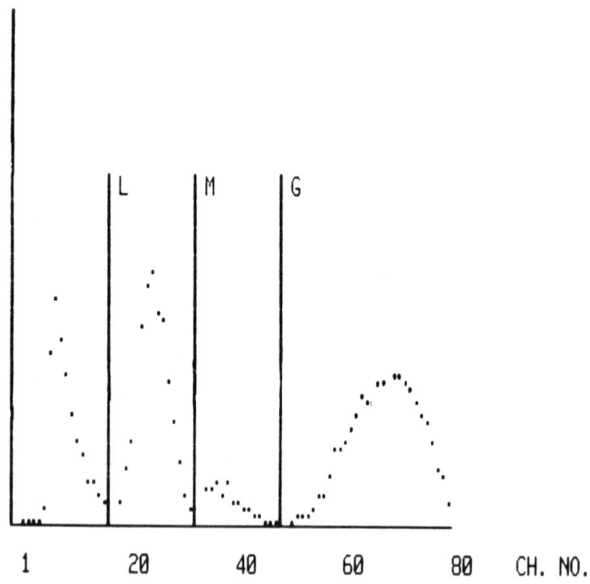

FIGURE 3. ORTHO ELT-8/WS WBC histogram from a normal leukocyte distribution with the following results: WBC 7.9 × 10³/mm³, 37.6% lymphocytes, 7.0% monocytes, and 55.4% granulocytes.

hematology laboratory. Currently, much time is spent by trained technicians in producing manual differentials of samples that turn out to be normal. Not only is this boring to the technicians, but it also limits the time that is available for the vital task of analyzing truly abnormal specimens. If the extended white cell parameter can actually eliminate some normal bloods from the manual differential workload, technicians would have more time to perform other important laboratory tasks and staff size could remain stable.

CLINICAL STUDY

An ORTHO ELT-8/WS was evaluated in a large metropolitan general hospital laboratory. The purpose of the evaluation was to compare the performance of ORTHO ELT-8/WS automated leukocyte subpopulation counts to the manual differential leukocyte counting method in a clinical setting.

All differential requests received by the laboratory during the first shift were included in the study. Specimens for analysis were obtained from a population of hematologically normal and abnormal individuals whose differential leukocyte counts covered a wide range. Five-hundred forty-five samples were analyzed over a four-week period.

Because ORTHO ELT-8/WS provides lymphocyte, monocyte and granu- locyte subpopulation results, manual differential leukocyte results were grouped accordingly. Manual lymphocyte counts include normal, atypical, and abnormal lymphocytes and plasma cells. Manual monocyte counts include normal monocytes and large mononuclear cells. Manual granulocyte counts include the various stages of neutrophils, eosinophils, and basophils.

INTERPRETATION OF ORTHO ELT-8/WS OUTPUT

FIGURE 3 displays the histogram and subpopulation counts from a normal leukocyte distribution, which will be compared to abnormal distributions to illustrate the flag scheme of ORTHO ELT-8/WS.

The "plus alert" (FIG. 4) is triggered when a WBC anomaly is detected. A manual differential leukocyte count is required when the + symbol is printed in place of WBC subpopulation results. A WBC anomaly is detected by the microprocessor when lymphocyte, monocyte, or granuloctye peaks are absent or overlap owing to abnormal leukocyte morphology. When a WBC anomaly is detected in which the debris peak overlaps the leukocyte distribution, both the WBC and the subpopulation results are flagged with the "plus alert." In the case of a WBC count "plus alert," an alternate WBC count method must also be performed.

The "star alert" (FIG. 5) occurs when WBC, lymphocyte, monocyte, or granulocyte results exceed the normal ranges set by the operator. A microscopic review of a differential slide is advised when the * symbol is printed to the right of the abnormal result. The "star alert" is also generated when the number of lymphocytes is greater than the number of granulocytes. A message will appear on the computer screen to indicate a lymphocyte/granulocyte reversal.

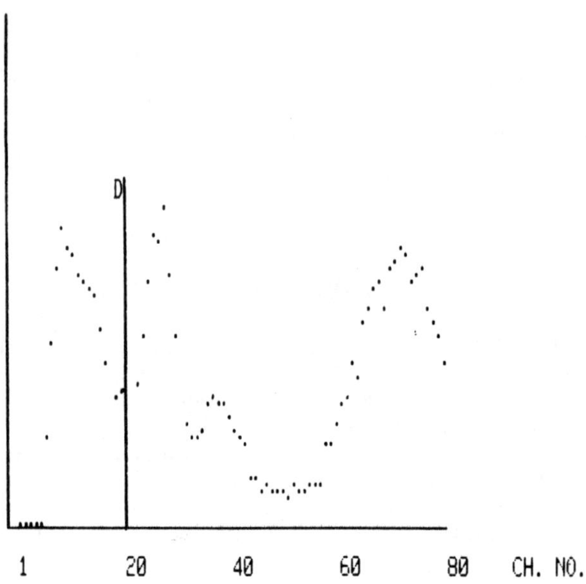

FIGURE 4. ORTHO ELT-8/WS WBC histogram from an abnormal sample detected as a WBC Anomaly with the following results: WBC $15.8 \times 10^3/mm^3$, + lymphocytes, + monocytes, and + granulocytes. The "+" denotes the plus alert flag.

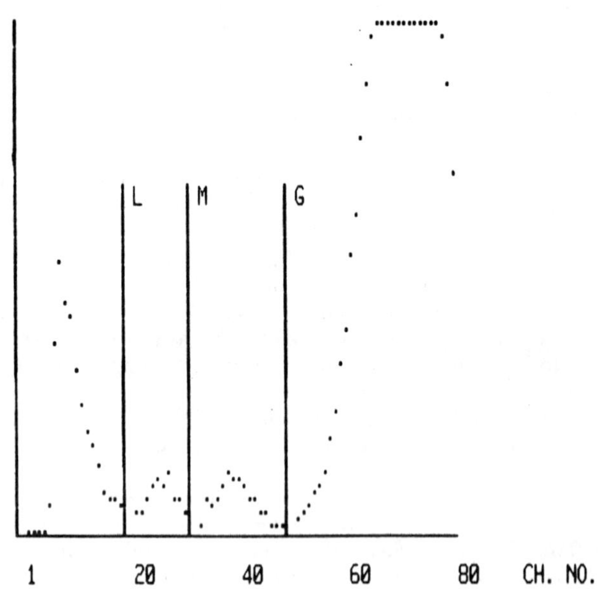

FIGURE 5. ORTHO ELT-8/WS WBC histogram from an abnormal leukocyte distribution with the following results: WBC $25.4 \times 10^3/mm^3$, 3.0* % lymphocytes, 3.8* % monocytes, and 93.2* % granulocytes. The "*" denotes the star alert flag.

CORRELATIONS

Results from the analysis of 500 specimens were included in the comparison between ORTHO ELT-8/WS and manual differential count methods. ORTHO ELT-8/WS flagged the remaining 45 specimens as WBC anomalies which caused them to be excluded from correlations.

Correlation of leukocyte subpopulation counts expressed as percentages are presented in FIGURES 6, 7, and 8 for lymphocytes, monocytes, and granulocytes, respectively. The correlation coefficients of .944, .809 and .941, respectively, indicate that ORTHO ELT-8/WS correlated well with manual results over wide count ranges.

RECOMMENDATIONS FOR CLINICAL UTILITY

An important use of ORTHO ELT-8/WS is in the reduction of the number of manual differential leukocyte counts performed on normal specimens. The extent of workload reduction depends on the volume of normal manual differentials

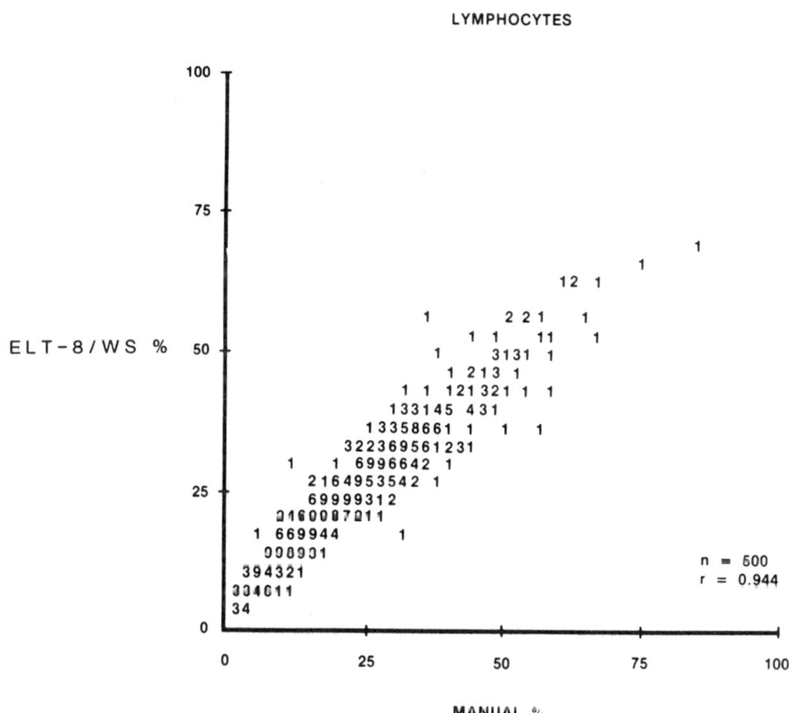

FIGURE 6. Correlation between ORTHO ELT-8/WS and manual lymphocyte percent for 500 samples. Correlation coefficient is 0.944. The numerals 1,2, ... 9 denote the number of data points at each coordinate.

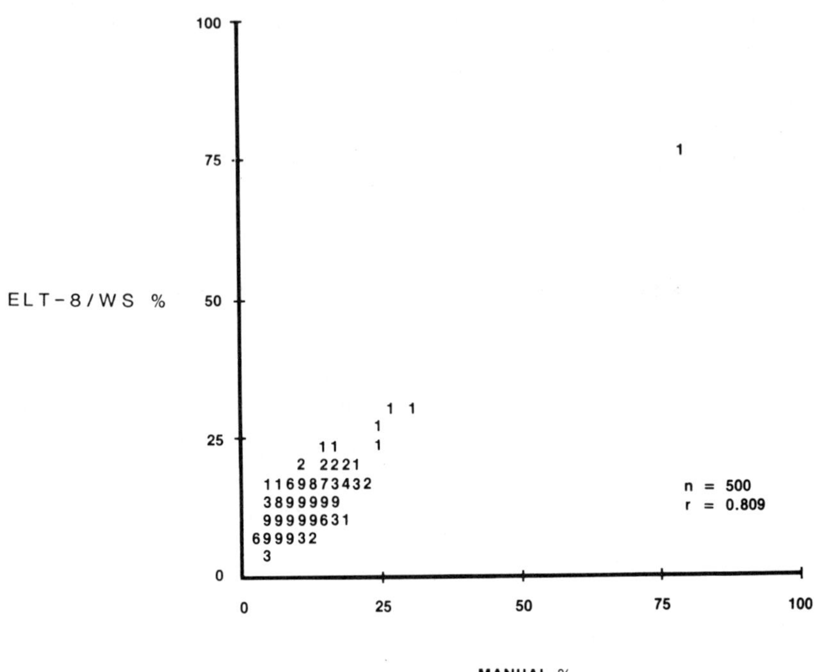

FIGURE 7. Correlation between ORTHO ELT-8/WS and manual monocyte percent for 500 samples. Correlation coefficient is 0.809. The numerals 1,2, ... 9 denote the number of data points at each coordinate.

currently performed, the correlation between the ORTHO ELT-8/WS and the manual methods, and the normal value ranges defined by the laboratory.[3] As stated previously, the correlations between the automated and manual results in this study were very good. The volume of normal manual differential leukocyte counts in this study was reduced by approximately 80%. Because volumes of normal manual differential leukocyte counts as well as normal value ranges vary from institution to institution, workload reduction will also vary. Very wide ORTHO ELT-8/WS alert ranges, set according to laboratory normal value ranges will reduce the normal manual differential leukocyte count workload tremendously. However, the consequence is an increased incidence of false normal results.[4] Carefully chosen normal value ranges for use as alert ranges by the automated system are essential for effective clinical use.

Once baseline data have been obtained and confirmed by the manual differential leukocyte count method, the automated method further reduces manual method workload when it is used as a monitor of patient leukocyte status.

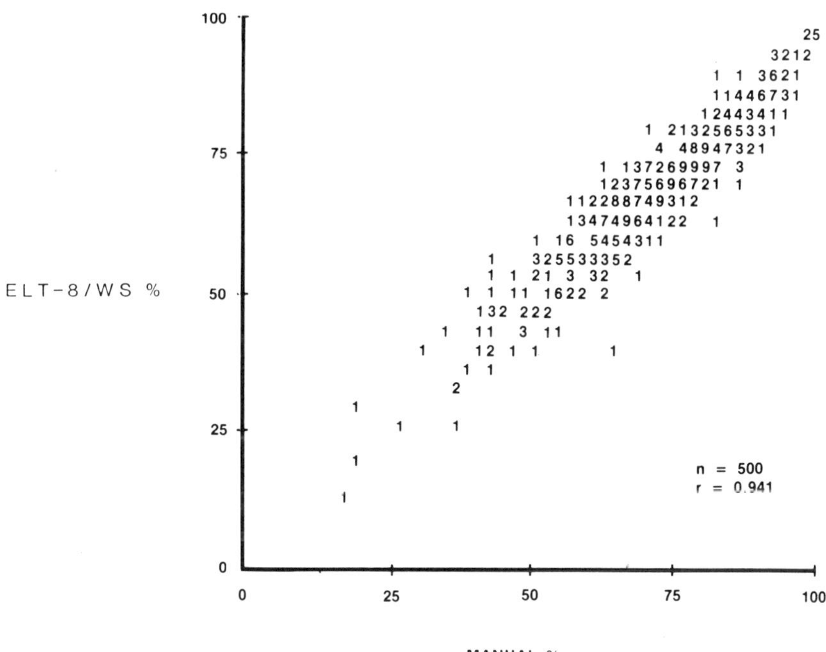

FIGURE 8. Correlation between ORTHO ELT-8/WS and manual granulocyte percent for 500 samples. Correlation coefficient is 0.941. The numerals 1,2, … 9 denote the number of data points at each coordinate.

Because patient leukocyte status can be continually monitored using ORTHO ELT-8/WS, fewer manual differential leukocyte counts are done. For added confidence, slides may be scanned microscopically. If a change in leukocyte status is reflected by ORTHO ELT-8/WS results, a complete manual differential leukocyte count is advised.

CONCLUSIONS

Hematology analyzers can effectively utilize a combination of narrow-angle forward scatter and wide-angle light scatter to accurately differentiate normal whole blood leukocytes into lymphocytes, monocytes, and granulocytes. Furthermore, this type of system can detect a majority of leukocyte abnormalities and flag them. This can result in a significant reduction of the manual differential workload in the clinical hematology laboratory.

SUMMARY

Newly introduced, extended leukocyte parameters on laser-based hematology analyzers permit normal samples to be readily identified allowing these samples to be excluded from manual analysis. This enables more hematology laboratory resources to be focused on abnormal specimens. A new reagent that lyses erythrocytes while leaving the optical properties of the leukocytes unaltered is used. A special optical bench measures both narrow-angle forward light scatter (an indicator of cell size) and wide-angle light scatter (an indicator of cell granularity). The two scatter measurements are combined to produce a histogram in which lymphocytes, monocytes, and granulocytes are clearly delineated. A microprocessor detects the separation between histogram peaks, sets the proper thresholds, produces the three-part count, and indicates abnormal samples on the basis of histogram peak position, shape, and resolution. In a clinical study correlations between instrument counts and manual leukocyte differentials were very good. The potential for reducing the number of manual differentials of normal samples is signficant.

REFERENCES

1. SALZMAN, G. C. 1982. Light scattering analysis of single cells. *In* Cell Analysis. N. Catsimpoolas, Ed. 1:111–143. Plenum Press. New York, NY.
2. SALZMAN, G. C., J. M. CROWELL, J. C. MARTIN, T. T. TRUJILLO, A. ROMERO, P. F. MULLANEY & P. M. LaBAUVE. 1975. Cell classification by laser light scattering: Identification and separation of unstained leukocytes. Acta Cytol. 19:374–377.
3. RICH, E. C., T. W. CROWSON & D. P. CONNELLY. 1983. Effectiveness of differential leukocyte count in case finding in the ambulatory care setting. J. Am. Med. Assoc. 249:633–636.
4. KOEPKE, J. A., Ed. 1977. Differential Leukocyte Counting. College of American Pathologists. Skokie, IL.

Methodological Considerations for Implementation of Lymphocyte Subset Analysis in a Clinical Reference Laboratory

K. A. MUIRHEAD,[a, c] P. K. WALLACE,[b] T.C. SCHMITT,[a]
R. L. FRESCATORE,[b] J. A. FRANCO,[b] AND P. K. HORAN[a]

[a]Smith Kline and French Laboratories
Philadelphia, Pennsylvania 19101

[b]SmithKline BioScience Laboratories
King of Prussia, Pennsylvania 19406

INTRODUCTION

Beginning with the distinction between T-cell and B-cell neoplasms, expanded by the development of hybridoma antibodies and publicized by the increasing incidence of Acquired Immune Deficiency Syndrome, applications of lymphocyte subset analysis in clinical diagnosis have grown dramatically in the past 10 years (see ref. 1-3 for general reviews). Flow cytometry, because of its speed, sensitivity, objectivity, and quantitative capability, has been increasingly preferred in the clinical research laboratory over traditional microscopic methods for subset analysis.[4-7] Along with the increasing number of diagnostic applications, there have arisen a wide variety of procedures for flow cytometric subset analysis. Some, but by no means all, of the significant variables which have been reported include sample collection medium,[8] storage medium and temperature,[9,10] cell preparation technique,[11,12] reagents and staining methodology,[13] light scatter gating parameters,[14] viability,[7,15] fixation method,[16,17] and type of data analysis.[13]

One reason for this plethora of procedures is that each variation represents a different compromise on the various aspects of cell preparation and/or analysis optimized for the purposes of a particular laboratory. Our goal was to adopt a method compatible with the operations of a large clinical reference laboratory and capable of analyzing 100 or more specimens a day. Since patients are not in-house, a major requirement is a sample collection/transportation/storage method which optimizes specimen stability. This allows shipment to a central location for analysis, repetition of analyses, and examination of additional markers suggested by initial analysis results without requiring a fresh specimen. A second major requirement is a time-efficient sample preparation method, needed because of the large number of specimens. In addition to simplicity and speed, we required that the sample preparation method selected give results comparable to those obtained using Ficoll-Hypaque (FH), the traditionally accepted method of lymphocyte purification. We have therefore compared lymphocyte subset fre-

[c]To whom reprint requests should be addressed (Mail Code L-102).

quencies obtained for a healthy reference population using FH and several variations on the lysis method. We have also compared the effects of storage media and conditions on the reproducibility of subset analysis of lymphocytes prepared by these methods.

For maximum efficiency of sample preparation, direct immunofluorescence staining is preferable to a two-step indirect staining procedure. However, this is only true provided that directly labeled reagents are available which are capable of giving baseline positive/negative resolution on the instrumentation used. Otherwise any time saved in preparation is likely to be offset by increased complexity, subjectivity, and time required for analysis of histograms where resolution is incomplete. For these reasons, we felt that criteria for choosing among different commercially available reagents were, in approximate order of importance (1) degree of positive/negative separation; (2) availability of appropriate controls for nonspecific staining; (3) stability and reproducibility; (4) insensitivity to method of cell preparation; and (5) price. We have compared a number of commercial T- and B-cell reagents in terms of the first four of these criteria. We also discuss our approach to instrument optimization, staining quality control, viability considerations, and biohazard containment strategies.

MATERIALS AND METHODS

Antibody Reagents

The following FITC-conjugated monoclonal antibodies used for direct staining procedures were provided by their manufacturers: OKT3, OKT4, OKT8, and OKT11 (Ortho Pharmaceutical Corp., Raritan, NJ); T4, T8, T11, and B1 (Coulter Electronics, Inc., Hialeah, FL); Leu2a, Leu3a, Leu3ab, Leu4, Leu12, LeuM3, and HLE (Becton Dickinson, Inc., Sunnyvale, CA). For comparison of direct versus indirect staining, unconjugated Leu2a, Leu3a, Leu4, OKT11, and T11 (same manufacturers as indicated above) were used as first step reagents and FITC-conjugated F(ab')₂ goat anti-mouse immunoglobulin (heavy and light chain specific, Cappel Laboratories, West Chester, PA) was used as the second step reagent. FITC-conjugated F(ab')₂ goat anti-human immunoglobulin (heavy and light chain specific) was also purchased from Cappel Laboratories.

Collection, Storage, and Preparation of Peripheral Blood

Peripheral blood (PBL) from apparently healthy volunteers was drawn into heparinized tubes (Becton Dickinson Systems, Rutherford, NJ) and either stored without further manipulation or mixed with an equal volume of transport medium. The transport medium consisted of RPMI with 25 mM HEPES plus glutamine and gentamicin (RPMI-HEPES; Grand Island Biological Co., Grand Island, NY). Specimens were stored at either 4°C or ambient temperature.

Peripheral leukocytes were prepared either by Ficoll-Hypaque (FH) density gradient centrifugation[11] or by lysis of whole blood.[12] For FH preparation, PBL was diluted 1:1 with Hanks buffered salt solution (Grand Island Biological Co.), layered over FH, and centrifuged at $400 \times g$ for 40 minutes at 20°C. Mononuclear cells at the supernatant-FH interface were removed and washed three times at 4°C with phosphate-buffered saline containing 1% bovine serum albumin (Sigma

Chemicals, St. Louis, MO) and 0.1% sodium azide (PBS-BSA-Az). The washed cells were resuspended at approximately $3-5 \times 10^6$ per ml prior to staining.

For preparation by the lysis technique, three variations were used. In Method A, the lysing reagent and protocol supplied by Ortho Pharmaceutical Corp. were used: after immunofluorescence staining, 0.1 ml of whole blood was mixed with 2 ml of lysing reagent (8.3 g NH_4Cl, 1 g $KHCO_3$, and 0.37 g Na_4EDTA per liter, pH 7.4) and incubated at room temperature for approximately 15 minutes. After lysis was complete, cells were washed once at 4°C with PBS-BSA-Az, resuspended in 0.5 ml of the same buffer, and kept on ice until analyzed. In method B, 1 ml of whole blood was mixed with 15 ml of modified lysing reagent (same composition as above plus 17 mM pH 7.2 tris-[hydroxyamino]methane) and incubated for 15 min at room temperature with inversion every 5 minutes. In method C, dextran sedimentation was used to obtain leukocyte enrichment prior to the lysis step: 3 ml of whole blood was mixed with 3 ml of 2% dextran (molecular weight 100,000–200,000; US Biochemical, Cleveland, OH) and allowed to sediment at 37°C for 15 min in silicone-coated vacutainer tubes (Becton Dickinson Systems, Rutherford, NJ). The leukocyte-enriched supernatant was removed, washed with PBS-BSA, and then lysed by method B. After lysis by methods B or C, cells were washed three times at 4°C with PBS-BSA-Az and resuspended in 1 ml of the same buffer for immunofluorescence staining.

In some experiments to determine the effect of removal of cytophilic antibody, cells were washd and incubated in RPMI-HEPES containing 1% BSA and 50 units/ml penicillin and streptomycin (Pfizer Inc., New York, NY) for 60 minutes at 37°C after FH or lysis.

Immunofluorescence Staining

For direct or indirect immunofluorescence staining, 100 µl of mononuclear or leukocyte cell preparation was mixed with 100 µl PBS-BSA-Az and the amount of monoclonal antibody recommended by the manufacturer and incubated for 30 minutes on ice. For indirect staining, cells were washed three times in cold PBS-BSA-Az and resuspended in 200 µl FITC-conjugated secondary antibody for another 30 minutes on ice. After either direct or indirect staining, all further manipulations were carried out on ice. Cells were then washed three times with PBS-BSA-Az and resuspended in 200 µl of PBS-BSA (without azide) and 20 µl propidium iodide (Sigma Chemicals; 50 µg/ml in PBS). All staining was carried out in either 15 ml conical centrifuge tubes (Corning Glass, Corning, NY) or 96-well microtiter plates (Costar, Cambridge, MA).

Flow Cytometric Analysis

Samples were analyzed on an EPICS V flow cytometer/sorter (Coulter Electronics, Inc., Hialeah, FL) using excitation power of 500 mW at 488 nm. Forward angle light scatter was collected through a neutral density filter (OD 2.0). Right angle scatter was reflected to one photomultiplier by a 488 nm dichroic mirror and collected through a 488 nm bandpass interference filter. A 515 nm longpass interference filter was used to prevent scattered laser light from reaching the fluorescence detectors. Propidium iodide ("red") fluorescence was reflected to a second photomultiplier by a 560 nm dichroic mirror and collected through a 590 nm longpass absorption filter. Antibody ("green") fluorescence was collected at a third photomultiplier through a 560 nm shortpass interference/515 nm longpass absorption filter combination. Antibody fluorescence was collected using a 3-decade logarithmic amplifier.

At the beginning of each day instrument performance was monitored by running 10 μm fluorescent particles (full bright fluorospheres, Fine Particles Division, Coulter Electronics, Inc.) using standard conditions for laser power, photomultiplier high voltage, and gain. Light scatter and fluorescence means and their associated coefficients of variation (CV) were required to fall within 10% of established standard values before cell samples were analyzed.

Staining with HLE (pan-leukocyte) antibody was used to assess the viability of each specimen. A two-parameter histogram of green (HLE) fluorescence versus red (propidium iodide) fluorescence was accumulated for 10,000 cells. Viability was calculated as the percentage of HLE-positive cells that were propidium iodide negative. If viability fell within acceptable limits (>85%, see DISCUSSION), forward angle and right angle scatter gates were then set to define the lymphocyte population.[14] Staining with LeuM3 (anti-monocyte) antibody was used to determine the number of small monocytes falling within the lymphocyte gates, generally less than 0.5% of the total number of lymphocytes.

For analysis of other surface markers, lymphocyte-gated log green fluorescence histograms were accumulated for 5–10,000 viable lymphocytes. Where baseline positive/negative separation was obtained, percent positive cells was determined by integration of the area under the positive peak. Where separation was incomplete, percent positive cells was determined by subtraction of an appropriate negative control histogram.

Experimental Protocol

Monoclonal reagents from three manufacturers were compared using techniques expected to give optimum results: freshly drawn specimens, FH purification, and indirect immunofluorescence staining. Positive/negative resolution obtained using direct versus indirect immunofluorescence staining was assessed using selected T- and B-cell markers. The effects of FH versus lysis preparation procedures and various collection and storage conditions were then compared for a wider variety of lymphocyte markers. Finally, reference data were accumulated for a larger number of individuals using the rapid preparation procedures defined by these studies, and the degree of individual variation over time was assessed.

RESULTS

Reagent Comparisons

Monoclonal lymphocyte subset reagents are available from a variety of commercial sources and we compared those obtained from Ortho Pharmaceuticals (OR), Becton Dickinson (BD), and Coulter Electronics (CEI). Reagents specific for helper/inducer T cells, suppressor/cytotoxic T cells, total T cells, and sheep red cell (SRBC) receptor bearing (T) cells were compared using FH prepared cells and indirect immunofluorescence staining. In general, reagents from different manufacturers gave very comparable results for the various

TABLE 1. Comparison of Commercially Available Monoclonal Reagents for T-Lymphocyte Subsets[a]

Cell Type Identified	Reagent Supplier		
	OP[b]	BD[c]	CEI[d]
Helper/inducer	47 ± 6	28 ± 6	43 ± 3
Suppressor/cytotoxic	16 ± 5	14 ± 7	13 ± 4
Total T cells (pan T)	63 ± 7	58 ± 9	NA[e]
Total T cells (SRBC receptor)	76 ± 3	NA[e]	80 ± 7

[a]Results expressed as mean % marker positive cells ± standard deviation ($n = 9$); cells from the same 3 individuals were stained in parallel with each manufacturer's reagents in 3 separate experiments. Indirect immunofluorescence techniques were used, with the same fluorescent secondary antibody being applied in each case.

[b]Ortho Pharmaceutical Corp.; reagents used were OKT4, OKT8, OKT3, and OKT11 (in descending order).

[c]Becton Dickinson, Inc.; reagents used were Leu3a, Leu2a, and Leu4 (in descending order).

[d]Coulter Electronics, Inc.; reagents used were T4, T8, and T11 (in descending order).

[e]Not available.

lymphocyte subsets (TABLE 1). One exception was the BD helper/inducer reagent Leu3a, which gave significantly lower results than similar reagents for the other two manufacturers ($p < .005$). Most reagents gave excellent positive/negative separation, with the exception of the CEI SRBC-receptor reagent, T11. While no significant difference in mean percent positive cells was observed for T11 and OKT11, the standard deviation for % T11 positive was significantly greater ($p < .05$) than that for % OKT11 positive.

Since the most rapid sample preparation method possible consistent with maintaining baseline positive/negative separation was desired, direct and indirect immunofluorescence staining procedures were compared. FH prepared cells from two individuals were stained with Leu2a, Leu3ab, and Leu4 for T cells and with anti-surface immunoglobulin (sIg) for B cells. In all cases the positive/negative separation was unimpaired and analysis by integration could be used for cells stained by either method. The ratio of positive cells obtained by direct staining to positive cells obtained by indirect staining was 0.98 ± 0.097 ($n = 7$), not significantly different from 1.0. A separate study comparing directly labeled T11 and OKT11 showed no significant difference in either the degree of positive/negative separation or the standard deviations obtained ($n = 8$, data not shown). All further studies were carried out using direct staining procedures; irrelevant antibodies of the same isotype and fluorescein·protein ratio as the monoclonal reagents were used as controls for nonspecific staining.

Comparison of FH and Lysis Cell Preparation Methods

Since lysis of erythrocytes from whole blood samples has been described as a more rapid method of preparing cells for subset analysis than FH separation,[12] we compared the effects of these two preparation methods (TABLE 2). All of the monoclonal reagents examined gave equivalent results whether cells were FH

TABLE 2. Comparison of Ficoll-Hypaque and Whole Blood Lysis Cell Preparation Methods[a]

Reagent Used	F-H	Lysis A
OKT4[b]	41 ± 8	42 ± 6
Leu3ab[b]	34 ± 8	34 ± 9
Leu2a[b]	24 ± 7	23 ± 8
Leu4[b]	64 ± 8	64 ± 8
OKT11[b]	74 ± 4	76 ± 6
Leu12[c]	14 ± 3	13 ± 2
B1[d]	14 ± 3	14 ± 3

[a]Results expressed as mean percent marker positive cells ± standard deviation ($n = 15$); specimens from 15 separate individuals were prepared in parallel using either FH or Lysis A technique. FH purified cells were stained after preparation; for the lysis technique, whole blood was stained and then lysed as in ref 13. Direct immunofluorescence staining was used in all cases.
[b]See TABLE 1 for manufacturer and specificity.
[c]Pan-B specificity (Becton Dickinson, Inc.), $n = 7$.
[d]Pan-B specificity (Coulter Electronics, Inc.), $n = 6$.

purified and then stained, or whether whole blood preparations were lysed after staining (Lysis A). However a polyclonal anti-immunoglobulin reagent (sIg) gave significantly higher B-cell percentages than the monoclonals for FH prepared cells (26% ± 3%, $n = 6$), apparently owing to staining of cytophilic immunoglobulin (see below).

Staining procedures carried out in a microtiter tray are faster and more convenient than those using multiple tubes for each specimen. However, the Lysis A procedure requires addition of a large volume of lysing reagent after staining, making it impractical to carry out in a microtiter tray. As an alternative, we tried lysis of a single larger sample of blood followed by aliquotting into microtiter wells for staining (Lysis B). We also examined the effects of dextran sedimentation prior to lysis (Lysis C), as a means of enriching for leukocytes and avoiding lysis of inconveniently large volumes of blood in cytopenic specimens. The results obtained using Lysis B and Lysis C are equivalent to each other (TABLE 3) and to the results obtained using FH or Lysis A (TABLE 2).

The role of cytophilic antibody in sIg staining of cells prepared by FH or the modified lysis method was also assessed (TABLE 4). Incubation for 1 hour at 37°C to remove cytophilic antibody significantly decreased the proportion of sIg positive cells in both FH preparation ($p < .002$) and Lysis C preparations ($p < .025$), but had no significant effect on the proportion of B1-positive cells ($p > 0.2$). After removal of cytophilic antibody, enumeration of B cells using sIg gave results comparable to those obtained using the monoclonal markers for either preparation method.

Comparison of Collection and Storage Conditions

As indicated above, maximum sample stability was a major consideration. The ability to use both FH and lysis methods for several days after specimen

TABLE 3. Comparison of Pre-Stain Lysis and Dextran/Lysis Cell Preparation Methods[a]

Reagent Used	Lysis B	Lysis C
OKT4[b]	47 ± 8	46 ± 5
Leu2a[b]	28 ± 5	28 ± 5
Leu4[b]	74 ± 8	76 ± 5
OKT11[b]	82 ± 9	84 ± 6
B1[b]	11 ± 5	11 ± 4

[a]Results expressed as mean % marker positive cells $\pm$ standard deviation ($n = 11$). Whole blood from 11 individuals was lysed either with (Lysis C) or without (Lysis B) prior dextran sedimentation (see MATERIALS AND METHODS). After lysis, cells were stained by direct immunofluorescence using the indicated reagents.

[b]Manufacturer and specificity as indicated in TABLE 2.

collection was also deemed important since poor lysis sometimes occurs in specimens with high reticulocyte counts, requiring the use of FH as an alternative method of preparation. Blood samples from three healthy individuals were drawn with heparin as the anticoagulant, then stored without further additions or with addition of an equal volume of buffered RPMI (see METHODS). Samples were prepared, stained, and analyzed both immediately and at 24-hour intervals for up to 4 days after collection. Previous experience indicated that after 24 hours cells stored without additions in heparin gave poor FH separations. Therefore only Lysis B was carried out for the full 4 days for cells stored in heparin without additional medium. Cells stored in RPMI-HEPES were analyzed using both preparation methods over the whole time course.

As shown in TABLE 5, excellent reproducibility was obtained for specimens stored for up to 3 days at room temperature in RPMI-HEPES regardless of sample preparation method, and for specimens stored in heparin that were prepared by Lysis B. Specimens stored in heparin for longer than 24 hours gave indistinct interfaces and poor reproducibility if prepared by FH (data not shown). Slight but not significant decreases in OKT11 ($0.1 < p < 0.2$) and Leu4 ($p > 0.25$) positives were noted on day 4 for samples prepared by FH even when specimens were stored in RPMI-HEPES. This decrease was not seen for cells prepared by

TABLE 4. Effect of Cytophilic Antibody on B-Cell Enumeration

Procedure[a]	Percent sIg Positive[b]	Percent B1 Positive[b]
FH, 0°C	27 ± 4	13 ± 4
FH, 37°C	11 ± 4	10 ± 4
Lysis C, 0°C	16 ± 2	10 ± 5
Lysis C, 37°C	9 ± 1	7 ± 3

[a]Cells were prepared in parallel by FH or Lysis C procedures (see MATERIALS AND METHODS); one aliquot from each preparation was incubated for 60 min at 37°C prior to staining while a duplicate aliquot was held on ice.

[b]Mean $\pm$ standard deviation ($n = 7$), analysis by integration.

TABLE 5. Effect of Storage Conditions on Lymphocyte Subset Enumeration

Reagent[b]	Storage/Preparation Method	% Positive Cells[a]			
		Day 1[c]	Day 2	Day 3	Day 4
OKT4	RPMI/FH	41 ± 8	39 ± 9	38 ± 6	35 ± 8
	RPMI/Lysis B	39 ± 8	39 ± 8	41 ± 9	42 ± 17
	HEP/Lysis B	39 ± 8	42 ± 8	40 ± 8	46 ± 13
OKT11	RPMI/FH	75 ± 3	71 ± 2	73 ± 4	62 ± 9
	RPMI/Lysis B	73 ± 2	74 ± 3	74 ± 2	73 ± 5
	HEP/Lysis B	73 ± 2	72 ± 7	76 ± 3	77 ± 4
Leu2a	RPMI/FH	24 ± 8	21 ± 3	21 ± 4	17 ± 9
	RPMI/Lysis B	28 ± 6	28 ± 6	28 ± 5	26 ± 5
	HEP/Lysis B	28 ± 6	25 ± 3	26 ± 4	29 ± 9
Leu 4	RPMI/FH	64 ± 6	59 ± 4	61 ± 8	49 ± 11
	RPMI/Lysis B	67 ± 4	66 ± 6	66 ± 6	68 ± 7
	HEP/Lysis B	67 ± 4	64 ± 6	66 ± 6	71 ± 5
Leu 12	RPMI/FH	16 ± 1	16 ± 1	12 ± 1	5 ± 3**
	RPMI/Lysis B	18 ± 2	15 ± 2	15 ± 1	5 ± 7**
	HEP/Lysis B	18 ± 2	16 ± 3	13 ± 1	1 ± 0**
Ratio[d]	RPMI/FH	1.8 ± 1.0	1.7 ± 0.7	1.7 ± 0.5	2.5 ± 1.4
	RPMI/Lysis B	1.5 ± 0.6	1.5 ± 0.4	1.5 ± 0.6	1.6 ± 0.6
	HEP/Lysis B	1.5 ± 0.6	1.7 ± 0.4	1.7 ± 0.8	1.8 ± 1.0

[a]Mean ± standard deviation ($n = 3$); ** = significant difference over time, ($p < .05$).
[b]Reagents and specificities as in TABLES 1 and 2.
[c]Day 1 = day of specimen collection.
[d]Percent OKT4 positive/percent Leu2a positive.

Lysis B regardless of storage medium. A significant decrease ($.01 < p < .025$) in Leu12 positive B cells was also observed on day 4 for all storage media and preparation methods.

It has previously been reported that storage at 4°C for 24 hours can affect helper/suppressor ratios.[10] However, we found no significant differences among cells prepared by Lysis B and stained with any of the T or B cell monoclonal markers after storage for 24 hours at 4°C in either RPMI-HEPES or heparin ($n = 5$, data not shown).

Summary of Normal Values and Individual Variability

Since "normal" for each laboratory really represents a reference population of healthy individuals that is sampled regularly, TABLE 6 summarizes data obtained over a 7-month period on our reference population. The 20 individuals represented in this table were sampled from a minimum of once to a maximum of 14 times over this period (approximately 10 ml per specimen). Samples were prepared by Lysis B or Lysis C, stained by direct immunofluorescence, and analyzed by integration of the discrete positive portion of each distribution.

TABLE 6. Cumulative Normal Values for Lymphocyte Subset Analysis[a]

Specimen Age (days)	n[c]	Viability[d]	OKT4	Leu2a	Leu4	OKT11	B1	Ratio[e]
			Percent Positive[b]					
1	41	96 ± 5 (78–99)	46 ± 7 (27–64)	29 ± 5 (20–39)	74 ± 6 (60–86)	80 ± 7 (61–90)	11 ± 3 (5–18)	1.6 ± 0.5 (0.8–3.0)
2	41	97 ± 2 (89–99)	46 ± 6 (34–61)	31 ± 6 (18–43)	76 ± 6 (66–88)	83 ± 6 (71–94)	11 ± 3 (3–19)	1.5 ± 0.5 (0.9–2.9)
3	20	96 ± 2 (90–99)	47 ± 5 (37–56)	32 ± 5 (21–38)	76 ± 4 (70–84)	81 ± 6 (72–93)	10 ± 4 (4–19)	1.5 ± 0.4 (1.1–2.7)
Overall	105	96 ± 3 (78–99)	46 ± 6 (27–64)	30 ± 5 (18–43)	75 ± 6 (60–88)	82 ± 6 (61–94)	11 ± 3 (3–19)	1.6 ± 0.5 (0.8–3.0)

[a]Specimens were collected in heparin and stored at room temperature for the indicated length of time (day 1 = day of specimen collection). Samples were prepared by the Lysis B or Lysis C method. Twenty healthy individuals were sampled from 1 to 14 times (approx. 10ml/sample) over a 7-month period.
[b]Mean percent marker positive cells ± standard deviation and (range).
[c]Number of samples analyzed.
[d]Percent pan-leukocyte antibody-positive cells that were propidium iodide negative.
[e]Percent OKT4 positive/percent Leu2a positive.

Results are classified by marker and by the number of days after a specimen was collected that the particular sample was stained and analyzed. Means for each marker remained unchanged for 3 days, and the range of variation did not change with sample age, indicating that measurement variability did not significantly increase with storage time.

Since some of the individuals represented in TABLE 6 were sampled a number of times over the 7-month period studied, we also determined the range of variation in subset frequencies measured in one individual as opposed to a group of people. Data for the most frequently sampled individual of TABLE 6 are shown in TABLE 7. Where several analyses were made on the same specimen at different ages (e.g., days 0, 1, 2), mean values are given. Comparison of TABLES 6 and 7 suggests that the variability of determinations on one individual over time is nearly as great as that seen over a group of individuals.

DISCUSSION

Most of the methodological comparisons presented here are motivated by the need to achieve a highly time-efficient preparation and staining procedure without compromising ease of analysis or sample stability. As shown in TABLE 1, most of the commercially available reagents examined gave equivalent results using FH purification and indirect immunofluorescence staining. Although the Leu3a reagent used in this series gave lower results than similar reagents from other manufacturers, this did not appear to be due to increased overlap of positive and negative cells since baseline separation was obtained for all the helper/ inducer reagents. Because all of the data in TABLE 1 were obtained using the same lot of reagent, it is possible that this discrepancy was due to the particular sample

TABLE 7. Subset Enumeration in a Single Individual Over Time

| Day Drawn[b] | Percent Positive[a] | | | | | Ratio[c] | Viability[d] | n^e |
	OKT4	Leu2a	Leu4	OKT11	B1			
1	43	32	76	76	14	1.3	98	1
4	36 ± 12	32 ± 1	77 ± 4	80 ± 0	12 ± 0	1.0 ± 0.4	93 ± 6	2
9	40 ± 1	36 ± 2	78 ± 9	79 ± 4	12 ± 2	1.1 ± 0.1	97 ± 2	3
23	44 ± 1	34 ± 2	80 ± 1	83 ± 1	11 ± 1	1.2 ± 0.1	97 ± 2	3
51	50 ± 2	33 ± 3	80 ± 2	84 ± 2	8 ± 1	1.5 ± 0.2	97 ± 1	3
100	42 ± 5	31 ± 4	77 ± 6	84 ± 5	9 ± 1	1.4 ± 0.1	87 ± 13	2
116	41 ± 4	30 ± 4	73 ± 11	74 ± 18	10 ± 1	1.2 ± 0.0	92 ± 4	2
128	55	29	71	86	9	1.9	99	1
134	45 ± 2	29 ± 1	80 ± 1	84 ± 2	10 ± 1	1.6 ± 0.2	98 ± 1	3
141	44	19	76	82	14	2.3	99	1
142	48	25	77	84	13	1.9	99	1
158	44 ± 2	23 ± 1	80 ± 2	86 ± 1	12 ± 1	1.9 ± 0.0	97 ± 3	2
165	42	32	77	83	14	1.3	97	1
180	43	24	73	85	11	1.8	98	1
Overall	44 ± 5	31 ± 5	78 ± 5	82 ± 5	11 ± 2	1.4 ± 0.4	96 ± 4	26
Range	27–55	19–38	65–88	62–87	7–14	0.8–2.3	79–99	

[a]Mean ± standard deviation indicated where a specimen was analyzed more than once (e.g., on days 1, 2, and 3 after collection). Samples were collected and prepared as indicated in TABLE 6. Manufacturers and specificities as in TABLE 2.

[b]Day 1 = day on which individual was first sampled.

[c]Percent OKT4 positive/percent Leu2a positive.

[d]Percent HLE-positive cells that were propidium iodide negative.

[e]Number of times a specimen was analyzed after collection.

of reagent. Later data using an improved Leu3ab reagent in a direct staining procedure gave results more consistent with other helper reagents (TABLE 2). Lot specific factors may also have been responsible for poor positive/negative separation observed with T11 in the study of TABLE 1. Subtraction of a secondary-antibody-only control was required to enumerate T11 positives as compared with simple integration for OKT11 positives, giving rise to a significantly larger standard deviation for % T11 positive although the means did not differ significantly. However, further comparison of T11 and OKT11 using directly labeled reagents showed no significant differences between them. Thus, although remarkably similar results are obtained using different reagent sources, crossover studies in going from one lot of reagent to another are clearly desirable as a quality control measure regardless of the particular reagents or sources chosen. Direct immunofluorescence staining was found to give results equivalent to indirect staining with a 50% reduction in staining time and no loss in positive/ negative resolution. Note, however, that this last conclusion was drawn based on use of the relatively high excitation intensities available in a laser-based flow cytometer. In systems where lower intensity excitation sources are used, indirect staining may be required to obtain adequate sensitivity and resolution.

Whole blood lysis methods were also evaluated as a means of decreasing sample preparation time. We found that cells prepared by the literature method, in which samples are stained prior to erythrocyte lysis (Lysis A), gave results equivalent to those prepared by FH for all of the T and B cell monoclonal markers examined (TABLE 2). Staining in microtiter wells improved sample preparation efficiency but required lysis before staining (Lysis B) rather than after (Lysis A). For monoclonal reagents, Lysis B and a variation utilizing dextran enrichment (Lysis C) gave results equivalent to one another (TABLE 3) and to FH or Lysis A (TABLE 2).

Use of a polyclonal anti-immunoglobulin reagent (sIg) for enumeration of B cells in FH preparations gave significantly higher results (26% positive) than did use of monoclonal reagents (13–14% positive). To assess the role of cytophilic antibody in sIg staining, the effect of incubation at 37°C was determined for cells prepared by FH or Lysis C (TABLE 4). The level of sIg staining was significantly decreased in Lysis C preparations but still gave B-cell estimates higher than those obtained using monoclonal reagents. Removal of cytophilic antibody significantly decreased the proportion of sIg positive cells for both FH preparations and Lysis C preparations. Neither removal of cytophilic antibody nor preparation method significantly affected B-cell numeration using the B1 monoclonal reagent.

In selecting sample storage medium and conditions, specimen stability and sample reproducibility were of paramount importance. We also wished to be able to use FH as an alternate preparation method in specimens where adequate lysis was not obtained (e.g., those with high reticulocyte counts). Previous experience indicated that, at times greater than 24 hours after specimen collection, specimens stored with RPMI-HEPES medium admixed in equal amounts with heparin anticoagulated whole blood gave better FH separations than those stored in heparin only. As shown in TABLE 5, equivalent results were obtained for up to three days for specimens stored at room temperature in heparin or RPMI-HEPES and then prepared by Lysis B. For FH preparation, RPMI-HEPES but not heparin storage medium gave acceptable reproducibility for up to 3 days. Thus storage in RPMI-HEPES medium at room temperature offers excellent analysis reproducibility and maximum flexibility in preparation procedure for up to 3 days after specimen collection. These results are in good agreement with recent studies on a wider variety of collection and storage media and conditions.[8] We have been unable to demonstrate any effect of storage at 4°C on helper/suppressor ratio for cells prepared by Lysis B, whether they are stored in heparin or in RPMI. Since previous studies on this effect were done using FH-prepared cells, this tends to confirm that differential losses can occur when using this cell preparation method.[10]

TABLES 6 and 7 summarize our cumulative normal values obtained over a 7 month period on 20 individuals, and compare the variability seen in one individual over that period of time with the variability among individuals. Although no formal analysis of variance has been done, it appears that the variability within one individual over time may be almost as large as that seen over 20 individuals. It is also of interest to note that one sample in TABLE 7 had a viability of 78% (one of the two samples from the specimen at 100 days). However, none of the marker values for that particular sample represented extremes of the overall range and most fell midrange. Thus, whatever the cause of the significantly decreased viability in this sample, it appeared not to have resulted in selective losses of particular lymphocyte subtypes. Nonetheless, there is a philosophical question as to what lower limit of sample viability is acceptable to avoid even the possibility of selective losses. We have somewhat arbitrarily set a lower limit of 85% total leukocyte viability as an indication that cell preparation and staining procedures are unlikely to have caused problems of this nature. If a sample has a less than acceptable viability, the specimen is reprocessed and restained; if viability remains unacceptable, a new specimen is obtained.

A brief comment on the concept of "normal" values is also in order. While all of the individuals included in these studies were healthy "normal" individuals, a cursory comparison of the suppressor values for FH prepared cells in TABLE 1 ($n = 9$) and 2 ($n = 15$) would leave one with the totally misleading impression that direct and indirect staining give discordant results. However, in TABLE 1, only three individuals were studied, and two of them had values at the low end of the range for suppressor cells. For the purpose of comparing reagents this was unimportant, since the different reagents were all run on the same individual. But for the purpose of establishing reference values for the laboratory, it is clear that it is the number of individuals rather than the number of replicates that is of prime importance.

As described in MATERIALS AND METHODS, optimal instrument performance is monitored by daily (or more frequent, as warranted) analysis of fluorescent 10 µ beads under standard running conditions. FIGURE 1 illustrates the variation in instrument performance noted in this manner over an arbitrarily chosen 4-week period. Although some fluctuations are inevitable, largely because laser power is difficult to set exactly reproducibly, any steady downward trend would be readily noticeable. This might not be the case if relative alignment alone (i.e., maximal mean and minimum CV obtainable on a given day) were used as a criterion of instrument performance. In addition to instrument optimization using standard spheres, an unstained sample and a cell sample stained with the appropriate negative control reagent are also important reference points. Unless the fluorescence distribution of the negative control matches that of the negative population in the marker stained population, short term instrument fluctuations and/or staining variability must be suspected. For this reason, we feel that the availability of an appropriate control reagent (same isotype and fluorescein:protein ratio as the marker in question) is almost as important a criterion in choosing reagents as good baseline separation. Comparison of the irrelevant antibody control with the unstained control also provides some indication of the degree of non-specific staining for each reagent.

Although all the specimens reported on in this study came from nominally healthy individuals, all patient specimens must be regarded as potentially pathogen-bearing and cell preparation, staining, and analysis procedures must be adjusted accordingly. We have chosen to do all preparation and staining in a level P2 biohazard containment hood. Gloves must be worn at all times when handling samples, and tubes or microtiter trays must be covered when removed from the hood (for example, when centrifuging). Because fixation does not allow later assessment of viability, we have chosen not to fix specimens after staining and before flow cytometric analysis. However, samples are transferred to the appropriate sample holders for flow cytometric analysis in the P2 hood after staining and handled the minimum amount necessary during the analysis. To avoid aerosols created by a jet-in-air configuration, a totally enclosed biohazard flow tip is used. Measurements are made in a square glass chamber from which the stream empties directly into a waste tube and thence into waste containers containing 10% final vol/vol Clorox. After sample analysis is completed, a sample of 10% Clorox is run through the system for 30 minutes to flush all the lines and the system is then flushed for a final 30 minutes with PBS.

In summary, we have discussed a number of methodological considerations in adapting lymphocyte subset analysis to the setting of a high volume clinical

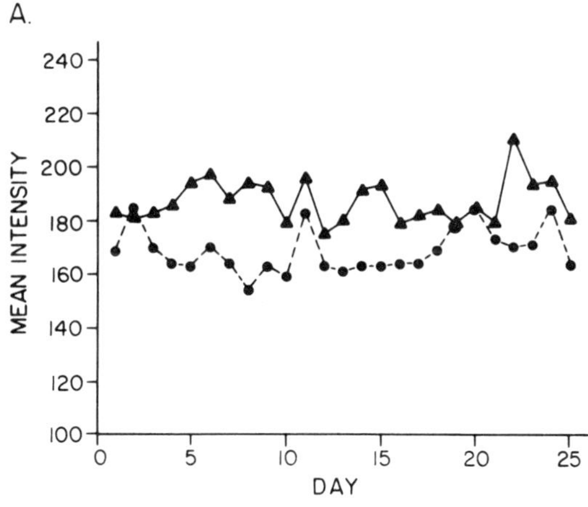

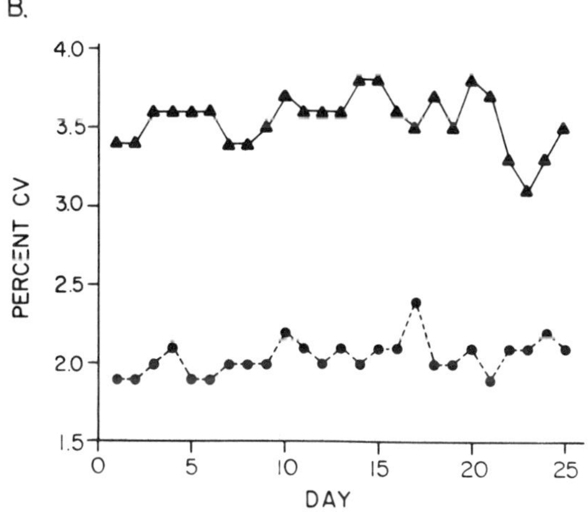

FIGURE 1. Use of fluorescent microspheres for instrument optimization. Fluorescent full bright spheres were run daily using standard instrument settings as described in MATERIALS AND METHODS. Mean green fluorescence (▲) and light scatter (●) intensities are plotted as a function of time in Panel A. Coefficients of variation for green fluorescence (▲) and light scatter (●) are similarly displayed in Panel B.

reference laboratory. With minor exceptions, all of the commercially available monoclonal T- and B-cell reagents examined give equivalent results. However, the exceptions suggest that lot specific variations in reagents should be monitored as a quality control measure. Direct immunofluorescence staining decreases sample preparation time without loss of positive/negative resolution. Whole blood lysis prior to staining allows handling of cells in microtiter wells, further increasing the efficiency of sample preparation. B-cell enumeration is sensitive to the combination of reagent and preparation method chosen. Incubation to remove cytophilic antibody is required after either FH or lysis preparation when sIg is used as a B-cell marker, though not for monoclonal markers. A dextran sedimentation step prior to lysis adds relatively little time to the total preparation procedure, does not alter subset analysis results, and can considerably decrease analysis times required for cytopenic specimens. Specimen collection in heparin and storage at room temperature in RPMI-HEPES offers excellent sample reproducibility for up to 3 days after specimen collection and a choice of FH or lysis preparation methods as needed.

We have also discussed a number of more philosophical questions, such as instrument optimization, methods of biohazard containment, appropriate controls, and what constitutes a set of "normal" values. Clearly, clinical flow cytometry will continue to have valuable diagnostic applications only so long as it is used critically.

SUMMARY

As the diagnostic utility of lymphocyte subset analysis has been recognized in the clinical research laboratory, a wide variety of reagents and cell preparation, staining and analysis methods have also been described. Methods that are perfectly suitable for analysis of smaller sample numbers in the biological or clinical research setting are not always appropriate and/or applicable in the setting of a high volume clinical reference laboratory. We describe here some of the specific considerations involved in choosing a method for flow cytometric analysis which minimizes sample preparation and data analysis time while maximizing sample stability, viability, and reproducibility. Monoclonal T- and B-cell reagents from three manufacturers were found to give equivalent results for a reference population of healthy individuals. This was true whether direct or indirect immunofluorescence staining was used and whether cells were prepared by Ficoll-Hypaque fractionation (FH) or by lysis of whole blood. When B cells were enumerated using a polyclonal anti-immunoglobulin reagent, less cytophilic immunoglobulin staining was present after lysis than after FH preparation. However, both preparation methods required additional incubation at 37°C to obtain results concordant with monoclonal B-cell reagents. Standard reagents were chosen on the basis of maximum positive/negative separation and the availability of appropriate negative controls. The effects of collection medium and storage conditions on sample stability and reproducibility of subset analysis were also assessed. Specimens collected in heparin and stored at room temperature in buffered medium gave reproducible results for 3 days after specimen collection, using either FH or lysis as the preparation method. General strategies for instrument optimization, quality control, and biohazard containment are also discussed.

ACKNOWLEDGMENTS

The authors would like to thank Drs. Don Griswold, Zdenka Jonak, and Nabil Hanna for helpful discussions. The skillful secretarial assistance of Ms. Judy Krause is also deeply appreciated.

REFERENCES

1. SCHLOSSMAN, S. F. & E. L. REINHERZ. 1984. Human T-cell subsets in health and disease. Spring Semin. Immunopathol. 7:9–18.
2. FOON, A. K., R. W. SCHROFF & R. P. GALE. 1982. Surface markers on leukemia and lymphoma cells: Recent advances. Blood 60:1–19.
3. THIEL, E. 1983. Monoclonal antibodies against differentiation antigens of lymphopoiesis. Blood 47:247–261.
4. COLVIN, R. B. 1984. Flow cytometric analysis of T-cells: diagnostic applications in transplantation. *In* Ann N. Y. Acad. Sci. 428:5–13.
5. LOVETT, E. J. III, B. SCHNITZER, D. F. KEREN, A. FLINT, J. L. HUDSON & K. D. MCCLATCHY. 1984. Application of flow cytometry to diagnostic pathology. Lab. Invest. 50:114–140.
6. MUIRHEAD, K. A. 1984. Applications of flow cytometry in clinical diagnosis. Tr. Anal. Chem. 3:107–111.
7. HORAN, P. K., E. M. LORD, D. RYAN, J. HEAL & T. C. SCHMITT. 1984. Applications of flow cytometry in clinical pathology. 1983. *In* Diagnostic Immunology: Technology Assessment and Quality Assurance. J. H. Rippey & R. M. Nakamura, Eds.:78–91. College of American Pathologists. Skokie, IL.
8. NICHOLSON, J. K. A., B. M. JONES, G. D. CROSS & J. S. MCDOUGAL. 1984. Comparison of T- and B-cell analyses on fresh and aged blood. J. Immunol. Methods. 73:29–40.
9. SHIELD, C. F., P. MARLETT, A. SMITH, L. GUNTER & G. GOLDSTEIN. 1983. Stability of human lymphocyte differentiation antigens when stored at room temperature. J. Immunol. Methods 62:347–352.
10. WEIBLEN, B. J., K. DEBELL, A. GIORGIO & C. R. VALERI. 1984. Monoclonal antibody testing of lymphocytes after overnight storage. J. Immunol. Methods 70:179–183.
11. BOYUM, A. 1968. Separation of lymphocytes from blood and bone marrow. Scand. J. Clin. Lab. Invest. 21 (Suppl. 97):77–106.
12. HOFFMAN, R. A., P. C. KUNG, W. P. HANSEN & G. GOLDSTEIN. 1980. Simple and rapid measurement of human T lymphocytes and their subclasses in peripheral blood. Proc. Natl. Acad. Sci. USA 77:4914–4917.
13. HOFFMAN, R. A. & W. P. HANSEN. 1981. Immunofluorescent analysis of blood cells by flow cytometry. Int. J. Immunopharmac. 3:249–254.
14. RITCHIE, A. W. S., R. A. GRAY & H. S. MICKLEM. 1983. Right angle scatter: A necessary parameter in flow cytometric analysis of human peripheral blood mononuclear cells. J. Immunol. Methods. 64:109–117.
15. LOKEN, M. R. & A. M. STALL. 1982. Flow cytometry as an analytical and preparative tool in immunology. J. Immunol. Methods 50:R85–R112.
16. LANIER, L. L. & N. L. WARNER. 1981. Paraformaldehyde fixation of hematopoietic cells for quantitative flow cytometry analysis. J. Immunol. Methods 47:25–30.
17. SCHROFF, R. W., C. D. BUCANA, R. A. KLEIN, M. M. FARRELL & A. C. MORGAN, Jr. 1984. Detection of intracytoplasmic antigens by flow cytometry. J. Immunol. Methods 70:167–177.

A New Immune Monitoring System for the Determination of Lymphoid Cell Subsets

ALEX M. SAUNDERS[a] AND CHIN-HAI CHANG

Becton Dickinson Monoclonal Center
2375 Garcia Avenue
Mountain View, California 94043

In recent years many studies dealing with the identification and function of populations of immune competent cells have been reported.[1-5] These studies have shown that the various immune functions are performed by particular cell populations and subsets. In some instances, functional cell subsets have been shown to have unique cell surface markers. Although some of these cell markers are not strictly limited to particular functional populations, it is operationally practical to use them to identify and count cell subsets.

Particular interest has recently been focused on the classification of functional lymphocyte subsets. One of the problems in such studies is separating monocytes from lymphocytes so that the two populations can be individually studied. Monocytes and lymphocytes are often confused because of their morphologic and physical similarity, resulting in a loss of accuracy and precision in test results. This problem is compounded by the differing responses of the two cell types to disease processes, and by the lack of correlation between the numbers of lymphocytes and monocytes.

In the past, the methods used to separate lymphocytes and monocytes for immune monitoring studies have depended on differences in cell size. When cells were segregated on the basis of electronic volume or light scattering, the larger cells were presumed to be monocytes. These methods of determining monocyte and lymphocyte concentrations have been shown to be inaccurate at best, and have been shown to be completely inaccurate in experiments using cytochemical markers.[6] These cytoenzymatic methods, developed by Schmalzl and Braunsteiner,[7] were later commercialized.[8,9] This enzymatic method of classification provides no information about lymphocyte subsets, however.

In research laboratories, immune monitoring procedures are most often based on separating mononuclear cells by differential sedimentation, a procedure first popularized by Böyum.[10] The yield and purity of cell preparations vary among laboratories, depending on operator skill. The most common procedures of this type, using Ficoll-Hypaque®, are time consuming and therefore not generally applicable to routine clinical laboratory use.

With the advent of monoclonal antibody technology, new approaches to immune monitoring and to cell subset determination have become possible. Monoclonal antibodies are highly specific research tools that can be conjugated with fluorochromes or other contrast-producing molecules. The use of fluoro-

[a]Present address: K.L.A. Instruments, 2051 Mission College Blvd., Santa Clara, CA 95054.

chrome-conjugated monoclonal antibodies in conjunction with instruments such as the flow cytometer has resulted in practical methods of evaluating subsets of immune competent cells. These methods are finding clinical applications in differential diagnosis and immune monitoring of a variety of disease states.

In this present study, a new immunomonitoring system for the analysis of leukocyte cell surface markers is described. Interference from monocytes is eliminated. The system combines monoclonal antibody technology and flow cytometry, and is applicable to both whole blood and mononuclear cell preparations. In this system, monocytes are labeled with two fluorescent dyes so that they can be differentiated from single color cells. The procedures, instrumentation, software, and results of the new immunomonitoring system are described in this report.

MATERIALS AND METHODS

Monoclonal Reagents

The phycoerythrin (PE) and fluorescein isothiocyanate (FITC) conjugated monoclonal antibodies, obtained from Becton Dickinson Monoclonal Center, (Mountain View, CA) used in this study are shown in TABLE 1.

Mixtures of Monoclonal Reagents

Co-titration of Anti-Leu-M3 PE and Anti-Leu-M3 FITC (Mixture a)

The proper co-titration of Anti-Leu-M3 PE and Anti-Leu-M3 FITC for all the PE- and FITC-conjugated reagents used in this study was found to be critical. Co-titration was accomplished by serially diluting Anti-Leu-M3 FITC together with approximately one-half the full strength of Anti-Leu-M3 PE supplied by the manufacturer. FIGURE 2 shows the result of a satisfactory titration in which the doubly labeled cells fall along a diagonal in the green (FITC) and red (PE) fluorescent dimensions. After this balance was achieved, the same Anti-Leu-M3 mixture, hereafter referred to as double Anti-Leu-M3, was used throughout the study. The titration must be repeated with each lot of Anti-Leu-M3 PE and Anti-Leu-M3 FITC.

Mixture of Anti-T, Anti-B, and Monocyte Monoclonal Reagents (Mixture b)

The following mixture was used for each test:

2.5 µl of Anti-Leu-2a PE;
2.5 µl of Anti-Leu-3a PE;
5 µl Anti-HLA-DR FITC;
−X µl of Anti-Leu-M3 PE plus Anti-Leu-M3 FITC (double Anti-Leu-M3; volume determined by co-titration).

TABLE 1.

Monoclonal Antibody	Clone	Specificity
Anti-Leu-M3 PE and Anti-Leu-M3 FITC[11]	P9	mature monocytes
Anti-Leu-2a PE[2,3]	SK1	cytotoxic/suppressor T cells
Anti-Leu-3a FITC[2,3]	SK3	helper/inducer T cells
Anti-Leu 2a PE and Anti-Leu-3a PE	SK1&SK3	pan T cell
Anti-Leu-7[12]	HNK-1	subset of natural killer cells (large granular lymphocytes)
Anti-HLA-DR FITC[13]	L203	B cells (also monocytes, activated T cells)

Mixture of Anti-T Suppressor Cells, Anti-T Helper Cell, and Monocyte Monoclonal Reagents (Mixture c)

Each test used the following:

2.5 µl Anti-Leu-3a FITC;
2.5 µl Anti-Leu-2a PE;
X µl double Anti-Leu-M3 (volume determined by co-titration).

Mixture of Anti-T Suppressor Cell, Natural Killer, and Monocyte Monoclonal Reagents (Mixture d)

The mixture for each test was as follows:

2.5 µl Anti-Leu-2a PE;
2.5 µl Anti-Leu-7 FITC;
X µl double Anti-Leu-M3 (volume determined by co-titration).

NOTE: Values given are for prepared mononuclear cells. Amounts are different for whole blood and may require further careful titration.

Specimen Collection and Preparation

All blood samples were collected in EDTA using Becton Dickinson Vacutainer® and processed within 8 hours of collection.

Instrumentation

When PE and FITC are excited simultaneously by a single light source, their emission spectra overlap significantly. These overlapping emission spectra must be separated by an electronic dual compensation network in order for singly and doubly labeled cells to be indentified. In this study, dual compensation was accomplished through the use of a FACS™ Analyzer equipped with a light scatter detector (Becton Dickinson FACS Systems, Sunnyvale, CA).

The FACS Analyzer is equipped with a mercury arc lamp. Excitation light is isolated at 485 ± 20 nm. The dichroic mirror had a transition point at 560 nm. The secondary filter for fluorescein analysis has peak wavelength at 530 ± 15 nm. The secondary filter for phycoerythrin has a peak wavelength of 575 ± 25 nm.

Data were collected in four parameter list mode including cell volume, right angle light scatter, green fluorescence, and red fluorescence. The data were analyzed with a Consort 30 computer system programmed to set gates and markers automatically on two-dimensional displays, and calculate the final results in percent lymphocytes independent of monocytes. This will become clearer in the results section.

Staining Procedure

Peripheral Blood Mononuclear Cells (PBMC)

PBMC were prepared from whole blood using a standard Ficoll-Hypaque (Pharmacia) laboratory procedure. The mononuclear cell layer was washed twice and re-suspended at 2×10^7 cells per ml in Hank's Balanced Salt Solution containing 0.1% sodium azide. Twenty-five microliter aliquots of this cell suspension were then added to test tubes containing mixtures of monoclonal reagents. The tubes were incubated for 30 minutes in a covered ice bath to prevent light from reaching the samples. Fifty microliter aliquots of 1% w/vol formaldehyde were then added to each tube, and they were incubated 5 minutes in the dark at 25°C. The cells were washed once with 3 ml of cold phosphate buffered saline (PBS). The supernatant was then aspirated, leaving behind 50 µl of fluid to avoid disturbing the pellet, and 0.5 ml of Analyzer sheath fluid (Ultracount Hematology Diluent, cat. no. 2483, Clay Adams Instrumentation) was added to each tube. The tubes were stored on ice in the dark until FACS analysis, which was performed within 6 hours of staining and according to the manufacturer's operating instructions. Fluorescing and nonfluorescing microbeads (Becton Dickinson FACS Systems) were used to calibrate the FACS Analyzer.

Whole Blood

Whole blood samples were incubated with various combinations of monoclonal reagents using a procedure similar to that outlined directly above. Immediately following the incubation step, the stained samples were incubated for at least 10 minutes in the dark with a proprietary lysing agent containing formaldehyde. The erythrocytes were therefore lysed as the leukocytes were fixed. Platelets were also fixed and preserved by this technique. The samples were then washed and suspended in PBS containing 1% formaldehyde. Fixed samples are stable for 3 days if refrigerated.

RESULTS

FIGURES 1 through 5 are dot plots obtained on the FACS Analyzer when preparations of PBMC from healthy donors were stained and analyzed. FIGURE 1 shows a dot plot of PBMC labeled with double Anti-Leu-M3 and analyzed on the

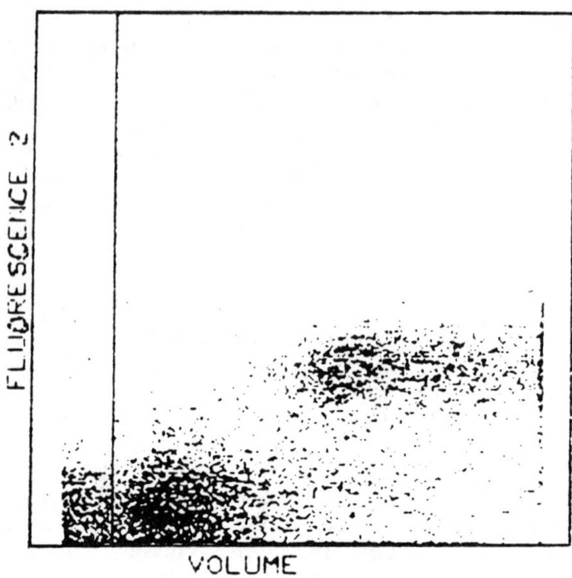

FIGURE 1. Dot plot of PBMC from a healthy donor labeled only with double Anti-Leu-M3 and analyzed with the FACS Analyzer plus light scatter detector on the basis of cell volume and fluorescence.

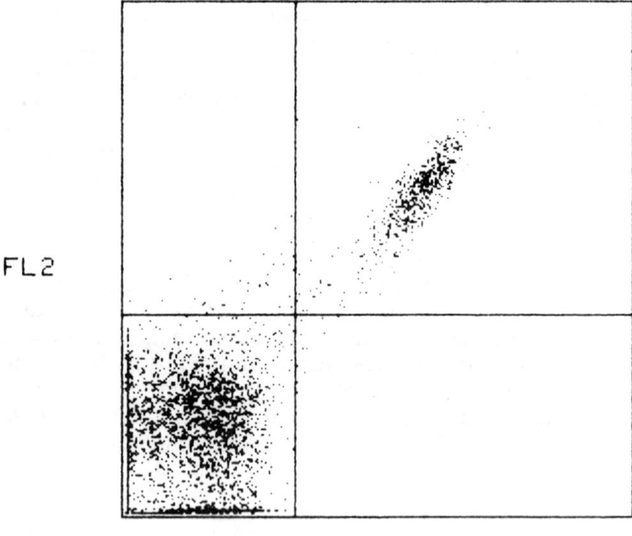

FIGURE 2. Dot plot of PBMC from a healthy donor labeled with double Anti-Leu-M3 and analyzed on the basis of red (PE) and green (FITC) fluorescence with the FACS Analyzer plus light scatter detector.

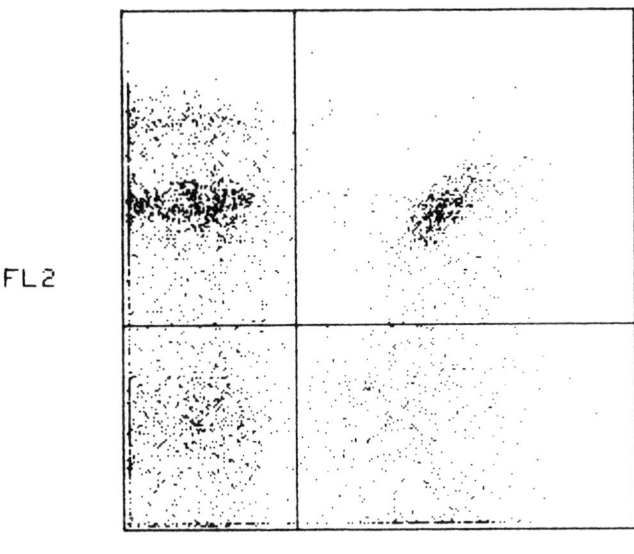

FIGURE 3. Dot plot of PBMC from a healthy donor labeled with anti-T (Anti-Leu-2,3 PE), B (Anti-HLA-DR FITC), and monocyte (double Anti-Leu-M3) reagents.

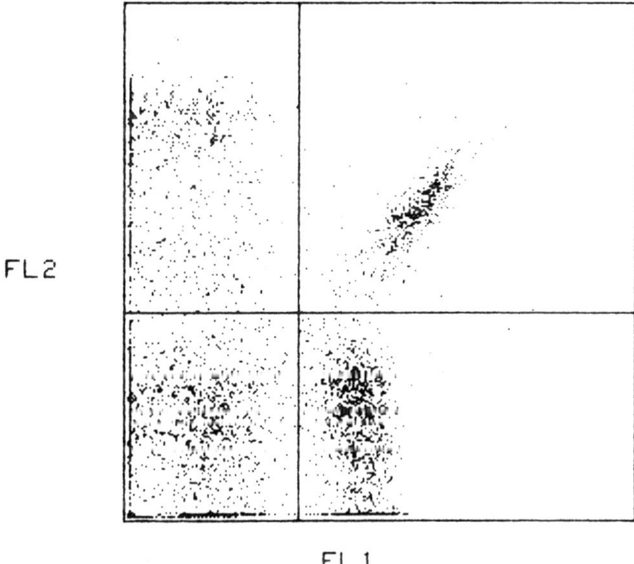

FIGURE 4. Dot plot of PBMC from a healthy donor labeled with double Anti-Leu-M3 (monocytes), Anti-Leu-2a PE (cytotoxic/suppressor T cells), and Anti-Leu-3a FITC (helper/inducer T cells) reagents.

basis of cell volume. The plot clearly shows that the volumes of the lymphocyte and monocyte populations overlap. If volume gates were set to exclude the majority of monocytes, a significant number of lymphocytes in the lower right quadrant would also be eliminated. The monocyte and lymphocyte populations therefore cannot be cleanly separated on the basis of cell volume alone.

FIGURE 2 is a dot plot showing that better resolution of monocytes and lymphocytes was obtained when PBMC were stained with double Anti-Leu-M3 and analyzed on the basis of their red (PE) and green (FITC) fluorescence. In this subsample, the monocytes and total lymphocytes were clearly segregated, and only the monocytes were doubly positive. The monocytes were found to represent 22.1% of the total PBMC, while the lymphocytes represented 72.5% of the total PBMC (TABLE 2). This subsample provides a reference count of monocytes for FIGURES 3 to 5.

FIGURE 3 is a dot plot illustrating the frequency of T cells, B cells, and monocytes, according to mixture b, in the PBMC preparation. The lower left section shows null cells which were not labeled by any of the reagents used. The upper right section contains all doubly labeled cells, including the monocytes, which may also be HLA-DR positive. TABLE 2 shows the percentages of the various cell populations contained in this subsample. The number of doubly labeled lymphocytes (here labeled TB for convenience) was obtained by automatic subtraction of the reference count of monocytes from FIGURE 2.

FIGURE 4 shows a dot plot of the results obtained when the PBMC subsample was labeled with Anti-Leu-2a PE, Anti-Leu-3a FITC, and double Anti-Leu-M3.

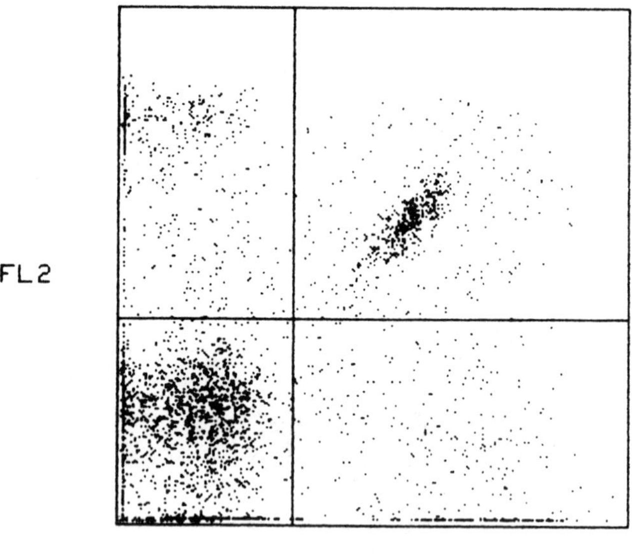

FIGURE 5. Dot plot of PBMC from a healthy donor labeled with double Anti-Leu-M3 (monocytes), Anti-Leu-2a PE (cytotoxic/suppressor T cells), and Anti-Leu-7 FITC (natural killer cells).

TABLE 2. Summary of the Data from Figures 2 through 5.

Figure No.	Cell Type	% Of Total PBMC[a]	% Of Lymphocytes[b]
2	monocytes	22.1	
(control)	lymphocytes	72.5	
3	T cells	48.5	62.2
	B cells	6.5	8.3
	TB cells	2.2	2.8
	null cells	20.8	26.7
4	Leu-2+ cells	18.9	24.3
	Leu-3+ cells	29.5	37.8
	Leu-2+3+ cells	3.3	4.3
5	Leu-2+ cells	16.6	21.4
	Leu-7+ cells	8.6	11.0
	Leu-2+7+ cells	3.2	4.1

[a]Total percent has PBMC as denominator.
[b]Lymphocyte percent has had monocytes subtracted from the denominator.

When the doubly labeled monocytes are subtracted, the results summarized in TABLE 2 show that 24.3% of the lymphocytes were Leu-2+ cytotoxic/suppressor T cells, 37.8% were Leu-3+ helper/inducer T cells, and 4.3% of the lymphocytes were doubly labeled with both Anti-Leu-2a PE and Anti-Leu-3a FITC.

FIGURE 5 is a dot plot of the results obtained when a PBMC subsample was labeled with Anti-Leu-2a PE, Anti-Leu-7 FITC, and double Anti-Leu-M3. The results are summarized in TABLE 2. When doubly labeled monocytes were automatically subtracted, 4.1% of the lymphocytes were found to be doubly positive Leu-2 + 7+ cells. Other investigators have reported that Leu-2 + 7+ cells are large granular lymphocytes with complex history and function.[14]

FIGURES 6 through 9 are dot plots illustrating the need of alternatives to size as the only criterion for eliminating monocytes from lymphocyte subclassification.

In FIGURE 6, incubated with mixture a, volume is plotted against fluorescence and the fluorescence positive cells are monocytes. One observes that no matter where the volume gate is set, some monocytes will be included or some lymphocytes will be excluded. The gate was arbitrarily set at the line in the figure.

FIGURES 7a and 7b (same cells as FIG. 6) show the effect of setting this gate. In FIGURE 7a, the gate is not set and all monocytes are evident as doubly stained events. In FIGURE 7b, where the gate has been set, there is a residue of monocytes representing about 15% of the original monocytes.

In FIGURE 8a and 8b, incubated with mixture c, the same experiment is shown. The residual monocytes are again apparent.

In FIGURES 9a and 9b, the double Anti-Leu-M3 was not used and therefore monocytes with weak Leu-3a reactions are seen in the lower right quadrant. When the gate is set, in FIGURE 9b, the influence of these monocytes is not apparent, but one knows they are a contamination in the lower right quadrant from the corollary experiment in FIGURE 8b. When the experiment represented by FIGURE 8b is not available, there will be no suspicion that a false count was obtained.

FIGURE 10 is a dot plot of FITC fluorescence and PE fluorescence in a

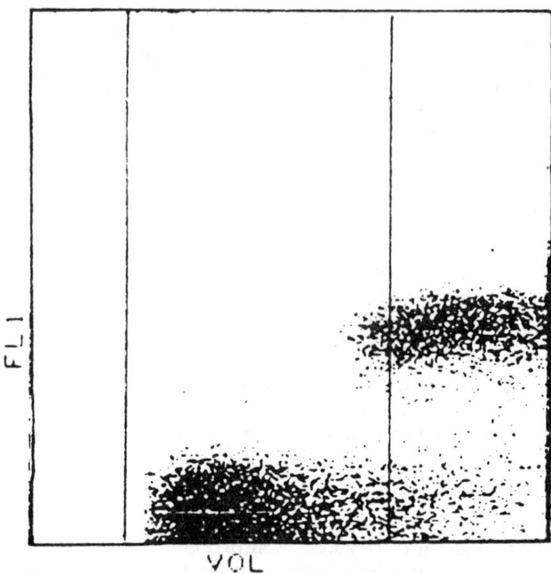

FIGURE 6. FIGURES 6–9 are all on the same normal blood sample prepared as mononuclear cells. This dot plot of volume and fluorescein fluorescence illustrates overlap of monocytes and lymphocytes in volume measurement. Monocytes are the upper, or fluorescent, cluster. The vertical line represents the gate used in Figures 7b, 8b, 9b.

subsample from an individual having an AIDS-related history and incubated with mixture d. The Leu-2 + 7+ cells are clearly a strong feature of this figure and form a distinct cluster. These double positive cells are counted separately from monocytes as indicated in TABLE 2. These double positive cells vary independently from the singly positive Leu-7+ cells, although both are often elevated in AIDS-related conditions.

FIGURE 11 is a dot plot showing the resolution of lymphocytes, monocytes, and granulocytes obtained after whole blood from a healthy donor was treated with the proprietary lysing agent. The lysed sample was analyzed by the FACS Analyzer using the volume and light scattering parameters. In FIGURE 11, the three cell populations can be readily distinguished. The lymphocytes and debris are clearly separated, thus permitting the gating box to be drawn around the lymphocytes. Our findings indicate that the resolution of the lymphocyte cluster that was obtained by using the proprietary lysing agent in our whole blood protocol is equivalent to the resolution obtained when purified PBMC are used.

DISCUSSION

Several individual subclasses of functionally distinct lymphocytes can be identified on the basis of antigenic determinants found on the cell surface. T cells,

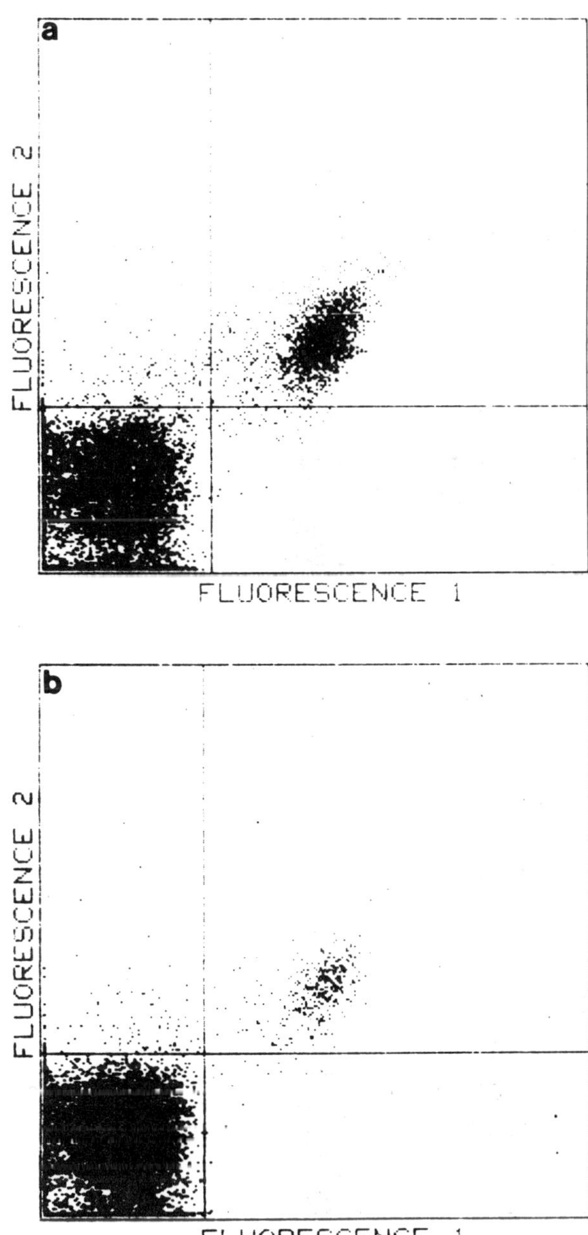

FIGURE 7. a, Ungated, and b, gated, subsample stained with antibody mixture a.

 ANNALS NEW YORK ACADEMY OF SCIENCES

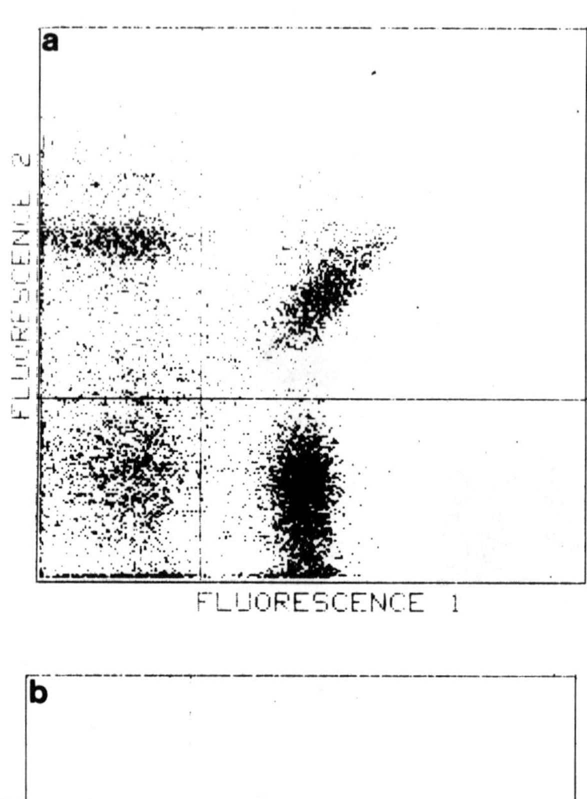

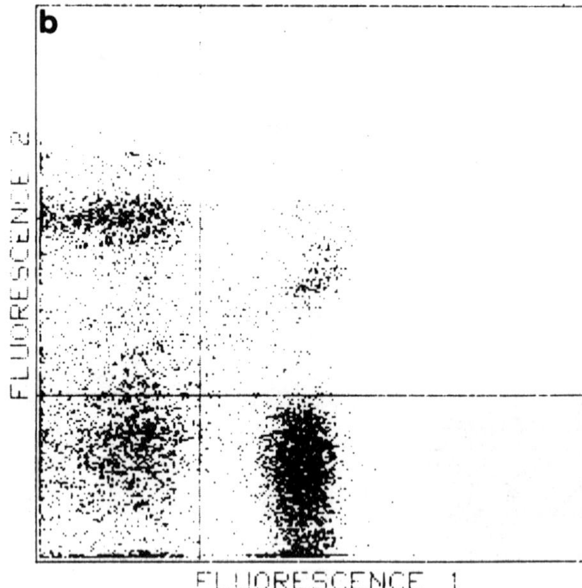

FIGURE 8. a, Ungated, and b, gated, subsample stained with mixture c.

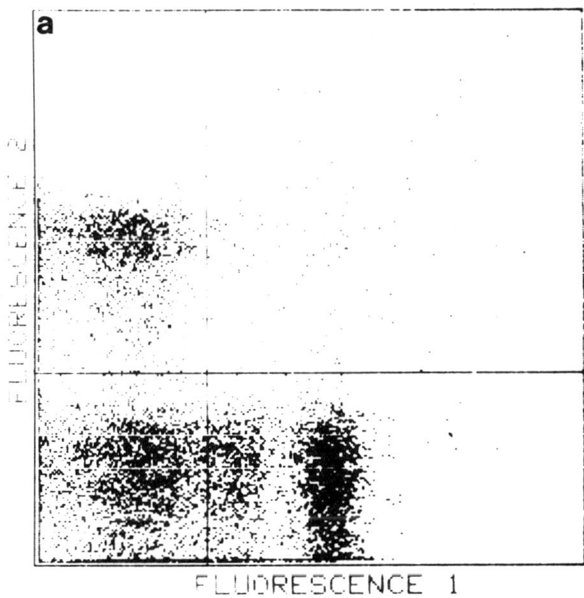

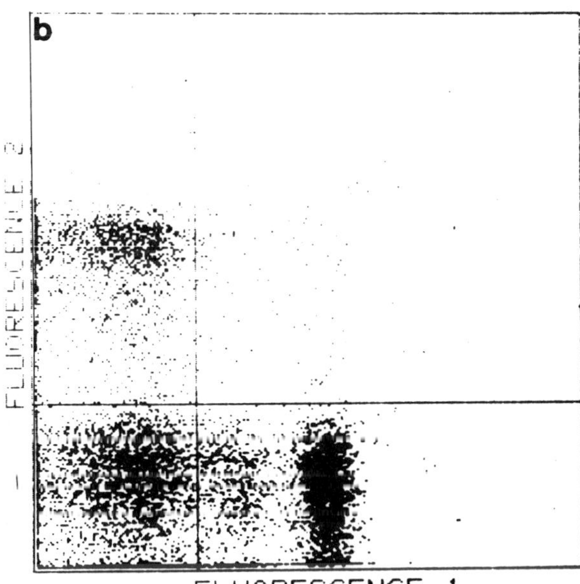

FIGURE 9. a, Ungated, and b, gated, subsample stained with Anti-Leu-3a FITC and Anti-Leu-2a PE but without Anti-Leu-M3 in the mixture.

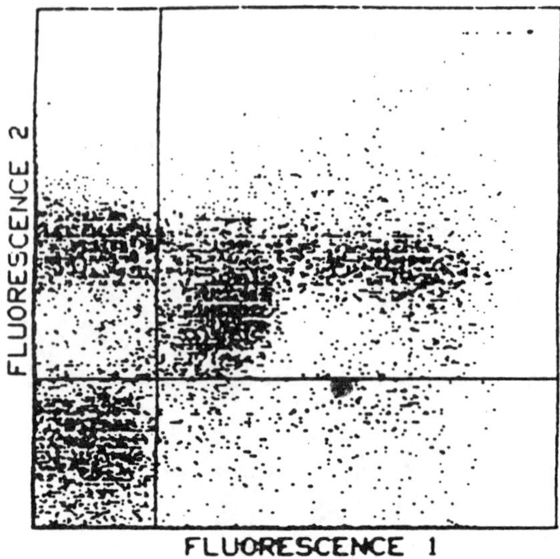

FIGURE 10. Dot plot of PBMC from an abnormal donor labeled with double Anti-Leu-M3, Anti-Leu-2a PE, and Anti-Leu-7 FITC. See text for description.

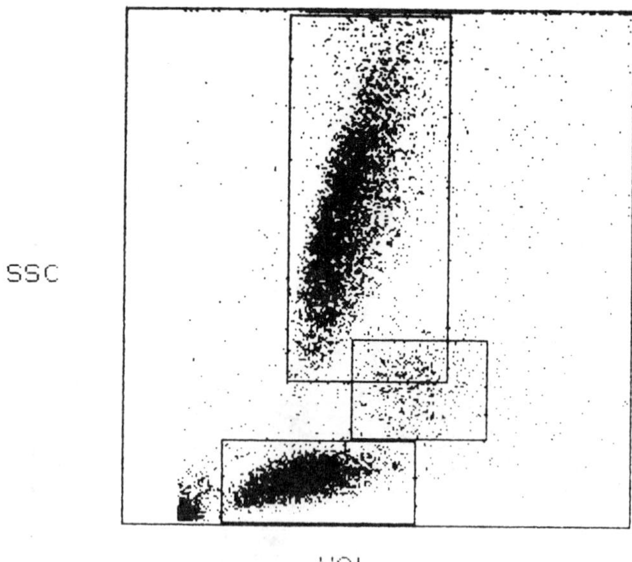

FIGURE 11. Dot plot of whole blood preparation treated with the proprietary lysing agent prior to analysis with the FACS Analyzer plus light scatter detector. Dot plot shows four clusters resolved. Granulocytes are in the upper central region, monocytes are just right of center as a diffuse cluster, lymphocytes are "gated" in the small rectangle and debris with platelets are to the left of the small rectangle.

for example, are divided into at least two subsets called helper/inducer T cells and cytotoxic/suppressor T cells. In humans, the former subset reacts specifically with Anti-Leu-3a, while the latter reacts with Anti-Leu-2a.[2-5] Identification of these T cell subsets and other lymphoid cell populations may have diagnostic or therapeutic significance in a variety of immunoregulated diseases.[15,16]

This report describes an improved method of monitoring cellular immune systems, and may facilitate studies of lymphoid diseases. In this new automated testing system, lymphoid cell subsets in whole blood or mononuclear cell preparations are identified and enumerated without interference from monocytes. The method is based on the use of monoclonal reagents conjugated with phycoerythrin and fluorescein isothiocyanate, and depends on analysis by a FACS Analyzer equipped with a light scatter detector.

In this system, each test sample is divided into several subsamples. In the first subsample, monocytes are doubly labeled with a highly specific anti-monocyte reagent, Anti-Leu-M3 conjugated with phycoerythrin and fluoroscein isothiocyanate. No other stain is used in the first subsample. As a result, only the monocytes are doubly positive, and the monocytes and lymphocytes are clearly segregated. This first subsample provides a reference count of monocytes for subsequent use in computer calculation of the remaining doubly labeled cells. Total lymphocytes for the denominator are calculated in the same way by subtracting monocytes from total PBMC.

In a second, third, or additional subsample, single color monoclonal reagents are used. Each of these other subsamples is also stained with the double color anti-monocyte reagent. A separate cell count and classification by color is made for each subsample. The reference count of monocytes obtained from the first subsample is subtracted from the doubly positive group in subsequent subsamples which may contain doubly positive lymphocytes bearing several of the surface markers under study. The doubly stained monocytes would interfere with accurate enumeration of these cells if they were not subtracted. As a result of the subtraction step, the lymphocyte subsets are enumerated as if monocytes were not present in the subsample. Note that there are two columns of figures in TABLES 2 and 3. These represent percent of a subset using all mononuclear cells as denominator (PBMC) or only lymphocytes as denominator.

TABLE 3. Summary of the Data from Patient of Figure 10

Test	Cell Type	% Of Total	% Of Lymphocytes
Control	monocytes	12.9	
	lymphocytes	81.4	
TBM	T cells	60.7	69.7
	D cells	5.1	5.8
	TB cells	7.9	9.1
	null cells	13.4	15.4
Leu23	2+ cells	52.8	60.6
	3+ cells	14.2	16.3
	2+3+ cells	4.6	5.3
Leu27	2+ cells	31.5	36.2
	7+ cells	14.8	17.0
	2+7+ cells	21.9	25.2

Conventional immunofluorescent techniques usually require the physical separation of lymphocytes from other leukocytes and erythrocytes by density gradient centrifugation as a preliminary step. This separation step eliminates the possibility that nonspecifically stained granulocytes might be counted as specifically stained lymphocytes. The necessity of separating the lymphocytes from other leukocytes is a serious impediment to rapid clinical analyses. In this present study, our findings indicated that the resolution of the lymphocyte cluster obtained by using the proprietary lysing agent was equivalent to that obtained by using Ficoll-Hypaque separated PBMC. The use of hemolyzed whole blood instead of PBMC in this immune monitoring system requires fewer steps, and therefore decreases the opportunities for technical errors in the clinical laboratory.

The precision of the immune monitoring system depends on enumeration of a statistically representative test sample. For a standard deviation of 1% or less, at least 10,000 cells per subsample must be counted. Samples of this size require an automated method of analysis. We have found that the FACS Analyzer equipped with a light scatter detector provides reliable, accurate test results.

In studies currently underway in several laboratories, we are using the automated immune monitoring system to determine the lymphocyte subsets in samples from patients with a variety of immune disorders, including acquired immune deficiency syndrome (AIDS), lympho-proliferative diseases, and transplantation-related immunosuppression. The results of these studies will be reported in subsequent publications.

ACKNOWLEDGMENTS

The authors wish to thank Patty Duncan, Ruth Isenberg, and Kathleen Rulon for their help in the preparation and typing of this manuscript, and Silvia Jost for her technical assistance.

REFERENCES

1. EVANS, R. L., H. LAZARUS, A. C. PENTA & S. F. SCHLOSSMAN. 1978. J. Immunol. **120**:1423–1428.
2. EVANS, R. L., D. W. WALL, C. D. PLATSOUCAS, F. P. SIEGAL, S. M. FIKRIG, C. M. TESTA & R. A. GOOD. 1981. Proc. Natl. Acad. Sci. USA **78**:544–548.
3. ENGLEMAN, E. G., C. J. BENIKE, E. GLICKMAN & R. L. EVANS. 1981. J. Exp. Med. **154**:193–198.
4. KOTZIN, B. L., C. J. BENIKE & E. G. ENGLEMAN. 1981. J. Immunol. **127**:931–935.
5. LEDBETTER, J. A., R. L. EVANS, M. LIPINSKI, C. CUNNINGHAM-RUNDLES, R. A. GOOD & L. A. HERZENBERG. 1981. J. Exp. Med. **153**:310–323.
6. ORNSTEIN, L., H. ARNSLEY & A. M. SAUNDERS. 1976. Blood Cells **2**:557–585.
7. SCHMALZL, F. & H. BRAUNSTEINER. 1970. Ser. Haematol. **3**:93–131.
8. SAUNDERS, A. M., W. GRONER & J. KUSNETZ. 1970. Adv. Automated Analysis **1**:20–26.
9. MANSBERG, H. P., A. M. SAUNDERS & W. GRONER. 1974. J. Histochem. Cytochem. **22**:711–724.
10. BOYUM, A. 1968. Scand. J. Clin. Lab. Invest. **21**:9–29.
11. DIMITRU-BONA, A., G. R. BURMESTER, S. J. WATERS & R. J. WINCHESTER. 1983. J. Immunol. **130**(1):145–152.

12. ABO, T. & C. M. BALEH. 1981. J. Immunol. **127**:1024–1029.
13. LAMPSON, L. A. & R. LEVY. 1980. J. Immunol. **125**:293–299.
14. LANIER, L. L., A. M. LEE, J. H. PHILLIPS, N. L. WARNER & G. F. BABCOCK. 1983. J. Immunol. **131**:1789–1796.
15. SCHROFF, R. W., R. P. GALE & J. L. FAHEY. 1982. UCLA Symposium on Molecular and Cellular Biology **24**:285–291.
16. SCHROFF, R. W., M. S. GOTTLIEB, H. E. PRINCE, L. L. CHAI & J. L. FAHEY. 1983. Clin. Immunol. Immunopathol. **27**:300–314.

Characteristics of Monoclonal Antibody Measurements in Human Peripheral Blood[a]

JERRY T. THORNTHWAITE, DANIEL SECKINGER,
PHYLLIS ROSENTHAL, AND ANTONIO VAZQUEZ

Oncology Laboratories
Department of Research Pathology
Cedars Medical Center
Miami, Florida 33136

INTRODUCTION

Advances in monoclonal antibody production and flow cytometry have revolutionized the evaluation of the immune system in patients. Our knowledge of how the immune system functions has been greatly facilitated by the subclassification of thymus (T) and bone marrow (B)–derived lymphocytes.[1,2] The enumeration of these cell types in peripheral blood is important in understanding immunoregulation in the management of leukemias,[3-5] renal transplant rejection,[6] autoimmune diseases,[6] viral infections,[7] and acquired immunodeficiency syndrome (AIDS).[8,9]

In this article, we will concentrate on some of the technical aspects of dual parameter fluorescence measurements, blood storage conditions, and the relationship between lymphocyte count and T-cell subpopulations. The data will be almost exclusively reported for helper (H) and suppressor (S) T lymphocytes. Recently, the commercial availability of monoclonal antibodies, the ability to analyze lymphocytes in peripheral blood without cell separation, and automation of the enumeration of lymphocyte subtypes using flow cytometry,[10-12] have greatly increased the reliability of these assays for clinical use. In order to minimize sample preparation and analysis times, we have developed an accurate, rapid flow cytometric dual immunofluorescence assay for simultaneously enumerating peripheral blood lymphocytes and their H-H and S-cell subpopulations.

Furthermore, the testing of lymphocyte subsets is a procedure that currently is not available in many clinical laboratories, thus necessitating that samples be mailed to specialized reference laboratories. Even in cases where the test is performed in house, one must know the stability of these samples so they can be analyzed without affecting the normal workflow of the laboratory. In this respect, we compared peripheral blood stored in several anticoagulants at room temperature or 4°C to determine the most stable storage conditions for T-cell phenotyping. Finally, lymphopenia, which may correlate with a reduced H/S ratio, has been proposed as a pre-screen for immune deficiency disorders. Data

[a]This work was supported in part by an equipment grant from the Chatlos Foundation through the Miami Cancer Institute.

will be presented comparing absolute lymphocyte counts and H/S ratios in control and patient populations.

METHODS

Specimen and Storage

Blood was drawn from healthy laboratory volunteers into three different anticoagulants: ethylenediaminetetracetic acid (EDTA), sodium heparin, and acid-citrate-dextrose (ACD). An aliquot of each sample was stored at room temperature (20°C) and another sample of each anticoagulant was refrigerated at 4°C.

Dual Immunofluorescence Analysis

The method for preparing the cells for immunofluorescent flow cytometric analysis was as follows. A 100 μl sample of whole blood was added to a 3 ml conical autoanalyzer cup (Technicon) on ice. A 5 μl aliquot of the fluorescein (FL)-labeled Leu-3a + 3b helper antibody (Becton-Dickinson) was added and slightly vortexed. Subsequently, a 10 μl aliquot of the B-Phycoerythrinated (B-PE)-labeled Leu 2a suppressor antibody was added and the sample vortexed again. The sample was left on ice for 20 minutes. After incubation, a 2 ml aliquot of 3-part diff lysing reagent was added to the cell suspension. After vortexing, the samples were left at room temperature for five minutes and then centrifuged at 400 × g for 9 min at 4°C. The cell pellet was resuspended in a phosphate-buffered isotonic saline solution, stored on ice, and analyzed within one hour.

Flow Cytometric Analysis

The samples were analyzed on an Ortho System 50 Cytofluorograf adapted with a System 30 flow cell and a 2150 computer system (Ortho Diagnostic Systems, Westwood, MA). Single human peripheral blood leukocytes were forced by laminar flow to travel single file through a 200 μm square orifice. Each cell was individually illuminated as it passed through a focused laser beam of 600 mw at 4880. At this intersection point, four parameters were measured on each cell. As shown in FIGURE 1a, the dual light scatter parameters measure the light scattered at forward angles (proportional to the cross sectional area of the cell or cell size) and at 90 degrees to the incident beam (proportional to the degree of heterogeneity or granularity of the cell) and are used to identify the cells as lymphocytes, monocytes, or granulocytes.[10-12] An analysis gate was established so that only cells with lymphocyte light scattering characteristics are analyzed for their F- and B-PE-fluorescent antibody content. For each sample 10,000 lymphocytes are analyzed at a rate of about 400 cells/second. A program was written in which only the lymphocyte region (FIG. 1a) of the light scatter distribution is analyzed for 10,000 lymphocytes, which may or may not bind the helper or suppressor antibodies. For each fluorescent distribution a clear region between the positively and negatively stained lymphocytes was established (FIGS. 1c,d; 2a,b). The

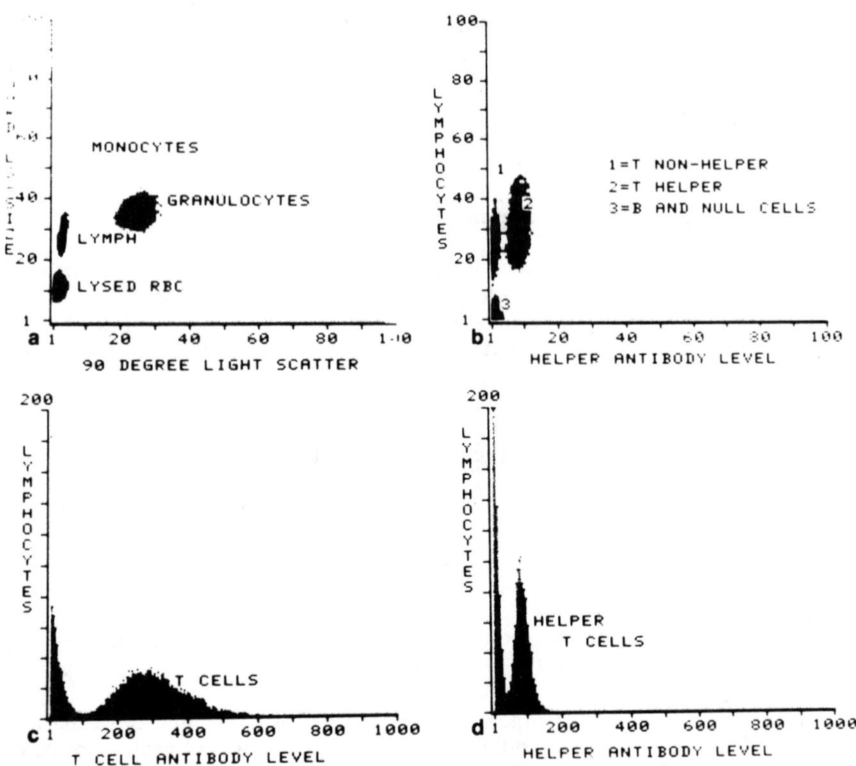

FIGURE 1. Multiparameter flow cytometric data: (a) Forward light scatter (*y*-axis) versus 90° scatter (*x*-axis) of human peripheral blood where the erythrocytes were lysed; (b) Dual fluorescent parameter analysis of lymphocytes stained with Leu-4 (T-Cell, *y*-axis) and Leu 3a + 3b (Helper cell, *x*-axis); (c) Fluorescent distributions of T-cells; and (d) Helper cells.

fluorescent distribution is determined by the percentage of positive cells, the peak and modal distributions, coefficient of variation (CV) of the curve, and skewness. The percentages of the helper and suppressor cells were analyzed in order to determine the mean and standard deviation (SD) for the samples.

RESULTS

Light Scatter Measurements

A light scatter profile for leukocytes from a typical experiment are shown in FIGURE 1a. The lymphocyte distribution is easily distinguished from the monocytes, granulocytes, and lysed erythrocytes based on their forward narrow-angle and right-angle light scatter properties regardless of the immunodeficiency state of the patients studies to date.

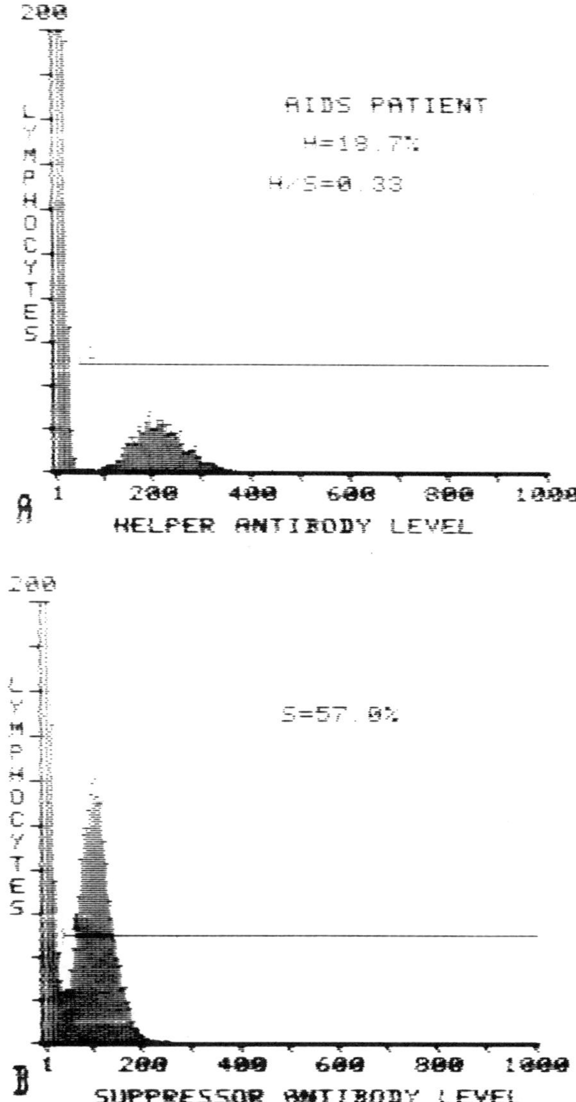

FIGURE 2. Nonfluorescent (left of region 1) and fluorescent (region 1) distributions of (a) Helper cells-fluorescein labeled Leu − 3a + 3b antibody and (b) Suppressor cells-B-phycoerythrin labeled Leu−2a antibody.

TABLE 1. Absolute Lymphocyte, Helper Cell, Suppressor Cell, and H/S Ratio Values for Controls (n = 167)

Type	Mean	± 1 SD	Range (Min-Max)
Absolute lymphocytes (/mm³)	2001	1333–2669	780–4450
Helper cells (%)	42.8	35.3–50.3	19.8–16.7
Suppressor cells (%)	21.6	15.2–28.0	5.1–48.1
H/S ratio	2.17	1.42–2.92	0.6–4.31

Dual Fluorescent Antibody Measurements

For analysis of the lymphocyte subpopulations by dual fluorescent antibody flow cytometry, it was necessary to first have a detectable signal and then to split the green and red fluorescent signals so they did not interfere with each other. FIGURE 1c,d reveals the fluorescent distributions of total T-lymphocytes (Fl-Leu 4) and the subpopulations of helper cells (B-PE-Leu 3a), respectively. FIGURE 1b shows the advantage of analyzing two antibodies simultaneously in which three lymphocyte subpopulations are clearly distinguished. FIGURE 2 shows the results from a control patient with a H/S ratio of 2.10. Both the FL-helper (FIG. 2a) and B-PE-suppressor (FIG. 2b) antibody stained cells can easily be distinguished from the non-fluorescent cells. Furthermore, less than 1% of the fluorescent cells are recorded in the other fluorescent channel distribution.[12] A summary of the

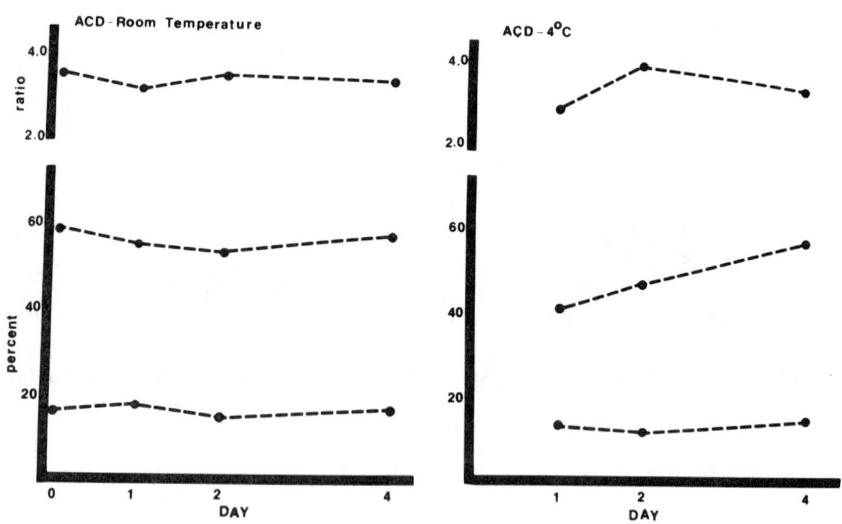

FIGURE 3. Changes in the percentages of helper and suppressor cells and their H/S ratio for ACD blood stored at room temperature (20°C) and 4°C.

absolute lymphocyte, helper cell, suppressor cell, and H/S ratio values for 167 control samples is shown in TABLE 1. These data are similar to the values published on 100 control samples.[15]

Stability as a Function of Anticoagulant and Temperature

FIGURES 3, 4, and 5 graphically show the typical effects of the anticoagulant used and temperature stored on the H/S cell populations. In FIGURE 3, the most stable anticoagulant and storage temperature conditions for preserving the receptor proteins on human peripheral blood lymphocytes are ACD and 20°C (room temperature). At 4°C, there is only a slight increase or decrease, depending on the control, in the percentages of helper cells over the 4 days. Heparin stored blood (FIG. 4) showed slight fluctuations in the H/S ratio at 20°C, while there was a significant decrease in the helper cells at 4°C. Starting at day 0, the effects of storing blood in EDTA caused in some controls dramatic changes in the helper cell populations in which the percentages of helper cells increased at 20°C and decreased at 4°C. In fact, the percentage of helper cells in freshly drawn EDTA tubes were low compared to the ACD or heparin samples which were equivalent. Relative to the helper cells, the suppressor cells showed few changes regardless of the anticoagulants or temperatures used. Equivalent results have been seen in other laboratory control subjects.[20]

H/S Ratios and Absolute Lymphocyte Counts (ALC)

Over 200 peripheral blood samples, which include controls and patients, have been analyzed for H/S cell phenotyping and absolute lymphocyte values. The

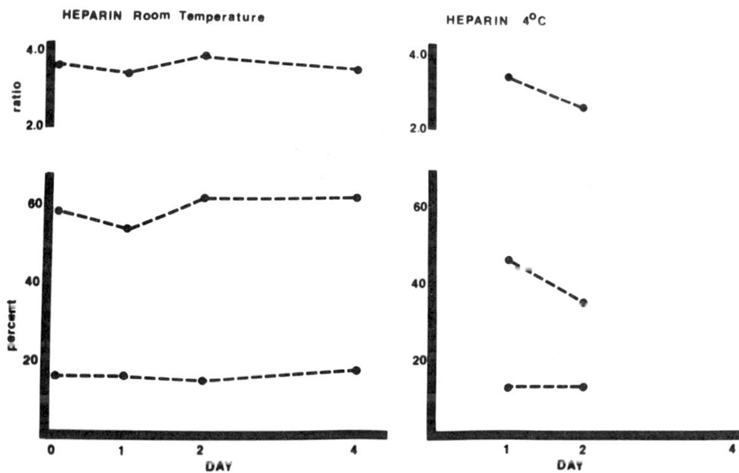

FIGURE 4. Changes in the percentages of helper and suppressor cells and their H/S ratio for heparin anticoagulated blood stored at room temperature (20°C) and 4°C.

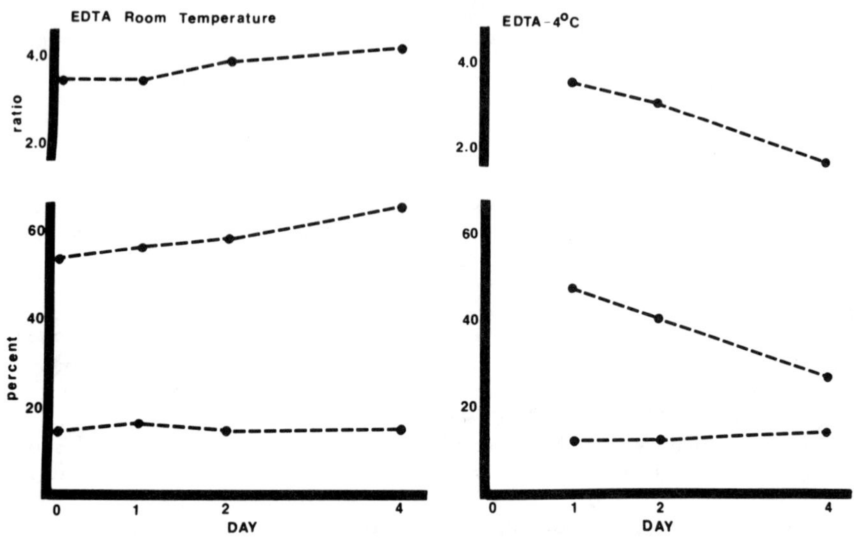

FIGURE 5. Changes in the percentages of helper and suppressor cells and their H/S ratio for EDTA blood stored at room temperature (20°C) and 4°C.

most interesting data are from the analyses of the controls and AIDS/Pre-AIDS (patients who do not fit all of the criteria for AIDS, but are strongly suspect) patient bloods. TABLES 1 and 2 summarize these populations of controls and patients, respectively. The mean ALC were 2001 and 1465/mm³ for the control and patient samples, respectively. However, the mean H/S ratios were considerably different in which the control ratio was 2.17, while the AIDS population was only 0.45.

In FIGURE 6, three graphs describe the changes in ALC and H/S ratio during a year period for three healthy laboratory volunteers. Except for the single data point in Control 2, the H/S Ratios remain fairly constant. The ALC, especially for Control 1, changed over time and in two of the controls (FIG. 6a,b). Lymphopenia (less than 1500 cells/mm³) could be determined depending on the time of analysis. In fact, as shown in FIGURE 7, patients absolute lymphocyte counts can

TABLE 2. Absolute Lymphocyte, Helper Cell, Suppressor Cell, and H/S Ratio Ratio Values for AIDS/pre-AIDS Patients (n = 31)

Type	Mean	± 1 SD	Range (Min-Max)
Absolute lymphocytes (/mm³)	1465	661–2269	76–3660
Helper cells (%)	18.4	9.1–27.7	2.7–35.1
Suppressor cells (%)	42.7	34.5–50.9	28.9–58.9
H/S ratio	0.45	0.17–0.73	0.08–1.16

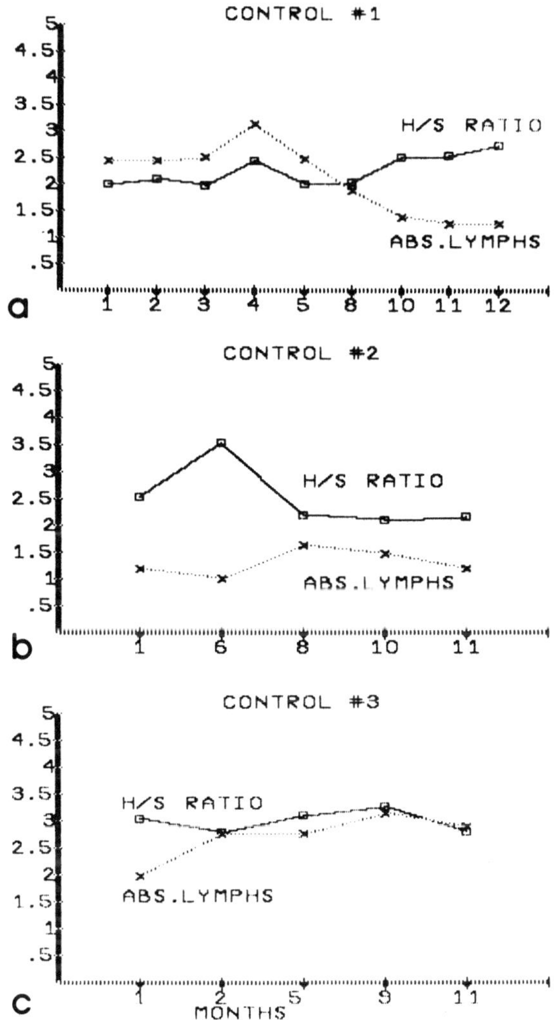

FIGURE 6. Changes in H/S ratio and absolute lymphocyte count as a function of time for three healthy laboratory staff members. Ordinate:absolute lymphocytes × 1000/mm³.

vary considerably over even a short time. On the other hand, the H/S remains stable even in patients. FIGURES 8 and 9 show graphically the distributions of ALC and H/S data in AIDS/pre-AIDS patients, respectively. Interestingly, the ALC distributions overlapped considerably with patients who were above or below the H/S ratio of 1.2 that has been determined to be the immunodeficiency threshold.[15] Conversely, the H/S ratio was below 1.2 for all of the AIDS/pre-AIDS patients studied (FIG. 9, TABLES 2 and 3). AIDS/pre-AIDS patient data are shown in TABLE 3 for the patients studied to date.

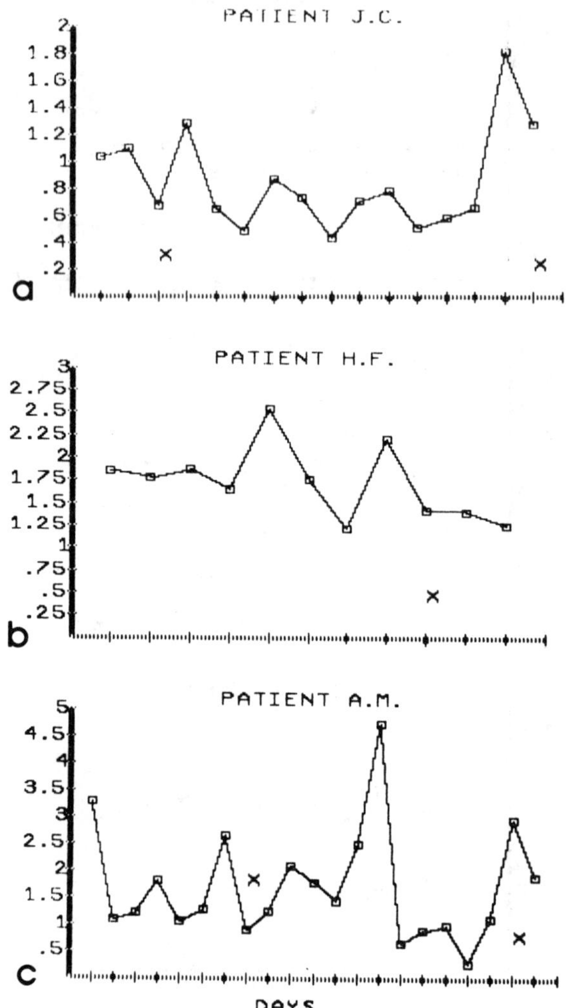

FIGURE 7. Changes in H/S ratio (marked by an x) and absolute lymphocyte count as a function of time for three patients. Ordinate:absolute lymphocytes $\times$ 1000/mm^3.

DISCUSSION

Monoclonal antibodies, especially helper and suppressor T-lymphocytes, have been instrumental in delineating T-cell imbalances in a variety of immunologic diseases such as chronic active hepatitis, autoimmune hemolytic anemia hemolytic anemia, glomerulonephritides, multiple sclerosis, leprosy, or AIDS.[1]

In this study we have shown that flow cytometric analysis of lysed whole

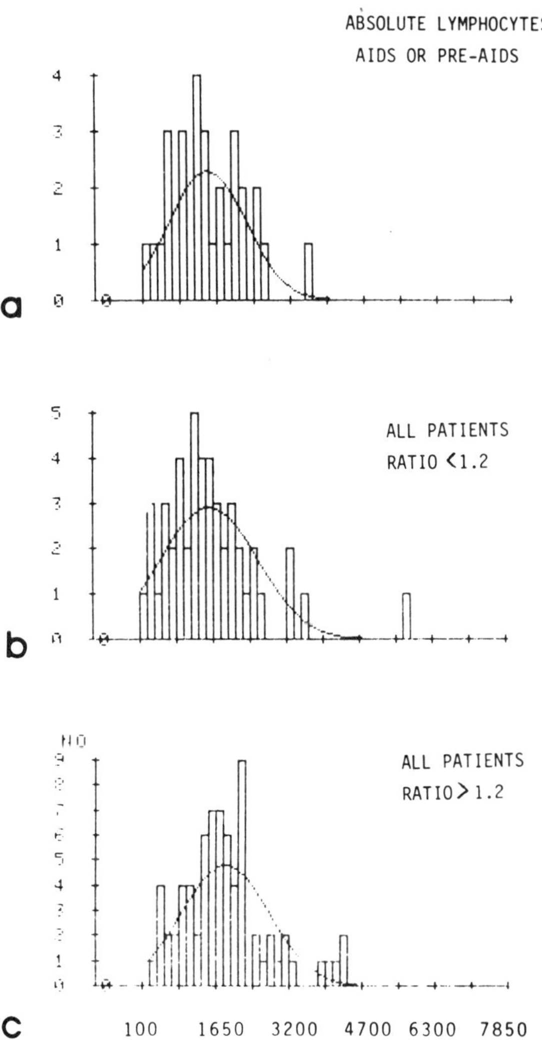

FIGURE 8. Absolute lymphocyte count distributions for (a) AIDS/pre-AIDS (n = 31) or patients other than AIDS with H/S ratios (b) below 1.2 (n = 17) or (c) above 1.2 (n = 68). x = lymphocytes/mm^3; y = patient number

blood,[10-12] using the forward and 90 degree light scattering parameters, can discriminate lymphocytes, monocytes, and granulocytes[10] and avoids problems inherent in cell separation. Furthermore, even though the lymphocyte cluster can easily be identified in our system, under certain disease states the cluster may be contaminated by monocytes and a few granulocytes.[13] However, this contamination is probably low (less than 5%) (refs. 11, 14, and 15, unpublished results) based on cell sorter studies. The percentage of lymphocytes in Region 1 of the

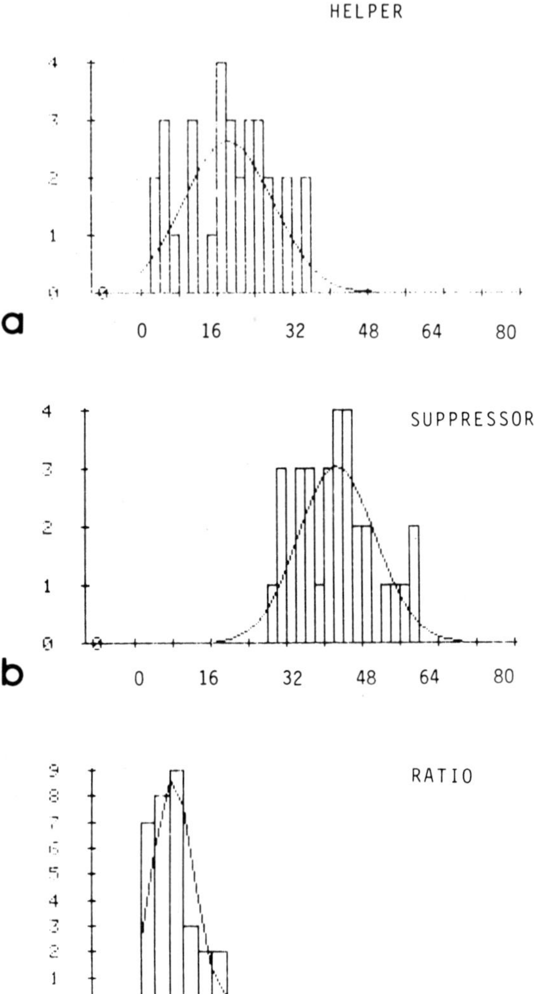

FIGURE 9. AIDS/pre-AIDS patients distributions for (a) Helper, (b) Suppressor, and (c) H/S ratio measurements. x = percentage (a, b) or ratio (c); y = patient number.

forward and 90 degree light scatter histograms (FIG. 1a) correlates well (r = 0.97) with the manual procedure over a wide range (2–99%) of lymphocyte concentrations.[11]

In order to enhance the clinical applicability of monoclonal antibody measurements of cells by flow cytometry, a simple method for simultaneously measuring two antibodies would decrease the preparation and analysis times. We

TABLE 3. AIDS/pre-AIDS, Helper/Suppressor, and Absolute Lymphocyte Count (ALC) Data[a]

AIDS/Pre-AIDS Patient No.	Helper (%)	Suppressor (%)	H/S Ratio	WBC	% Lymphocytes	ALC
1	19.9	41.2	0.48	4100	24	984
2	31.8	35.0	0.91	4400	30	1320
3	23.7	31.1	0.76	6600	31	2046
4	3.1	37.9	0.08	5100	12	612
5	19.3	52.4	0.37	5500	23	1265
6	17.5	57.3	0.31	3800	25	950
7	22.6	40.6	0.56	3700	2	74
8	16.5	44.5	0.37	4600	45	2070
9	4.4	44.3	0.10	4200	13	546
10	6.2	37.2	0.17	7700	8	616
11	10.8	58.1	0.19	9900	17	1683
12	25.8	43.5	0.59	6100	23	1403
13	24.0	44.3	0.54	4300	37	1591
14	18.4	43.3	0.42	6800	36	2448
15	35.1	31.6	1.11	4500	58	2610
16	27.9	38.0	0.73	6100	35	2135
17	16.3	47.2	0.35	3900	36	1404
18	14.7	42.2	0.35	7200	34	2448
19	11.9	45.9	0.26	9500	7	665
20	10.9	40.8	0.27	2200	45	990
21	16.6	48.3	0.34	6100	60	3660
22	34.8	30.1	1.16	6600	26	1716
23	20.3	46.7	0.43	5600	25	1400
24	26.6	43.3	0.61	6400	20	1280
25	20.0	48.5	0.41	4700	28	1316
26	23.8	55.5	0.43	4400	50	2200
27	30.2	36.6	0.83	5200	37	1924
28	2.8	34.2	0.08	7400	3	222
29	5.8	35.7	0.16	ND[b]	ND	ND
30	4.0	28.9	0.14	19700	2	394
31	25.2	58.9	0.43	17900	11	1969

[a]See TABLE 2 and FIGURE 8a.
[b]ND-Not Determined.

describe such a method in which whole blood is mixed with helper (Leu 3a + 3b) FL-conjugated antibody[16,17] and suppressor (Leu 2a) B-PE-conjugated antibody.[10,19] As shown in FIGURES 1 and 2, B-PE antibody conjugates can be combined with F-labeled antibodies for two color analyses. It is possible to simultaneously excite fluorescein and phycoerythrin using the 4880 Å argon-ion laser line. The emission maxima of fluorescein (530 nm) and phycoerythrin (600 nm) can be separated from each other so there is minimal crosstalk between the green and channels of the flow cytometer.[12] Therefore, simultaneous enumeration of helper/suppressor cells can be accomplished on patients as graphically shown in FIGURES 1 and 2. Furthermore, virtually any dual antibody combination may be used. FIGURE 1b illustrates the application of the dual antibody labeling technique using H/S or helper T-lymphocyte antibodies. For example, as shown in an earlier report,[12] the H/S, B/T, activated suppressor, and natural killer cells can be determined each in a single analysis run on the flow cytometer.

There are significant alterations as a function of time in the percentages of helper and suppressor lymphocytes depending on the storage anticoagulant time and temperature.

The most stable conditions were observed for blood stored at 20°C in ACD. No significant alterations in the percentages of helper and suppressor cells were observed for up to four days after blood was drawn. By day 4, however, the light scatter distributions, while resolvable for the 3-part differential, show significant populations of cells appearing between the lymphocyte and granulocyte distributions.[20] A comparison, using citrate-phosphate dextrose (CPD) or ACD as anticoagulants at room temperature, showed less clumping of leukocytes with ACD,[21] whereas another report[22] stated that ACD caused more aggregation of leukocytes than CPD. Although CPD was not used in this study, the results with ACD show consistent percentages of lymphocytes during the storage periods which suggest cell clumping is minimal.

Storage of blood at 4°C for as little as 24 hours, regardless of the anticoagulant used, resulted in significant alterations in the light scatter patterns.[20] The most apparent change was seen with EDTA at 4°C in which there is an increase of the skewness to the right for the granulocyte 90 degree scatter distributions with little change in the forward angle light scatter measurements.

While the percentages of T-helper cells in EDTA decreased significantly during cold storage and increased at 20°C, the percentages of T-suppressor cells do not change significantly regardless of the anticoagulant or temperature. Therefore, the suppressor cell population is not affected by cold storage or anticoagulant type; this lack of affect could be due to either the suppressor cells not migrating from the lymphocyte light scatter distribution and/or to their cell-surface receptor antibody affinity not being as sensitive to cold storage as helper cells. Finally, Grunow et al.[23] showed that the relative percentage of T-cells in CPD-anticoagulated whole blood samples stored at 4°C for 24 hours dropped 20%.

While it has been proposed that prescreening of immunodeficiency can be accomplished using absolute lymphocyte counts,[24,25] our data show lymphopenia (<1500 lymphocytes/mm³) to be a poor indicator of immunodeficiency for several reasons. First of all, ALC can vary significantly in normal subjects (FIG. 6) as well as patients (FIG. 7). Conversely, the H/S ratio remains consistent over a long period. Secondly, although lymphopenia as a trend is seen with AIDS/Pre-AIDS patients, the measurement is by no means consistent (FIG. 8, TABLES 2 and 3). The H/S ratios for all 31 AIDS/Pre-AIDS patients, on the other hand, were all below the minimal normal ratio value of 1.2.[20]

CONCLUSION

The data presented here describe a method for whole blood dual antibody fluorescent labeling without concern for the purification of the lymphocytes population before flow cytometric analysis. It is hoped that this preparative method will speed up and make more economical the current methods of determining lymphocyte subsets, which, with a knowledge of the complete clinical assessment, will add to the battery of tests that are used to define an immunosuppressive state. Furthermore, the phenotypic determinations of helper and suppressor lymphocytes are influenced by some interaction of anticoagulant, storage time, and temperature. These studies show that either ACD or heparin is an adequate anticoagulant for the storage of human peripheral blood for the analysis of helper/suppressor T-lymphocyte subpopulations. From these studies, storage at room temperature was preferred to that at 4°C. These data show the feasibility of shipping peripheral blood as shown by the stability of the ACD samples for 3 days at room temperature. Finally, lymphopenia is not a good prescreen for immunodeficiency in patients or blood donors. This measurement varies considerably in control and patient populations. All AIDS/pre-AIDS patients had abnormally low H/S ratios. For AIDS, new tests (e.g., antibody against human T-cell leukemic virus[26-29] or interleukin-2,[30]) are needed to screen more clearly for potential AIDS victims. Work is currently in progress to develop such tests.

SUMMARY

Three areas of monoclonal antibody measurements using flow cytometry have been presented. These include a description of a dual immunofluorescent method for measuring two antibodies simultaneously, the effects of blood storage on enumeration of helper (H) and suppressor (S) cells, and the relationship between absolute lymphocyte count and H/S ratio in both control and AIDS patients.

These studies reveal that a dual immunofluorescent labeling method is useful for enumerating lymphocytes from peripheral blood which bear the helper, suppressor and/or thymus-derived (T) cell receptors. Fluorescein (FL)-conjugated Leu-3a + 3b antibodies were used to enumerate helper T-lymphocytes, while the B-phycoerythrin (B-PE)-conjugated Leu-2a antibodies were utilized for enumerating suppressor T-lymphocytes. Dual immunofluorescently stained lymphocytes, prepared from whole blood, were analyzed by flow cytometry. Two light scatter parameters, (forward and 90 degree scatter) were used to define the lysed erythrocyte, lymphocyte, monocyte, and granulocyte populations. Only the lymphocytes were analyzed for dual immunofluorescence activity. The helper and suppressor distributions from 167 control patients were as follows: The average percentage ± SD of the helper and suppressor cells were 42.8 ± 7.5 and 21.6 ± 6.4, respectively. The H/S ratio was 2.17 ± .75. These studies show that the H/S ratio can be determined in a single preparative sample and analyzed by dual immunofluorescence in a single flow cytometric analysis even though the H/S ratio may vary from normal during a disease condition. The dual immunofluorescent assay enables one to correlate the activities of two antibodies against cell surface receptors and allows the measurement of a large number of samples in a minimal time.

This study also compared the effects of anticoagulant, storage time, and temperature on the phenotypic determination of the percentages of helper and suppressor T-lymphocytes in human peripheral blood. Blood was drawn in ACD, heparin, and EDTA and stored for up to 4 days at room temperature or 4°C. Phenotypic determination of helper/suppressor lymphocytes was most stable for ACD or heparinized blood at room temperature. Marked changes were observed in the percentages of helper cells at 4°C, whereas the percentages of suppressor cells did not change appreciably regardless of the anticoagulant storage time or temperature.

Finally, the relationship between ALC and the H/S ratio in control and AIDS patients was determined. The ALC varied considerably in both control and patient populations as a function of time. Conversely, the H/S ratio remained constant. Furthermore, the analysis of 31 AIDS/pre-AIDS patients showed large fluctuations in the ALC, whereas the H/S ratio remained below the minimal normal H/S ratio value of 1.2.

ACKNOWLEDGMENTS

We thank the Medical Media Center for preparation of the figures and the Library for invaluable assistance.

REFERENCES

1. BACH, M. A. & J.-F. BACH. 1981. The use of monoclonal anti-T antibodies to study T cell imbalances in human disease. Clin. Exp. Immunol. **45**:449–456.
2. FAIRBANKS, T. R. 1980. Current status of lymphocyte subpopulation testing in humans. Am. J. Med. Technol. **46**:471–475.
3. GUPTA, S. & R. A. GOOD. 1980. Markers of human subpopulations in primary immunodeficiency and lymphoproliferative disorders. Sem. Hematol. **17**:1–25.
4. KERSEY, J. H. & K. J. GAJL-PECZALSKA. 1975. T and B lymphocytes in humans. Am. J. Pathol. **81**:446–457.
5. PANDOLFI, F., G. DeROSSI, et al. 1982. Immunologic evaluation of T chronic lymphocyte leukemia cells: Correlation among phenotype, functional activites and morphology. Blood **59**:668–695.
6. REINHERZ, E. L. & S. F. SCHLOSSMAN. 1980. The differenation and function of human T lymphocytes. Cell **19**:821–825.
7. SCHROFF, R. W., K. A. FOON, R. J. BILLING & J. L. FAHEY. 1982. Immunologic classification of lymphocyte leukemias based on monoclonal antibody–defined cell surface antigens. Blood **59**:207–215.
8. MIIDVA, D., V. MRTHUR, R. W. ENLOW, et al. 1982. Opportunistic infections and an immune deficiency in homosexual men. Ann. Int. Med. **96**:700–704.
9. FOLLANSBEE, S. E., D. F. BUSCH, C. B. WOFSY, et al. An outbreak of pneumocystis carinii pneumonia in homosexual men. Ann. Int. Med. **96**:705–713.
10. SALZMAN, G. C., J. M. CROWELL, J. C. MARIN, et al. 1975. Classification by laser light scattering: Identification and separation of unstained leukocytes. Acta Cytol. **19**:374–377.
11. IP, S. H., C. W. RITTERSHAUS, K. W. HEALY, C. C. STRUZZIERO, R. A. HOFFMAN & P. W. HANSEN. 1982. Rapid enumeration of T-lymphocytes by a flow cytometric immunofluorescent method. Clin. Chem. **28**:1905–1909.
12. THORNTHWAITE, J. T., D. SECKINGER, E. V. SUGARBAKER, P. K. ROSENTHAL & D. A. VAZQUEZ. 1986. Dual immunofluorescent analysis of human peripheral blood lymphocytes. Am. J. Clin. Pathol. In press.
13. IWATANI, Y. N. AMINO, H. MORI, et al. 1982. Effects of various isolation methods for human peripheral lymphocytes on T cell subsets determined in a fluorescent activated cell sorter (FACS), and demonstrations of a sex difference of suppressor/cytotoxic T cells. J. Immunol. Methods **54**:31–42.
14. HOFFMAN, R. A., P. C. KUNG, W. P. HANSEN & G. GOLDSTEIN. 1980. Simple and rapid measurement of human T lymphocytes and their subclasses in peripheral blood. Proc. Natl. Acad. Sci. USA **77**:4914–4917.
15. HOFFMAN, R. A. & W. P. HANSEN. 1981. Immunofluorescence analysis of blood cells by flow cytometry. Int. J. Immunopharmacol. **3**:249–254.

16. EVANS, R. L., D. W. WALL & C. D. PLATSOUCAS. 1981. Thymus-dependent membrane antigens in man: Inhibition of cell-mediated lympholysis by monclonal antibodies to the TH2 antigen. Proc. Natl. Acad. Sci. USA **78**:544–549.

17. KOTZIN, B. L., C. J. BENIKE & E. G. ENGLEMAN. 1981. Induction of immunoglobulin secreting cells in the allogeneic mixed leukocyte reaction: Regulation by helper and suppressor lymphocyte subsets in man. J. Immunol. **27**:931–937.

18. GLAZER, A. N. & C. S. HIXSON. 1977. Submit structure and chromophore composition of rhodophytan phycoerythrins, porphyridium cruentum B-phycoerythrin and b-phycoerythrin. J. Biol. Chem. **252**:32–42.

19. VERNON, T. O., A. N. GLAZER & L. STRYER. 1982. Fluorescent phycobiliprotein conjugates for analysis of cells and molecules. J. Cell. Biol. **93**:981–986.

20. THORNTHWAITE, J. T., P. K. ROSENTHAL, D. A. VAZQUEZ & D. SECKINGER. 1984. The effects of anticoagulant and temperature on the measurements of helper and suppressor cells. Diag. Immunol. **2**:167–174.

21. LUCETE, R. M. 1973. Influence of ACD and CPD anticoagulants on white cell aggregations. Transfusion **13**:435–438.

22. WALDMAN, A. A. & C. SHANDER. 1980. Human blood components I. Preparation and characteristics of leukocyte concentrates from single units of human blood. Transfusion **20**:384–392.

23. GRUNOW, J. E., A. R. LUBET, M. J. FERGUSON & M. E. GAULDEN. 1976. Perferential decrease in thymus dependent lymphocytes during storage at 4°C in anticoagulant. Transfusion **16**:610–615.

24. METROKA, C. E., S. CUNNINGHAM, M. S. POLLACK, J. A. SONNABEND, J. M. DAVIS, B. GORDON, R. D. FERNANDEZ & J. MOURADIAN. 1983. Generalized lymphadenopathy in homosexual men. Ann. Int. Med. **99**:585–591.

25. IOACHIM, H. L., C. W. LERNER, M. L. TAPPER. 1983. The lymphoid lesions associated with the acquired immunodeficiency syndrome. Am. J. Surg. Pathol. 7:543–553.

26. POPOVIC, M., M. G. SARNGADHARAN, E. READ & R. C. GALLO. 1984. Detection, isolation, and continuous production of cytopathic retroviruses (HTLV-III), from patients with AIDS and pre-AIDS. Science **224**:497–500.

27. GALLO, R. C., S. Z. SALAHUDDIN, M. POPOVIC, G. M. SHEARER, M. KAPLAN, B. F. HAYNES, T. J. PAIKER, R. REDFIELD, J. OLESKE, B. SAFAI, G. WHITE, P. FOSTER & P. D. MARKHAM. 1984. Frequent detection and isolation of cytopathic retroviruses (HTLV-III) from patients with AIDS and at risk for AIDS. Science **224**:500–503.

28. SCHÜPBACH, J., M. POPOVIC, R. V. GILDEN, M. A. GONDA, M. G. SARNGADHARAN & R. C. GALLO. 1984. Serological analysis of a subgroup of human T-lymphotropic retroviruses (HTLV-III) associated with AIDS. Science **224**:503–505.

29. SARNGADHARAN, M. G., M. POPOVIC, L. BRUCH, J. SCHÜPBACH & R. C. GALLO. 1984. Antibodies reactive with human T-lymphotropic retroviruses (HTLV-III) in the serum of patients with AIDS. Science **224**:506–508.

30. GILLIS, S. & K. A. SMITH. 1977. Long-term culture of tumor-specific cytotoxic T cells. Nature **268**:154–157.

Analysis of Neoplastic Leukocyte Surface Antigens in Unfractionated Blood[a]

RAUL C. BRAYLAN, NEAL A. BENSON,
BARBARA A. BENSON, AND VIRGINIA A. NOURSE

Department of Pathology
University of Florida College of Medicine
Gainesville, Florida 32610

INTRODUCTION

Conventional diagnosis of leukemias in peripheral blood is based on numerical and morphologic abnormalities of leukocytes. A modern approach to the characterization of leukemic cells is the analysis of cell surface antigens using leukocyte subset-specific monoclonal antibodies.[1-7] This analysis is helpful because leukemic cells may express differentiation and other antigens that are not usually present in normal mature circulating leukocytes. In most instances the initial diagnosis of leukemia is not difficult since large numbers of neoplastic cells can be easily detected by microscopy. However, samples in which the leukemic cells represent only a small fraction of the entire leukocyte population pose a special diagnostic problem. In such cases, the interpretation of morphologic changes of cells in blood smears or the microscopic analysis of immunostains may be extremely difficult.

In 1975, Salzman *et al.*[8] demonstrated that by using flow cytometry (FCM), unfixed, unstained human leukocytes can be separated into distinct populations of leukocytes corresponding to lymphocytes, monocytes, and granulocytes based only on the measurement of laser light simultaneously scattered by each cell at two different angles (dual scatter). The identity of the various cellular groups was determined by electronic cell sorting and subsequent cytologic examination. Hansen *et al.*[9] have more recently confirmed the earlier observations. Utilizing this information, Hoffman *et al.*[10] described a technique to enumerate T-lymphocyte subsets in buffy coats or whole blood samples. Lymphocytes were identified by FCM using dual scatter detectors and their surface antigens were analyzed by immunofluorescence using monoclonal antibodies against T-cell differentiation antigens. In this paper we show that a similar approach can be used to detect leukemic or lymphoma cells in unfractionated blood samples based on the observation that circulating neoplastic cells result in changes in the light scatter and/or antigenic patterns of normal blood leukocytes.

[a]Supported in part by grant CA-23393, National Cancer Institute.

MATERIALS AND METHODS

Sample Preparation

Leukocyte surface antigens were labeled using mouse monoclonal antibodies according to the method of Hoffman *et al.*[10] with the following modifications: For each antigen investigated, 100 µl of heparinized blood were mixed with 100 µl of phosphate-buffered saline (PBS), pH 7.4 and 5 µl of the appropriate antibody were added to the mixture. Control samples were exposed to normal mouse immunoglobulin at equivalent concentrations. After a 15 minute incubation on ice, 3 ml of a red cell lysing solution (8.29 g NH_4Cl, 1.0 g $KHCO_3$, and 37 mg disodium EDTA per liter of distilled water) were added. After 10 minutes at room temperature, samples were centrifuged, washed once in PBS, and 100 µl of a 1:100 dilution of fluorescein (FITC)-labeled sheep anti-mouse antibody (Cappel, West Chester, PA) were added. Following a 15 minute incubation on ice, the cells were washed and resuspended in 0.5 ml of PBS. Between 10^4 and 2×10^4 cells were analyzed. Analysis of cell surface immunoglobulins required a modified technique to remove serum immunoglobulins. One ml of heparinized blood was mixed with 5 ml of RPMI-1640 medium (GIBCO, Grand Island, NY), centrifuged, and the pellet was exposed to the red cell lysing solution for 15 minutes at room temperature. After centrifugation, the cells were resuspended in 5 ml of warmed medium and incubated for 15 minutes at 37°C in a water bath to remove cytophilic immunoglobulin. After centrifugation, the cells were exposed to 1:100 dilution of the appropriate monoclonal antibody against immunoglobulin light chain (Seward, London, UK). Following a 15 minute incubation on ice and washings in cold medium, the cells were exposed to FITC-sheep anti-mouse antibody at 1:100 dilution for 15 minutes, washed and resuspended in PBS. For the detection of small numbers of monoclonal B-cells we analyzed 2×10^5 cells stained with each anti-light chain antibody. Before analysis all preparations were filtered through a 44 µm-pore nylon mesh.

Flow Cytometric Analysis

A FACS II (Becton Dickinson FACS Systems, Sunnyvale, CA) equipped with additional electronic components to provide three-parameter correlated measurements[11] was used for analysis. The light source was the 488 nm line of a 4 watt argon ion laser regulated at 400 mW constant light output. The optical filters for the green fluorescence detector (FITC) included two 500 nm long-pass filters to block scattered laser light, one 520 nm, 3-cavity bandpass, and one 515 nm long-pass filter. A 50 µm diameter nozzle was used. Forward low-angle light scatter was measured using the standard light scatter detector of the instrument. Right-angle light scatter was measured using a detector positioned at a 90° angle with respect to the laser beam without optical filters.

Data Analysis

Data acquisition was accomplished via a parallel link to a specially designed multiplexer capable of sending sequentially up to three digitized measurements per cell (forward angle scatter, right-angle scatter, and green fluorescence) from the flow cytometer analog-to-digital converters to a 9845T desktop computer (Hewlett-Packard, Fort Collins, CO). Computer software was developed in our laboratory to provide control of data acquisition. Using this program, a "window" is positioned around the leukocyte cluster of interest and the fluorescence signal from the cells within the window is acquired and stored. This program was designed to allow an operator to select subpopulations of cells based upon their forward and right-angle light scatter and to store their fluorescence distributions for later enumeration of labeled cells. The program functions in two modes: "on-line," acquiring data from the flow cytometer and retaining it in memory while an

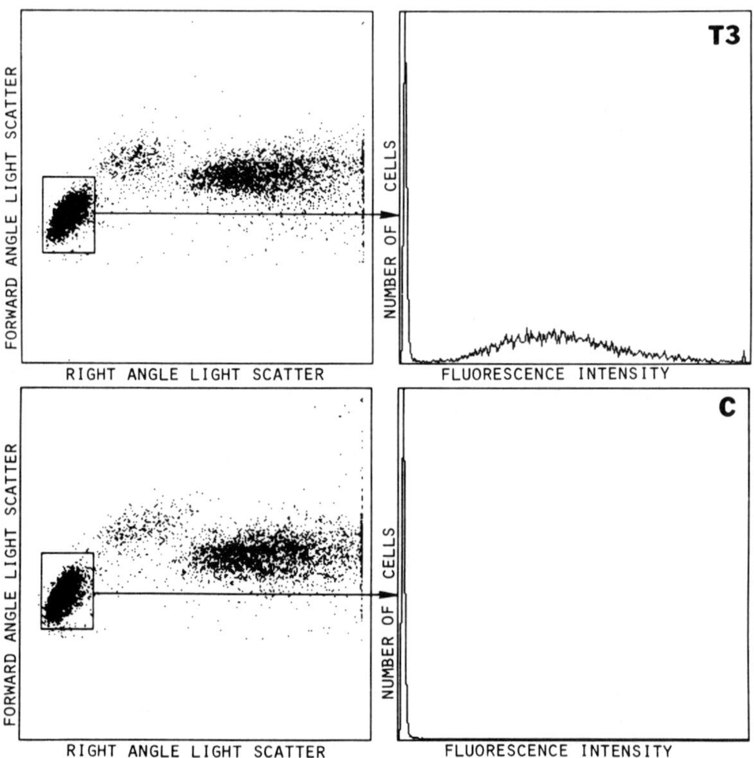

FIGURE 1. Measurement of T-cells in whole blood preparations using dual scatter analysis. Normal blood was exposed to anti-T3 monoclonal antibody (T3) (upper charts) or to normal mouse IgG as a control (C) (lower chart) followed by FITC-labeled sheep anti-mouse antibody. Red cells were lysed. The left charts show "dot plots" produced by the forward (vertical axis) and right angle (horizontal axis) light scatter of leukocytes. Each dot represents a single cell. The clusters in the windows represent lymphocytes. The other two clusters correspond to monocytes (middle) and granulocytes (right). Platelets and residual red cell ghosts fall below a size threshold. Histograms to the right show the fluorescence distribution of cells in the lymphocyte windows. In this experiment 70.3% of the lymphocytes are T3 positive. Other windows may be selected for analysis of populations of interest.

operator selects populations of interest and stores only their fluorescence distributions; and "off-line," reading previously acquired flow cytometric data from disk storage, instead of from the flow cytometer during actual data acquisition. The two light scatter measurements are displayed on the computer screen as points on a 256 × 256 element matrix (dot-plot) on a cell-by-cell basis. While examining the dot-plot, an operator can vary the density of the cells on the screen and designate up to four subpopulations for fluorescence distribution analysis by enclosing each in a rectangle (window) drawn on the screen using a cursor. The limits of each window can be altered, if desired, by the operator to accommodate sample-to-sample variations. Display of the fluorescence distributions of any of the four windows is selectable, and following display, the distributions may be stored on disk and a hard-copy may be obtained.

RESULTS

The correlated measurement of forward and right-angle light scatter signals from normal leukocytes is shown in FIGURE 1. Lymphocytes, monocytes, and granulocytes form discrete clusters easily discernable on the computer screen. The fluorescence distribution of lymphocytes labeled with anti-T antibody (T3) was obtained by placing a "window" around the appropriate cell cluster and the percentage of labeled cells was calculated by comparison with a control sample (C). We tested the reproducibility of this method using blood from 10 normal volunteers (TABLE 1). The samples were divided into two aliquots. One aliquot was

TABLE 1. FCM Analysis of T3-Lymphocytes in Whole Blood (Percentage of Total Lymphocytes)

Donor	Day	Technician 1 Window 1	Window 2	Technician 2 Window 1	Window 2	Technician 3 Window 1	Window 2	Technician 4 Window 1	Window 2
1	1	68.6	68.7	69.1	68.8	68.6	68.7	69.0	68.8
	2	67.5	66.7	67.2	66.9	66.9	67.6	67.1	67.3
2	1	74.0	73.9	74.3	74.5	74.1	74.2	74.2	74.6
	2	73.0	72.9	73.4	73.2	73.1	73.0	73.5	73.5
3	1	74.8	74.5	75.0	74.7	75.1	75.0	75.3	75.5
	2	74.8	74.8	75.2	75.4	75.0	75.3	75.0	76.3
4	1	66.5	66.1	66.3	66.5	66.3	66.3	66.4	67.0
	2	66.0	65.8	65.9	66.4	65.6	65.8	66.6	66.4
5	1	81.5	81.2	81.5	81.9	81.6	81.6	81.9	81.7
	2	81.7	82.0	82.3	81.7	81.6	81.9	82.1	82.3
6	1	79.2	78.9	79.3	79.6	79.6	79.4	79.4	79.4
	2	80.1	80.0	80.1	80.1	79.9	80.0	80.1	80.4
7	1	67.9	67.4	68.0	68.1	67.7	67.1	67.9	68.0
	2	70.2	70.3	70.5	71.0	70.4	70.2	70.5	70.5
8	1	76.3	76.1	76.4	76.3	76.2	76.3	76.3	76.4
	2	77.2	76.8	77.6	77.1	77.1	77.1	77.3	77.4
9	1	70.3	70.3	70.5	70.7	70.3	70.3	70.4	70.6
	2	67.9	67.5	67.4	67.6	67.7	67.8	68.8	68.1
10	1	73.6	73.6	73.7	73.7	73.7	73.7	73.7	74.3
	2	73.5	72.5	72.9	73.2	72.1	72.3	73.3	73.1

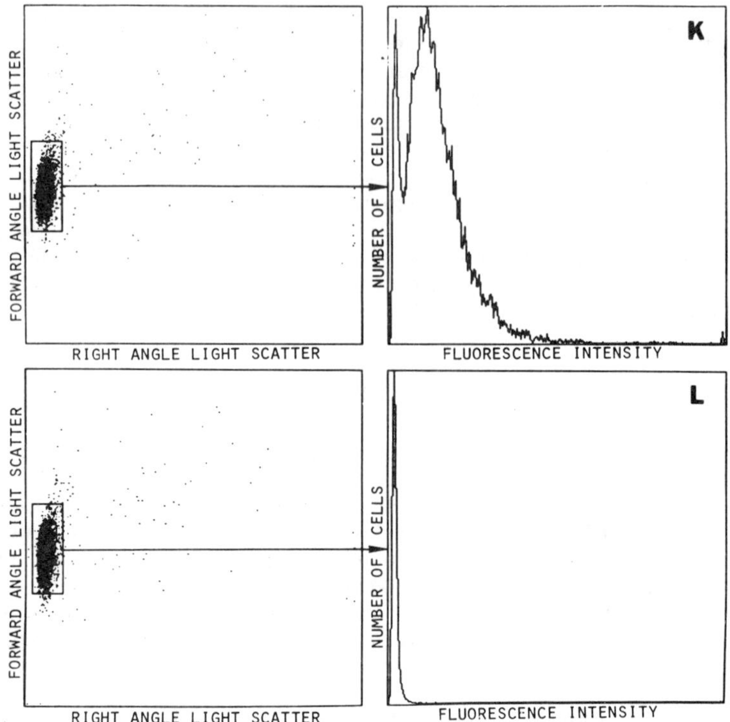

FIGURE 2. Blood from a patient with chronic lymphocytic leukemia. Almost all cells are in the lymphocytic area. Analysis of cell surface immunoglobulins shows that most cells express kappa (K) but not lambda (L) immunoglobulin light chains indicating that they are a monoclonal B-cell population.

prepared on the day the blood was drawn (Day 1). The other sample was kept at 4°C and prepared the next day (Day 2). T-lymphocytes were labeled with OKT3 monoclonal antibody (Ortho Diagnostic Systems, Westwood, MA). After the samples were run and the data stored, the analysis was performed by four laboratory technicians who were asked to select a lymphocyte window on the computer screen in each of the samples on two different occasions. The results show that the biologic variation among different donors was greater than the analytical variation of samples from each donor.

In leukemic samples in which the neoplastic cells did not differ morphologically from normal leukocytes (e.g. chronic leukemias), abnormal light scatter signals were not detected. In these cases, however, the presence of neoplastic cells in any of the normal clusters increased the density and/or altered the normal distribution of cells bearing various differentiation antigens within the cluster. FIGURE 2 illustrates an example of chronic lymphocytic leukemia, a B-cell leukemia. Virtually all cells analyzed fell in the lymphocytic cluster and the analysis of immunofluorescence revealed that most cells expressed a single light chain (kappa) indicating that they were monoclonal B-lymphocytes.

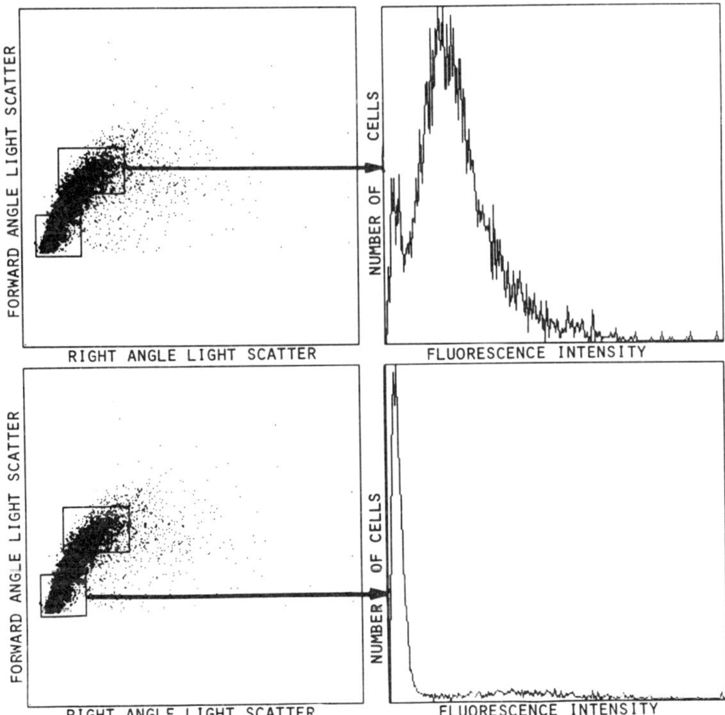

FIGURE 3. Analysis of leukocytes in a patient with acute leukemia. Light scatter analysis shows an elongated lymphocyte cluster resulting from the presence of numerous cells displaying higher light signals (upper window) than normal lymphocytes (lower window). Immunofluorescence analysis after exposure to J5, an antibody that recognizes an antigen present in acute lymphoblastic leukemia of the common type, shows that most cells in the upper window are positive while much fewer cells are fluorescent in the lower window.

In acute leukemias or in certain lymphomas with peripheral blood involvement, neoplastic cells have an abnormal morphology and are designated on blood smears as "blasts" (e.g., myeloblasts, lymphoblasts). Because of their abnormal morphology, these blasts form aberrant clusters on dual light scatter analysis. The analysis of fluorescence of cells in these clusters provided information on the antigenic profile of the leukemic cells. FIGURE 3 represents the analysis of leukocytes of a patient with possible acute leukemia. The conventional examination of blood smears revealed the presence of blasts of unknown nature. Dual light scatter analysis of blood leukocytes revealed an abnormally elongated lymphocyte cluster with a portion of the cells producing higher scatter signals than normal lymphocytes. Phenotype analysis of the cells in the abnormal cluster demonstrated that the majority of cells expressed J5 (Coulter Immunology, Hialeah, FL), an antigen present in common acute lymphocytic leukemia cells and rarely expressed in normal hemopoietic cells.[12] In FIGURE 4, blood of a patient with lymphoma was analyzed in this manner. On conventional smears, blasts of

 ANNALS NEW YORK ACADEMY OF SCIENCES

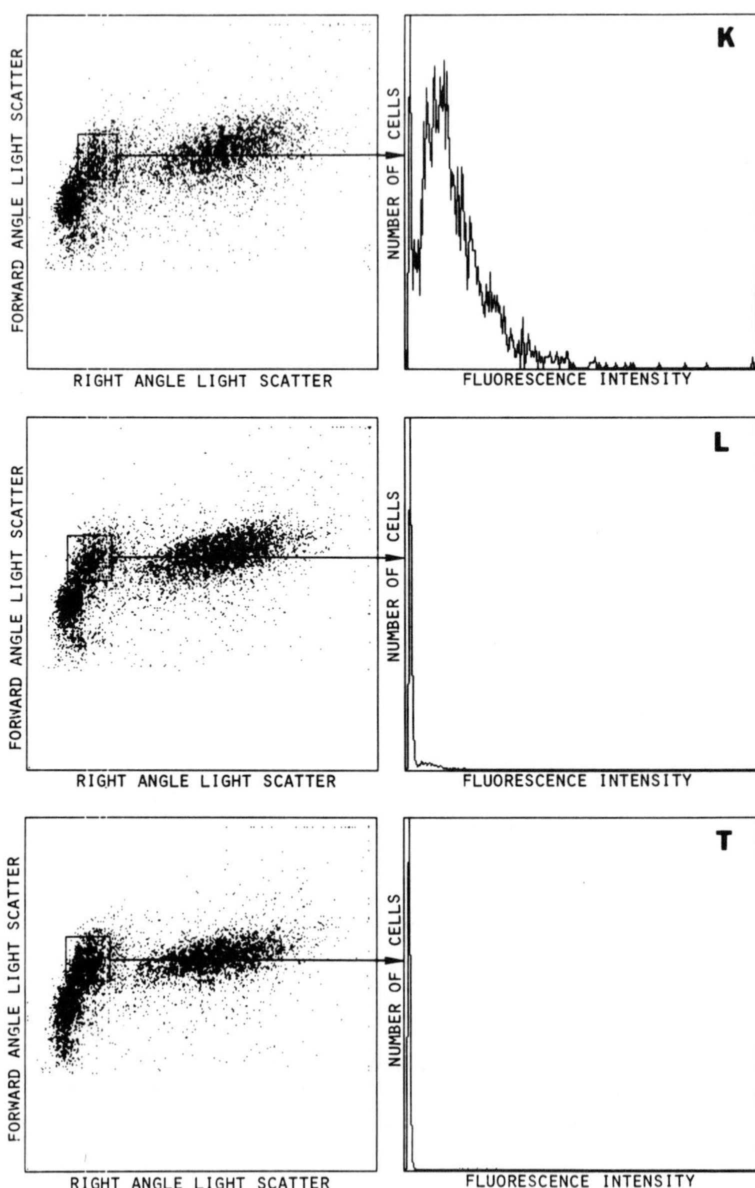

FIGURE 4. Identification of abnormal cells in peripheral blood of a patient with lymphoma. Blood contained "blasts" when examined by conventional cytologic methods. The windows enclose an abnormal cluster corresponding to the blast population. Analysis of fluorescence shows that virtually all cells in the blast window express kappa light chain (K). Lambda-bearing cells (L) are few and no T-cells (T) are seen.

uncertain nature were observed. Dual light scatter analysis revealed the presence of abnormal signals between the normal lymphocyte and monocyte clusters. A window placed around the abnormal cluster demonstrated that virtually all cells in the cluster expressed kappa (K) immunoglobulin. Lambda (L)-positive cells were scarce and T-cells were not present. In contrast, the cluster corresponding to normal lymphocytes contained mostly T-cells and very few immunoglobulin-bearing cells (data not shown). These observations demonstrated that the abnormal circulating cells detected by dual scatter analysis were monoclonal B-cells representing circulating lymphoma cells.

The measurement of small numbers of circulating monoclonal B-cells in patients with B-cell lymphomas in whom no leukemic cells were detectable by conventional hematologic techniques required a more complex analysis. In these cases neoplastic cells may be present in very low numbers so that no overt change in light scatter signals or in antigenic distribution is readily detectable. However, the presence of the neoplastic monoclonal B-cells should alter the normal balance between circulating kappa- and lambda-bearing lymphocytes. For analysis of these samples we applied the method of Ault *et al.*[13] This method uses a "D" value to measure the magnitude of the difference between kappa and lambda fluorescence distributions. The D value is derived from the Kolmogorov-Smirnov test[14] and represents the maximum difference between the kappa and lambda cumulative frequency distributions expressed as a percentage of the total cells analyzed. In 18 normal individuals studied in our laboratory, the D value was 3.1 ± 1.7 (SD). To determine the sensitivity of the technique, in two separate experiments we added known numbers of cells from B-cell lymphomas to normal unfractionated blood. TABLE 2 shows the results. In both experiments the D values exceeded the normal range when total leukocytes contained approximately 10% or more neoplastic cells. We also studied 6 patients with proven B-cell lymphoma and no evidence of blood involvement by conventional leukocyte counting or smear examination. The results shown in TABLE 3 suggest an imbalance of the kappa/lambda ratio in three cases.

TABLE 2. Sensitivity of D-Value

	% Lymphoma Cells Added	D-Value
Experiment 1	100	94
	50	46
	20	11
	10	6
	2	4
	0	5
Experiment 2	52	47
	15	9
	8	6
	4	2
	1	2
	0	4
Normal D-Value ($n = 18$)	3.1 + 3.4 (2 SD)	

TABLE 3. D-Values in Blood of Patients with B-cell Lymphomas

Diagnosis	D-Value
Diffuse mixed	3.9
Follicular large cell	0.9
Diffuse large cell	4.3
Follicular and diffuse mixed	9.1
Small lymphocytic	11.2
Small lymphocytic	10.2
Normal D-Value (n = 18)	3.1 + 3.4 (2 SD)

DISCUSSION

The analysis of surface antigens on leukemic cells using monoclonal antibodies against specific leukocyte differentiation antigens is becoming increasingly important in the diagnosis and classification of leukemias.[1-7] FCM is the method of choice for the analysis of leukocyte antigens because it provides rapid, objective and quantitative data on a large number of cells. In the past, measurements of immunofluorescence of normal leukocytes by FCM have been made in conjunction with forward (low angle) light scatter analysis which allows the measurement of fluorescence to be made on cells within a desired range.[15] However, since lymphocytes, monocytes, and granulocytes partially overlap in size, forward light scatter alone cannot discriminate well among the various types of leukocytes. Earlier studies showed that different types of leukocytes can be identified on the basis of the intensity of light scattered by the cells at two different angles (forward and right-angle) with respect to the light beam.[8] While forward low-angle light scatter relates mainly to size, right-angle light scatter relates to internal cell components such as cytoplasmic granules.[9] In this study we demonstrated that dual light scatter analysis can be used as an adjunct to immunofluorescence not only for the identification of normal leukocyte subsets but also for the rapid detection and characterization of human leukemia/lymphoma cells in peripheral blood.

The advantage of this approach is that it provides the means for rapidly establishing a direct relationship between structural and antigenic properties of a large number of leukocytes. The methodology that we used is extremely simple. The sample preparation is rapid, it uses a small amount of blood and does not require time-consuming blood fractionation, physical purification of cells or special cell identification procedures. Furthermore, the data can be stored permanently and retrieved as necessary for further interactive analysis.

The blood of most patients with B-cell lymphomas does not appear to contain neoplastic cells when examined by conventional cytological techniques.[16] However, when the distribution of cell surface immunoglobulins was analyzed, a high percentage of these patients with normal leukocyte counts showed an imbalance in the normal ratio between kappa- and lambda-bearing circulating cells (clonal excess).[13,17] Furthermore, this imbalance paralleled the changes observed in the lymphomatous lymph nodes from the same patients[17] strongly suggesting involvement of blood by neoplastic cells. These studies were performed on Ficoll-Hypaque separated mononuclear blood cells. Monocytes were a major source of nonspecific immunofluorescence and prevented accurate interpretation of results.[17]

In our study we used whole blood preparations and dual light scatter lymphocyte isolation for the quantitation of kappa and lambda-bearing lymphocytes thereby eliminating monocytes solely on the basis of their light scatter properties. Applying Ault's method[13] of calculating kappa/lambda imbalance, our results showed a similar sensitivity in the detection of monoclonal B-cells. Also, in three of six patients with B-cell lymphoma, our analysis suggested the presence of circulating neoplastic cells which were not detectable by conventional hematologic methods.

In summary, the technique described in this paper can be applied to both acute and chronic leukemias and is especially useful in the detection of low numbers of circulating neoplastic cells. The ability to detect a small number of leukemic or lymphoma cells is of particular importance not only in the initial diagnosis of these diseases but also in treated patients in whom the presence of residual or recurrent neoplasia may have important prognostic or therapeutic implications. At present, the number of monoclonal antibodies reacting with the various types of leukemic cells is limited. With increasing availability and improved specificity of immunologic reagents, a methodology such as the one described in this study will allow more sensitive and rapid detection of circulating neoplastic leukocytes.

SUMMARY

The detection of leukemic cells in peripheral blood is based on cytologic and cytochemical methods. Recently, the characterization of leukemic cells has been improved by the analysis of cell surface antigens. Abundant leukemic cells are relatively easy to identify. Small numbers of circulating leukemic cells, however, may be difficult to recognize by conventional or immunological techniques. This is of particular importance in treated patients in whom the presence of low levels of circulating leukemic cells may have considerable clinical relevance. We used multiparameter correlated flow cytometric analysis to detect and characterize human leukemia/lymphoma cells in small samples of unfractionated peripheral blood. Leukocyte surface antigens were labeled with monoclonal antibodies and simultaneous forward and right-angle light scatter and fluorescence signals were measured from each cell. An interactive computer program was written that permitted the two light scatter measurements to be displayed on the screen on a cell-by-cell basis (dot-plot). Subpopulations of interest could then be selected on the dot-plot for analysis of their fluorescence distribution. On the basis of their dual scatter properties and their antigenic profile, neoplastic leukocytes were recognized even in cases in which leukemic cells were infrequent and not detectable by conventional cytologic methods.

REFERENCES

1. RITZ, J., J. M. PESANDO, J. NOTIS-MCCONARTY, H. LAZARUS & S. F. SCHLOSSMAN. 1980. A monoclonal antibody to human acute lymphoblastic leukaemia antigen. Nature **283**:583–585.

2. REINHERZ, E. L., P. C. KUNG, G. GOLDSTEIN, R. H. LEVY & S. F. SCHLOSSMAN. 1980. Discrete stages of human intrathymic differentiation: Analysis of normal thymo-

cytes and leukemic lymphoblasts of T-cell lineage. Proc. Natl. Acad. Sci. USA 77:1588–1592.

3. SALLAN, S. E., J. RITZ, J. PESANDO, R. GELBER, C. O'BRIEN, S. HITCHCOCK, F. CORAL & S. F. SCHLOSSMAN. 1980. Cell surface antigens: Prognostic implications in childhood acute lymphoblastic leukemia. Blood 55:395–402.

4. SCHROFF, R. W., K. A. FOON, R. J. BILLING & J. L. FAHEY. 1982. Immunologic classification of lymphocytic leukemias based on monoclonal antibody-defined cell surface antigens. Blood 59:207–215.

5. BERNSTEIN, I. D., R. G. ANDREWS, S. F. COHEN & B. E. McMASTER. 1982. Normal and malignant human myelocytic and monocytic cells identified by monoclonal antibodies. J. Immunol. 128:876–881.

6. MALCOM, A. J., R. C. SHIPMAN & J. G. LEVY. 1982. Detection of a tumor-associated antigen on the surface of human myelogenous leukemia cells. J. Immunol. 128:2599–2603.

7. GRIFFIN, J. D., R. J. MAYER, H. J. WEINSTEIN, D. S. ROSENTHAL, F. S. CORAL, R. P. BEVERIDGE & S. F. SCHLOSSMAN. 1983. Surface marker analysis of acute myeloblastic leukemia: Identification of differentiation-associated phenotypes. Blood 62:557–563.

8. SALZMAN, G. C., J. M. CROWELL, J. C. MARTIN, T. T. TRUJILLO, A. ROMERO, P. F. MULLANEY & P. M. LaBAUVE. 1975. Cell classification by laser light scattering identification and separation of unstained leukocytes. Acta Cytologica 19:374–377.

9. HANSEN, W. P., R. A. HOFFMAN & K. W. HEALEY. 1982. Light scatter as an adjunct to cellular immunofluorescence in flow cytometric systems. J. Clin. Immunol. 2(Suppl):32–41.

10. HOFFMAN, R. A., P. C. KUNG, W. P. HANSEN & G. GOLDSTEIN. 1980. Simple and rapid measurement of human T lymphocytes and their subclasses in peripheral blood. Proc. Natl. Acad. Sci. USA 77:4914–4917.

11. BRAYLAN, R. C., N. A. BENSON, V. NOURSE & H. KRUTH. 1982. Correlated analysis of cellular DNA, membrane antigens and light scatter of human lymphoid cells. Cytometry 2:337–343.

12. RITZ, J., L. M. NADLER, A. K. BHAN, J. NOTIS-McCONARTY, J. M. PESANDO & S. F. SCHLOSSMAN. 1981. Expression of common acute lymphoblastic leukemia antigen (CALLA) by lymphomas of B-cell and T-cell lineage. Blood 58:648–652.

13. AULT, K. A. 1979. Detection of small numbers of monoclonal B lymphocytes in the blood of patients with lymphoma. N. Engl. J. Med. 300:1401–1405.

14. YOUNG, I. T. 1977. Proof without prejudice: Use of the Kolmogorov-Smirnov test for the analysis of histograms from flow systems and other sources. J. Histochem. Cytochem. 25:935–941.

15. SLEASE, R. B., R. WISTAR, JR., & I. SCHER. 1979. Surface immunoglobulin density on human peripheral blood mononuclear cells. Blood 54:72–87.

16. COME, S. E., E. S. JAFFE, J. C. ANDERSEN, R. B. MANN, B. L. JOHNSON, V. T. DE VITA, JR. & R. C. YOUNG. 1980. Non-Hodgkin's lymphomas in leukemic phase: Clinico-pathologic correlations. Am. J. Med. 69:667–672.

17. LIGLER, F. S., R. G. SMITH, J. R. KETTMAN, J. A. HERNANDEZ, J. B. HIMES, E. S. VITETTA, J. W. UHR & E. P. FRENKEL. 1980. Detection of tumor cells in the peripheral blood of nonleukemic patients with B-cell lymphoma: Analysis of "clonal excess." Blood 55:792–801.

The Use of Flow Cytometry in the Diagnosis and Biological Characterization of the Non-Hodgkin's Lymphomas

STANLEY E. SHACKNEY[a]

Laboratory of Theoretical and Physical Biology
National Institute of Child Health and Human Development
National Institutes of Health
Bethesda, Maryland 20205

and

Division of Medical Oncology
Allegheny General Hospital
Pittsburgh, Pennsylvania 15212

INTRODUCTION

The non-Hodgkin's lymphomas have proved to be especially well-suited for study by means of flow cytometry (FCM). Malignant lymphoid tissues can be disaggregated into suspensions of intact, whole single cells with relative ease. Like other malignancies of the hematopoietic system, the lymphomas often have distinctive enzymatic and membrane surface markers that lend themselves readily to FCM studies. Finally, the lymphomas have distinctive cytomorphologic features that can be correlated with their clinical behavior; these cytomorphologic features are amenable to flow cytometric analysis.

To date, we have analyzed over 250 cases of human lymphoma, both at our own institution and in collaboration with Drs. Robert Lukes and John Parker at the University of Southern California. I would like to focus here on two aspects of our flow cytometry studies, namely, the detection of aneuploidy in the lymphomas and the estimation of cell proliferative rates in these malignancies.

ANEUPLOIDY IN THE LYMPHOMAS

Many human tumors consist of clonal proliferations of cytogentically abnormal cells.[1] In contrast, normal host cells are invariably diploid. Thus, the finding of aneuploidy in a clinical specimen constitutes strong evidence for the diagnosis of neoplasia.

---

[a]Address reprint requests to Division of Medical Oncology, Allegheny General Hospital, 320 East North Avenue, Pittsburgh, PA 15212.

Technical Considerations: Choice of Fluorescent Stain

We began our ploidy studies in the mid-1970s with the acriflavine-Feulgen stain. The drawbacks of this stain were extensive cell loss during the staining procedure (up to 90%), a high coefficient of variation of the DNA measurement (6%–7%), and only fair reproducibility of the position of the G_1 peak of normal cells. Another major disadvantage was that cells deteriorated in the fixative over the course of time. The major advantage of the stain was its compatibility with formalin fixation; the formalin fixative provided for better electronic cell volume measurements than ethanol.

Mithramycin, on the other hand, required ethanol fixation, which tended to degrade the electronic cell volume signal, but had other advantages. When we compared mithramycin with propidium iodide, both produced comparable coefficients of variation of the measurement (range 2%–4%), but we found mithramycin to be superior to propidium iodide with regard to reproducibility of the position of the G_1 peak of normal cells. Specifically, when freshly fixed, normal lymphocytes were split into multiple separate aliquots that were stained at the same time with mithramycin and measured sequentially, the variations in mean G_1 peak position were in the range of ± 3 percent. The small degree of intersample staining variability has permitted the use of a separate diploid reference standard consisting of normal lymphocytes run before and after each clinical sample.[2] With propidium iodide staining, the variability among sequentially measured (separately stained) aliquots of the same cell suspension was in the range of ± 7 percent. To minimize the effects of such staining variability with propidium iodide, an internal standard can be used (e.g., nucleated chicken erythrocytes admixed with the clinical sample[3,4]). Nonetheless, this variability may be a potential source of error in the clinical detection of aneuploidy with propidium iodide (see below). Various considerations in the choice of fluorescent stains for the detection of aneuploidy by FCM are listed in TABLE 1.

Staining Artifacts

During the course of our studies with mithramycin as a DNA stain, we have come to recognize a variety of staining artifacts that are associated with this dye.[5]

TABLE 1. Considerations in the Choice of Fluorescent Stain for the Detection of Aneuploidy by FCM

1. Magnitude of the coefficient of variation of the measurement (G_1 peak width/height).
2. Reproducibility of the ploidy index[a] (variation in G_1 peak position).
3. Idiosyncratic staining artifacts.
4. Ease and rapidity of staining.
5. Compatibility with other stains for multi-parameter studies—absorption/emission characteristics.
6. Compatibility with other stains for multi-parameter studies—fixative requirements.
7. Stability of cells in fixature over time.

$$^a\text{Ploidy index} = \frac{\text{tumor } G_1 \text{ peak channel}}{\text{diploid reference } G_1 \text{ peak channel}}$$

Two of these are of potential importance in the clinical detection of aneuploidy in the lymphomas.

There is a progressive increase in mithramycin fluorescence with increasing storage time of cells in ethanol at 4°C. This fixation time-dependent hyperchromatism can produce spurious shifts in the position of the G_1 peak of the sample and the diploid reference cells where they are stored in fixative for different lengths of time. We have found that this hyperchromatic shift proceeds more rapidly in fixative at room temperature and is essentially complete within 2 hours at 37°C. We routinely incubate both the diploid reference standard and the clinical specimen in fixative at 37°C for 2 hours prior to staining and analysis. This incubation has virtually eliminated artifactual fixation time-dependent shifts in the G_1 peak in clinical samples, without adversely affecting the detectability of true aneuploidy.

Our experience with the lymphomas to date would indicate the following:

1. Virtually every clinical sample contains at least some diploid cells that provide an internal diploid reference.
2. The position of the endogenous diploid G_1 peak in clinical samples corresponds with that of the external diploid reference with a high degree of reproducibility.
3. True hypodiploidy detected by flow cytometry must be rare in the lymphomas.

The second mithramycin staining artifact that is of potential importance in the detection of aneuploidy is a small spurious hyperdiploid peak that is often present in the DNA histogram of circulating mononuclear cells of normal individuals. This spurious side peak disappears after incubation of the cells in fixative at 37°C for 2 hours prior to staining. True aneuploidy is not affected by this incubation procedure.

Criteria for the Detection of Aneuploidy

There are at least three criteria in common use for the detection of flow cytometry (TABLE 2). We do not consider all of these criteria to be equally reliable.

As noted above, with mithramycin staining we have found at least some diploid cells in every clinical sample. In all samples (normal and neoplastic) containing a single G_1 peak, the mean ploidy index was 0.998 ± 0.015 (standard deviation). Ninety-five percent of cases fell within 3% of the mean (2 standard deviations), and all cases fell within 5.5% of the mean.[6] Thus, our experience does not support the validity of criterion 3 in TABLE 2 for the detection of aneuploidy.

TABLE 2. Criteria for Aneuploidy

1. Multiple G_1 peaks
2. Single G_1 peak normal peak position, increased CV[a]
3. Single G_1 peak, abnormal peak position, normal CV

[a]CV, coefficient of variation.

TABLE 3. Aneuploidy in the Lymphomas in Relation to Criteria for Aneuploidy

Author	Stain	Normal Variation in G_1 Peak Position[a]	Overall Frequency of Aneuploidy	Frequency of Aneuploidy	
				Criterion 3 Only[b]	Criterion 1 Only[b]
Shackney et al.[6]	Mithramycin	Diploid peak of normal tissues: ±3%	70/220 (31%)	0	57/220 (26%)
Diamond & Braylan[7]	Propidium iodide	±5.7%	22/30 (73%)	3/30 (10%)	7/30 (23%)
Costa et al.[8]	Propidium iodide	±11.7%	45/74 (61%)	34/74 (46%)	11/74 (15%)

[a]±Two standard deviations.
[b]Criteria defined in TABLE 3.

When comparing reported frequencies of aneuploidy in the lymphomas among different published series of cases, it should be kept in mind that there are differences in DNA stains used and that the reproducibility of the G_1 peak position may vary from stain to stain and from series to series. Some investigators consider criterion 3 in TABLE 2 as valid for the detection of aneuploidy by FCM, while others do not. A summary of pertinent data from several recently published studies is given in TABLE 3.

Clearly, the best criterion for aneuploidy by FCM is the presence of one or more distinct, abnormally positioned G_1 peaks in addition to a normally positioned G_1 peak in the same clinical specimen. The frequency of aneuploidy by this criterion alone varies from 15% to 26% in published series (TABLE 3). Our own experience indicates that the frequencies of aneuploidy by this criterion differ in B- and T-cell lymphomas, and differ within histologic subtypes of the B-cell lymphomas. The highest frequency of aneuploidy is observed among the large B-cell lymphomas (72%) and lowest among the indolent T-cell lymphomas (7%). These data are summarized in TABLE 4.

A G_1 peak with a high coefficient of variation (CV) may represent the overlap of closely spaced diploid and aneuploid peaks; however, a high CV can also be the result of instrument malfunction (e.g., misaligned optics, an unstable sample stream, or an aging laser plasma tube). Thus, a high CV of the G_1 peak in the clinical sample should be compared to that of a normal diploid reference standard run just before and/or just after the sample of interest. It is also worthwhile to stain and analyze a separate aliquot from the same clinical sample to confirm that the high CV of the G_1 peak is a reproducible finding.

Several techniques are available for resolving aneuploid and diploid G_1 peaks with single G_1 peaks with high CVs are encountered. Kruth et al. have reported that minimally aneuploid G_1 peaks can be detected in studies employing combined cell DNA and immunologic surface marker measurements.[7] Schuette et al.[8] have recently developed a mathematical method for resolving closely spaced diploid and aneuploid peaks and multiple aneuploid peaks when such peaks are present in the data.

TABLE 4. Aneuploidy by Flow Cytometry in the Non-Hodgkin's Lymphomas[6]

	Aneuploidy (*n*)	Criteria 1 and 2 %
B-cell lymphomas	(62/174)	36
Large B-cell lymphomas	(18/25)	72
B-cell lymphomas of intermediate cell size	(13/20)	65
Small B-cell lymphomas	(25/85)	29
Small B-cell lymphomas with maturation/ differentiation	(6/44)	14
T-cell lymphomas	(8/46)	17
Aggressive T-cell lymphomas	(7/30)	23
Indolent T-cell lymphomas	(1/16)	7

The Clinical Significance of Aneuploidy by Flow Cytometry

To date, we have studied two cases (one a lymph node, the other a sample of ascites fluid) that were diagnosed pathologically as benign granulomatous processes in which distinct aneuploid G_1 peaks were found by FCM. It is not clear to us whether these cases represented (a) benign clonal proliferations, (b) premalignant clonal proliferations, (c) early frankly malignant clonal proliferations, or (d) mistaken pathologic diagnoses.

We have adopted the practice of reporting findings of aneuploid G_1 peaks in lymphoid tissues as evidence for *neoplasia*, i.e., new clonal growths. New clonal growths may be clinically benign or malignant. When other diagnostic features are present (e.g., two or more aneuploid peaks or an aneuploid population consisting of abnormally large cells with a high S fraction), we suggest that these findings are *consistent with malignancy*. Of course, the absence of demonstrable aneuploidy by FCM does not rule out neoplasia or malignancy.

THE S FRACTION BY FLOW CYTOMETRY

The fraction of cells in the S phase of the cell cycle can be determined from the DNA histogram by any one of several methods that have been developed for this purpose. We have devised a graphical method that can be used in the presence of aneuploidy.[9] Using this method, all samples that consist of nonseparable mixtures of diploid and aneuploid populations are subject to analysis; failure to include such samples would introduce significant biases, since the aneuploid populations tend to have higher S fractions than the diploid ones.[6,10] We have also carried out multiparameter studies that have permitted the separate analysis of aneuploid and diploid subpopulations that have high S fractions.

In our studies, we have found that the large B-cell lymphomas have significantly higher S fractions than the small B-cell lymphomas or the T-cell lymphomas.[6] Furthermore, within cell populations of a given ploidy class, the subpopulations consisting of larger cells have been shown to have higher S fractions than the subpopulations of smaller cells.[10]

It is generally safe to assume that a high S fraction implies a high proliferative rate. In theory, both the duration of S and the duration of the total cell cycle time could be prolonged, in which case a high S fraction would *not* reflect a high proliferative rate; however, if the duration of S were prolonged, then one might expect that the average rate of DNA synthesis during S would be low. In contrast, if high S fractions were always associated with short cell cycle times, then high rates of DNA synthesis throughout most of S would be expected. In order to distinguish these two possibilities, we have carried out [3]HTdR labeling studies on some of the same human lymphomas that we have studied by flow cytometry, and we have performed image analysis studies on Feulgen-stained radioautographs of these cells. Although the number of cases analyzed to date is small, our experience so far would indicate the following:

1. It is clear that lymphomas with high S fractions by FCM also have high [3]HTdR labeling indices, and that mean grain counts per cell are also high in these cases. Lymphomas with low S fractions by FCM have low [3]HTdR labeling indices and low mean grain counts per cell when all cases are processed under the same experimental conditions. This would tend to confirm the premise that high S fractions by FCM imply high rates of cell proliferation.
2. Within individual cases of lymphoma, large cells in mid-S have higher grain counts than small cells in mid-S. This observation supports earlier FCM observations that smaller cells have lower proliferative rates than large cells within the same population.[10]

We have carried out multiparameter studies in the non-Hodgkin's lymphomas in which we have found repeatedly that B-cell lymphomas often contain mixtures of diploid and aneuploid subpopulations, where the aneuploid subpopulations consist of larger cells and have higher S fractions.[6,10] This would suggest that the lymphomas evolve by clonal selection, where clones with progressively higher proliferative rates overgrow abnormal clones with lower proliferative rates and normal cells. It would appear that in the B-cell lymphomas clones with higher proliferative rates also produce cells of larger size.

These observations may have far-reaching implications with regard to lymphoma classification and with regard to our understanding of the clinical and pathologic evolution from so-called favorable histologies to unfavorable histologies over time in individual patients. At the present time, it would seem reasonable to suppose that histopathologic transformation of the lymphomas represents a manifestation of clonal evolution. Future studies will test this hypothesis, and perhaps, they may provide insights into the underlying genetic mechanisms of clonal evolution in tumors.

SUMMARY

The detection of aneuploidy and the estimation of the fraction of cells in S in DNA histograms from patients with human non-Hodgkin's lymphoma are reviewed. Karyotype studies and DNA histograms each have advantages and disadvantages in the detection and monitoring of aneuploidy. The choice of fluorescent stain, staining artifacts, and the criteria for the detection of aneuploidy by FCM must be considered carefully. In general, the B-cell lymphomas are more frequently aneuploid by FCM than the T-cell lymphomas. When determining S fractions in the lymphomas, care must be taken not to exclude the aneuploid cases on methodological grounds; these cases generally have the highest S fractions and their exclusion would bias the data. Multiparameter studies have shown that the most aneuploid component of a mixed clinical sample generally has the highest S fraction in the sample, favoring the concept of clonal selection and clonal evolution of tumors.

REFERENCES

1. NOWELL, P. C. 1976. The clonal evolution of tumor cell populations. Science **193**:23–28.
2. SHACKNEY, S. E., B. W. ERICKSON & K. S. SKRAMSTAD. 1979. The T lymphocyte as a diploid reference standard for flow cytometry. Cancer Res. **39**:4418–4422.
3. COSTA, A., G. MAZZINI, G. DELBINO & R. SILVESTRINI. 1981. DNA content and kinetic characteristics of non-Hodgkin's lymphoma: Determined by flow cytometry and autoradiography. Cytometry **2**:185–188.
4. DIAMOND, L. W. & R. C. BRAYLAN. 1980. Flow analysis of DNA content and cell size in non-Hodgkin's lymphoma. Cancer Res. **40**:703–712.
5. CUNNINGHAM, R. E., K. S. SKRAMSTAD, A. E. NEWBURGER & S. E. SHACKNEY. 1982. Artifacts associated with mithramycin fluorescence in the clinical detection and quantitation of aneuploidy by flow cytometry. J. Histochem. Cytochem. **30**:317–322.
6. SHACKNEY, S. E., A. M. LEVINE, R. I. FISHER, P. NICHOLS, E. JAFFE, W. H. SCHUETTE, R. SIMON, C. A. SMITH, S. I. OCCHIPINTI, J. W. PARKER, J. COSSMAN, R. C. YOUNG & R. J. LUKES. 1984. The biology of tumor growth in the non-Hodgkin's lymphomas: A dual parameter study of 220 cases. J. Clin. Invest. **73**:1202–1214.
7. KRUTH, H. S., R. C. BRAYLAN, N. A. BENSON & V. A. NOURSE. 1981. Simultaneous analysis of DNA and cell surface immunoglobulin in human B-cell lymphomas by flow cytometry. Cancer Res. **41**:4895–4899.
8. SCHUETTE, W. H., S. E. SHACKNEY, M. MACCOLLUM & C. A. SMITH. 1983. A high resolution method for DNA histogram analysis that is suitable for the detection by flow cytometry of multiple closely spaced aneuploid cell populations in clinical samples. Cytometry **3**:376–386.
9. RITCH, P. S., S. E. SHACKNEY, W. H. SCHUETTE & C. A. SMITH. 1983. A practical graphical method for estimating the fraction of cells in S in DNA histograms from clinical tumor cells containing aneuploid cell populations. Cytometry **4**;66–74.
10. SHACKNEY, S. E., K. S. SKRAMSTAD, R. E. CUNNINGHAM, D. J. DUGAS, T. L. LINCOLN & R. J. LUKES. 1980. Dual parameter flow cytometry studies in human lymphomas. J. Clin. Invest. **66**:1281–1294.

Comparison of Morphologic Features and Mitotic Rate to Cytometrically Determined DNA Content of Poorly Differentiated Lymphocytic Lymphomas

MARGARITA PALUTKE,[a] BERTRAM SCHNITZER,[b]
DEBRA DRESNER,[a] PAMELA TABACZKA,[a]
VORAVIT RATANATHARATHORN,[c]
LEOPOLDO EISENBERG,[d] AND DAVID TENENBAUM[a,e]

[a]Department of Pathology
Section of Hematopathology
Berman Memorial Laboratories
Wayne State University School of Medicine
Detroit, Michigan 48201

[b]Department of Pathology
University of Michigan
Ann Arbor, Michigan 48104

[c]Department of Oncology
Wayne State University School of Medicine
Detroit, Michigan 48201

[d]Departments of Hematology and Oncology
Sinai Hospital
Detroit, Michigan 48235

INTRODUCTION

Several groups of investigators[1-3] have demonstrated that DNA content analysis of lymphomas of low, intermediate, and high grade malignancy[4,5] showed good correlation between S phase values and aggressiveness of the lymphomas. There was considerable overlap of values, however, between intermediate and high grade lymphomas. As part of a larger study of more than 80 cases of poorly differentiated lymphocytic lymphomas (PDL) we were able to perform DNA histograms of 22 cases. Working within the Rappaport classification,[6] we wanted to determined how morphologic criteria such as major cell type, number of

[e]No longer at Wayne State University School of Medicine; Present Affiliation: U.S. Army, Tank Automotive Command, Warren, MI 48397.

transformed cells, blasts, and numbers of mitotic figures correlated with S phase values. For this we chose a group of poorly differentiated lymphocytic lymphomas of different morphologic subtypes.

MATERIALS AND METHODS

Light Microscopic Studies

The fresh cut surface of a lymph node was gently touched to clean microscope slides. These imprints were air dried and stained with Leishman's stain. Transverse sections of the lymph node (2–3 mm in thickness) were fixed in B5 fixative for routine histologic processing. Four to five micron sections were stained with hematoxylin-eosin (H&E), periodic acid Schiff (PAS), and methyl green pyronin (MGP). For the present study special attention was paid to the growth pattern (nodular and/or diffuse), the morphologic characteristic of the major tumor cell population, the number of large transformed lymphocytes and blasts, and number of mitoses per high-power field. The latter was determined by counting mitoses in 10 high-power fields.

Immunologic Studies

The methods for preparing mononuclear cell suspensions and evaluation of surface immunoglobulin and other antigenic markers and receptors for unsensitized sheep red cells, complement (C), and immunoglobulin (Fc) have previously been described.[1-9] Cytocentrifuge preparations of all cell suspensions and the rosette tests were made using a Shandon-Elliot cytospin and were stained with Leishman's stain.

DNA Analysis

In all cases DNA histogram analysis was performed using cells obtained from teased lymphoid tissue and peripheral blood Ficoll-Hypaque monolayers. All cells were fixed in 70% ethanol prior to testing. Chicken erythrocyte nuclei were used as an external standard for the histograms. The nuclei were prepared by washing the whole chicken cells with $CaCl_2$ (0.332 g/L saline) and lysing the cell membrane with saponin (50 mg/100 ml saline). Free nuclei were then rewashed with $CaCl_2$ and fixed in 70% ethanol until ready to use. Sample cell concentrations were adjusted to 4×10^6 cells/ml to which 1.2×10^5 cells/ml of the chicken erythrocyte nuclei were added.[1] Treatment with RNase (1 mg/ml) at 37°C for 30 minutes followed to eliminate double-stranded RNA. The nuclear DNA was stained with propidium iodide (5 mg/100 ml) in 0.1% sodium citrate at 4°C for 20 minutes.[10] Prior to analysis, each sample was filtered through a 37 μm nylon mesh to remove cellular debris then sonicated to minimize clumping.

Analysis on 12 cases was performed on a 128 channel Coulter TPS-1 flow cytometer using the 388 nm line of a 5W argon ion laser at 500 mW constant light output for fluorescent excitation. A 590 nm long pass dichroic filter was placed in front of the red photomultiplier tube to allow the passage of red fluorescent

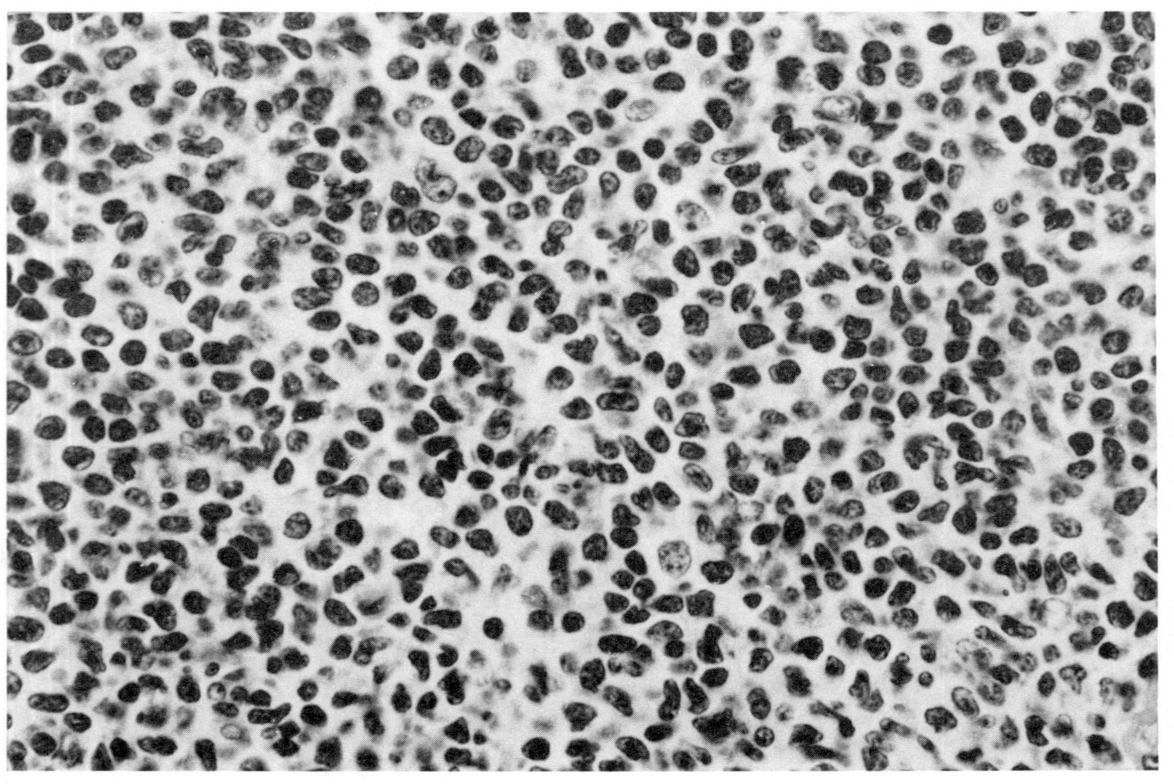

FIGURE 1. PDL of small, cleaved follicular center lymphocytes (Group 1). Few large, transformed lymphocytes and very few mitoses. H&E; × 540.

emissions for the DNA measurements. Data were analyzed on an Amdahl 470V/6 and 470V/8 computer using the DNA histogram program kindly supplied by Dr. Philip Dean of Livermore Laboratories in California.[11]

Analysis on 10 cases was performed on a 256 channel FACS 440 (Becton Dickinson, Sunnyvale, CA) flow cytometry system using 100nW of constant light output from the same laser. A 488 nm blocking and 575/25 nm narrow band pass filter system was placed in front of the red photomultiplier tube to detect the emissions. Data analysis was performed on a PDP-11/23 microcomputer Consort 40 (Becton Dickinson, Sunnyvale, CA) using a similar program also supplied by Dr. Philip Dean.

RESULTS

Light Microscopic Findings

The 22 cases were divided into three histological groups: (1) nodular PDL composed predominantly of small, cleaved lymphocytes[12] (11 cases); (2) follicular mantle zone lymphoma and those of intermediate differentiation[13-15] (6 cases); and (3) "blastic" PDL (5 cases) in which blasts were more numerous than cleaved lymphocytes.[16] Mitoses in Group 1 ranged from 0.1 (FIG. 1) to 5.7/HPF (FIG. 2) (mean = 2.5/HPF) and the percentage of large cells or blasts correlated well with the number of mitoses (TABLE 1). In Group 2 the number of mitoses was low in five cases: 0.2 − 1.1/HPF (mean = 0.7/HPF), but the percentage of large cells was high because these cells were present in remnants of normal follicle centers and in malignant pseudo-follicular proliferation centers (FIG. 3). Case 17 in this group differed from the other five cases in that both the mitotic rate (4.8/HPF) and the percentage of cells in S phase (10%) were considerably higher. Cytologic examination of the lymph node (FIG. 4) and blood revealed many prolymphocytes in addition to the usual mixture of round and small cleaved lymphocytes, characteristic of this type of lymphoma. Mitoses were seen predominantly in cells of the same size as the prolymphocytes rather than in large cells. In Group 3, blasts predominated although a minor cell population consisted of cleaved lymphocytes (FIG. 5). This group had the highest mitotic rate (4.5 − 8.8/HPF; Mean = 6.1/HPF).

Immunologic Findings

As seen in TABLE 1, all the cases were of B-cell type. The percentage of T cells varied from 7 to 37 percent. They were considered a nonneoplastic component of the lymphoid tissue and their numbers did not appear to influence the size of the S phase compartment.

Cell Cycle Analysis

The correlation of the percentage of cells in the S phase of the cell cycle and number of mitoses and types of tumor cells are given in TABLE 1. Control values are given in TABLE 2. A DNA histogram of a lymph node with benign reactive follicular hyperplasia is seen in FIGURE 6.

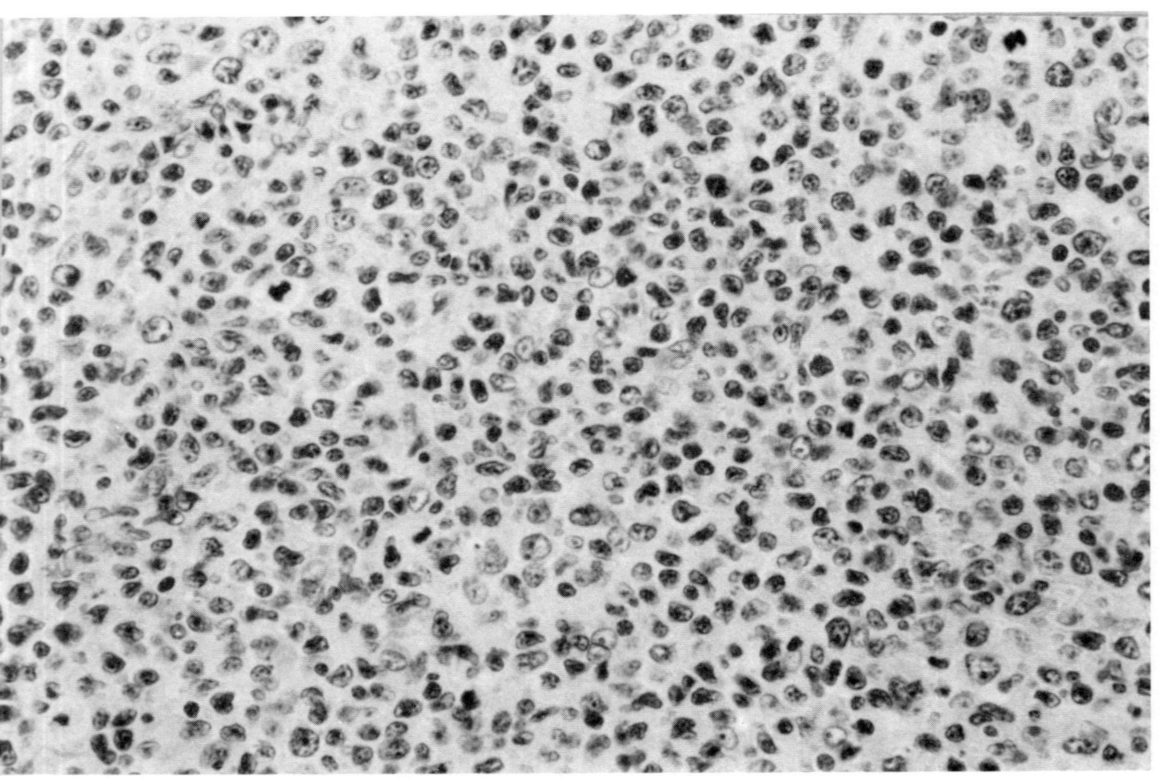

FIGURE 2. PDL from Group 1 with 20–30% large, transformed lymphocytes and high mitotic rate. H&E; × 400.

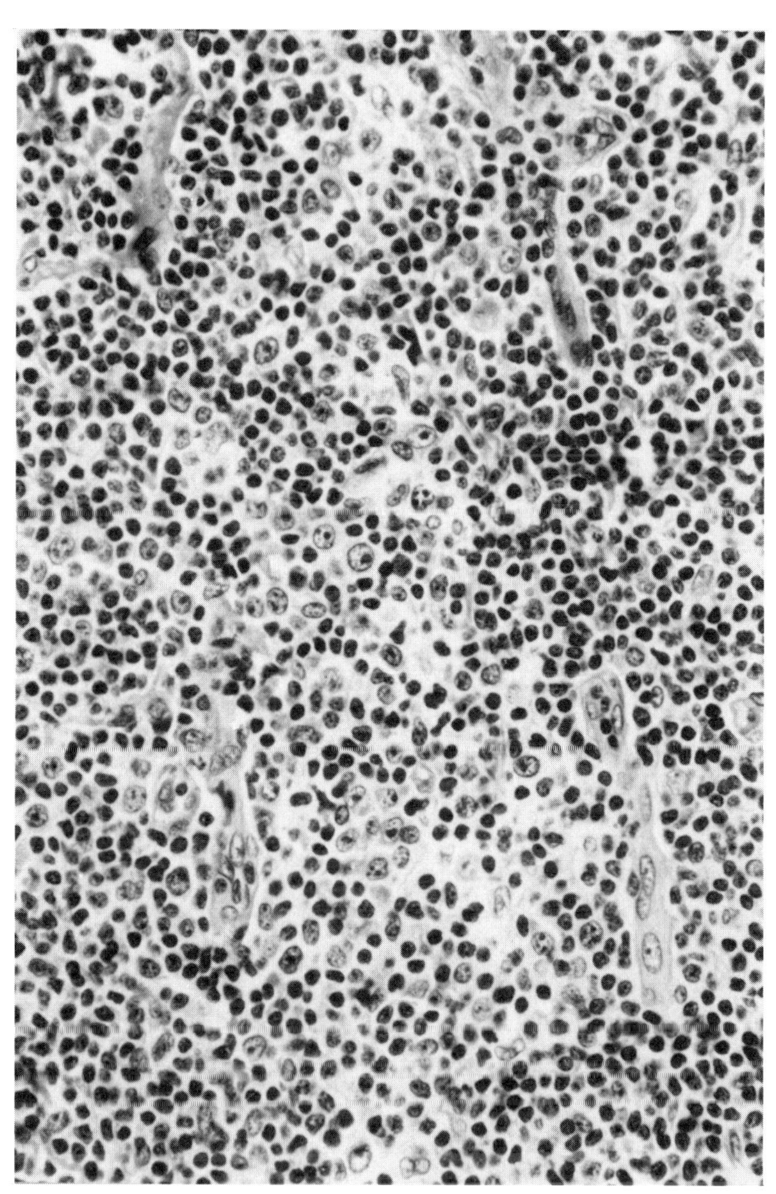

FIGURE 3. Lymphoma of intermediate differentiation (Group 2). Many large cells in an area of pseudofollicular transformation centers. H&E; × 400.

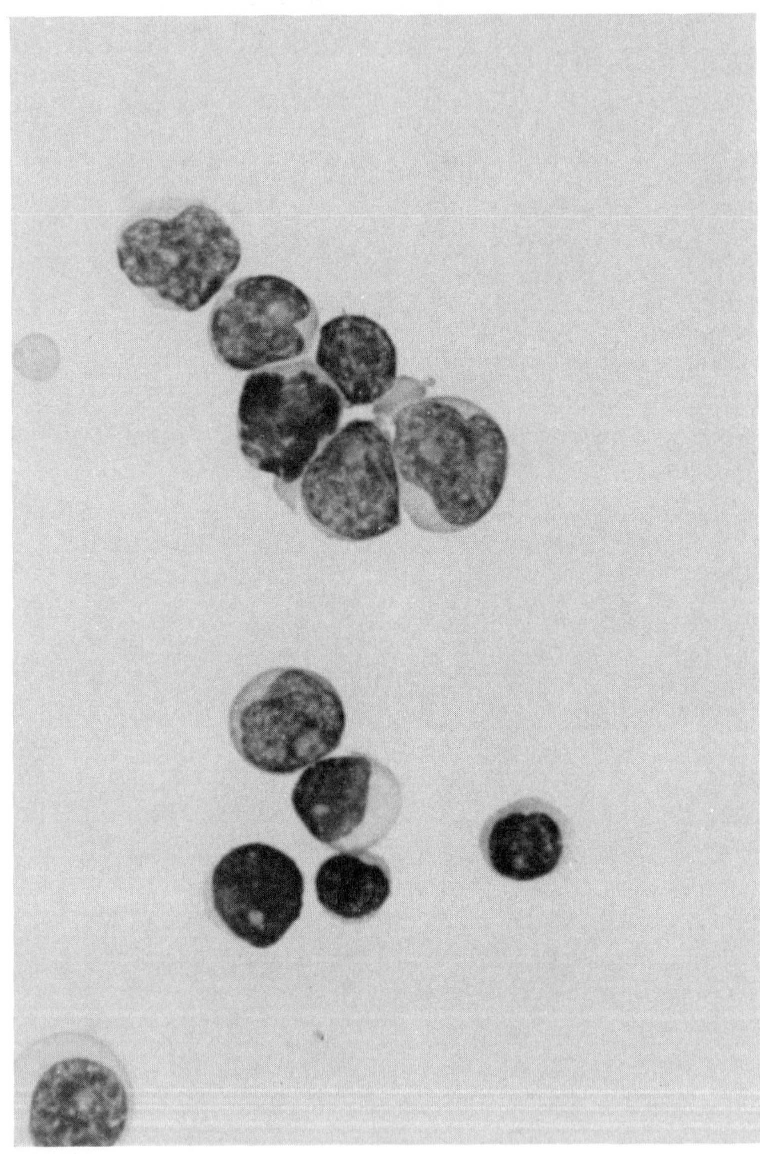

FIGURE 4. Cytocentrifuge preparation of cell suspension from lymph node of Case 17 (Group 2). There are many prolymphocytes in addition to the usual mixture of small, round, and cleaved lymphocytes. Note mitotic figure in a cell the size of a prolymphocyte. Leishman's stain; × 1008.

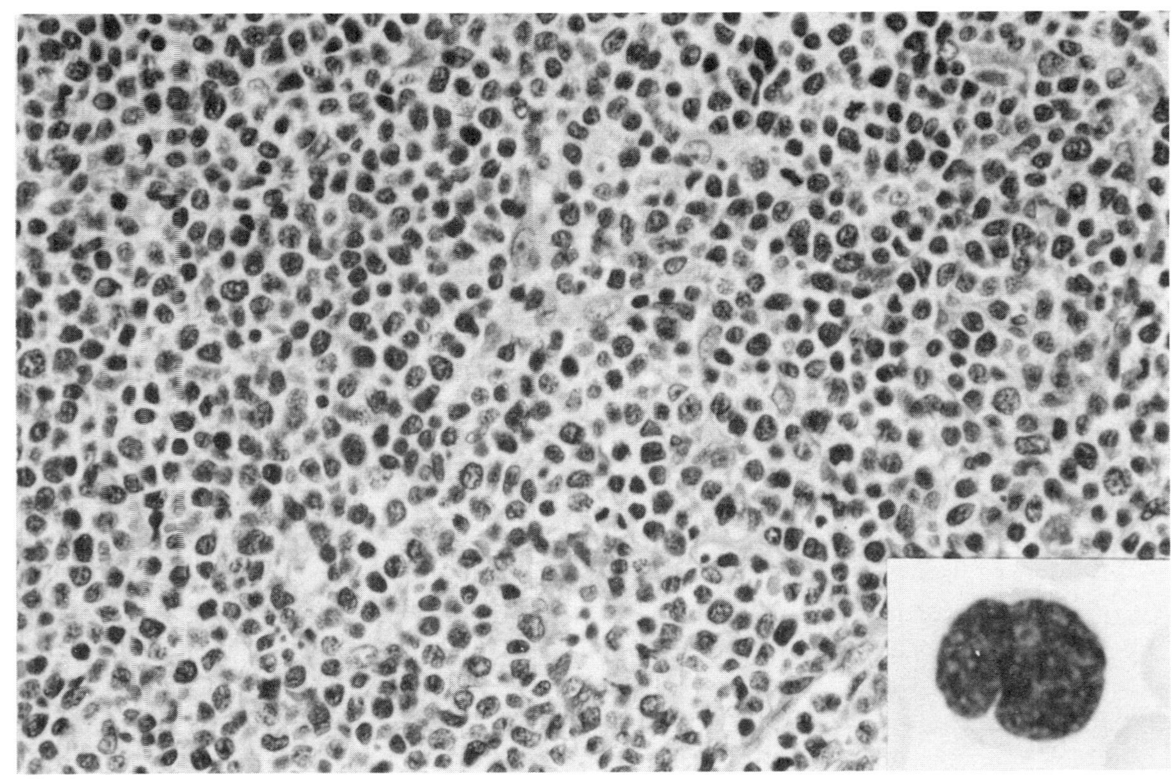

FIGURE 5. "Blastic variant" of PDL (Group 3). Blasts may have round or irregular nuclei. High motitic rate. H&E; × 400. *Insert:* Similar blasts are seen in peripheral blcod. Case 21. Leishman's stain; × 1600.

TABLE 1. Correlation of Histologic Criteria and Cell Cycle Analysis

Subtype of PDL	Case Number	Instrument	Immunology Surface Markers	% T Cells
1. Small Cleaved Follicular Center Cell	1	C	IgM,L,C	27
	2	C	IgM,L,D,C'	13
	3	BD	82% B_1^+ cells	18
	4	C	Fc,C'	35
	5	BD	IgM,K,Fc,C'	37
	6	BD	IgM,K	25
	7	C	IgM,K	18
	8	BD	IgM,L,C'	20
	9	BD	IgM,K,C,Fc	37
	10	C	IgM,L,C'	34
	11	C	IgM,L,C'	23
2. Follicular Mantle Zone — Intermediate	12	C	IgM,L,Fc	10
	13	C	IgM,K	18
	14	BD	IgM,L,D,C'	5
	15	BD	IgM,K,C'	19
	16	BD	IgM,K	26
	17	BD	IgM,L	27
3. "Blastic" PDL	18	C	IgM,L	17
	19	C	Fc	7
	20	C	IgM,L,C'	19
	21	C	IgM,K	
	22	BD	IgM,L	22

In Group 1 there was excellent correlation between percentage of large cells or blasts, number of mitoses and percentage of cells in S phase ($1.0 - 5.0\%$; Mean = 2.5%) (FIG. 7). In Group 2 the percentage of cells in S phase ($3.2 - 6.0\%$; Mean = 4.4%) correlated with the size of the large cell component but not with number of mitoses except for one case (case 17) (FIG. 8). In Group 3 there was a better correlation of the S phase ($9.7 - 22.3\%$; Mean = 16%) with the number of blasts than with the number of mitoses (FIG. 9).

Cases 19 and 22 showed two G_0/G_1 peaks (FIGS. 10 and 11) representing aneuploidy. The coefficient of variation (CV) of several other cases in all three groups was unusually high and could be a result of aneuploidy.

DISCUSSION

A correlation between the percentage of cells in the S phase of the cell cycle and clinical behavior of lymphocytic lymphomas of low, intermediate, and high grade

| Histologic Criteria | | | | | DNA Analysis | | | |
Nodular	Diffuse	Large Cells (%)	% Blasts (%)	Mitoses / High Field	Cells in G_0G_1 Phase (%)	Cells in S Phase (%)	DNA Ratio	CV
X		<10		0.1	97.9	1.5	4.26	4.45
X		<10		0.3	97.3	1.4		5.76
X		<10		0.6	96.5	2.2	2.82	3.70
X		<10		1.6	97.4	1.6		3.33
X		10–20		1.7	95.0	2.0	4.10	3.90
X		10–20		1.7	96.0	1.0	2.65	3.20
X		<10		2.2	95.9	3.3	2.70	3.44
X	X	10–20		3.0	95.0	3.0	2.74	3.90
X		20–30		4.7	90.0	5.0	3.33	3.90
X	X	<10	10–20	5.6	97.0	4.5	3.00	5.73
X	X	10–20	10	5.7	95.3	4.7	3.07	5.15
X		20–30		0.7	92.4	4.9	3.00	5.04
faintly	X	20–30		0.2	97.1	3.2	3.14	3.53
faintly	X	20–30		0.7	88.9	4.4	2.94	3.70
faintly	X	10–20		1.1	90.0	6.0	2.71	3.80
faintly	X	20–30		0.9	93.0	3.4	2.50	3.86
faintly	X	10–20[a]		4.8	90.0	10.0	2.38	5.60
	X		pred	6.5	90.0	9.7	2.79	5.15
X	X		pred	5.7	84.6	NC[b]	NC	NC
X			pred	8.8	84.4	15.6		7.72
X	X		pred	5.1	71.1	22.3	2.90	5.63[c]
X	X	<10	>50	4.5	NC	NC[b]	NC	NC

[a]Many prolymphocytes. [b]Not calculated owing to aneuploidy.
[c]Done on blood.

malignancies has been described.[1-3] The study by Diamond *et al.*[1] demonstrated that the range of S phase values is quite broad in the intermediate (4.9 − 21.5%) and high grade (15.4 − 31.5%) groups.

Our study of 22 cases of poorly differentiated lymphocytic lymphomas, including six cases of lymphocytic lymphoma of intermediate differentiation, was aimed at determining whether histologic features had any direct correlation to the percentage of cells in S phase.

The 22 cases were divided into three groups: (1) small, cleaved follicular center cell lymphomas;[12] (2) mantle zone lymphomas and lymphomas of intermediate differentiation;[13-15] and (3) "blastic" type.[16]

In Group 1 there was a good correlation between the mitotic rate (.1 − 5.7%; Mean − 2.5%), the number of large, transformed round cells and/or blasts (< 10 − 30%), and the percentage of cells in S phase (1.0 − 5.0; Mean = 2.5%).

TABLE 2. Control Values for DNA Analysis During Time of This Study

	Percentage of Cells				
	G_0/G_1	S	G_2/M	CV	DNA Ratio
From Coulter TPS-1					
Peripheral bloods (11)	98.43 ± 0.74	0.82 ± 0.37	0.75 ± 0.55	3.27 ± 0.37	2.93 ± 0.09
Bone marrows (3)	88.55 ± 1.53	6.70 ± 0.82	4.75 ± 0.94	3.72 ± 0.25	2.84 ± 0.08
Reactive lymph nodes (4)	95.63 ± 1.50	2.98 ± 1.26	1.37 ± 1.27	4.07 ± 0.77	2.80 ± 0.05
Tonsils (6)	92.45 ± 1.87	3.66 ± 0.64	3.83 ± 1.54	3.38 ± 0.40	2.86 ± 0.04
Spleens (2)	94.96 ± 1.21	2.93 ± 1.80	2.12 ± 0.60	3.62 ± 0.25	2.96 ± 0.06
Total (26)					2.88 ± 0.06
From FACS 440					
Peripheral bloods (10)	99.30 ± 0.41	0.70 ± 0.46	0.00 ± 0.00	3.09 ± 0.30	2.98 ± 0.05
Reactive lymph nodes (7)	96.83 ± 1.14	1.67 ± 0.75	1.67 ± 0.75	3.42 ± 0.16	2.82 ± 0.07
Tonsils (4)	87.75 ± 2.16	8.75 ± 1.48	3.50 ± 1.12	3.13 ± 0.11	2.87 ± 0.05
Total (21)					2.89 ± 0.06

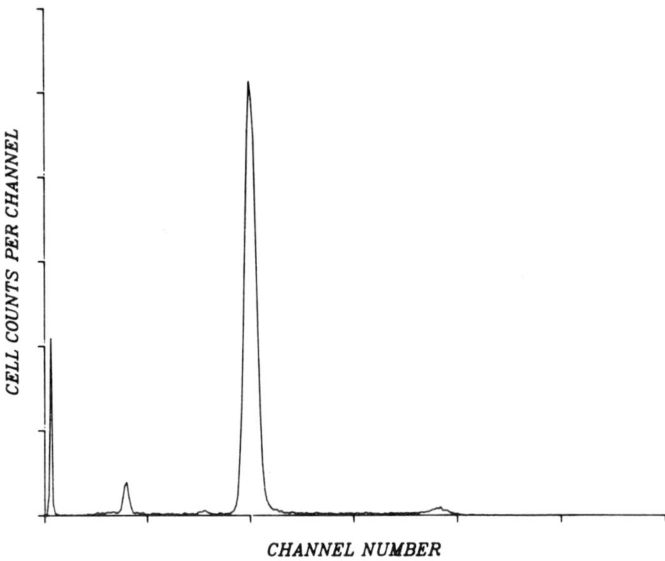

FIGURE 6. DNA histogram of lymph node showing benign follicular hyperplasia.

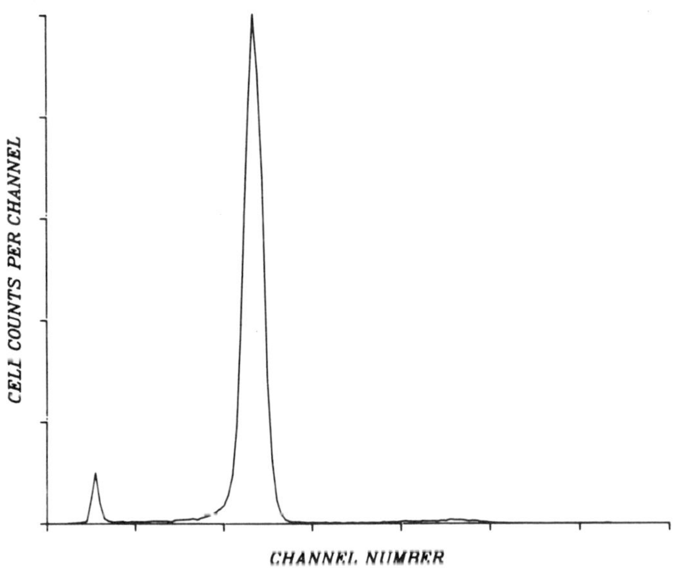

FIGURE 7. DNA histogram of PDL from Group 1 with small percentage of cells in S phase, 1.5%.

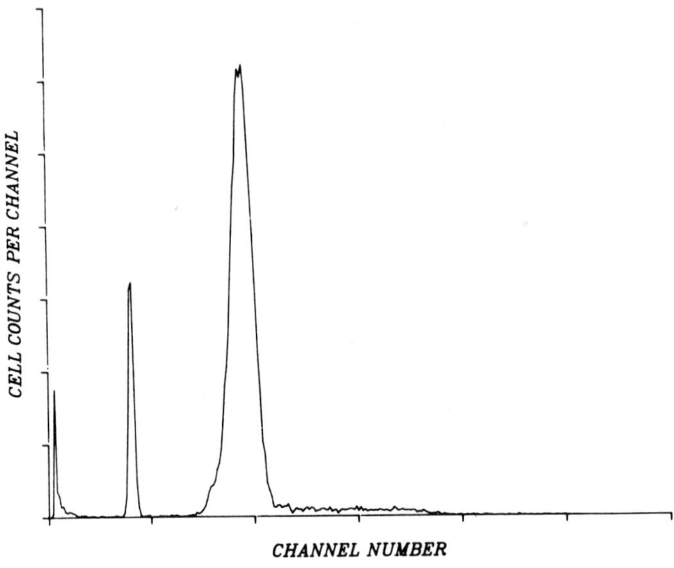

FIGURE 8. DNA histogram of lymphoma of intermediate differentiation with unusually high S phase (10%), Case 17.

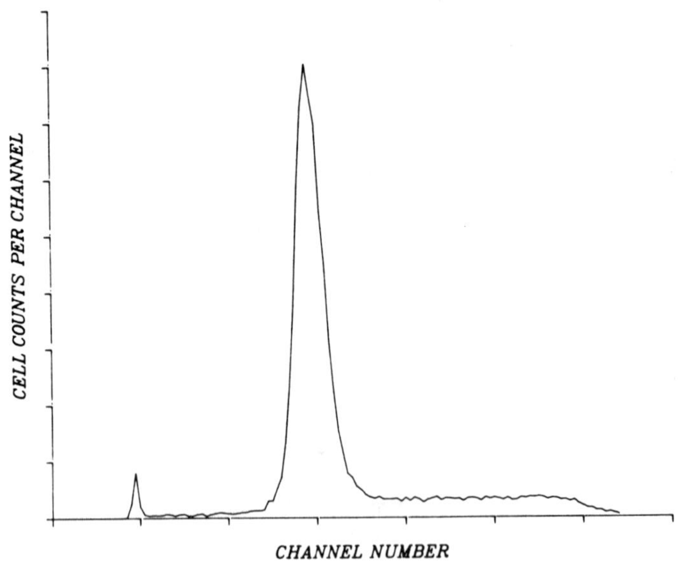

FIGURE 9. DNA histogram of "blastic" variant (Group 3). Case 21. S phase = 22.3%.

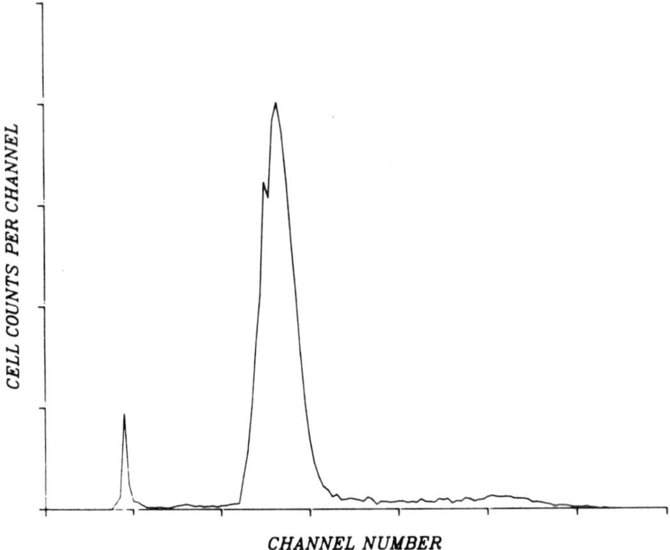

FIGURE 10. DNA histogram of "blastic" variant (Group 3). Case 19. Two distinct G_0/G_1 peaks are noted. S phase is also high, but was not calculated.

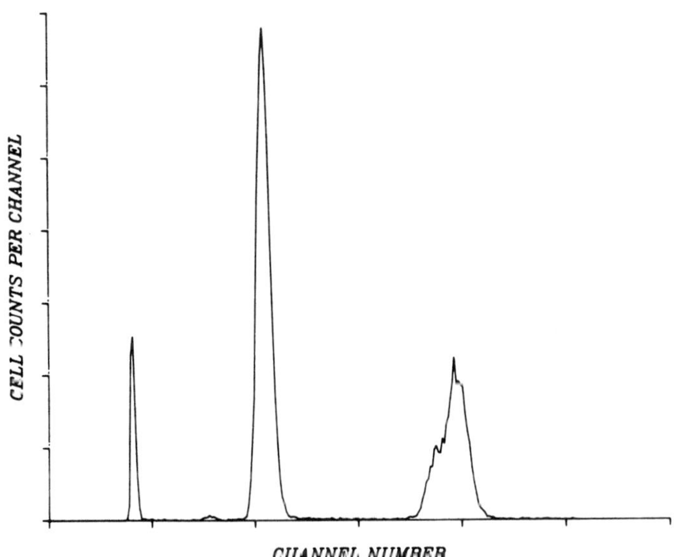

FIGURE 11. DNA histogram of "blastic" variant (Group 3). Case 22. Most of the tumor cells are near tetraploid. Normal T cells arc 22% in this case and are presumed to be part of the diploid G_0/G_1 population.

Lymphomas of intermediate differentiation and their variant, mantle zone lymphoma, (Group 2) are characterized by a mixture of small cleaved and mature small round lymphocytes and pseudofollicular proliferation centers composed of large, transformed lymphocytes and remnants of normal follicular centers which also contain large, transformed lymphocytes. There was no correlation between mitotic rate (0.2 − 1.1%; Mean = 0.7%) and the S phase component (3.24 − 6.0%; Mean = 4.4%). The high S phase component may be due to a large number of large transformed cells, which may have a prolonged S phase. In one of these cases placed in this group, Case 17, the number of mitoses (4.8/HPF) was considerably higher than in the other five and the percentage of cells in S phase (10%) was also higher. In this case a large number of partially transformed nucleolated cells, prolymphocytes, were noted and mitoses were present in cells of similar size. This case probably represents an accelerated phase of this type of lymphoma.

In Group 3 the "blastic" PDLs had high mitotic rates (4.5 − 8.8/HPF; Mean = 6.1/HPF) and high S phases (9.7 − 22.3%; Mean = 16%). Compared to Group 1, the S phase component was considerably higher relative to the mitotic rate and more consistent with the number of blasts in the lesion. The two cases of distinct aneuploidy also occurred in this group. This group of lymphomas was considered to have arisen from follicular center cells as evidenced by the presence of small, cleaved cells and nodular growth patterns in four of the five cases. Nodular PDL of small, cleaved follicular center cell type (Group 1) may undergo transformation into lymphomas consisting of large transformed lymphocytes or blasts.[16] These blasts, because of their small size, may be more difficult to distinguish from the cleaved lymphocytes than are large transformed cells and thus a transformation of a nodular PDL to a more aggressive lymphoma may be undetected by the pathologist. Some of the PDLs may be this variant at initial presentation. Three of the five patients in the "blastic" group died within two years. The other two have had a much more aggressive course than the patients in the other two groups.

In summary, in nodular PDLs composed of small cleaved lymphocytes and in "blastic" PDLs there appears to be a correlation between mitotic rate and the number of blasts or large transformed lymphocytes. This does not appear to be the case in lymphomas of intermediate differentiation where relatively high S values do not correspond to the mitotic rate but may reflect the presence of large transformed cells. This group of cases, however, is quite small and further studies are needed. Nodular PDL is considered a lymphoma of low grade malignancy. It is important to recognize "blastic" transformation or a "blastic" variant of PDL. Because the prognosis of patients with this form of PDL is considerably worse,[16] they should be considered to lymphomas of intermediate or high grade malignancy.[4,5]

SUMMARY

A direct correlation between the percentage of cells in S phase of the cell cycle and the clinical behavior of lymphocytic lymphomas of low, intermediate, and high grade malignancy has been described. The histopathologist has used the mitotic rate and other morphologic criteria such as size of cells and nuclear characteristics as predictors/indicators of the aggressiveness of a tumor. We compared the S phase values of 22 cases of poorly differentiated lymphocytic lymphoma (PDL) of the B cell type, using flow cytometric measurement of DNA content, to morphologic features and mitotic rate (MR). The 22 cases were divided into 3 histologic groups: (1) nodular PDL composed of small, cleaved lymphocytes (11 cases); (2) follicular mantle zone lymphoma and those of intermediate differentiation (6 cases); and (3) "blastic" PDL (5 cases). In Group 1 there was excellent correlation of MR, percentage of cells in S phase, and proportion of large cells (transformed lymphocytes) per high power field (HPF). In Group 2, this correlation was not found between MR and percentage of cells in S phase in five of the six cases. The high S phase in this group did correlate with the large proportion of large cells found primarily in pseudofollicular proliferation centers and in remnants of true follicular centers. These cells may have a prolonged S phase and thus fewer mitoses were seen. In Group 3, although both MR and S phases were high, a direct correlation between them as noted in the Group 1 cases was not seen, but an excellent correlation of the high S phase and the number of blasts was present. The fact that three of the five patients in this group died rapidly (within less than 2 years of presentation) and the two survivors were experiencing rapid progression of disease, supports the concept that this group represents a clearly different, more aggressive subclass of PDL.

REFERENCES

1. DIAMOND, L.W., B.N. NATHWANI & H. RAPPAPORT. 1982. Flow cytometry in the diagnosis and classification of malignant lymphoma and leukemia. Cancer **50**(6):1122–1135.
2. DIAMOND, L.W. & R.C. BRAYLAN. 1980. Flow analysis of DNA content and cell size in non-Hodgkin's lymphoma. Cancer Res. **40**(3):703–712.
3. SHACKNEY, S.E., K.S. SKRAMSTAD, R.E. CUNNINGHAM, D.J. DUGAS, T.L. LINCOLN & R.J. LUKES. 1980. Dual parameter flow cytometry studies in human lymphomas. J. Clin. Invest. **66**(6):1281–1294.
4. ROSENBERG, S.A., C.W. BERARD, B.W. BROWN, JR., *et al.* 1982. National Cancer Institute sponsored study of classifications of non-Hodgkin's lymphomas: Summary and description of a working formulation for clinical usage. Cancer **49**(10):2112–2135.
5. DORFMAN, R.E., J.S. BURKE & C.W. BERARD. 1981. A new working formulation on non-Hodgkin's lymphomas: Background, recommendations, histologic criteria, and relationship to other classifications. *In* Advances in Malignant Lymphomas. Proceedings of the Third Annual Bristol Meyers Symposium on Cancer Research. Academic Press. New York, N.Y.
6. RAPPAPORT, M. 1966. Tumors of the hematopoietic system. *In* Atlas of Tumor Pathology. Section 3, Fasc. **8**:98–101. Armed Forces Institute of Pathology. Washington, D.C.
7. PALUTKE, M., P. KHILANANI & R. WEISE. 1976. Immunologic and electron-microscopic characteristics of a case of immunoblastic lymphadenopathy. Am. J. Clin. Pathol. **65**(6):929–941.
8. PALUTKE, M., D.J. PATT, R.W. WEISE, C. VARADACHARI, R. WYLIN, C.R. BISHOP & P.M. TABACZKA. 1977. T cell leukemia-lymphoma in young adults. Am. J. Clin. Pathol. **68**(4):429–439.
9. PALUTKE, M., L. EISENBERG, J. KAPLAN, M. HUSSAIN, K. KITHIER, P. TABACZKA, I. MIRCHANDANI & D. TENENBAUM. 1983. Natural killer and suppressor T cell chronic lymphocytic leukemia. Blood **62**(3):627–634.
10. KRISHAN, A. 1977. Flow cytometry: Long term storage of propidium iodide/citrate-stained material. Stain Technol. **52**(6):339–343.
11. DEAN, P.N. & J.H. JETT. 1974. Mathematical analysis of DNA distribution derived from flow cytometry. J. Cell Biol. **60**(2):523–527.
12. LUKES, R.J., C.R. TAYLOR, J.W. PARKER, T.L. LINCOLN, P.K. PATTENGALE & B.H. TINDLE.

1978. A morphologic and immunologic surface marker study of 299 cases of non-Hodgkin's lymphomas and related leukemias. Am. J. Pathol. **90**(2):461–486.

13. PALUTKE, M., L. EISENBERG, I. MIRCHANDANI, P. TABACZKA & M. HUSSAIN. 1982. Malignant lymphoma of small cleaved lymphocytes of the follicular mantle zone. Blood **59**(2):317–322.

14. WEISENBURGER, D.D., H. KIM & H. RAPPAPORT. 1982. Mantle-zone lymphoma: A follicular variant of intermediate lymphocytic lymphoma. Cancer **49**(7):1429–1438.

15. MANN, R.B., E.S. JAFFE & C.W. BERARD. 1979. Malignant lymphomas—A conceptual understanding of morphologic diversity. Am. J. Pathol. **94**(1):105–192.

16. COME, S.E., E.S. JAFFE, J.C. ANDERSEN, R.B. MANN, B.L. JOHNSON, V.T. DEVITA JR. & R.C. YOUNG. 1980. Non-Hodgkin's lymphomas in leukemic phase: Clinicopathologic correlations. Am. J. Med. **69**(5):667–674.

Lymphocyte Subsets in Lymph Node Hyperplasias and B Cell Neoplasms as Determined by Fluoresceinated Antibodies and Flow Cytometry

SALLY E. SELF, NICHOLAS M. BURDASH,
ANNE D. PONZIO, AND MARIANO F. LAVIA

Department of Laboratory Medicine
Medical University of South Carolina
Charleston, South Carolina 29425

In an attempt to see how monoclonal antibody staining and cell flow cytometry could be applied to lymphocyte suspensions made from surgically obtained tissues, 21 hyperplastic lymph nodes and 19 B cell neoplasms were analyzed. Information was obtained concerning percentages of lymphocyte subsets in different patterns of B cell hyperplasia and the correlation of helper/suppressor ratios in lymph nodes as compared to peripheral blood. In B cell neoplasms the immunotype of the neoplasm was identified and the composition of residual nonneoplastic/-reactive lymphocytes was determined. The expression of differentiation and activation antigens, e.g. transferrin receptor and OKT10, in certain cases was also determined. In some cases of B cell neoplasm, peripheral blood was also analyzed and assessment of blood involvement was made as well as correlation of the distribution of lymphocyte subsets between the tissue compartment and blood compartment.

MATERIALS AND METHODS

Specimen Preparation

Tissue samples were obtained immediately after surgical removal and placed in McCoy's medium (Grand Island Biological Co. Laboratories, Grand Island, NY) for transport to the laboratory. Samples were minced very finely using a small pair of scissors, and the cells in suspension were placed in a plastic test tube. Cells were washed several times in phosphate-buffered saline with 0.2% bovine serum albumin and 0.01% sodium azide (PBS); all centrifugation steps were at $600 \times g$ for 3 minutes. After the last wash, cells were resuspended in 2 ml PBS and filtered through fine nylon mesh to remove clumps. A cell count and viability study were performed utilizing trypan blue dye exclusion. A viability of 70% was required to insure there was a minimum of nonspecific staining due to dead cells and to minimize the possibility of selective cell death of various populations. The cell concentration was adjusted to 1×10^7/ml. Cytocentrifuge preparations were routinely made for all specimens. These were examined along with the histologic section of the node to insure that representative cells were examined.

Staining Procedures

Samples were examined for antigens reactive with commercially available monoclonal antibodies OKT3, 4, 6, 8, 10, 11, OKM1 (Ortho Pharmaceutical Corporation, Raritan, NJ), Leu-M3, HLA-DR, CALLA, Leucocyte, Leu-12, Transferrin Receptor (Becton Dickinson Monoclonal Center Inc., Mountain View, CA), and with fluorescein conjugated goat or burro $F(ab')_2$ antisera to human kappa, lambda, mu, delta, and gamma chains (Kallestad Laboratories Inc., Austin, TX).

For those monoclonal antibodies which were already fluorescein conjugated, 100 μl of specimen cells at 1×10^7/ml were added to a tube containing 100 μl of PBS and 5–10 μl of the appropriate antibody, then incubated on ice for 30 min. Specimen negative controls consisted of sample and PBS but no antibody. Any potential contaminating red cells were lysed with 1–2 ml of Ortho lysing reagent. Samples were centrifuged 3 min at $600 \times g$ and resuspended in 1.5 ml PBS.

Unlabeled monoclonal antibodies were utilized as described above, except that samples were washed once in PBS following the lysis step. In the second incubation, 100 μl of appropriately diluted goat $F(ab')_2$ anti-mouse IgG fluorescein conjugate (TAGO Inc., Burlingame, CA) and 40 μl of AB-negative serum were added to the cell pellets, mixed, and incubated on ice for 30 minutes. Samples were washed once and resuspended in 1.5 ml PBS. Specimen negative controls consisted of sample and PBS in the primary incubation and goat anti-mouse conjugate and AB-negative serum in the second incubation.

Surface immunoglobulin staining was performed as follows. One hundred μl of sample cells at 1×10^7/ml were lysed, washed twice in PBS, resuspended in 1.5 ml PBS, and incubated at 37°C for 30 min to remove cytophilic antibodies. Samples were pelleted, 50 μl of appropriately titered conjugates were added, tubes were incubated in the dark for 30 min on ice, washed twice in PBS, and resuspended in 1.5 ml PBS. Specimen negative controls consisted of incubation with fluorescein conjugated goat $F(ab')_2$ antisera to mouse IgG.

All samples were resuspended in PBS and analyzed within 24 h; if any delay beyond 24 h was expected, cells were fixed in 2% paraformaldehyde overnight and resuspended in PBS the following day. To insure reliability of the monoclonal antibodies and flow cytometer, peripheral blood from a normal individual was run with each patient specimen. A monoclonal marker specific for leukocytes was also run as a control and was always positive. In addition, the OKM1 and Leu-M3 markers for monocytes were run and in all cases were found in very low percentages. Analyses for OKT6 markers were also performed and were negligible in all cases.

Analysis

Cells were analyzed with an Ortho Specrum III (Ortho Diagnostic Systems, Inc., Westwood, MA).[1,2] As with whole blood, cells from nodes were gated on the basis of forward scatter versus 90° light scatter as seen in FIGURE 1. Generally 2000–8000 lymph node cells were examined with each monoclonal antibody. The

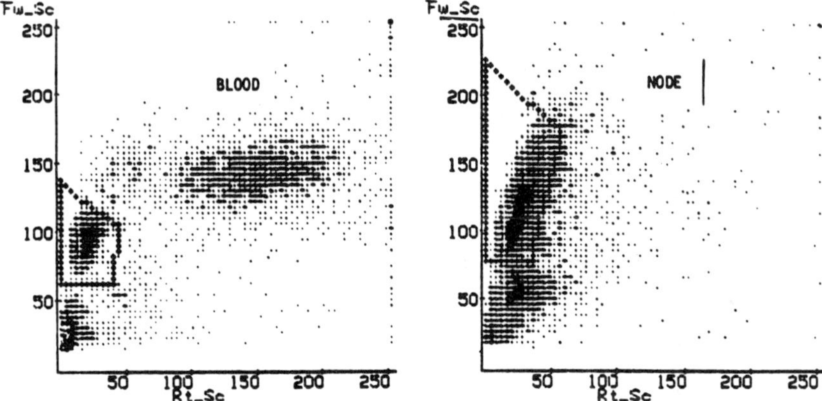

FIGURE 1. Cytograms of peripheral blood and lymph node suspension demonstrating gating of lymphocytes.

net percentage of cells reacting to an antibody was the total percentage of fluorescent cells with antibody minus the percentage of reactive cells in the control tube.

RESULTS

Hyperplasias

Twenty one lymph nodes with the histologic picture of reactive hyperplasia were analyzed. The fluorescent histograms of the tissue cell suspensions were similar to those seen in normal peripheral blood for the respective antibody, indicating similar staining intensity, as seen in FIGURE 2. The results of lymphocyte subset quantitation in hyperplastic lymph nodes with the mean and range of each marker are shown in TABLE 1. Lymph node cells which stained positively for the HLA-DR antigen had a mean percentage of 58. In general the percentages of HLA-DR cells were higher than the B cell percentages as determined by surface immunoglobulin or specific B cell antigens. The kappa/lambda ratio of surface immunoglobulin ranged from 1.2 to 2.6 similar to our normal peripheral blood range of 1.4 to 2.7 (not shown). The OKT4/OKT8 ratio had a broad range of 0.6 to 8.5. Two nodes had 4/8 values less than 1.0. One of these with a ratio of 0.6 came from a patient with systemic lupus erythematosus on dilantin therapy for epilepsy, and suffering from an Epstein-Barr virus infection. The other node with a ratio of 0.9 came from a patient with suspected AIDS and suspected toxoplasmosis. Seven nodes had a 4/8 ratio of 3.0 or greater. One of these, with a 4/8 ratio of 8.5, had the typical histological finding of dermatopathic lymphadenitis as seen in FIGURE 3. Five nodes, with 4/8 ratios of 3.0, 3.3, 5.1, 6.4, and 8.1, had prominent paracortical nodules typical of activation of the T cell area of the nodes as seen in FIGURE 4.

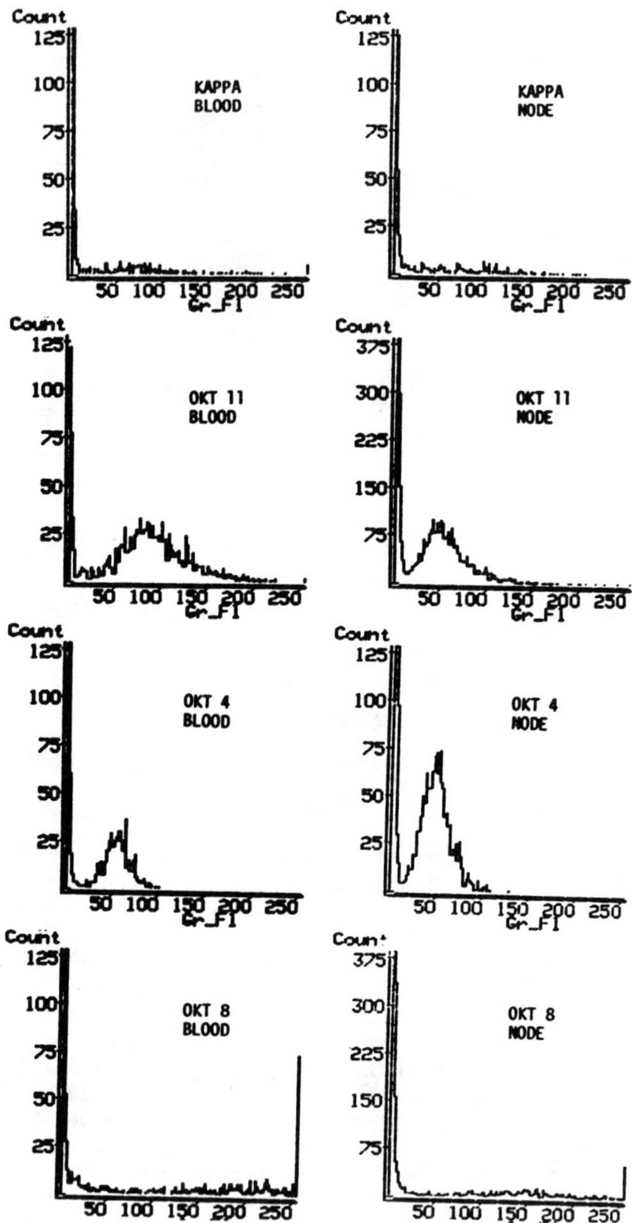

FIGURE 2. Fluorescent histograms of peripheral blood lymphocytes compared with suspensions made from hyperplastic lymph nodes demonstrating similar fluorescent intensities for the respective antibodies. The abscissa, denoting fluorescence intensity, is on a linear scale.

TABLE 1. Results of Lymphocyte Subset Analyses in Hyperplastic Lymph Nodes

Marker	No. Cases	Mean (%)	Range (%)
Mu	20	24	9–46
Delta	19	19	7–44
Gamma	20	13	2–30
Kappa	21	20	9–39
Lambda	21	13	4–26
OKT3	19	55	31–81
OKT11	21	57	33–83
OKT4	21	39	21–68
OKT8	21	17	7–41
OKT6	19	1.5	0–5
HLA-DR	20	58	12–86
Leu-12	11	41	16–73
Transferrin Receptor	14	17	1–51
4/8	21	3.0	0.6–8.5
K/λ	21	1.7	1.2–2.6

In seven cases, peripheral blood lymphocytes were analyzed at the same time as the lymph node. There appeared to be no relationship between blood and node 4/8 ratios as seen in TABLE 2. However, in all cases the 4/8 ratio in the node was higher than that in the blood.

B Cell Neoplasms

Portions of 14 lymph nodes, two spleens, one tonsil, one pharyngeal mass, and one lower abdominal mass with B cell lymphoma involvement were analyzed. In 10 cases, peripheral blood was also available for examination. The lymphomas were classified according to the NIH working formulation.[3]

There were three cases of small lymphocytic lymphoma. The results of marker studies are shown in TABLE 3. In all three cases, peripheral blood was studied and

TABLE 2. Blood and Node OKT4/OKT8 Ratios in Hyperplasia

Blood	Node
0.6	1.3
1.5	2.3
1.2	2.5
2.5	8.1
0.7	1.8
2.1	3.3
0.4	0.9

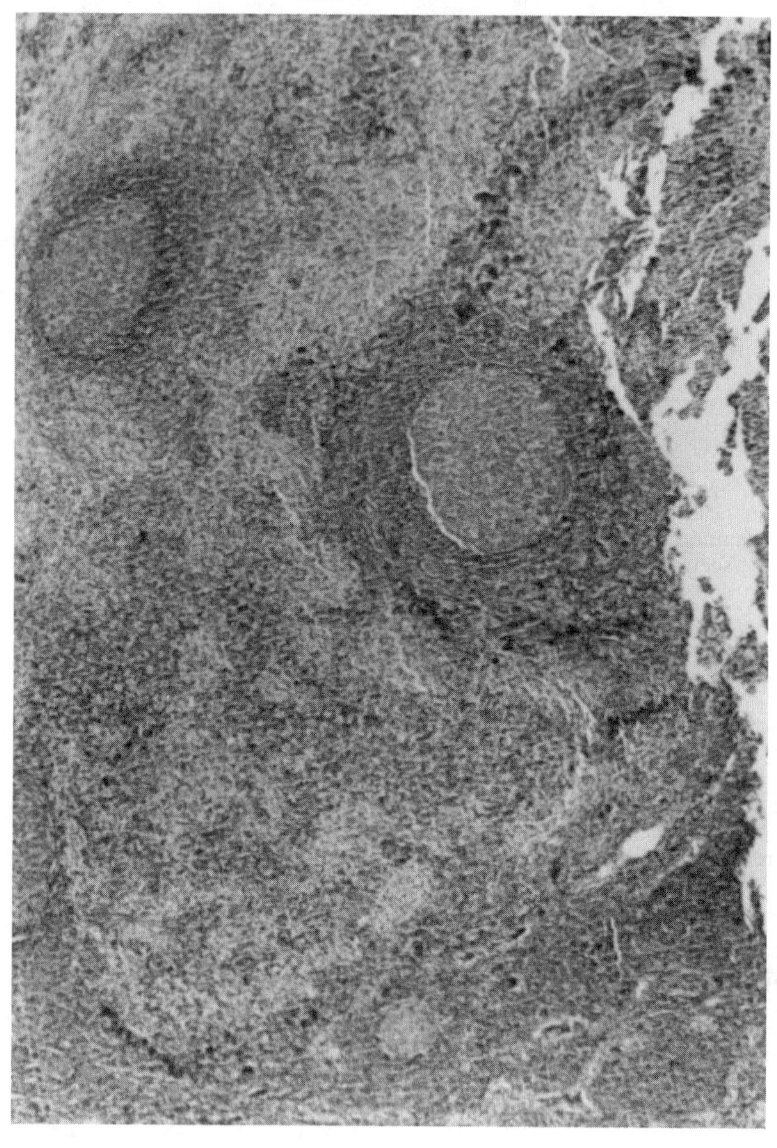

FIGURE 3. Histologic section of lymph node demonstrating dermatopathic lymphadenitis. Original magnification ×40; shown at 90% of original size.

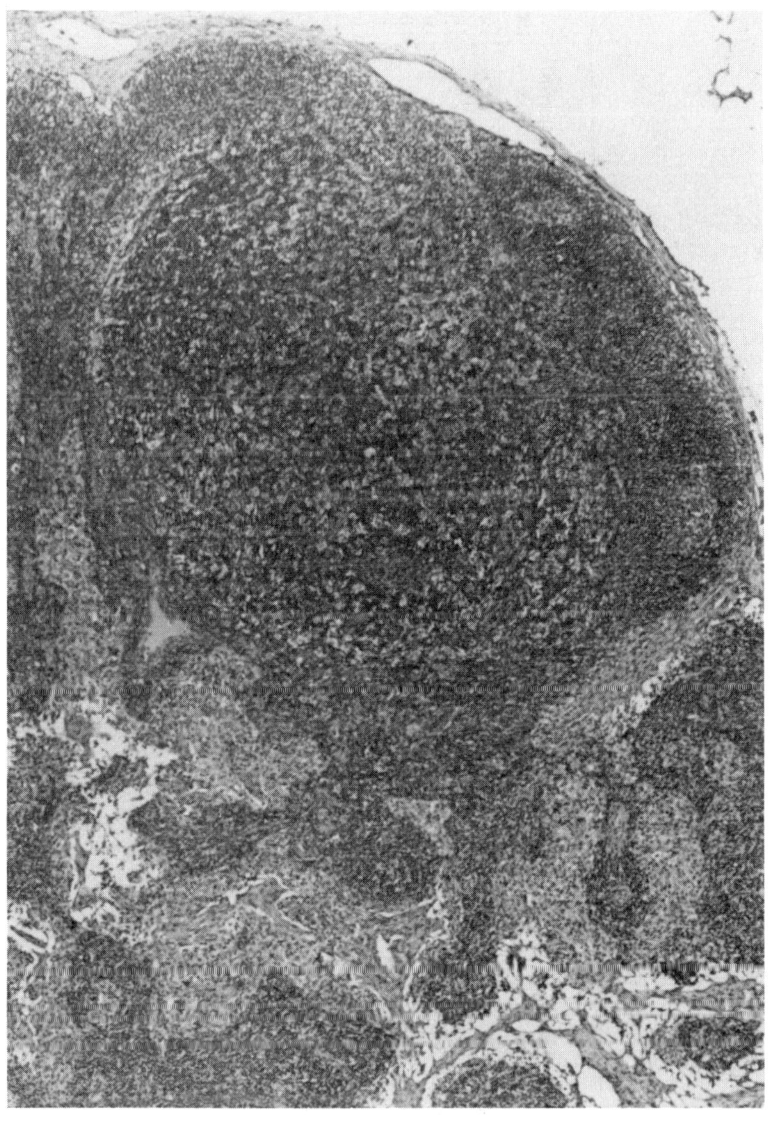

FIGURE 4. Histologic section of lymph node demonstrating paracortical hyperplasia. Original magnification ×40; shown at 90% of original size.

TABLE 3. Percentage of Lymphocyte Subsets in Small Lymphocytic Lymphomas

		Surface Immunoglobulin					OKT Markers							Transferrin	
		μ	δ	γ	κ	λ	3	11	4	8	10	HLA-DR	CALLA	Receptor	4/8
Case 1	Node	0.2	0.1	1.4	0.2	7	21	21	13	11		87			1.2
	Blood[a]	2	0.1	0.2	2	0.7	41	53	23	26		53			0.9
Case 2	Node	0.3	1	53	2	39	14	17	15	5	0.2	97	0	2	3.0
	Blood[b]	4	0.3	49	2	37	4	5	4	1	0.6	95	0	0.6	4.0
Case 3	Note	17	27	0	58	0	13	16	10	5	19	96	0.4	3	2.0
	Blood[c]	24	25	3	59	0	6	7	3	3	3	84	0.1	0.7	1.0

[a]Case 1, WBC $6.5 \times 10^3/mm^3$, 47% lymphocytes.
[b]Case 2, WBC $12.4 \times 10^3/mm^3$, 87% lymphocytes.
[c]Case 3, WBC $31.8 \times 10^3/mm^3$, 62% lymphocytes.

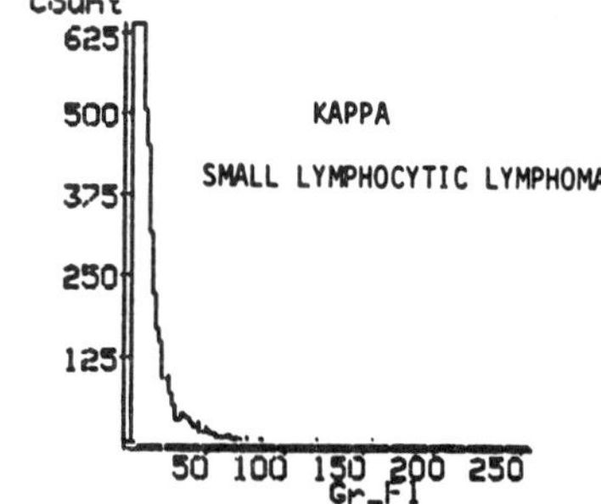

FIGURE 5. Fluorescent histogram of a cell suspension of a small lymphocytic lymphoma (Case 3) stained for kappa surface immunoglobulin and demonstrating low fluorescence intensity.

peripheral blood involvement by the neoplasm was demonstrated. The morphology of the lymphocytes in the peripherial blood was typical of chronic lymphocytic leukemia (CLL). In Case 1, although there was no detectable expression of monotypic surface immunoglobulin, T cell markers were abnormally low while HLA-DR markers were significantly elevated. In Case 2, lambda light chains and gamma heavy chains were expressed, while in Case 3, kappa light chains along with mu and delta heavy chains were expressed. The intensity of fluorescence of surface immunoglobulin was faint as seen in the fluorescence histogram shown in FIGURE 5. The low mean channel of fluorescence is analogous to the faint fluorescence appreciated visually when one examines cases of CLL under the fluorescent microscope. The percentage of T cells as determined by OKT11 was low in all cases, 16 to 21% in the nodes and 7 to 53% in the blood. The OKT4 and OKT8 populations were low and therefore changes in their percentages were inconsequential. The HLA-DR marker was present on the majority of cells in all samples examined, as would be expected in a B cell neoplasm. Because of unavailability at the time, anti B cell monoclonal antibodies were not run, although it is recognized that they would be helpful in detecting malignant B cells when the surface immunoglobulin fluorescence is at a low level. In both cases in which it was run, the Transferrin Receptor antibody was essentially negative.

The categories of follicular small cleaved and follicular mixed lymphomas were combined for the purposes of this study. This was done as a result of disagreements between referring pathologists and the authors regarding categorization of some cases. The combination is felt to be justifiable because of the similarity in prognosis and the close relationship of the neoplasms. Four cases fell into this group, with the results seen in TABLE 4. Immunoglobulins on the surface of neoplastic cells demonstrated a variety of immunotypes. In Cases 6 and 7, peripheral blood was also analyzed. There was no obvious involvement of the blood in Case 6, however Case 7 showed evidence of blood involvement by the presence of malignant lymphocytes of the same immunotype as in the node. The percentage of T cells, as determined by OKT11, ranged from 25% to 55% in these follicular neoplasms. The 4/8 ratio of node T cells ranged from 3.0 to 5.6. In Case 6 where these markers were done both in the node and in the blood, the ratio was 3.0 in the node and 1.0 in the peripheral blood, again demonstrating the lack of relationship between the two compartments. CALLA staining was done in two cases and was positive in one case, but negative in the other.

Four cases of diffuse mixed B cell lymphoma were analyzed. The results are

TABLE 4. Percentage of Lymphocyte Subsets in Follicular Small Cleaved and Follicular

		Surface Immunoglobulin					OKT Markers					HLA-DR	CALLA	Transferrin Receptor	4/8
		μ	δ	γ	κ	λ	3	11	4	8	10				
Case 4	Node	40	16	14	4	35	61	55	54	14					3.9
Case 5	Spleen	68		1	66	0.5		32	23	6		73			3.8
Case 6	Node	4	4	45	4	3	23	25	18	6	58	81	39	4	3.0
	Blood	13	7	22	17	12	72	81	40	39	22	18	0.9	35	1.0
Case 7	Node	58	30	1	0.6	57	30	32	28	5		83	2	9	5.6
	Blood	56	68	1	0.3	69		3				94			

TABLE 5. Percentages of Lymphocyte Subsets in Diffuse Mixed Lymphomas

		Surface Immunoglobulin					OKT Markers					HLA-DR	CALLA	Transferrin Receptor	4/8
		μ	δ	γ	κ	λ	3	11	4	8	10				
Case 8	Node	10	14	0.7	23	0.1	56	61	47	17	54	64		2	2.8
Case 9	Node	25	42	12	5	39	48	49	33	12		56			2.8
	Blood	11	20	1	0.7	14	25	25	9	21		62			0.4
Case 10	Node	0	0.1	63	0.6	30	4	4	1	3		99	0.1	12	0.3
	Blood	2	1	63	4	40	15	16	11	5		74	0.2	6	2.2
Case 11	Node	98	19	11	1	96	25	58	29	21		99		92	1.4

shown in TABLE 5. Once again a variety of surface immunoglobulin phenotypes were expressed by the malignant cells. Percentage of T cells, as determined by OKT11, ranged from 4% to 61%. The 4/8 ratio in these nodes ranged from 0.3 to 2.8. CALLA was negative in both cases analyzed for this antigen. The transferrin receptor was low (2% and 12%) in two cases and high (92%) in one case. In Cases 9 and 10, in which peripheral blood was examined, there was involvement of B cells having the same immunophenotype as the node. There was no relationship between the 4/8 ratios of the nodes and the corresponding bloods.

TABLE 6 shows five cases of diffuse large cell lymphomas. In Case 12, surface immunoglobulins were undetectable. Case 13 had a kappa, mu immunophenotype, while Cases 14 and 15 had lambda, mu immunophenotypes. Although only a few cells were available for analysis in Case 16, light chain determinations indicated kappa monotypism. In Cases 14–16 peripheral blood was also analyzed and showed no involvement.

Three cases of diffuse small noncleaved lymphomas were analyzed as seen in TABLE 7. The histology of Case 17 was that of classical Burkitt's lymphoma, while Cases 18 and 19 were of undifferentiated type. Cases 17 and 19 demonstrated kappa, mu monotypism, while in Case 18, a pharyngeal mass, only enough cells were obtained to demonstrate kappa monotypism. The T cell percentages as determined by OKT3 and OKT11 ranged from 5% to 29%.

DISCUSSION

Although the analysis of lymph nodes by flow cytometry and fluoresceinated antibodies has certain drawbacks, namely loss of architectural and cytologic details and differential cell survival in suspension, these can be overcome in part by careful correlation with histologic sections and cytocentrifuge preparations made from cell suspensions. The advantages of flow cytometry include inherent objectivity, quantitation, and information concerning the intensity of fluorescence. In addition, it is quite easy to analyze peripheral blood samples in the same manner as tissue cell suspensions, allowing correlation between findings in blood and tissue.

The analysis of lymph node hyperplasias has provided guidelines against which parameters from neoplastic nodes can be judged. For example, the kappa/lambda ratios in those B cell lymphomas which expressed surface light chains were clearly outside the range of 1.2–2.6 determined for hyperplastic nodes. However, surface immunoglobulin light chain monotypism alone does not make a diagnosis of malignant lymphoma as shown by Levy et al.[4] who presented 12 cases of reactive lymphoid hyperplasia with monotypic surface immunoglobulin. Histologic correlation is crucial to the diagnosis of malignant lymphoma. The range and means of T cell percentages in hyperplastic nodes in this study (33–83%, mean 57%) is similar to the 50–82% with a mean of 63% obtained by Knowles et al.[5] The range of 4/8 ratios in this study (0.6 to 8.5) is broader than that of Knowles et al. (1.0 to 6.2), although the means are similar (3.0 compared to 3.4). The high 4/8 ratios in this study were seen in lymph nodes showing prominent paracortical hyperplasia. The lack of correlation between 4/8 ratios in blood and in lymph nodes implies that no inference can be made about the T cell subsets in one compartment by analyzing the subsets in the other compartment.

TABLE 6. Percentages of Lymphocyte Subsets in Diffuse Large Cell Lymphomas

		Surface Immunoglobulin					OKT Markers					HLA-DR	CALLA	Transferrin Receptor	4/8
		μ	δ	γ	κ	λ	3	11	4	8	10				
Case 12	Node	1	1	2	4	2	16	20	13	4		43			3.3
Case 13	Spleen	31	2	7	25	4	59	56	27	20					1.4
Case 14	Node	95	0.4	3	1	89	9	22	7	63	88	94	0.4	78	0.1
	Blood	4	4	2	5	5	85	89	33	53	10	6		1	0.6
Case 15	Node	90	2	1	0	88	6	7	2	5	21	100	0.8	20	0.4
	Blood	2	28	0.5	19	9		76							
Case 16	Abdominal mass				65	2		76							
	Blood				10	4	88	90	53	30					1.8

TABLE 7. Percentages of Lymphocyte Subsets in Diffuse Small Non-Cleaved (Burkitt's and Undifferentiated) Lymphomas

		Surface Immunoglobulin					OKT Markers							Transferrin	
		μ	δ	γ	κ	λ	3	11	4	8	10	HLA-DR	CALLA	Receptor	4/8
Case 17	Node	79	0.7	1	76	0.3	5	5	3	3		92			1.0
Case 18	Pharyngeal mass				87	2	2	9							
Case 19	Tonsil	87	16	0.4	88	0	17	29	11	10		99	23	84	1.1

The difficulties associated with the application of flow cytometry to the diagnosis of malignant lymphomas have been pointed out in the excellent review by Lovett *et al.*[6] However, as shown in this study, in most cases of B cell lymphoma, surface immunoglobulin monotypism can readily be identified. The highest percentages of T cells in B cell lymphomas were seen in the follicular lymphomas and diffuse mixed lymphomas, although Knowles *et al.*[5] in a larger study found no obvious correlation between percentage of T cells and morphologic category. In both this study and that of Knowles *et al.*, there was no correlation between 4/8 ratios and morphologic categories.

High percentages of cells staining positively for the Transferrin Receptor were seen in one of three cases of diffuse mixed lymphoma, one of two cases of diffuse large cell lymphoma, and one case of diffuse small noncleaved (non-Burkitt's) lymphoma. Aisenberg *et al.*[7] found one half of diffuse histiocytic (large cell) lymphomas positive for this marker, while other types of lymphomas were generally deficient in this marker. OKT10 staining in this study was variable and no significant correlation can be made from these few cases. Aisenberg *et al.* also found OKT10 positivity in most cases of diffuse histiocytic (large cell) lymphoma as well as in many other subtypes.

CALLA positivity was seen in one of the two follicular lymphomas analyzed for this marker. Anderson *et al.*[8] found 69% of patients with nodular lymphoma expressed this antigen. One case of diffuse small noncleaved (non-Burkitt's) lymphoma expressed this antigen in the present study, while all other lymphomas tested for this antigen were negative.

In summary, cell flow cytometry is a useful ancillary technique in understanding cell populations in lymph node hyperplasias and B cell neoplasms. In the current study, marker analysis was most helpful in the differential diagnosis of flamboyant follicular hyperplasia and follicular lymphoma (Cases 6 and 7) and in the diagnosis of extranodal lymphoma (Cases 16 and 18). Objectivity, analysis of large numbers of cells, and rapidity of processing provide the major advantages of this technique. The analysis of solid tissue suspensions by flow cytometry coupled with microscopic examination shows great potential for the detection and categorization of certain lymphomas.

ACKNOWLEDGMENTS

The authors are grateful to A. Julian Garvin, M.D., for his participation in the microscopic examinations and to Donna H. Runey for her excellent technical assistance.

REFERENCES

1. Hoffman, R. A. & W. P. Hansen. 1981. Immunofluorescent analysis of blood cells by flow cytometry. Immunopharmacology **3**:249–254.
2. Ip, S. H., C. W. Ritterhaus, K. W. Healey, C. C. Struzziero, R. A. Hoffman & W. P. Hansen. 1982. Rapid enumeration of T lymphocytes by a flow-cytometric immunofluorescence method. Clin. Chem. **28**:1905–1909.
3. Nathwani, B. N. & C. D. Winberg. 1983. Non-Hodgkin's lymphomas: An appraisal of the "Working Formulation" of non-Hodgkin's lymphomas for clinical use. *In* Malignant Lymphomas: A Pathology Annual Monograph. A. C. Sommers & P. P. Rosen, Eds.:1–63. Appleton-Century-Crofts. Norwalk, CT.

4. LEVY, N., J. NELSON, P. MEYER, R. J. LUKES & J. W. PARKER. 1983. Reactive lymphoid hyperplasia with single class (monoclonal) surface immunoglobulin. Am. J. Clin. Pathol. **80**(3):300–308.

5. KNOWLES, D. M., J. P. HALPER & F. A. JAKOBIEC. 1984. T-lymphocyte subpopulations in B-cell-derived-non-Hodgkin's lymphomas and Hodgkin's disease. Cancer **54**(4):644–651.

6. LOVETT, E. J., B. SCHNITZER, D. F. KEREN, A. FLINT, J. L. HUDSON & K. D. McCLATCHEY. 1984. Application of flow cytometry to diagnostic pathology. Lab Invest. **50**(2):115–140.

7. AISENBERG, A. C., B. M. WILKES & N. L. HARRIS. 1983. Monoclonal antibody studies in non-Hodgkin's lymphoma. Blood **61**(3):469–475.

8. ANDERSON, K. C., M. P. BATES, B. L. SLAUGHENHOUPT, G. S. PINKUS, S. F. SCHLOSSMAN & L. M. NADLER. 1984. Expression of human B cell-associated antigens on leukemias and lymphomas: A model of human B cell differentiation. Blood **63**(6):1424–1433.

Surface Immunoglobulin Light Chain Expression in Pre-B Cell Leukemias[a]

BENJAMIN KOZINER, JANET STAVNEZER,
AYAD AL-KATIB, DAVID GEBHARD,
ABRAHAM MITTELMAN, MICHAEL ANDREEFF, AND
BAYARD D. CLARKSON

Memorial Sloan Kettering Cancer Center
New York, New York 10021

INTRODUCTION

Our understanding of cellular differentiation is being rapidly increased by the development of monoclonal antibodies and by modern techniques of molecular genetic analysis. In particular, our knowledge of the phenotype, i.e. the cell lineage and state of differentiation, of acute lymphoblastic leukemia (ALL) and the blastic phase of chronic myelogenous leukemia (BL-CML) are undergoing revision. Most cases of ALL have lymphoblasts that were thought to fall into the non-B, non-T ("null") category, but have recently been shown to be within the B-cell differentiative compartment. An early study, using cytoplasmic immunoglobulin (cIg) μ heavy chain expression as a marker for pre-B-cells, indicated that approximately 18% of "null" ALL cases were of the pre-B phenotype, i.e. they expressed cIg μ chain but did not express surface immunoglobulin (sIg).[1] Monoclonal antibodies raised by immunizing mice with pre-B-cell lines have been useful in further elucidating the B-cell lineage of some ALL cases. Both B1[2] and BA2,[3] which are pre-B cell markers defined by such monoclonal antibodies, have been clearly demonstrated on ALL cells. Moreover, analysis of immunoglobulin (Ig) gene rearrangements in ALL cells conclusively showed a commitment to the B-cell pathway in all 25 cases that were examined.[4,5] It has recently been reported that the leukemic cells from a patient with null "common" ALL were induced to proceed along the B-differentiative pathway so as to acquire a B-associated antigen (BA2).[5] These cells showed decreased expression of terminal deoxynucleotidyl transferase (TdT) and expressed cIg and ultimately sIg.

Similarly to ALL, about one-third of patients with BL-CML have been found to have leukemic cells with pre-B lymphoid characteristics. Cytoplasmic μ chain, but not sIg, has been detected in the blast cells from some patients with BL-CML[6] who also expressed high levels of TdT and were positive for the Philadelphia chromosome. Ig heavy chain genes were rearranged in BL-CML cells from seven and Ig light chain genes were rearranged in three of eight patients examined by Bakhshi *et al.,*[7] providing further evidence of their commitment to differentiate along the B-cell lineage.

[a]This work was parly supported by grants CA-08748, CA-20194, CA-05826, CA-29564 from the National Cancer Institute, and AI 17558 from the National Institutes of Health.

In the course of investigating the expression of Ig light chains by flow cytometry in patients with B-cell neoplasias to measure "clonal excess" (CE), which is indicative of monoclonality, we turned our attention to cases of acute leukemia. The analysis of CE is based on the work of Ligler *et al.*[8] who identified neoplastic B cells by an excess of cells bearing one light chain type on the cell surface at discrete levels of fluorescence intensity, as measured on a fluorescence-activated cell sorter (FACS). CE was defined as the value obtained applying the formula ($\%\kappa^+ - \%\lambda^+$) cells/total number of light chain bearing cells for every 10 channels of fluorescence intensity. This work attested to the capacity of flow cytometry to detect small numbers of neoplastic B cells that cannot otherwise be detected by regular light or fluorescence microscopy. By using this methodology, we hereby document the finding of identifiable CE in cells from four patients with BL-CML, three with ALL, one with leukemic lymphoblastic lymphoma (LBL), and in the pre-B-cell line REH, thereby demonstrating that these leukemias and the REH cell line contain cells which are further differentiated along the B-cell lineage than was heretofore believed.

MATERIALS AND METHODS

Cell Preparation

Mononuclear cells from heparinized samples of peripheral blood (PB) and bone marrow (BM) aspirates were isolated by density centrifugation.[9] Viability by trypan blue dye exclusion always exceeded 90%. One $\times 10^6$ cells were directly stained with burro anti-κ and anti-λ antibodies purified by ion-exchange chromatography and conjugated with FITC (Meloy Laboratories, Inc., Springfield, VA) as previously described.[9] Saturating titers of antibody were determined by fluorescence microscopy. Immunofluorescence analysis was performed on a Cytofluorograf System 30-L (Ortho Diagnostic Systems, Westwood, MA), equipped with a Lexcel argon ion laser (operated at 500 mw laser power) and careful light-scatter gating for the lymphoid population.

CE Analysis by Flow Cytometry

CE analysis was performed by the previously described method of Ligler *et al.*[8] FIGURE 1 is a graphic example of a normal control blood sample and the resultant CE analysis. The upper figure is the flow cytometric frequency distributions of κ versus λ light-chain-bearing cells. The x-axis is channel of increasing fluorescence intensity and the y-axis is frequency. Flow cytometric analysis shows the typical heterogeneous distribution of fluorescence intensity in nearly equal proportions for both κ and λ light-chain-bearing cells. Dissecting this curve by CE analysis yields the bottom figure. Again, the x-axis is channels of increasing fluorescence intensity while the y-axis is now the percent difference of κ^+ cells versus λ^+ cells for each 10-channel unit of intensity. A value greater than zero indicates a majority of κ-bearing cells, a value less than zero indicates a majority of λ-bearing cells. Analysis of CE for a population of 44 normal control blood samples indicated that 95% confidence limits for this population ranged from a high of .258 on the κ side to a low of $-.114$ on the λ side. This analysis agrees with the accepted ratio of 2:1 κ to λ light chain expression. A series of at least six

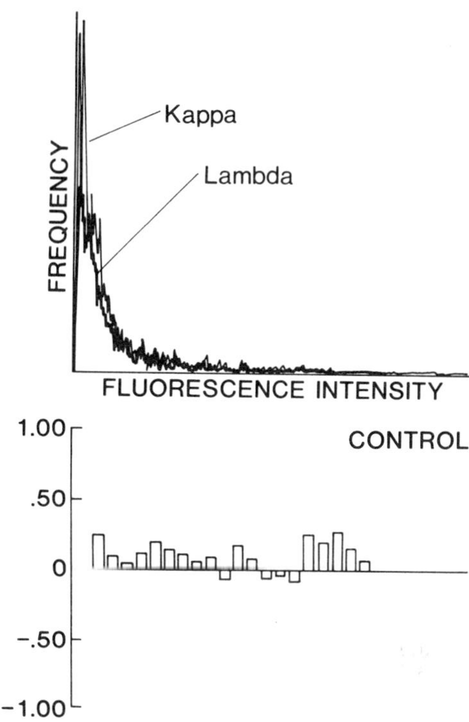

FIGURE 1. The upper figure shows the immunofluorescence profile for surface κ and λ light chain bearing cells in a normal control. The lower figure represents the actual clonal excess calculation of the profile.

consecutive 10-channel readings exceeding .4 on the κ side or −.4 on the λ side was considered as positive CE for the respective light chain immunoglobulin.

Cell Sorting and Dual Labeling by Flow Cytometry

Fluorescence-activated cell-sorting was performed on a FACS IV (Becton-Dickinson FACS System, Sunnyvale, CA.) DNA/"common" ALL antigen (CALLA) determination and DNA/κ measurements were carried out on an Ortho FC200 Flow Cytometer with a 50 mW argon ion laser and an interactive computer system, as previously described [10] Staining for DNA and CALLA antigen was carried out on density-separated cells that were washed with buffer (PBS with 0.2% BSA) and incubated with 200 μl of CALLA antibody (Coulter Clone J5, reconstituted with 0.5 ml of sterile distilled H_2O of which 10 μl was diluted in 390 μl of buffer) for 30 minutes at 4°C. After washing twice with buffer, the supernatant was decanted and 200 μl of FITC labeled goat-anti-mouse antibody (Cappel Laboratories, West Chester, PA), diluted 1:40, was added. After incubation for 30 minutes in an ice bath, the cells were washed twice, fixed in 70% ethanol overnight, washed twice in buffer and incubated with 250 μl of RNase (RASE, Worthington Inc., 500 U/ml) at 37°C for 30 minutes. Then 250 μl of propidium iodide (5 μg/ml in 1.12% Na citrate buffer) was added, and after 30 minutes fluorescence was measured in the Ortho FC 200 Flow Cytometer.

Staining for *DNA and* κ *light chain* was carried out as above: the κ-FITC stained cells were fixed, then treated with RNase, and counterstained with propidium iodide. DNA ploidy was determined by dividing the $G_{0/1}$ DNA content of the sample by that of control (DNA diploid) lymphocytes and by comparison with admixed and simultaneously stained normal lymphocytes in accordance with the procedures outlined in the "Convention on Nomenclature for DNA Cytometry".[11] Differences in DNA content of ⩾ 5% can be detected. "2.0c" denotes diploid DNA stemline.

Immunoperoxidase Staining for Intracytoplasmic μ Heavy Chain

Cytocentrifuge slides obtained from cell suspensions, were fixed in acetone for 5 minutes. Endogenous peroxidase was blocked by incubating the slides with 3% H_2O_2 in PBS. The slides were then subjected to the immunoperoxidase staining procedure for the μ heavy chain as described before[12] using the Dako PAP Kit (Dako Corporation, Santa Barbara, CA).

Analysis of Genomic DNA

High molecular-weight DNA, prepared from leukemic cells expressing CE, was digested to completion (using an internal marker DNA to determine when the digestion was complete) with *Bam*HI or *Eco*RI restriction endonucleases, size-fractionated by agarose-gel electrophoresis, and transferred to nitrocellulose filters.[13,14] Nick-translated DNA fragments containing human Ig genes were hybridized with the nitrocellulose blots in 50% formamide, 6 × SSC, 0.02% bovine serum albumin, 0.02% polyvinylpyrrolidine, 0.02% Ficoll, 100 μg *E. coli* DNA per ml, 5% dextran sulfate (Pharmacia) for about 18 h at 41°C. The blots were washed[14] and visualized by autoradiography.

Analysis of RNA

Total cell RNA was isolated either by SDS-phenol-chloroform extraction, followed by digestion with DNase, or by the guanidium isothiocyanate-cesium chloride method.[15] For analysis by RNA blotting, total cell RNA (35 μg) or poly(A) + RNA was denatured with glyoxal and DMSO, and electrophoresed on a 1% agarose gel in 10 mM phosphate, pH 7.[16] To be sure that the RNA was intact and that the proper quantity was actually loaded onto the gel, the gel was photographed after staining with ethidium bromide. The RNA was transferred to diphenylthioether paper[17] and hybridized to Cκ, Cλ, and Cμ probes as described for the DNA blots, except that yeast RNA (1 mg/ml) replaced *E. coli* DNA. The blots were washed as described.[14]

Probes for Nucleic Acid Hybridization

Cloned human Ig gene probes were kindly provided by P. Leder, J. Ravetch, G. Hollis, S. Korsmeyer, and P. Hieter. The probe for the Cκ gene was a 2.5 kb *Eco*RI fragment.[18] The probe for Cλ genes was a 0.8 kb *Bam-Hind*III fragment

TABLE 1. Clinical Characteristics of the Eight Leukemic Patients Expressing Light Chain Clonal Excess in PB or BM Cells[a]

Case	Age/Sex	DX	WBC (Cells $\times 10^3/mm^3$)	Therapy	BM		PB	
					BL (%)	Lymph (%)	BL (%)	Lymph (%)
1	41/M	BL-CML	94.7	no	76	20.5	75	8
2	24/F	Acl-CML	20.5	no	2	2	0	5
3	35/M	BL-CML	43.0	yes	98.5	1	39	43
4	29/F	BL-CML	75.0	yes	NA	NA	31	12
5	15/F	ALL	0.3	no	78	11	2	82
6	22/M	ALL	77.2	yes	28	70	0	84
7	7/M	ALL	32.1	no	82.5	7	17	41
8	15/M	LBL	3.5	yes	73	7	0	24

[a]Abbreviations: BM = bone marrow; PB = peripheral blood; BL = blasts; BL-CML = blastic phase, chronic myeloid leukemia; Acl-CML = accelerated phase, chronic myeloid leukemia; NA = not available; ALL = acute lymphoblastic leukemia; LBL = lymphoblastic lymphoma.

containing the Mcg C λ gene.[19] This probe detects a number of Cλ genes. The probe for the Cμ gene was a 1.3 kb *Eco*RI fragment.[20]

RESULTS

Clinical Characteristics

Thirty-two patients presenting with acute leukemia (22 cases), CML (5 cases), or LBL (5 cases) were analyzed for CE by flow cytometry. Of the patients with acute leukemia, 11 had ALL, 9 were nonlymphoid, and 2 remained categorized as "undifferentiated." Identifiable CE was found in 8 patients, including 4 cases of BL-CML, 3 "null" ALL, and 1 LBL in leukemic phase (TABLE 1).

Determination of CE and DNA by Flow Cytometry

All four BL-CML patients were positive for Philadelphia chromosome with patient 1 having 55 chromosomes and double Philadelphia chromosomes. His cells were TdT$^+$ and reacted with monoclonal antibodies against the J-5, BA1 BL1, and BL2 antigens. Patient 2 was TdT$^+$ in PB and BM, had low RNA content, hyperdiploidy, and was diagnosed as having an accelerated phase of CML.

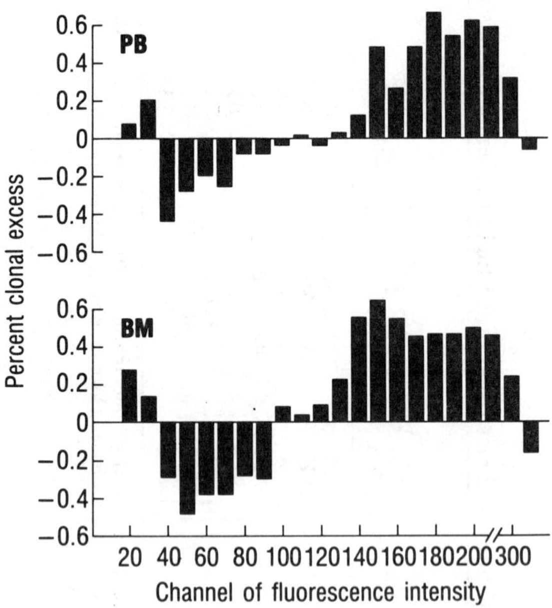

FIGURE 2. Graphic display of clonal excess calculation in patient 1 with blastic CML showing predominance of κ$^+$ cells in channels of brighter fluorescence (140–300).

Patient 3 was TdT$^-$ in PB and BM and was diagnosed as having BL-CML. Patient 4 was Philadelphia chromosome$^+$, TdT$^-$, and initially presented in blast crisis. All four CML patients had demonstrable κCE. PB and BM were concurrently studied in patients 1 and 3: CE was found in both tissues in patient 1 (FIG. 2) but only in the BM from patient 3.

κ$^+$ PB cells from patient 1, estimated to amount to approximately 20% of the entire cell suspension, were isolated by fluorescence-activated cell sorting and subsequently analyzed for DNA aneuploidy by simultaneous two-color flow cytometry of (green = fluorescein) and DNA (red = propidium iodide). κ$^+$ Cells were shown to be diploid whereas the κ$^-$ cells had diploid and hyperdiploid DNA content (FIG. 3). Simultaneous staining for DNA and CALLA (FIG. 4) showed evidence for a proliferating, hyperdiploid, CALLA$^+$, κ$^-$ stemline; diploid cells were CALLA$^-$ but κ$^+$. This patient was analyzed twice more while receiving therapy. The PB κCE was observed to diminish to normal levels.

In patient 2, 69% and 73% of PB and BM cells were aneuploid, respectively, while κCE was detected in the BM only. In patient 3, aneuploidy was not detected in either PB or BM, while κCE was detected only in the BM. Patient 4 was not so analyzed. In conclusion, there was no correlation between κCE expression and aneuploidy. Only the BM of the three ALL patients showed a definite CE (2λ, 1κ), while no clear CE was detected in two concurrently studied PBs. The LBL patient

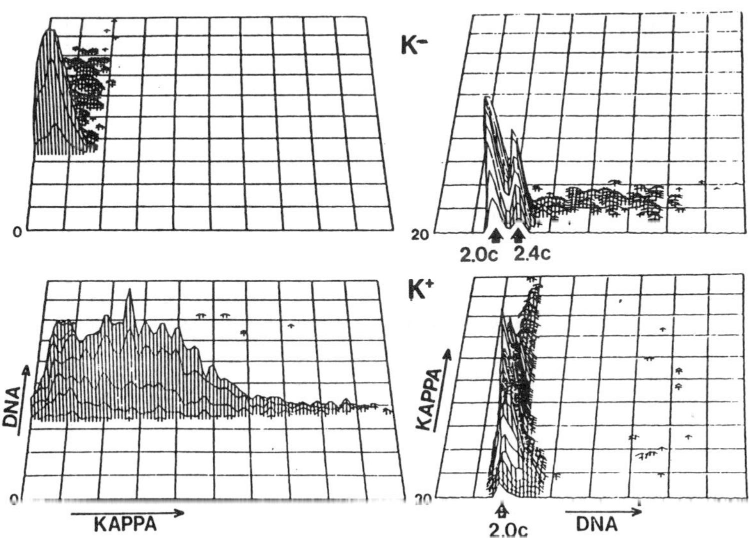

FIGURE 3. DNA/κ light chain flow cytometry of bone marrow cells in patient 1. Cells were stained for immunoglobulin light chain, sorted into κ$^+$ and κ$^-$ populations, and fixed and counterstained for DNA with propidium iodide after RNase treatment. κ(FITC green fluorescence) and red pulse width (RPW) were measured simultaneously. The upper two histograms show the combined DNA/κ measurement of the κ$^-$ population for (left) and DNA (right) axis. Cell suspension was entirely depleted of κ$^+$ cells and were diploid (2.0c) and hyperdiploid (2.4c) with proliferation. The lower two histograms show κ$^+$ cells (left). They have diploid (2.2c) DNA content only. κ$^+$ cells are, therefore, not part of the hyperdiploid population.

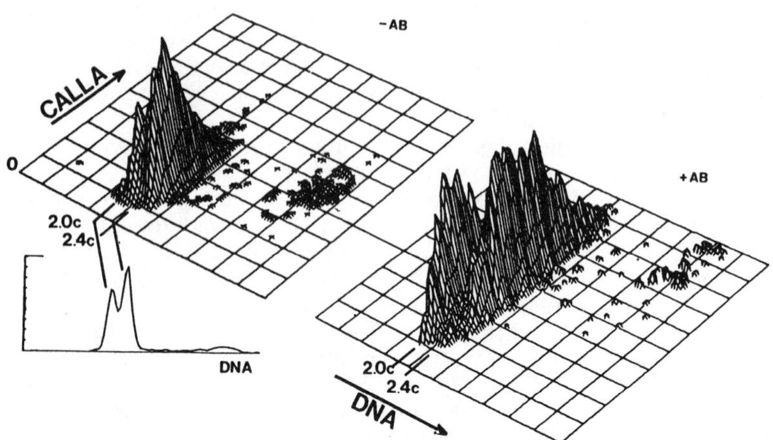

FIGURE 4. DNA/CALLA flow cytometry of bone marrow cells in patient 1. Cells were stained with CALLA (+AB) or nonspecific ascites (−AB), then treated with GAM-FITC, fixed, treated with RNase, and counterstained with propidium iodide for DNA. Red (DNA) and green (FITC) fluorescence and red pulse width (RPW) were simultaneously measured. Computer generated three-dimensional histogram on the left (−AB) shows nonspecific green fluorescence in the absence of CALLA as primary antibody. Histogram on the right shows CALLA fluorescence when antibody is present. DNA histogram shows the presence of diploid (2.0c) and hyperdiploid (2.4c) cells. Hyperdiploid cells expressed CALLA, but were very heterogeneous, with some cells having negative (background) FITC fluorescence.

had detectable λCE in the BM, but the PB was not analyzed. None of the ALL or LBL patients possessed an aneuploid stemline detectable by flow cytometry.

Cytoplasmic μ Heavy Chain by Immunoperoxidase Method

Evaluable smears were obtained from six of the eight patients that had CE; three of these six cases (2, 7, and 8) showed positive staining for cIg μ (> 10% of the cells staining positively). The other three cases were negative (1, 4, and 6).

Search for Ig Gene Rearrangements in BL-CML Cells

As B-cells that express sIg would be expected to contain rearranged Ig genes, we examined the configuration of the Ig heavy and light chain genes in BM cells from patients 1 and 3 by DNA blotting experiments. High molecular-weight DNA was digested with the restriction endonucleases, *Bam*HI or *Eco*RI, electrophoresed on agarose gels, and blotted to nitrocellulose;[13] the blots were then annealed with nick-translated probes for Ig μ, κ, or λ chain genes. As shown in FIGURE 5 (in the blot annealed with Cμ) cells from both patients 1 and 3 contained two rearranged Cμ genes; some of the BM cells from patient 1 also contained a μ

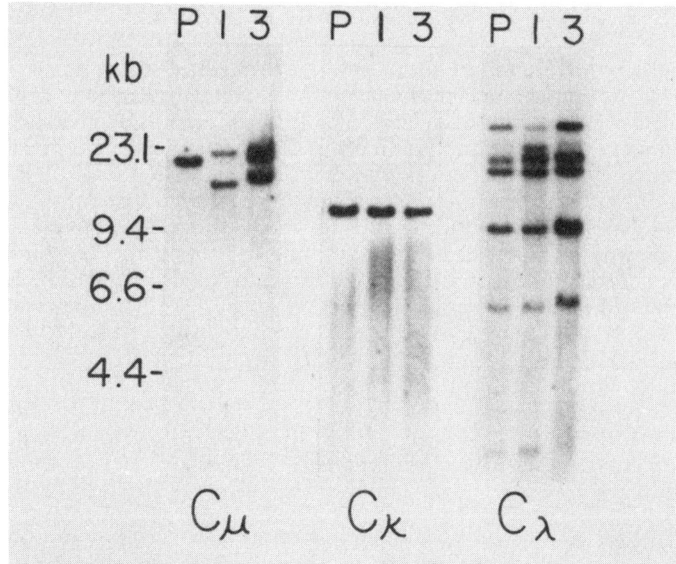

FIGURE 5. Blots of BL-CML DNAs annealed with Ig genes. Ten μg of genomic DNA from placenta (lanes P) and from BM cells from patient 1 (lanes 1) and from patient 3 (lanes 3) were digested with *Bam*HI (blots annealed with Cμ and Cκ) or *Eco*RI (blot annealed with Cλ), electrophoresed in an 0.8% agarose gel, blotted to nitrocellulose and annealed with nick-translated Cμ, Cκ or Cλ genes (indicated below each blot). The sizes (in kb) of fragments of λ bacteriophage DNA, digested with *Hin*d III, which were electrophoresed in a parallel lane are indicated on the left. The additional Cλ band (~ 24 kb) in patient 1 occurs because λ genes demonstrate restriction fragment polymorphism among individuals, and not because of the rearrangement of λ genes as found in B cells that express λ chains.[19]

gene fragment migrating at the size of the germline μ fragment present in DNA from placenta (lane P), a non-Ig-producing tissue. Although approximately 20% of the BM cells from these patients expressed κCE, we could not detect rearranged κ chain genes in these cells (lanes 1, 3 in blot annealed with Cκ). No rearranged λ genes were detected. The presence of three fragments in approximately equivalent amounts in the BM cells from patient 3 indicates that there were at least two different clones of cells present, one or both of which may be the blastic population.

To explain our inability to detect rearranged κ genes, it is possible that only a portion of the patients' BM cells contained rearranged κ genes. If the BL-CML cells were undergoing κ gene rearrangement, different cells would contain different κ gene configurations, and many of the cells would still contain κ genes in the germline configuration. Alternatively, it is possible that the κ+ cells in the BM were a polyclonal normal cell population reacting against the leukemia. This hypothesis seems improbable because these normal cells would not be expected to demonstrate a κCE, as such an antigen-stimulated B cell population would be more mature than the leukemic cells and should express both κ and λ chains.

CE Expression by a "Common" ALL Cell Line (REH)

In order to provide a more direct assay for Ig light chain expression, it was important to determine whether the leukemic cells that expressed light chain CE contained Ig light chain RNA. The BM or PB cells from the leukemic patients could not be used for an examination of Ig RNA because even after purification by density centrifugation these cells were contaminated to a level of about 10% with non-leukemic cells whose light chain mRNA would interfere. This was confirmed by an analysis of normal PB in which we detected both κ and λRNA (data not shown). Thus, we decided to examine Ig light chain RNA and Ig light chain genes in a cell line that was derived from a leukemic patient. The REH cell line was derived from a "common" ALL case and was originally considered to have a non-T, non-B phenotype, displaying reactivity with CALLA antibody and a high level of TdT enzymatic activity.[21,22] Further phenotypic characterization of this cell line by us has revealed expression of pre-B cell antigens (BL1 and BL2) as detected by monoclonal antibodies,[23] and reactivity with antibodies recognizing HLA-DR and DC/DS specificities.[24] In addition to cytoplasmic μ heavy chain expression, detected by immunoperoxidase technique, flow cytometric determination of sIg revealed λCE in approximately 35% of dimly staining cells (FIG. 6).

Analysis of the configuration of the Ig light chain genes in REH cells by DNA blotting revealed that both κ genes had been deleted (FIG. 7a, lane 2) and that λ genes had been rearranged (FIG. 7b, lane 2) relative to the λ genes present in non-Ig-producing cells (placenta, lane 1), as expected in λ light chain[+] B cells. When total cell RNA (FIG. 8) or poly (A) + RNA (not shown) from REH cells was examined for the presence of Ig RNAs by RNA blotting, the results of the CE assay were further supported. Lane 1 in FIGURE 8a shows the results of the hybridization of 35μg of total cell RNA with the Cλ probe. A 1.1-kb poly (A) + RNA, the size of λRNA, was detected. When this blot was melted and rehybridized with the Cκ probe, no κ RNA was detected in REH cells (FIG. 8b, lane 1). As Ig λ chains were detected on the surface of REH cells it would be predicted that REH cells synthesize μ mRNA and μ heavy chains, since the expression of sIg generally appears to require synthesis of light and heavy chains.[25] When total

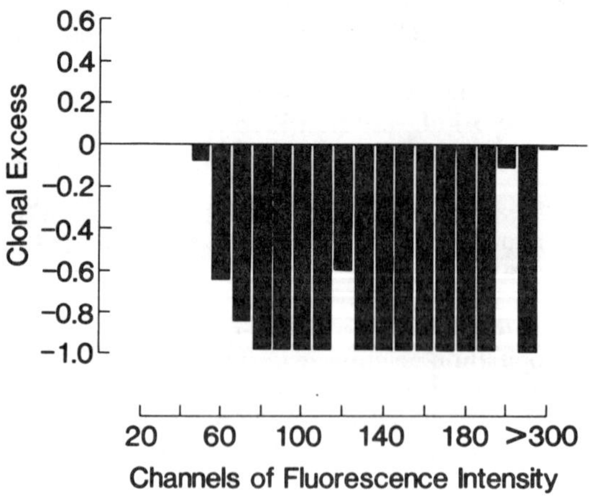

FIGURE 6. Graphic display of clonal excess calculation in REH cell line showing predominance of λ light-chain-bearing cells.

FIGURE 7. Blots of REH cell DNAs annealed with Ig light chain genes.

a. *Cκ genes*: Ten μg of genomic DNA from placenta (lane 1), and from REH cells (lane 2) were digested with *Bam*HI, electrophoresed in an agarose gel, and blotted to nitrocellulose.[12,13] The blot was annealed with the Cκ probe.

b. *Cλ genes*: Ten μg of *Eco*RI-digested DNA from placenta (lane 1) or from REH cells (lane 2), annealed with the Cλ probe. The arrows indicate the two rearranged Cλ genes present in REH cells. The bands migrating at 25 kb and larger, which were present in REH cells but not in placenta DNA (from another individual), probably differed in size owing to restriction fragment polymorphism among individuals.[19] The two λ bands indicated by arrows do not correspond in size to any of the known Cλ germ line fragments.[19]

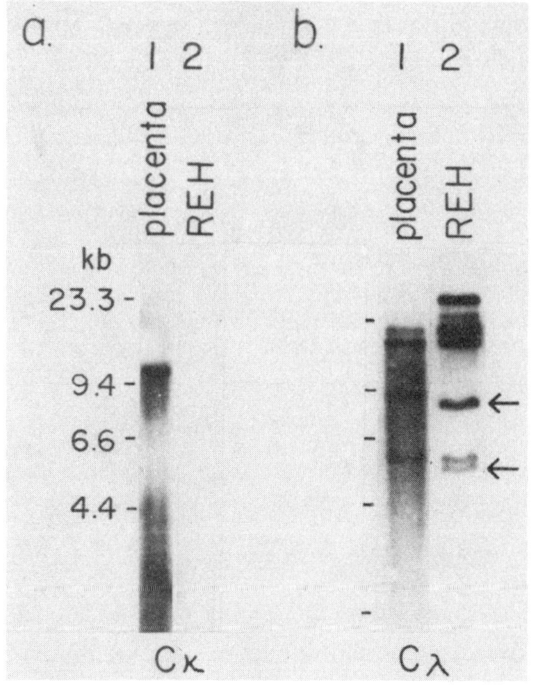

cell RNA from REH cells was hybridized with the Cμ probe (FIG. 8c, lane 1) it was found that REH cells do indeed contain RNA which hybridized with the Cμ probe. The μ RNAs in REH cells corresponded in size with μ RNAs detected in PB lymphocytes (lane 3).

To determine whether REH cells had the potential to differentiate along the B-cell pathway, REH cells were treated with the phorbol ester 12-O-tetradecanoyl-phorbal-13-acetate (TPA) for 6 days at a concentration of 10^{-9} M. RNA was extracted from treated and untreated cells and analyzed by RNA blotting. TPA was found to cause approximately a fivefold increase in the amount of λ chain RNA (FIG. 8a, lane 2, compared with lane 1), and approximately a twofold increase in the amount of μ chain RNA (FIG. 8c, lane 2). The amount of λ chain mRNA present after induction was about one half as much as that present in a λ-producing EBV-transformed B-cell line (FIG. 8a, lane 3). No κ RNA was detected (FIG. 8b, lane 2). Together these results indicate that REH is a B-cell line (not a pre-B-cell line) which can be induced to differentiate along the B-cell pathway.

DISCUSSION

Since expression of sIg light chains normally requires differentiation past the pre-B-cell stage, its detection in patients with pre-B-cell leukemia could be explained by the maturation of some of the leukemic cells beyond the pre-B stage. Nadler *et*

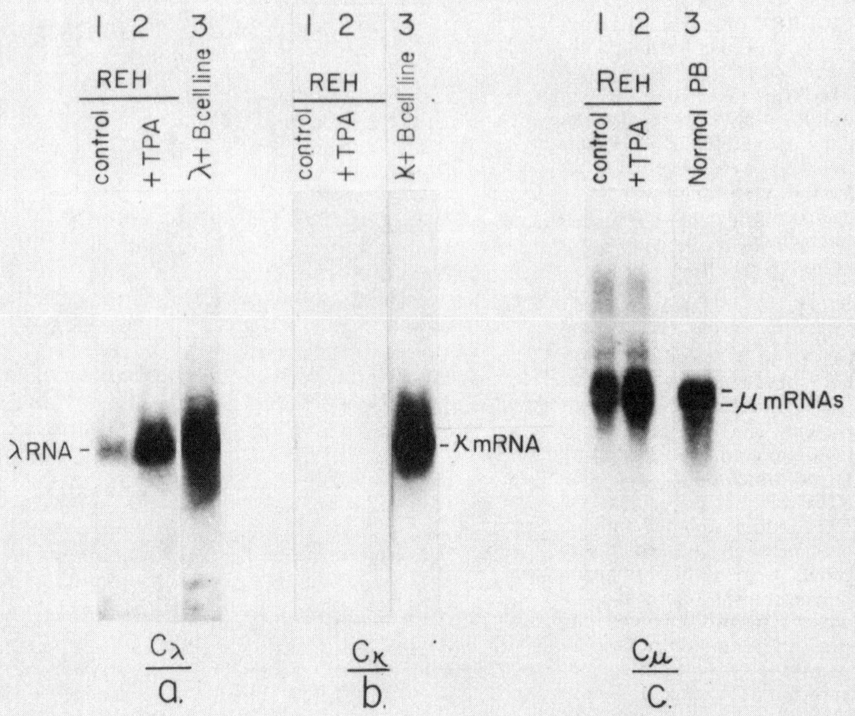

FIGURE 8. Blots of RNA from the REH cell line hybridized with a:Cλ; b:Cκ; c:Cμ probes.

a. Each lane contained 35 μg of total cell RNA, which was denatured, electrophoresed, and blotted as described in METHODS. The RNAs were from REH cells (lane 1), REH cells treated with TPA for 6 days (lane 2), and a λ-producing EBV-transformed human lymphoblastoid cell line (lane 3). The blot was hybridized with the Cλ gene probe (150×10^6 cpm), washed, and exposed to X-ray film for 13 hours. The position of 1.1 kb λ RNA is indicated.

b. The lanes contain: 35 μg of total cell RNA from REH cells, uninduced (lane 1); 35 μg total cell RNA from REH cells induced with TPA (lane 2); and 1 μg of poly (A) + RNA from a κ-producing EBV-transformed human lymphoblastoid cell line (lane 3). The blot was hybridized with the Cκ probe (150×10^6 cpm), washed and exposed to X-ray film for 8 hours. The position of κRNA (1.1 kb) is indicated.

c. The lanes contain: 35 μg total cell line RNA from REH cells, uninduced (lane 1); 35 μg of total cell RNA from REH cells induced with TPA (lane 2); and 16 μg of total cell RNA from normal PB cells (lane 3). The blots were hybridized with a genomic DNA fragment (35×10^6 cpm) coding for the human Cμ gene.[20] Exposure to X-ray film was carried out for 7 days.

al.[26] found that expression of the B1 antigen, which is present on normal B-lymphocytes and in 70% of the cases of Ia+, CALLA+, non T-cell ALL, can be induced in a portion of the remaining 30% of these non-T-cell ALL cases (mostly in those which are CALLA+) by the use of culture media conditioned by leukocytes treated with TPA or PHA. In some cases of Ia+, CALLA+, B1−, cIg μ−, non T-cell ALL, TPA could induce expression of both B1 and cIg μ. Nadler *et al.* proposed that within any one patient with non T-cell ALL there is a heterogeneous spectrum of cells at various stages of early B-cell differentiation including: (1) pre-B-cells (Ia[+] CALLA[+] B1[−] cIg μ[−]); (2) intermediate pre-B-cells (Ia[+] CALLA[+] B1[+] cIg μ[−]; and (3) true pre-B-cells (Ia[+] CALLA[+] B1[+] cIg μ[+]). More recently, Cossman *et al.*[5] reported that TPA treatment of the "null" lymphoblasts of a young patient with Philadelphia chromosome[−] ALL resulted in the serial progression of these lymphoblasts through pre-B and B-cell stages *in vitro*. As early as 24 hours after incubation with TPA, μ chain was synthesized in all of the cells, and sIg (μ and κ chains) were detected on a fraction of the leukemic cells. The rapidity of the appearance of the Igs argues that they were products of previously rearranged μ and κ genes. Moreover, since less than 30% of the TPA-treated leukemic cells expressed sIg, although all lymphoblasts contained cIg μ heavy chain, the leukemic cells appeared to be developmentally heterogeneous.

The existence of intratumor heterogeneity is strongly supported by the evidence that blast crisis in CML usually results from the evolution of a genetically altered clone derived from the Philadelphia chromosome[+] cells,[6] i.e. additional chromosome abnormalities are observed in the Philadelphia chromosome[+] cells as blast crisis evolves, only to disappear upon remission. Intratumor phenotypic heterogeneity, particularly with regard to which Ig chains are expressed, may be due to the occurrence of Ig gene rearrangements during maturation of the leukemic cells. Bakhshi *et al.*[7] presented evidence indicating that two different blast crises of CML within the same patient contained Ig λ genes in two different configurations, although the Ig H chain genes were identical, suggesting that different blast crises were derived from a cell which was capable of undergoing λ chain gene rearrangement. The expression of Ig ultimately depends on a stochastic process of site-specific recombinational events[27] in which Ig chain gene formation precedes that of light chain genes.[28] Because they found that Ig heavy chain gene rearrangements were relatively common in non-T-cell ALL[4] and in BL-CML,[7] whereas light chain gene rearrangements were less common, Korsmeyer *et al.* proposed[4] that precursor lymphoid B cells might be the most frequent population involved in non-T-cell ALL.

The results of our analysis of the REH cell line, which was derived from "null" ALL blasts, differ from those of Korsmeyer *et al.*[4] on the same cell line, as they detected no cIg μ chain nor sIg expression, whereas we detected cIg μ chain, μ RNA (of the size of μ mRNA), sIg λ chain and λ RNA (of the size of λ mRNA). They did detect, however, cIg λ chain and secreted λ chain, and like us, rearranged Ig μ and λ genes. As we have not exchanged cell lines we cannot explain the discrepancy.

It is surprising that in patient 1 the sorted κ[+] population had a diploid stemline and that κ CE was not associated with aneuploidy in any patient as determined by flow cytometry. It cannot be excluded that maturation to κ[+] is possible in some rare diploid cells arising by some form of clonal evolution, although this process usually involves addition, not loss, of DNA and chromosomes. Our finding of CE in cell populations with a predominance of "null" lymphoblasts might represent a mature variant clone. It is also possible that the CE observed in CML cases is present in clonally restricted cells not obviously blastic by morphology. Alternatively, sIg light chain could be expressed in cells reacting to the tumor although these cells should not be clonally restricted to express either κ or λ. Demonstration of any antitumor activity in this population would have implications for therapeutic management. However, the finding of λ

CE in a significant proportion of REH cells is a strong argument that sIg light chain can be expressed in the neoplastic cells in "null" ALL.

In conclusion, by using the very sensitive CE method, we have been able to detect sIg light chain expression in a significant fraction of leukemias that were previously classified to be at the pre-B stage, and also in the REH cell line, which was derived from "null" lymphoblasts. Although clinical correlation in a larger group of patients will be required, the detection of CE in pre B-cell leukemias is a finding that may ultimately influence both diagnosis and prognosis by recognizing a leukemic subtype that might exhibit a particular clinical behavior and responsiveness to therapy.

SUMMARY

Cytofluorographic analysis of surface immunoglobulin (sIg) light chain clonal excess (CE), defined as $(\%\kappa^+ - \%\lambda^+)/(\%\kappa^+ + \%\lambda^+)$ cells per discrete level of fluorescence intensity, was carried out on mononuclear cells of 32 leukemic patients. Eight demonstrated sIg light chain CE, including four blastic chronic myeloid leukemias (BL-CML), three "null" acute lymphoblastic leukemias (ALL), and one leukemic lymphoblastic lymphoma. Six of the leukemias demonstrated a κCE and two had a λCE. Sorted κ^+ PB cells from a BL-CML patient were shown to have a diploid DNA stem line and to bear the "common" ALL antigen. To provide further support for our finding of the expression of sIg light chains in ALL, we studied the REH cell line, derived from a "common" ALL patient and found cytoplasmic μ heavy chain and surface Ig λ CE. Nucleic acid blotting experiments on REH revealed that both κ genes had been deleted and that λ genes had been rearranged, as expected in B cells expressing λ light chains. Moreover, REH cells contained μ and λ RNA. When REH cells were treated with TPA the amount of μ chain RNA increased by approximately fivefold and the amount of λ chain RNA increased by approximately twofold. The finding of sIg light chain in pre-B cell leukemias and in the REH cell line, suggests that these leukemic cells are further differentiated along the B-cell lineage than was previously believed.

ACKNOWLEDGMENTS

We thank Ms. Olga Kekish and Edith Espirito for excellent technical assistance, and Dr. Phil Leder and his colleagues for DNA probes.

REFERENCES

1. VOGLER, L. B., W. M. CRIST, D. E. BOCKMAN, E. R. PEARL, A. R. LAWTON & M. D. COOPER. 1978. Pre-B-cell leukemia: A new phenotype of childhood lymphoblastic leukemia. N. Engl. J. Med. **298**:872.

2. NADLER, L. M., R. STASHENKO, J. RITZ, R. HARDY, J. M. PESANDO & S. F. SCHLOSSMAN. 1981. A unique cell surface antigen identifying lymphoid malignancies of B-cell origin. J. Clin. Invest. **67**:134.

3. KERSEY, J. H., T. W. LeBIEN, C. S. ABRAMSON, R. NEWMAN, R. SUTHERLAND & M. GREAVES. 1981. A human leukemia associated and lymphohemopoietic progenitor cell surface structure identified with monoclonal antibody. J. Exp. Med. **153**:726.

4. KORSMEYER, S. J., A. ARNOLD, A. BAKHSHI, J. V. RAVETCH U. SIEBENLIST, P. A. HIETER, S. O. SHARROW, T. W. LEBIEN, J. H. KERSEY, D. G. POPLACK, P. LEDER & T. A. WALDMANN. 1983. Immunoglobulin gene rearrangement and cell surface antigen expression in acute lymphocytic leukemias of T and B-cell precursor origins. J. Clin. Invest. **71**:301.

5. COSSMAN, J., L. M. NECKERS, A. ARNOLD & S. J. KORSMEYER. 1982. Induction of differentiation in a case of common acute lymphoblastic leukemia. N. Engl. J. Med. **307**:1251.

6. KOEFFLER, H. P. & D. W. GOLDE. 1981. Chronic myelogenous leukemia: New concepts. N. Engl. J. Med. **304**:1201.

7. BAKHSHI, A., J. MINOWADA, A. ARNOLD, J. COSSMAN, J. P. JENSEN, J. WHANG-PENG, T. A. WALDMANN & S. J. KORSMEYER. 1983. Lymphoid blast crises of chronic myelogenous leukemia represent stages in the development of B-cell precursors. N. Engl. J. Med. **309**:826.

8. LIGLER, F. S., E. S. VITETTA, R. G. SMITH, J. B. HIMES, J. W. UHR, E. P. FRENKEL & J. R. KETTMAN. 1979. An immunologic approach for the detection of tumor cells in the peripheral blood of patients with malignant lymphoma: Implication for the diagnosis of minimal disease. J. Immunol. **123**:1123.

9. KOZINER, B., S. KEMPIN, S. PASSE, T. GEE, R. A. GOOD & B. D. CLARKSON. 1980. Characterization of B-cell leukemias: A tentative immunomorphological scheme. Blood **56**:815.

10. ANDREEFF, M, J. D. BECK, Z. DARZYNKIEWICZ, F. TRAGANOS, S. GUPTA, M. R. MELAMED & R. A. GOOD. 1978. RNA content of human lymphocyte subpopulations. Proc. Natl. Acad. Sci. USA **75**:1938.

11. HIDDEMANN, W., L. SCHUMANN, M. ANDREEFF, B. BARLOGIE, C. L. HERMAN, R. C. LEIF, B. H. MAYALL, R. F. MURPHY & A. A. BANDBERG. 1984. Convention on Nomenclature for DNA Cytometry. Cytometry **5**:445.

12. MASON, D. Y., C. FARRELL & C. R. TAYLOR. 1975. The detection of intracellular antigens in human leukocytes by immunoperoxidase staining. Br. J. Haematol. **31**:361.

13. SOUTHERN, E. M. 1975. Detection of specific sequences among DNA fragments separated by gel electrophoresis. J. Mol. Biol. **98**:503.

14. STAVNEZER, J., K. B. MARCU, S. SIRLIN, B. ALHADEFF & U. HAMMERLING. 1982. Rearrangements and deletions of immunoglobulin heavy chain genes in the double-producing B-cell lymphoma I 29. Mol. Cell. Biol. **2**:1002.

15. GLISIN, V., R. CRKVENJAKOV & C. BYUS. 1974. Ribonucleic acid isolated by cesium chloride centrifugation. Biochemistry **13**:2633.

16. MCMASTER, J. K. & G. G. CARMICHAEL. 1977. Analysis of single and double-stranded nucleic acids on polyacrylamide and agarose gels by using glyoxal and acridine orange. Proc. Natl. Acad. Sci. USA **74**:4835.

17. SEED, B. 1982. Diazotizable arylamine cellulose papers for the coupling and hybridization of nucleic acids. Nucl. Acids Res. **10**:1799.

18. HIETER, P. A., E. E. MAX, J. G. SEIDMAN, J. F. MAIZEL & P. LEDER. 1980. Cloned human and mouse kappa immunoglobulin constant and J region genes conserve homology in functional segments. Cell **22**:197.

19. HIETER, P. A., G. F. HOLLIS, S. J. KORSMEYER, T. A. WALDMANN & P. LEDER. 1981. The clustered arrangement of immunoglobulin lambda light chain constant region genes in man. Nature **294**:536.

20. RAVETCH, J. V., U. SIEBELIST, S. J. KORSMEYER, T. A. WALDMANN, & P. LEDER. 1981. Structure of the human immunoglobulin locus: Characterization of embryonic and rearranged J and D genes. Cell **27**:583.

21. ROSENFELD, C., A. GOUTNER, C. CHOQUET, A. M. VENUAT, B. KAJIBANDA, J. L. PICO & M. F. GREAVES. 1977. Phenotypic characterization of a unique none T-, non B- acute lymphoblastic leukemia cell line. Nature **267**:841.

22. LEBIEN, T. W., F. J. BOLLUM, W. G. YASMINEH & J. H. KERSEY. 1982. Phorbol ester-induced differentiation of a non T-, non B- leukemic cell line: Model for human lyphoid progenitor cell development. J. Immunol. **128**:1316.

23. AL-KATIB, A., C. Y. WANG, T. S. GEE & B. KOZINER. 1982. Characterization of non T

cell lymphoid neoplasia by a panel of monoclonal antibodies to B-cell differentiation antigens. Blood **60**:(#498) 143a.

24. AL-KATIB, A. & B. KOZINER. 1984. Leu-10 antigen distribution in leukemic disorders: Correlation with HLA-DR expression. Brit. J. Hematol. **57**:373.

25. WALL, R. & M. KUEHL. 1983. Biosynthesis and regulation of immunoglobulins. Ann. Rev. Immunol. **1**:393.

26. NADLER, L. M., J. RITZ, M. P. BATES, E. K. PARK, K. C. ANDERSON, S. E. SALLAN & S. F. SCHLOSSMAN. 1982. Induction of human B-cell antigens in non T-cell acute lymphoblastic leukemia. J. Clin. Invest. **70**:433.

27. COLECLOUGH, C., P. P. PERRY, K. KARJALAINEN & M. WEIGERT. 1981. Aberrant rearrangements contribute significantly to the allelic exclusion of immunoglobulin gene expression. Nature **290**:372.

28. KORSMEYER, S. J., P. A. HIETER, J. V. RAVETCH, D. G. POPLACK, T. A. WALDMANN & P. LEDER. 1981. Developmental hierarchy of immunoglobulin gene rearrangments in human leukemic pre-B cells. Proc. Natl. Acad. Sci. USA **78**:7096.

Frequency and Clinical Significance
of DNA Aneuploidy in
Acute Leukemia

W. HIDDEMANN,[a] B. WÖRMANN[a], J. RITTER,[b] E. THIEL,[d]
W. GÖHDE,[c] B. LAHME,[a] G. HENZE,[e] G. SCHELLONG,[b]
H. RIEHM,[f] AND TH. BÜCHNER[a]

*Departments of [a]Internal Medicine, [b]Pediatrics, and [c]Radiology
University of Münster
4400 Munster, Federal Republic of Germany*

*[d]Institute of Hematology
Gesselschaft für Strahlen- und Umweltforschung
Munich, Federal Republic of Germany*

*[e]Department of Pediatrics
University of Berlin
Berlin, Federal Republic of Germany*

*[f]Department of Pediatrics
University of Hannover
Hannover, Federal Republic of Germany*

INTRODUCTION

After the introduction of flow cytometry (FCM) to hematology by Büchner *et al.*[1,2] and shortly thereafter by Melamed and co-workers,[3] the majority of FCM studies was predominantly undertaken to elucidate the effect of cytostatic drugs on the cell cycle. Especially in acute leukemias, these investigations aimed at optimizing the combination of currently used cytostatic agents according to cell kinetic findings.[4-8]

In conjunction with these analyses of the cellular DNA content, aneuploidies, as defined by an aberrant DNA content, were described for the first time.[1,9] The overall frequency of detectable DNA aneuploidies, however, did not exceed 10% in these first investigations.[9,10]

Based on technical improvements of the applied flow systems, such as the development of a sheath flow glass chamber,[11] the use of a high resolution multichannel analyzer, and the rigid application of standardized reference measurements,[12,13] our group first reported DNA aneuploidies in acute leukemia at a frequency of 40%.[12,14,15]

This high frequency of aneuploid DNA stemlines could be confirmed by the present investigation on children and adults with acute leukemias. The present

---

[a]Address correspondence to: W. H., Dept. Internal Medicine, University Münster, Albert-Schweitzer-Str. 33, 4400 Münster, FRG.

TABLE 1. Overview of the Patients Analyzed for the Present Study

Patients	Acute Myeloid Leukemia	Acute Lymphoblastic Leukemia
Adults	125	91
Children	50	180

study also revealed correlations between the immunologic and morphologic subtypes of leukemias and a tendency towards longer remissions for patients with DNA aneuploidy in the childhood BFM studies AML 78 and ALL 79/81 as well as in the adult AML trial 78/81.

MATERIAL AND METHODS

For the present study, 446 patients were recruited from the Department of Internal Medicine of the University of Münster (adults with AML and ALL), the Departments of Pediatrics of the Universities of Münster, Berlin, and Giessen (children with AML), and the institutions participating in the multicenter BFM trials on childhood ALL and the German multicenter studies on ALL in adults (TABLE 1).

The diagnosis was based in all cases on the criteria defined by the FAB working committee[16] and additional cytochemical stainings including PAS, POX, alpha-naphtyl-acetate esterase, and acid phosphatase reactions.

While in ALL the FAB classification was not found useful for the further evaluation, the AML subgroups M 1–M 6 were included for subsequent analyses. In ALL, the immunologic subgroups were evaluated instead, as defined by polyclonal antibodies against T and B cells and by E-rosette formation in the childhood BFM study 79/81 discriminating between T, B, and non-T/non-B ALL whereas more refined analyses were carried out in the subsequent BFM trials and the adult ALL studies providing the means to distinguish the major subgroups T, B, C, and Null-ALL.[17–20]

In all patients the following parameters were evaluated in addition: age, sex, WBC, response to induction therapy, and duration of the first complete remission.

All patients were treated uniformly in the corresponding clinical trials, which

TABLE 2. Treatment Protocols for the Corresponding Patient Samples

Acute myeloid leukemia	
Adults	AML studies 78/81 and 81-[21–23]
Children	BFM studies 78 and 83[24–26]
Acute lymphoblastic leukemia	
Adults	Multicenter BMFT trials[20,27]
Children	BFM studies 79/81 and 81/83[18,28]

have been published in detail elsewhere (TABLE 2). The response to therapy was assessed according to CALGB criteria.[29,30]

In all patients, measurements of the cellular DNA content were carried out prior to therapy on heparinized samples from bone marrow aspirates or biopsies and/or peripheral blood. All specimens were subjected to a density gradient separation over Hypaque Ficoll (density 1.078 g/ml) for 20 min at room temperature. The white cell layer was removed, resuspended in HBSS, washed twice, and subsequently fixed in 96% ice-cold ethanol. Prior to staining the cells were centrifuged again at 1.000 g for 10 min, the pellet was resuspended in a 0.5% pepsin-HCl solution (pepsin Merck 1.000 U/g) for 5–8 min and stained by ethidium bromide and mithramycin in combination.[31,32] After 30 min of staining the cells were measured on a modified ICP 11 hooked up to a 1024 channel multichannel analyzer.

The quantitation of the cell cycle phases was carried out according to Göhde[33] with an additional correction for cell clumping.[13,34]

For the detection of DNA aneuploidies and the determination of DNA stemlines all samples were mixed with diploid mononuclear cells from normal blood donors at two different concentrations.[12,13] The appearance of a second $G_{0/1}$ peak and its variation according to the ratio between sample versus reference cells was considered representative for a DNA aneuploidy (FIGURE 1).[35] The degree of DNA aneuploidy was defined by the DNA index:[35,36] (rel. DNA content of $G_{0/1}$ cells of the sample) ÷ (rel. DNA content $G_{0/1}$ reference cells).

Statistical analyses included the modified chi-square test for groups of independent variables. Differences between mean values were calculated by the t-test.

The method of Kaplan Meier was applied for the evaluation of life tables with differences being assessed by the log rank test.

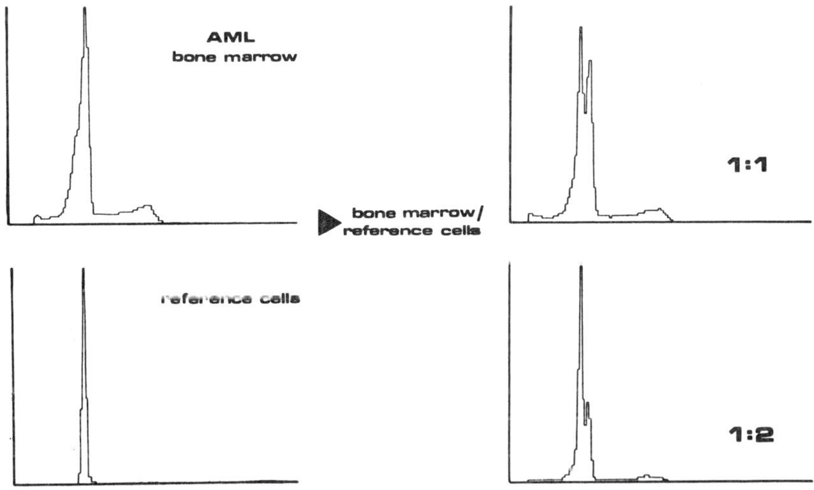

FIGURE 1. Reference measurements with diploid control cells at two concentrations in a patient's bone marrow sample revealing a DNA aneuploidy.

TABLE 3. Frequency of DNA Aneuploidies in Acute Leukemias

	N	DNA Aneuploidy
AML		
Adults	125	50 (40.0%)
Children	50	19 (38.0%)
ALL		
Adults	91	34 (37.4%)
Children	180	72 (40.0%)

RESULTS

DNA aneuploidies were identified in 175 of the 446 cases analyzed thus accounting for an incidence of 39.2% (TABLE 3).

DNA Aneuploidy in AML

From the 175 patients with AML, 10 had a history of preleukemic syndromes and were therefore excluded from the further analysis.

In the remaining group of 165 cases, aneuploid DNA stemlines were detected at a similar frequency in children (19 of 50, or 38%) and adults (48 of 115, or 41.7%) with an overall incidence of 40.6% (TABLE 4).

Biclonal leukemias were found in four patients, all of them being over 50 years of age.

Excluding these four extraordinary cases DNA indices ranged from 0.93–1.20 with a median of 1.10.

No differences in the frequency and degree of DNA aneuploidies were observed between male and female patients.

Analysis of the FAB subgroups, however, revealed a significantly lower frequency of DNA aneuploidies in M 1 leukemias as compared to the other morphologic subtypes ($p < 0.05$) (TABLE 5).

In addition, DNA indices were significantly lower in M 1 and M 2 leukemias as compared to M 4 and M 5 cases ($p < 0.05$) (TABLE 5, FIGURE 2).

TABLE 4. Incidence and Degree of DNA Aneuploidies in Children and Adults with AML

	N	DNA Aneuploidy	DNA Index Range Median
Adults			
de novo AML	115	48 (41.7%)	0.66–2.10
			1.10
AML after preleukemia	10	2	1.09–1.30
Children	50	19 (38.0%)	1.06–1.18
			1.10

TABLE 5. Incidence and Degree of DNA Aneuploidies for the AML FAB Subgroups M 1 to M 6

FAB Subtype	N	DNA Aneuploidy	Median DNA Index
M 1	30	7 (23.3%)[a]	1.08[b]
M 2	50	19 (38.0%)	1.07[b]
M 3	6	2	
M 4	36	16 (44.4%)	1.12
M 5	27	14 (51.9%)	1.11
M 6	11	4 (36.4%)	
Unclassified	1	1	

[a]M 1 versus M 2–M 6, $p < 0.05$.
[b]M 1 + M 2 versus M 4 + 5, $p < 0.05$.

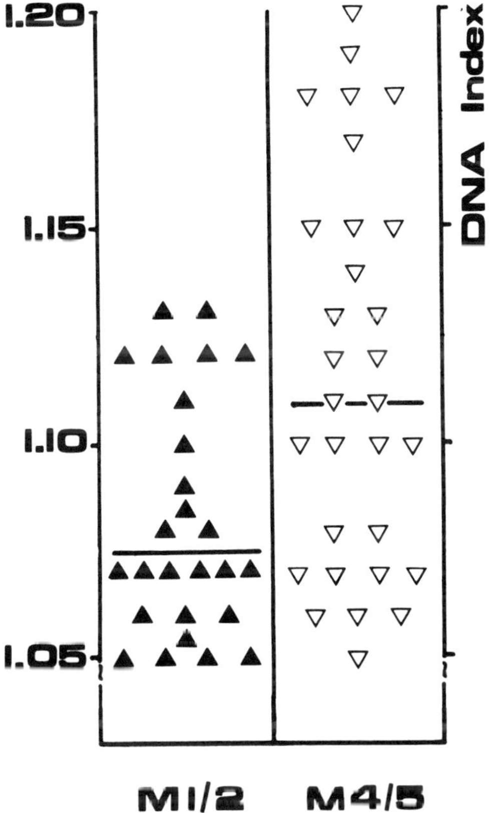

FIGURE 2. Distribution of DNA indices for the AML subgroups M 1 + M 2 versus M 4 + M 5.

While in adult patients no differences in the WBC were detected between patients with and without DNA aneuploidy, significantly lower WBC were found in children with DNA aneuploidy as compared to childhood AML cases without aneuploid DNA stemlines ($p < 0.05$) (FIGURE 3).

The prognostic significance of DNA aneuploidy was assessed for the incidence of complete remissions after induction therapy as well as for remission duration.

Both in children and adults with AML no differences in remission rates were observed between cases with and without DNA aneuploidies.

For remission duration, however, a similar tendency towards a higher percentage of long term remissions in patients with DNA aneuploidies emerged from the comparison of children and adults with and without aneuploid DNA stemlines in the two studies with the longest observation time (adult AML trial 78/81, BFM AML 78). The differences were not statistically significant as evaluated by the log rank test.

DNA Aneuploidy in ALL

The frequency of DNA aneuploidy in ALL was 39.1%, i.e. 106 from 271 patients expressed aneuploid DNA stemlines. As in AML no differences in the

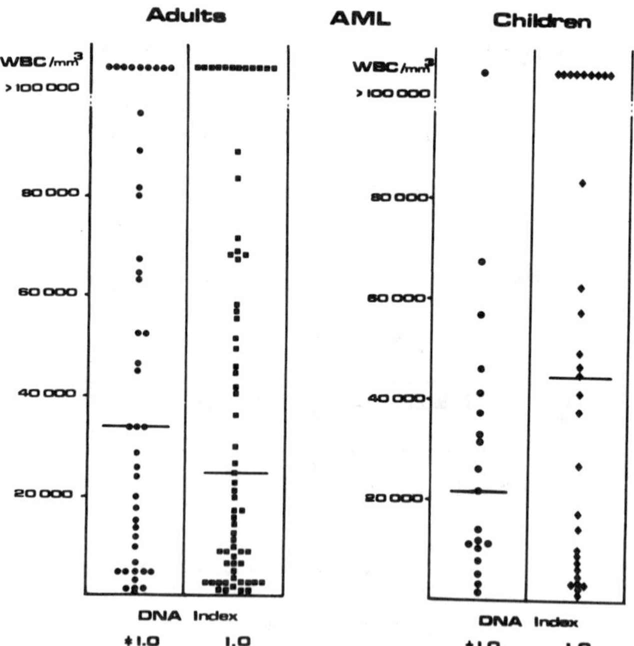

FIGURE 3. Distribution of WBC in children and adults with AML and with (DNA index ≠ 1.0) or without DNA aneuploidy (DNA index 1.0).

TABLE 6. Incidence and Degree of DNA Aneuploidies in Children and Adults with ALL

	N	DNA Aneuploidy	DNA Index Range Median
Adults	91	34 (37.4%)	0.88–1.83 1.14
Children	180	72 (40.0%)	0.85–2.00 1.16

incidence of DNA aneuploidies were revealed between children (72 of 180, or 40.0%) and adults (34 of 91, or 37.4%) with ALL (TABLE 6). Five patients had biclonal leukemias. DNA indices ranged from 0.85–2.0 with a median of 1.16. In both children and adults, more DNA aneuploidies were identified in female than in male patients (46.0 versus 34.2%, $p < 0.05$).

The analysis of DNA aneuploidy in relation to the immunologically defined ALL subgroups was hampered by the fact that the C-ALL antigen was not determined for the majority of childhood ALL patients in the BFM study 78/81. Since the frequency of Null-ALL is below 10% in children, however, the group of non-T/non-B ALL consists almost exclusively of C-ALL cases. Therefore, in childhood ALL non-T/non-B ALL was considered equivalent to C-ALL and was analyzed together with adult C-ALL patients.

On this basis a significantly higher frequency of DNA aneuploidies was revealed for non-T/non-B and C-ALL cases as compared to T-ALL and Null-ALL ($p < 0.05$) (TABLE 7).

Furthermore, the degree of DNA aneuploidy as expressed by the DNA index was higher in non-T/non-B and C-ALL than in T-ALL and Null-ALL ($p < 0.05$).

Within the group of childhood C-ALL and non-T/non-B ALL a significantly lower WBC was revealed for children with DNA aneuploidy as compared to cases without aneuploid DNA stemlines with median values of 29,592/mm^3 versus 54,371/mm^3 ($p < 0.05$).

A similar tendency was observed for adult patients as well without being statistically significant, however.

From the evaluation of the treatment results, no differences in the response to

TABLE 7. Incidence and Degree of DNA Aneuploidies for the Immunologic ALL Subgroups

Immunologic Subtype	N	DNA Aneuploidy	Median DNA Index
C-ALL non-T/non-B ALL	174	76 (44.4%)[a]	1.18[a]
T-ALL	40	9 (22.5%)	1.10
B-ALL	7	2	
Null ALL	25	7 (28.6%)	1.09

[a]C-ALL + non-T/non-B ALL versus T-ALL and Null ALL ($p < 0.05$).

the induction therapy emerged between patients with and without DNA aneuploidy in children or in adults.

Hampered by the small number of evaluable patients and a short observation time, no differences in remission duration were found in adult patients with and without aneuploid DNA stemlines.

In the childhood ALL study BFM 79/81, however, a tendency towards a higher proportion of long term remissions in children with DNA aneuploidy was observed with an accumulative proportion of patients in complete remission at five years of 0.91 as compared to 0.66 for patients without aneuploid DNA stemlines ($p = .053$; FIGURE 4).

Because of the relatively small number of cases analyzed in the BFM study 79/81 ($N = 62$) the correlation of DNA aneuploidy with other parameters of prognostic relevance such as risk factor, age, and WBC could not be assessed.

In the BFM study 81/83 a similar difference between patients with and without DNA aneuploidy is not shown at the present time.

DISCUSSION

Aberrations of the karyotype represent one of the longest known phenomena of malignant cells[37-39] and have proven a highly specific marker for the detection of cancer cells by cytogenetic analyses and FCM measurements of the cellular DNA content.[11,14,40-45]

Especially in acute leukemia, cytogenetic evaluations have greatly contributed to the present knowledge about the nature and classification of hematologic malignancies.[41,42,46-48]

The application of cytogenetic techniques for routine clinical investigations is limited, however, by the laborious and time-consuming processing of samples

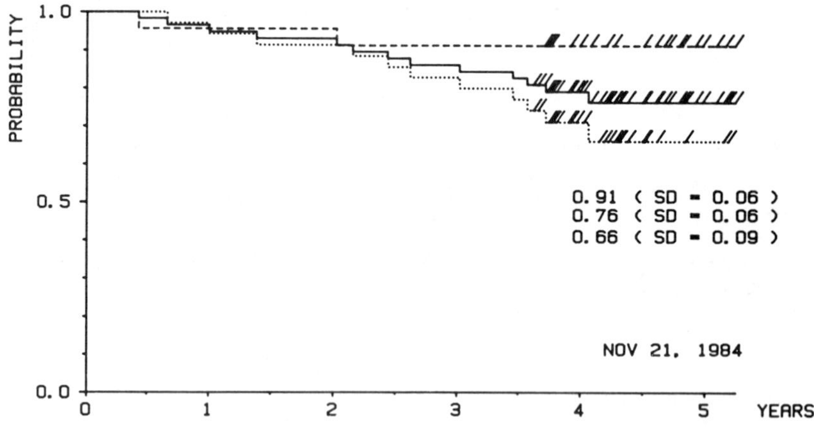

FIGURE 4. Remission duration in the childhood ALL study BFM 79/81 for patients with (dotted line) and without DNA aneuploidy (dashed line) $p = .053$; (solid line = total group).

and by the requirement of cell growth *in vitro* to gain evaluable metaphases.

FCM, on the other hand, provides a rapid means for the assessment of the cellular DNA content independent on cell proliferation but is restricted in its sensitivity by the fact that only numeric chromosome aberrations lead to a corresponding change in the cellular DNA content. Hence, structural abnormalities such as balanced translocations, which are not accompanied by a deviation of the total cellular DNA content, are not detectable by FCM.

Still, in acute leukemias DNA aneuploidies were identified in previous investigations in 8–30% of cases[36,45,49,50] and have proven to be of prognostic significance in childhood ALL.[50]

Based on high resolution FCM measurements and the application of standardized reference measurements, our group first reported DNA aneuploidies in acute leukemias at a frequency of 40%.[12,14,15] It was the aim of the present study to confirm this incidence of aneuploid DNA stemlines on a larger number of patients and to assess the clinical and prognostic significance of DNA aneuploidies in acute leukemias.

In the present study DNA aneuploidies were identified in 175 of 446 patients with acute leukemia thus accounting for a frequency of 39.2%.

These data are in good agreement with the previous results reported by our group and were recently confirmed by Andreeff *et al.* for childhood ALL.[51]

In spite of the limited comparability between FCM analyses and cytogenetic techniques the frequency of 39.2% DNA aneuploidies corresponds fairly well with the frequency of numeric chromosome aberrations identified by karyotype evaluations.[42,46,52-55]

Similarly, the lower rate of chromosome abnormalities in M 1 and M 2 leukemias as compared to the subtypes M 4 and M 5[56] is in agreement with the present study in which DNA aneuploidies were significantly less frequent in the AML subgroup M 1, while the degree of DNA aneuploidies as expressed by the DNA index was significantly lower for the subtypes M 1 and M2 as compared to M 4 and M 5.

The relevance of DNA aneuploidies for the prognosis of patients with AML was assessed separately for the response to induction therapy and the duration of the first complete remission. Neither in children nor in adults were differences observed in the rates of complete remission, early death, or non-response between patients with or without aneuploid DNA stemlines.

For remission duration, however, a tendency towards a higher percentage of long remissions was revealed from the two studies with the longest observation time, being the AML BFM study 78 in children and the AML study 78/81 in adults. The small numbers of patients analyzed, however, require very cautious analysis of these findings and do not provide a definite conclusion about the prognostic significance of DNA aneuploidies in AML, yet.

These findings are in agreement, though, with previous reports on longer remissions in children with ALL and DNA aneuploidy,[50] which was also found in cytogenetic studies indicating a higher proportion of long remissions in childhood ALL patients with high numeric chromosome aberrations in comparison to cases with pseudodiploid or diploid karotypes.[46,55,57,58]

A corresponding tendency is shown by the results of the BFM study 79/81 revealing a proportion of continuous complete remissions at five years of 0.91 for children with DNA aneuploidy as compared to 0.66 for cases without aneuploid DNA stemline ($p = 0.053$). In the subsequent BFM study ALL 81/83 no difference is detectable at the present time. This might be due to the shorter observation time.

Because of the small number of cases evaluated and the short observation period, no differences are revealed either for adult patients with ALL.

Both in children and in adults with ALL, however, the frequency of DNA aneuploidies differed significantly between the immunologic subtypes with a higher frequency of aneuploid DNA stemlines in non-T/non-B and C-ALL versus T- and Null-ALL ($p < 0.05$).

Equivalent data were obtained from cytogenetic studies.[46,53,59,60]

The present data indicate that DNA aneuploidies are identified by FCM analyses in a large proportion of children and adults with ALL and AML. The incidence and the degree of DNA aneuploidies were found to correlate with immunologic and morphologic leukemia subtypes. Further follow-up and the evaluation of larger numbers of patients will show if the identification of aneuploid DNA stemlines is also of prognostic significance in acute leukemias.

SUMMARY

Analyses of the cellular DNA content were carried out in 446 patients with newly diagnosed acute myeloid (AML) and lymphoblastic (ALL) leukemias in order to assess the frequency of DNA aneuploidies and its relation to immunologic and morphologic subtypes as well as its prognostic relevance.

Based on high resolution FCM analyses and standardized reference measurements, DNA aneuploidies were identified at a similar frequency in children and adults with AML (38.0% and 40.0%) and ALL (40.0% and 37.4%). In AML aneuploid DNA stemlines were significantly less frequent in FAB-M 1 cases as compared to the other morphologic subtypes ($p < 0.05$), whereas the degree of DNA aneuploidy was significantly lower in M 1 and M 2 leukemias as compared to the M 4 and M 5 subgroups ($p < 0.05$).

In ALL non-T/non-B and C-ALL revealed a higher frequency of DNA aneuploidies than T- and Null-ALL cases ($p < 0.05$).

No differences in the response to induction therapy were found between patients with and without DNA aneuploidy in children or adults with AML or ALL. In the childhood ALL trial BFM 79/81, however, a significantly higher frequency of long term remissions was observed in children with DNA aneuploidy (0.91 versus 0.66 at five years, $p = 0.053$). A similar though not significant tendency was also revealed from the AML studies BFM 78 in children and 78/81 in adults. In the subsequent studies these differences could not be confirmed at present, possibly because of the considerably shorter observation time.

REFERENCES

1. BÜCHNER, T., W. DITTRICH & W. GÖHDE. 1971. Die Impulscytophotometrie in der hämatologischen Cytologie. Klin. Wschr. **49**:1090–1092.
2. BÜCHNER, T., W. DITTRICH & W. GÖHDE. 1971. Impulscytophotometrie von Blut- und Knochenmarkszellen. Verh. Dtsch. Ges. Inn. Med. **77**:416–418.
3. MELAMED, M. R., L. R. ADAMS, F. TRAGANOS, A. ZIMRING & L. A. KAMENSTKY. 1972. Acridine Orange metachromasia for characterization of leukocytes in leukemia, lymphoma and other neoplasms. Cancer **29**:1361–1368.
4. BÜCHNER, T. 1974. Impulszytophotometrie in der Hämatologie. Blut **28**:1–7.
5. HILLEN, H., J. WESSELS & C. HAANEN. 1975. Bone-marrow-proliferation patterns in

acute myeloblastic leukaemia determined by pulse cytophotometry. Lancet i:609–611.

6. DOSIK, G. M., B. BARLOGIE, W. GÖHDE, D. JOHNSTON, J. L. TEKELL & B. DREWINKO. 1980. Flow cytometry of DNA content in human bone marrow: A critical reappraisal. Blood **55**:734–740.

7. HIDDEMANN, W., T. BÜCHNER, M. ANDREEFF, B. WÖRMANN, M. R. MELAMED & B. D. CLARKSON. 1982. Cell kinetics in acute leukemia—A critical reevaluation based on new data. Cancer **50**:250–258.

8. BARLOGIE, B., A. M. MADDOX, D. A. JOHNSTON, M. N. RABER, B. DREWINKO, M. J. KEATING & E. J. FREIREICH. 1983. Quantitative cytology in leukemia research. Blood Cells **9**:35–55.

9. BARLOGIE, B., T. BÜCHNER, J. S. HART, M. J. AHEARN & E. J. FREIRICH. 1975. Aneuploidy as seen in the DNA-histogram of acute leukemia during the course of therapy. *In* 1st Int. Symp. Pulsecytophotometry. C. A. M. Haanen, H. F. P. Hillen & J. H. C. Wessels, Eds.: 299–305. European Press Medikon. Ghent.

10. HIDDEMANN, W. 1978. Zellkinetische Untersuchungen mit Hilfe der Impulscytophotometrie bei akuten Leukämien. Dissertation. University of Münster.

11. GÖHDE, W., J. SCHUMANN, T. BÜCHNER, F. OTTO & B. BARLOGIE. 1979. Pulse cytophotometry: Application in tumor cell biology and clinical oncology. *In* Flow Cytometry and Sorting. M. R. Melamed, P. F. Mullaney & M. L. Mendelsohn, Eds.: 599–620. John Wiley & Sons, Inc. New York.

12. HIDDEMANN, W., B. WÖRMANN, J. RITTER, G. HENZE, H. J. LANGERMANN, U. KAUFMANN, G. SCHELLONG, H. RIEHM & T. BÜCHNER. 1982. Diagnostik von Aneuploidien bei akuten Leukämien mittels Impulszytophotometrie (ICP): Häufigkeit und klinische Relevanz. Verh. Dtsch. Ges. Inn. Med. **88**:934–937.

13. WÖRMANN, B. 1982. Intensivierte Induktionstherapie der akuten myeloischen Leukämie des Erwachsenen: Prognostische Bedeutung einiger zellulärer und zellkinetischer Parameter. Dissertation. University of Münster.

14. HIDDEMANN, W., B. WÖRMANN, J. RITTER, H. J. KLEINEMEIER, D. VON BASSEWITZ, A. ROESSNER, K. M. MÜLLER & T. BÜCHNER. 1983. DNA aneuploidy—a highly specific marker for cancer detection. Proc. Am. Soc. Clin. Oncol. **2**:7.

15. WÖRMANN, B., W. HIDDEMANN, J. RITTER, G. HENZE, H. J. LANGERMANN, U. KAUFMANN, G. SCHELLONG, H. RIEHM & T. BÜCHNER. 1983. DNA aneuploidy in childhood ALL-incidence and relation to prognostic factors. Proc. Am. Ass. Cancer Res. **24**:161.

16. BENNETT, J. M., D. CATOVSKY, M. T. DANIEL, G. FLANDRIN, D. A. G. GALTON, H. R. GRALNICK & C. SULTAN. 1976. Proposals for the classification of the acute leukaemias. Br. J. Haematol. **33**:451–458.

17. THIEL, E., H. RODT, D. HUHN, B. NETZEL, H. GROSSE-WILDE, K. GANESHAGURU & S. THIERFELDER. 1980. Multimarker classification of acute lymphoblastic leukemia: Evidence for further T subgroups and evaluation of their clinical significance. Blood **56**:759–772.

18. HENZE, G., H. J. LANGERMANN, R. FENGLER, M. BRANDEIS, K. G. EVERS, H. GADNER, L. HINDERFELD, A. JOBKE, B. KORNHUBER, F. LAMPERT, U. LASSON, R. LUDWIG, S. MÜLLER-WEIHRICH, M. NEIDHARDT, G. NESSLER, D. NIETHAMMER, M. RISTER, J. RITTER, A. SCHAAFF, G. SCHELLONG, B. STOLLMANN, J. TREUNER, W. WAHLEN, P. WEINEL, H. WEHINGER & H. RIEHM. 1982. Therapiestudie BFM 79/81 zur Behandlung der akuten lymphoblastischen Leukämie bei Kindern und Jugendlichen: intensivierte Reinduktionstherapie für Patientengruppen mit unterschiedlichem Rezidivrisiko. Klin. Pädiat. **194**:195–203.

19. THIEL, E. 1983. Monoclonal antibodies against differentiation antigens of lymphopoiesis. Blut **47**:247–261.

20. HOELZER, D., E. THIEL, H. LÖFFLER, H. BODENSTEIN, L. PLAUMANN, T. BÜCHNER, D. URBANITZ, P. KOCH, H. HEIMPEL, R. ENGELHARDT, U. MÜLLER, F. C. WENDT, H. SODOMANN, H. RÜHL, F. HERRMANN, W. KABOTH, H. DIETZFELBINGER, H. PRALLE, C. LUNSCKEN, K. P. HELLRIEGEL, S. SPORS, M. NOWROUSIAN, J. FISCHER, H. H. FÜLLE, P. MITROU, M. PFREUNDSCHUH, C. GÖRG, B. EMMERICH, W. QUEISSER, P. MEYER, L.

LABEDZKI, U. ESSERS, H. KÖNIG, K. MAINZER, D. FRITZE, D. MESSERER & T. ZWINGERS. 1984. Intensified therapy in acute lymphoblastic and acute undifferentiated leukemia in adults. Blood **64**:38–47.

21. BÜCHNER, T., D. URBANITZ, W. HIDDEMANN, D. KAMANABROO, R. MEISTER, L. BALLEISEN, U. DELVOS, E. M. LAGREZE, U. SCHMITZ-HUEBNER, H. SCHULTE & J. VAN DE LOO. 1979. Intensification of remission induction therapy for acute nonlymphocytic leukemia (ANLL). Blut **39**:133–140.

22. BÜCHNER, T., D. URBANITZ, B. EMMERICH, J. T. FISCHER, H. H. FÜLLE, A. HEINECKE, D. K. HOSSFELD, K. M. KOEPPEN, L. LABEDZKI, H. LÖFFLER, M. R. NOWROUSIAN, M. PFREUNDSCHUH, H. PRALLE, H. RÜHL & F. C. WENDT. 1982. For the AML Cooperative Group: Multicentre study on intensified remission induction therapy for acute myeloid leukemia. Leukemia Res. **6**:827–831.

23. BÜCHNER, T., D. URBANITZ, H. BRÜCHER, A. HEINECKE, W. HIDDEMANN, H. RÜHL, H. SCHULTE & F. WENDT. 1984. Chemotherapie der akuten myeloischen Leukämie des Erwachsenen. *In* Therapie der akuten Leukämien. Th. Büchner, D. Urbanitz & J. van de Loo Eds.: 59–72. Springer Verlag. Berlin.

24. CREUTZIG, U., J. RITTER, H. J. LANGERMANN, H. RIEHM, G. HENZE, D. NIETHAMMER, H. JÜRGENS, B. STOLLMANN, U. LASSON, H. KABISCH, W. WAHLEN, H. LÖFFLER & G. SCHELLONG. 1983. Akute myeloische Leukämie bei Kindern: Ergebnisse der kooperativen Therapiestudie BFM-78 nach 3 3/4 Jahren. Klin. Pädiat. **195**:152–160.

25. SCHELLONG, G., U. CREUTZIG & J. RITTER. 1984. Therapie der akuten myeloischen Leukämie bei Kindern. *In* Therapie der Akuten Leukämien. Th. Büchner, D. Urbanitz & J. van de Loo, Eds.: 73–94. Springer Verlag. Berlin.

26. CREUTZIG, U., J. RITTER, H. RIEHM, H. J. LANGERMANN, G. HENZE, H. KABISCH, D. NIETHAMMER, H. JÜRGENS, B. STOLLMANN, U. LASSON, U. KAUFMANN, H. LÖFFLER & G. SCHELLONG. 1985. Improved treatment results in childhood acute myelogenous leukemia: a report of the German cooperative study AML-BFM-78. Blood **65**:298–304.

27. HOELZER, D., E. THIEL, H. LÖFFLER, H. BODENSTEIN, L. PLAUMANN, TH. BÜCHNER, D. URBANITZ, P. KOCH, H. HEIMPEL, R. ENGELHARDT, U. MÜLLER, F. C. WENDT, H. SODOMANN, H. RÜHL, F. HERRMANN, W. KABOTH, H. DIETZFELBINGER, H. PRALLE, CH. LUNSCKEN, K. P. HELLRIEGEL, S. SPORS, M. NOWROUSIAN, J. FISCHER, H. H. FÜLLE, P. MITROU, M. PFREUNDSCHUH, CH. GÖRG, B. EMMERICH, W. QUEISSER, P. MEYER, L. LABEDZKI, U. ESSERS, H. KÖNIG, K. MAINZER, D. FRITZE, D. MESSERER & TH. ZWINGERS. 1983. Multizentrische Therapiestudie Akute Lymphatische Leukämie (ALL) und Akute Undifferenzierte Leukämie (AUL) des Erwachsenen. Verh. Dtsch. KrebsGes. **4**:723–731.

28. RIEHM, H. J. 1984. Therapie der akuten lymphoblastischen Leukämie des Kindes. *In* Therapie der Akuten Leukämien. Th. Büchner, D. Urbanitz & J. van de Loo, Eds.:51–58. Springer Verlag. Berlin.

29. OHNUMA, T., F. ROSNER, R. N. LEVY, J. CUTTNER, J. H. MOON, R. T. SILVER, J. BLOM, G. FALKSON, R. BURNINGHAM, O. GLIDEWELL & J. F. HOLLAND. 1971. Treatment of adult leukemia with L-asparaginase. Cancer Chemother. Rep. **55**:269–275.

30. YATES, J., O. GLIDEWELL, P. WIERNIK, M. R. COOPER, D. STEINBERG, H. DOSIK, R. LEVY, C. HOAGLAND, P. HENRY, A. GOTTLIEB, C. CORNELL, J. BERENBERG, J. L. HUTCHISON, P. RAICH, N. NISSEN, R. R. ELLISON, R. FRELICK, G. W. JAMES, G. FALKSON, R. T. SILVER, F. HAURANI, M. GREEN, E. HENDERSON, L. LEONE & J. F. HOLLAND. 1982. Cytosine arabinoside with daunorubicin or adriamycin for therapy of acute myelocytic leukemia: a CALGB study. Blood **60**:454–462.

31. ZANTE, J., J. SCHUMANN, B. BARLOGIE, W. GÖHDE & T. BÜCHNER. 1976. New preparating and staining procedures for specific and rapid analysis of DNA-distributions. *In* 2nd Int. Symp. Pulsecytophotometry. W. Göhde, J. Schumann & T. Büchner, Eds.: 97–106. European Press Medikon. Ghent.

32. BARLOGIE, B., G. SPITZER, J. S. HART, D. A. JOHNSTON, T. BÜCHNER, J. SCHUMANN & B. DREWINKO. 1976. DNA histogram analysis of human hematopoietic cells. Blood **48**:245–258.

33. GÖHDE, W. 1973. Zellzyklusanalysen mit dem Impulscytophotometer. Habilitations-schrift. Münster.

34. BECK, H. P. 1980. Evaluation of flow cytometric data of human tumours. Correction procedures for background and cell aggregations. Cell Tissue Kinet. **13**:173–181.

35. HIDDEMANN, W., J. SCHUMANN, M. ANDREEFF, B. BARLOGIE, C. J. HERMAN, R. C. LEIF, B. H. MAYALL, R. F. MURPHY & A. A. SANDBERG. 1984. Convention on nomenclature for DNA cytometry. Cytometry **5**:445–446.

36. BARLOGIE, B., W. HITTELMAN, G. SPITZER, J. H. TRUJILLO, J. S. HART, L. SMALLWOOD & B. DREWINKO. 1977. Correlation of DNA distribution abnormalities with cytogenetic findings in human adult leukemia and lymphoma. Cancer Res. **37**:4400–4407.

37. ARNOLD, J. 1879. Beobachtungen über Kernteilungen in den Zellen der Geschwülste. Virchows Arch. Pathol. Anat. **78**:279–301.

38. VON HANSEMANN, D. 1890. Über asymetrische Zellteilung in Epithelkrebsen und deren biologische Bedeutung. Arch. Patho. Anat. Physiol. **119**:299–326.

39. BOVERI, T. 1914. Zur Frage der Entstehung maligner Tumoren. 1–64. G. Fischer Verlag. Jena.

40. NOWELL, P. C. 1974. Chromosome changes and the clonal evolution of cancer. *In* Chromosomes and Cancer. J. German, Eds.:267–285. Wiley & Sons Inc. New York.

41. SANDBERG, A. A. 1980. The chromosomes in human cancer and leukemia. Elsevier North-Holland. New York.

42. HOSSFELD, D. K. 1985. Zytogenetik maligner Erkrankungen. *In* Klinische Onkologie. R. Gross & C. G. Schmidt, Eds. Thieme Verlag. (In press.)

43. BARLOGIE, B., B. DREWINKO, J. SCHUMANN, W. GÖHDE, G. DOSIK, D. A. JOHNSTON & E. J. FREIREICH. 1980. Cellular DNA content as a marker of neoplasia in man. Am. J. Med. **69**:195–203.

44. SCHUMANN, J., H. TILKORN, W. GÖHDE, F. EHRING & C. STRAUB. 1981. Zytogenetik maligner Melanome. Der Hautarzt **32** (Suppl. V):62–66.

45. BARLOGIE, B., M. N. RABER, J. SCHUMANN, T. S. JOHNSON, B. DREWINKO, D. E. SWARTZENDRUBER, W. GÖHDE & M. ANDREEFF. 1983. Flow cytometry in clinical cancer research. Cancer Res. **43**:3982–3997.

46. The Third International Workshop on Chromosomes in Leukemia. 1981. Cancer Genet. Cytogenet. **4**:96–137.

47. Fourth International Workshop on Chromosomes in Leukemia. 1984. Cancer Genet. Cytogenet. **11**:251–360.

48. BLOOMFIELD, C. 1985. The clinical usefulness of chromosome abnormalities in acute leukemia. *In* Tumor Aneuploidy. Th. Büchner, C. Bloomfield, W. Hiddemann, D. Hossfeld & J. Schuman, Eds.:13–24. Springer Verlag. Berlin.

49. ANDREEFF, M., Z. DARZYNKIEWICZ, T. K. SHARPLESS, B. D. CLARKSON & M. R. MELAMED. 1980. Discrimination of human leukemia subtypes by flow cytometric analysis of cellular DNA and RNA. Blood **55**:282–293.

50. LOOK, A. T., S. L. MELVIN, D. L. WILLIAMS, G. M. BRODEUR, G. V. DAHL, D. K. KALWINSKY, S. B. MURPHY & A. M. MAUER. 1982. Aneuploidy and percentage of S-phase cells determined by flow cytometry correlate with cell phenotype in childhood acute leukemia. Blood **60**:959–967.

51. ANDREEFF, M., A. REDNER, S. THONGPRASERT, B. EAGLE, P. STEINHERZ, D. MILLER & M. R. MELAMED. 1985. Multiparameter flow cytometry for determination of ploidy, proliferation and differentiation in acute leukemia: treatment effects and prognostic value. *In* Tumor Aneuploidy. Th. Büchner, C. Bloomfield, W. Hiddemann, D. Hossfeld & J. Schumann, Eds.:81–106. Springer Verlag. Berlin.

52. ROWLEY, J. D. 1978. The cytogenetics of acute leukemia. Clin. Haematol. **7**:385–406.

53. ARTHUR, D. C., C. D. BLOOMFIELD, L. L. LINDQUIST, B. A. PETERSON & M. E. NESBIT. 1981. Chromosome abnormalities in acute lymphoblastic leukemia (ALL): Frequency and clinical implications. Proc. Am. Ass. Cancer Res. **22**:345.

54. KANEKO, Y., J. D. ROWLEY, H. S. MAURER, D. VARIAKOJIS & J. M. MOOHR. 1982. Chromosome pattern in childhood acute non-lymphocytic leukemia (ANLL). Blood **60**:389–399.

55. WILLIAMS, D. L., A. TSIATIS, G. M. BRODEUR, A. T. LOOK, S. L. MELVIN, W. P. BOWMAN, D. K. KALWINSKY, G. RIVERA & G. V. DAHL. 1982. Prognostic importance of chromosome number in 136 untreated children with acute lymphoblastic leukemia. Blood 60:864–871.
56. ROWLEY, J. D., G. ALIMENA, O. M. GARSON, A. HAGEMEIJER, F. MITELMAN & E. L. PRIGOGINA. 1982. A collaborative study of the relationship of the morphological type of acute non-lymphocytic leukemia with patient age and karotype. Blood 59:1013–1022.
57. SWANSBURY, G. J., L. M. SECKER-WALKER, S. D. LAWLER, R. M. HARDISTY, S. E. SALLAN, O. M. GARSON & M. SAKURAI. 1981. Chromosomal findings in acute lymphoblastic leukaemia of childhood: an independent prognostic factor. Lancet ii:249–250.
58. SECKER-WALKER, L. M., G. J. SWANSBURY, R. M. HARDISTY, S. E. SALLAN, O. M. GARSON, M. SAKURAI & S. D. LAWLER. 1982. Cytogenetics of acute lymphoblastic leukaemia in children as a factor in the prediction of long-term survival. Br. J. Haematol. 52:389–399.
59. OSHIMURA, M., A. FREEMAN & A. SANDBERG. 1977. Chromosomes and causation of human cancer and leukemia XXVI. Banding studies in acute lymphoblastic leukemia. Cancer 40:1161–1172.
60. BLOOMFIELD, C. D., L. L. LINDQUIST, D. ARTHUR, R. W. MCKENNA, T. W. LEBIEN, B. A. PETERSON & M. E. NESBIT. 1981. Chromosomal abnormalities in acute lymphoblastic leukemia. Cancer Res. 41:4838–4843.

Detection of Central Nervous System Relapse in Acute Leukemia by Multiparameter Flow Cytometry of DNA, RNA, and CALLA[a]

ARLENE REDNER, MYRON R. MELAMED, AND
MICHAEL ANDREEFF

*Memorial Sloan-Kettering Cancer Center
New York, New York 10021*

INTRODUCTION

Leukemic involvement of the central nervous system may be difficult to diagnose by conventional cytology and to distinguish from reactive or inflammatory processes, particularly if immature or atypical cells are few.[1] Yet the leukemic cells in 40% of pediatric patients with acute lymphoblastic leukemia (ALL) have abnormal deoxyribonucleic acid (DNA) content, which is diagnostic of the disease, when measured by flow cytometry[2] while other features, including ribonucleic acid (RNA) content or the presence of common ALL antigen (CALLA) differentiate subtypes of leukemia.[3] These measurements offer possible new means of identifying leukemic cells. In this present study flow cytometry of cerebrospinal fluid (CSF) cells was performed on a series of pediatric patients with clinical evidence of central nervous system (CNS) leukemia or lymphoma, extending our previous study of DNA measurements in fifteen patients with central nervous system leukemia,[4] and now including simultaneous measurements of DNA and RNA, and DNA and common ALL antigen (CALLA).

MATERIALS AND METHODS

Flow cytometry of spinal fluid cells was performed on ninety-one patients. Thirty had a clinical diagnosis of central nervous system leukemia, for which they were entered into appropriate treatment protocols, and they had Wright-Giemsa cytospin preparation of spinal fluids showing blasts, monocytes, or unclassified cells. Sixty one patients, who had leukemia or lymphoma, but no evidence of CNS involvement, had 104 specimens of CSF examined by flow cytometry. In addition, peripheral blood and bone marrow specimens from all patients were examined by flow cytometry.

Heparinized bone marrow and peripheral blood samples were processed using a Ficoll Hypaque gradient technique to concentrate nucleated cells as

[a]Aided in part by grants from National Cancer Institute, The Department of Health and Human Services, CA29564, and CA20914. Dr. Redner is a Fellow of the Leukemia Society of America.

previously described.[3] Cerebrospinal fluid was (7 ml) obtained by lumber puncture and divided into four aliquots for differential court, chemistry, cytology, and flow cytometry.

Samples for cytology were preserved with an equal volume of 50% ethanol and then prepared by centrifugal cytology and Papanicolaou stain. Differential counts were performed on unfixed cytospin preparations stained with Wright-Giemsa.

Samples for flow cytometry (3 ml) were centrifuged at $1000\,g$ for 5 minutes and the pellet resuspended in Hanks Balanced Salt Solution (HBSS) to be stained in suspension with acridine orange (AO). The AO staining procedure has been described elsewhere.[5] In brief, 0.2 ml of the cell suspension in HBSS is mixed with 0.4 ml of 0.05 N HCl, 0.15 N NaCl, and 0.1% Triton X-100. After 30 seconds, 1.2 ml of AO staining solution is added (0.2 M Na_2HPO_4–citric acid buffer at pH 6.0, 1 mM EDTA-Na, 0.15 N NaCl, and 6 µg/ml acridine orange). Chromatographically purified AO (Polyscience, Inc. Warrington, PA) is used. Pretreatment of the cells with Triton X-100 makes them permeable to the dye and at low pH nucleic acids remain insoluble.[5] These staining conditions permit AO to intercalate into double helical DNA and fluoresce orthochromatically green in blue light. Double-stranded RNA is denatured in the presence of EDTA.[6] The single-stranded RNA complexes with acridine orange and forms micelles that fluoresce metachromatically red.[7]

Simultaneous measurements of green (DNA) and red (RNA) fluorescence, and of green pulse width (to distinguish single cells from cell doublets and other aggregates[8]) were performed using a compluter-interfaced research cytofluorograf (modified model FC 200, Ortho Instruments, Westwood, MA). Up to 5,000 cells per sample were measured. The data were stored in a Nova 1220 minicomputer (Data General, Southwood, MA) and analyzed with computer programs developed by Sharpless.[9]

DNA index was determined from the median value of DNA for $G_{0/1}$ cells of the sample compared to the median DNA of G_0 lymphocytes used as control (normal lymphocyte DNA index = 1.0). Samples and control lymphocytes were first measured separately, then mixed, stained, and measured simultaneously.[10] The RNA content of the cells in $G_{0/1}$ is expressed as the RNA index (RI $G_{0/1}$), which is defined as the mean RNA content of $G_{0/1}$ cells of the sample $\times$ 10 divided by the median RNA content of control G_0 lymphocytes. The RNA indices for "G_0" (RI G_0) and "G_1" (RI G_1) were calculated with "G_0" cells defined as cells with RNA content equal to or less than control lymphocytes and "G_1" cells with unit DNA content and RNA greater than control lymphocytes.

Simultaneous staining of CALLA surface antigen and DNA was performed on another portion of the samples using reconstituted lyophilized monoclonal antibody (Coulter J5 reagent) and propidium iodide (PI). Aliquots of 10 µl were stored at −80°C, and diluted prior to use with 390 µl phosphate-buffered saline with .2% bovine serum albumin. The cells (10^6 cells/ml) were washed, 200 µl of primary antibody was added, the cells were incubated at 4°C for thirty minutes, and then washed twice again. The supernatant was decanted and 200 µl of FITC-conjugated secondary antibody (Goat anti-mouse IgG (F(ab')$_2$, Cappel) was added in a 1:40 dilution. The sample was then incubated in an ice bath for 30 minutes, washed twice with PBS, followed by fixation with 70% ethanol for at least one hour, and then washed twice with HBSS. RNase (Worthington) was diluted with 1.12% Na citrate buffer pH 8.4 to 500 U/ml, 250 µl was added to 50 µl of cell suspension and incubated at 37°C for minutes. 250 µl propidium iodide (PI) (50 µg/ml in 1.12% Na citrate buffer) was finally added and the stain

TABLE 1. CNS Leukemia with Abnormal DNA Content

Case Number	Diagnosis	DI^a	% Cells				RNA Index	
			Abnormal DI	$G_{0/1}$	"G_1"	S	$G_{0/1}$	"G_1"
1	NHL^b	1.90	67.5	97	—	3	29.2	—
2	ALL^c	1.10	80.1	98	3	1	10.8	14.9
3	ALL	1.10	95.1	98	27	2	11.5	13.4
4	ALL	1.15	4.5	90	18	9	11.5	14.6
		.55	81.5	—	—	13	5.4	—
5	ALL	1.55	97.2	98	90	1	15.3	15.5
6	ALL	1.20	77.0	89	44	5	13.4	15.9
7	ALL	1.25	98.2	96	46	3	12.9	13.9
8	ALL	1.20	61.7	99	44	1	13.3	14.9
9	ALL	1.05	68.7	99	4	1	11.0	15.1
10	ALL	1.35	94.6	99	96	1	14.3	14.5
11	ALL	1.15	100	98	12	2	15.5	15.9
12	ALL	2.1	82.8	98	—	1	19.3	—

[a] DI = DNA index.
[b] NHL = Non Hodgkin's lymphoma.
[c] ALL = Acute lymphoblastic leukemia.

permitted to equilibrate at room temperature for 30 minutes.[11,12] Measurements of green fluorescence (FITC-conjugated antibody to CALLA), red fluorescence (DNA stained by PI), and pulse width were carried out on an Ortho FC-200 flow cytometer, and analyzed, as described above.

RESULTS

DNA/RNA Flow Cytometry: CNS Relapse

Flow cytometry of spinal fluid was performed on thirty patients with acute leukemia (ALL) or lymphoma (NHL) who had a clinical diagnosis of central nervous system involvement. All were in relapse except for two patients (#2 and 25) who had CNS involvement at the initial diagnosis of ALL.

Differential cell counts were performed on Wright-Giemsa stained cytospins from 27/30 patients with central nervous system relapse. Papanicolaou-stained cytocentrifuge slides were prepared on 22/30 patients, with 10 patients having positive cytological results, nine negative, and three interpreted as suspicious (TABLE 3).

With Abnormal DNA Index

Abnormal DNA stemlines were detected in a total of twelve patients. Coefficient of variation (CV) was 3.42 ± 1.3% for DNA measurements (TABLE 1). The DNA indices (DI) ranged from .55 to 2.1, (DI 1.0 = diploid), and the percentage of cells with abnormal stemline ranged from 36 to 100%, S-phase ranged from 1 to 13%, RNA indices of $G_{0/1}$ ranged from 5.4 to 29.2. One patient

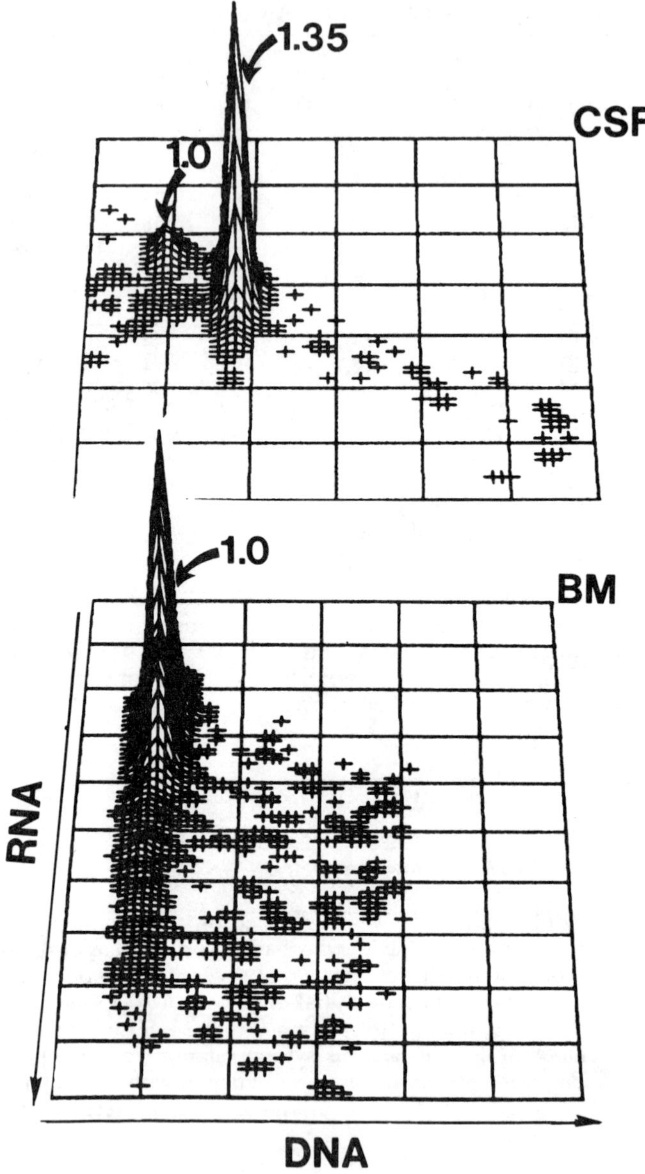

FIGURE 1. Computer drawn DNA-RNA histogram of CSF and bone marrow cells of a patient (# 10 TABLE 1) with an aneuploid (DI = 1.35) CSF relapse. At diagnosis of CSF relapse, the CSF (upper panel) contains 94.6% of cells with a DI = 1.35. The bone marrow specimen (lower panel) obtained on the same day shows 100% diploid (DI = 1.0) and no DNA-aneuploid cells.

(#4) had two abnormal DNA stemlines present in the spinal fluid. One other patient (#10; see also FIGURE 1) had a DI of 1.35 in the spinal fluid at the time of central nervous system relapse, with 94.6% of the cells in the spinal fluid having the abnormal DNA stemline, while bone marrow aspirate at that time had no cells with abnormal DNA content.

With Normal DNA Index

Eighteen patients had DNA diploid cells in spinal fluid at the time of diagnosis of central nervous system involvement (DNA index = 1.0; CV = 3.98 ± 1.2%). The S-phase fraction ranged from 0 to 7% except for two patients who had markedly elevated S fractions of 33 and 52%, both with Non-Hodgkin's lymphoma (FIGURE 2). RNA indices of "$G_{0/1}$" varied from 8.9 to 30.7 (TABLE 2). Ten of the 18 patients with diploid DNA leukemic cells (or lymphoma) had high RNA that served as a marker for CNS involvement. An example of a patient with ANLL and diploid DNA is shown in FIGURE 3. This patient had simultaneous CSF and bone marrow relapse, and leukemia cells in both the spinal fluid and the bone marrow biopsy had high RNA content. In three patients of this group, repeat flow cytometry was performed after intrathecal chemotherapy and the cells with increased RNA content were eradicated.

TABLE 2. CNS Leukemia with Normal DNA Content

Case Number	Diagnosis	DI[a]	% Cells				RNA Index	
			Abnormal DI	$G_{0/1}$	"G_1"	S	$G_{0/1}$	"G_1"
13	ALL[b]	1.0	0	98	—	—	10.2	—
14	NHL[c]	1.0	0	64	57	33	9.9	31.2
15	ALL	1.0	0	95	—	5	10.0	—
16	ALL	1.0	0	96	0	3	10.0	—
17	APL[d]	1.0	0	97	69	3	18.7	21.7
18	ALL	1.0	0	98	86	1	14.7	15.3
19	NHL	1.0	0	32	28	52	22.1	23.8
20	AML[e]	1.0	0	99	37	1	12.9	16.7
21	AMOL[f]	1.0	0	99	87	1	30.7	33.3
22	ALL	1.0	0	98	2	1	10.6	15.8
23	ALL	1.0	0	97	9	3	10.5	10.8
24	ALL	1.0	0	100	34	0	11.6	14.0
25	ALL	1.0	0	99	9	1	10.7	13.2
26	ALL	1.0	0	98	0	2	8.9	—
27	AML	1.0	0	97	54	2	14.9	17.8
28	ALL	1.0	0	97	67	2	12.1	14.0
29	ALL	1.0	0	93	4	5	9.4	13.3
30	ALL	1.0	0	95	62.1	4	13.8	15.3

[a]DI = DNA index.
[b]ALL – Acute lymphoblastic leukemia.
[c]NHL = Non Hodgkin's lymphoma.
[d]APL = Acute promyelocytic leukemia.
[e]AML = Acute myeloblastic leukemia.
[f]AMOL = Acute monocytic leukemia.

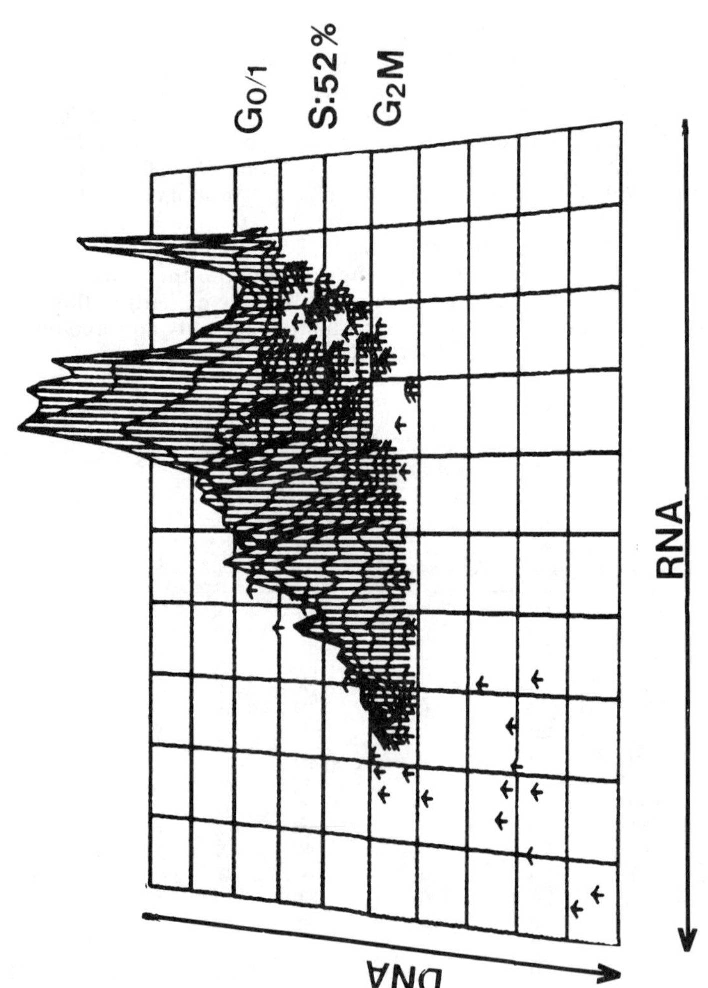

FIGURE 2. Computer drawn DNA-RNA histograms of CSF cells of patient with NHL (#19 TABLE 2). The sample has a DI = 1.0, but S-phase is 52% and RNA index G0/1 is 22.1.

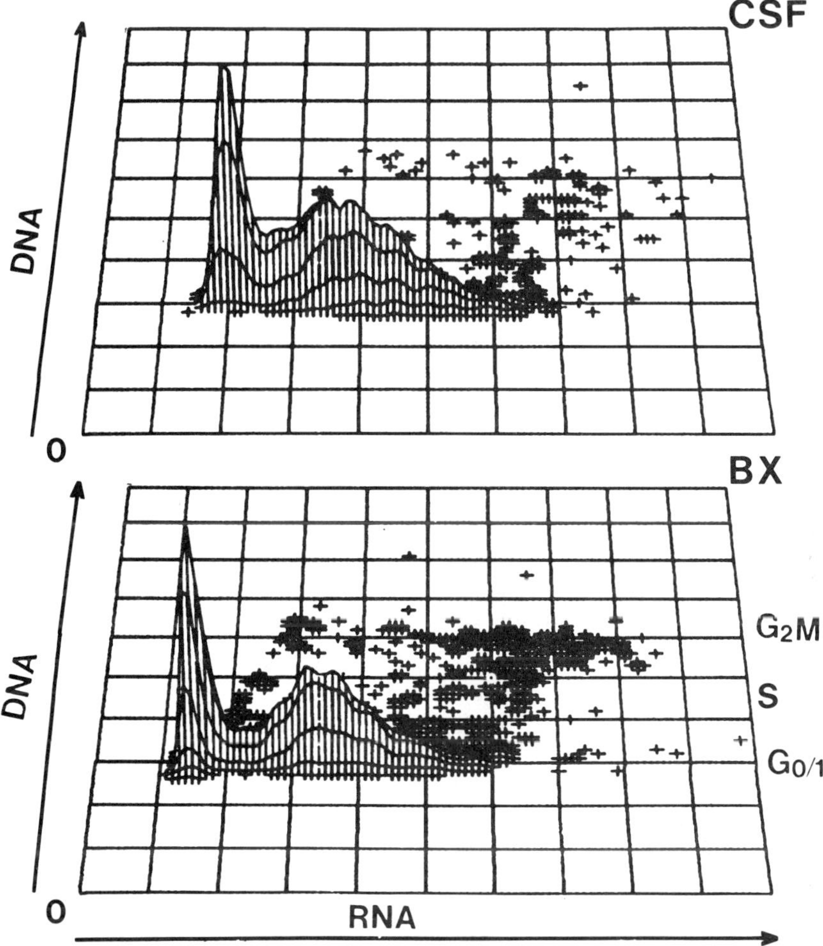

FIGURE 3. Computer drawn DNA-RNA histograms of CSF and bone marrow biopsy cells of a patient with acute promyelocytic leukemia who had a simultaneous CNS and bone marrow relapse. CSF (upper panel) shows 69% of the cells containing high RNA content (RI G_1 = 21.7). A bone marrow biopsy specimen obtained on the same day shows 55% of the cells with high RNA (RI G_1 = 20.7).

With CALLA Expression

Spinal fluid from five patients were examined for expression of common ALL antigen (CALLA) since the majority of pediatric leukemias are known to be CALLA positive. The specimens were labeled with the monoclonal antibody to CALLA, as described, and counterstained for DNA with propidium iodide. The presence of cells labeled by the antibody was considered indicative of CNS involvement with CALLA-positive leukemia (FIGURE 4). The combined DNA-CALLA antibody stain identifies CALLA-positive cells in CSF and allows correlation of antigen expression with cell cycle and DNA ploidy.

TABLE 3. Spinal Fluid Flow Cytometry Cell Count and Cell Morphology in Patients with Clinical Diagnosis of CNS Leukemia

Case Number	DI[a]	Hematology Cell Count/mm^3	Hematology Differential	Cytology
1	1.90	10 RBC 5 WBC	95% blasts	ND[b]
2	1.10	310 RBC 5 WBC	60% blasts	SUSP[d]
3	1.10	58 RBC 700 WBC	100% mono[c]	POS[e]
4	1.15 .55	65 RBC 262 WBC	100% blasts	POS
5	1.55	0 RBC 1230 WBC	ND	POS
6	1.20	0 RBC 865 WBC	100% blasts	ND
7	1.25	0 RBC 310 WBC	100% uncl mono[f]	POS
8	1.20	0 RBC 11 WBC	100% uncl mono	NEG[g]
9	1.05	0 RBC 64 WBC	100% uncl mono	ND
10	1.35	0 RBC 107 WBC	100% uncl mono	NEG
11	1.15	1,000 WBC	ND	POS
12	2.1	10 RBC 109 WBC	95% uncl mono	POS
13	1.0	5 RBC 75 WBC	90% blasts	ND
14	1.0	0 RBC 57 WBC	100% blasts	ND
15	1.0	6 RBC 190 WBC	100% blasts	POS
16	1.0	0 RBC 207 WBC	90% blasts	ND
17	1.0	3 RBC 22 WBC	30% blasts	NEG
18	1.0	0 RBC 60 WBC	100% blasts	NEG
19	1.0	0 RBC 1 WBC	21 blasts	NEG

TABLE 3. (*continued*)

Case Number	DI[a]	Hematology Cell Count/mm^3	Differential	Cytology
20	1.0	25 RBC 50 WBC	90% uncl mono	POS
21	1.0	0 RBC 8 WBC	80% blasts	NEG
22	1.0	30 RBC 949 WBC	100% blasts	ND
23	1.0	300 RBC 56 WBC	100% mono	SUSP
24	1.0	0 RBC 65 WBC	100% blasts	ND
25	1.0	165 RBC 18 WBC	65% uncl mono	SUSP
26	1.0	86 RBC 27 WBC	ND	POS
27	1.0	176 RBC 2 WBC	20% blasts	NEG
28	1.0	0 RBC 128 WBC	100% uncl mono	POS
29	1.0	0 RBC 15 WBC	40% blasts	NEG
30	1.0	91 RBC 18 WBC	95% blasts	NEG

[a]DI = DNA index.
[b]ND = not done.
[c]Mono = monocytes.
[d]SUSP = suspicious.
[e]POS = positive.
[f]Uncl mono = unclassified monocytes.
[g]NEG = negative.

DNA/RNA Flow Cytometry: No Leukemic CNS Involvement

Flow cytometry was performed on 98 samples of spinal fluid from 56 patients with leukemia or lymphoma and no evidence of central nervous system involvement clinically, by Wright-Giemsa cytospin or Papanicolaou cytology. These comprised 78 samples on patients in remission, 12 samples at diagnosis, and 8 samples at relapse. All these samples had DNA diploid cells only (mean CV = 3.6 ± 1.1). Cell cycle distribution and RNA indices were near normal for peripheral blood lymphocytes (TABLE 4).

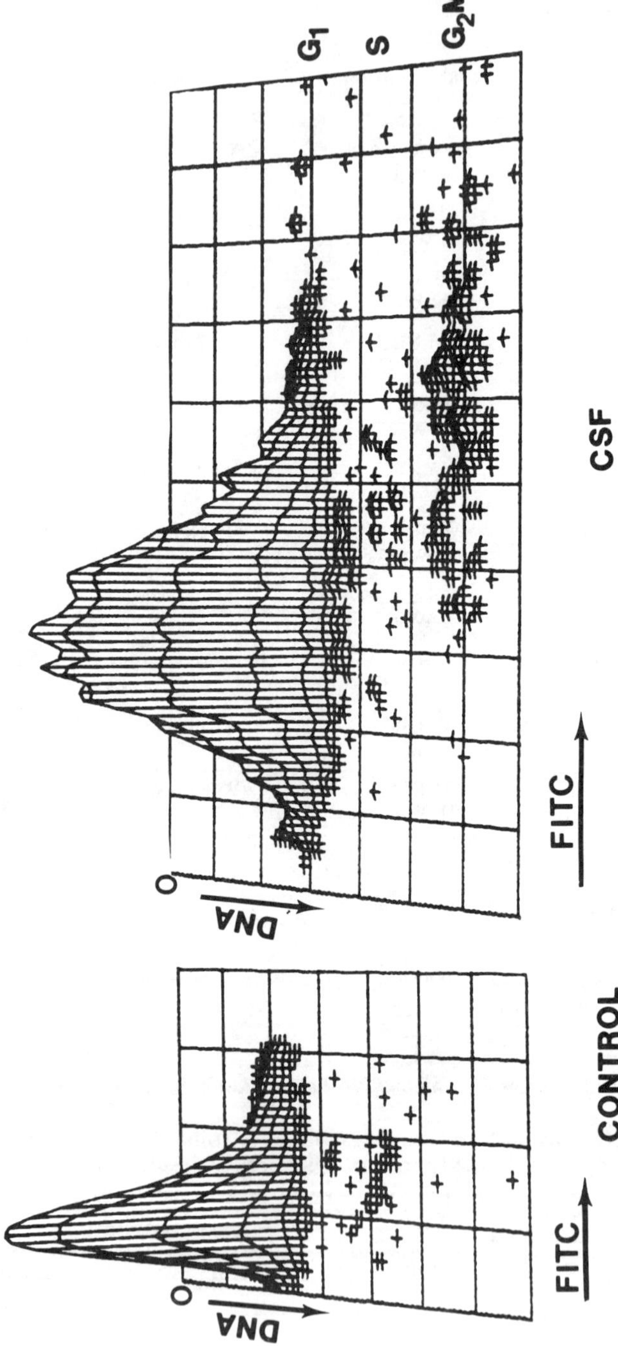

FIGURE 4. Computer drawn CALLA/DNA histogram of a spinal fluid sample of a patient with CALLA positive ALL. DI = 1.1 in the left panel; normal lymphocytes stained with antibody to CALLA define negative range of fluorescence. The right panel shows intense CALLA positively of CSF cells indicating presence of CALLA positive leukemic cells in the spinal fluid.

TABLE 4. Flow Cytometry of Spinal Fluid in Patients without CNS Leukemia

Parameter	Mean ± Standard Deviation	Range
% cells in $G_{0/1}$	98.12 ± 1.85	90–100%
% cells in G_0	94.42 ± 5.94	74–100%
% cells in G_1	3.71 ± 4.90	0–23%
% cells in S	1.68 ± 1.68	0–9%
RI $G_{0/1}$	9.44 ± .76	7.8–11.3
RI G_0	9.21 ± .64	7.8–10.8
RI G_1	11.4 ± 7.08	0–22
CV	3.6 ± 1.1	2.0–7.0

RI = RNA index. CV = Coefficient of Variation of DNA content. G_0, G_1, $G_{0/1}$, S = corresponding phases of cell cycle.

DNA/RNA Flow Cytometry as the only Indication of CNS Involvement

An abnormal DNA stemline was detected in spinal fluid of six patients who did not have central nervous system involvement clinically by Wright-Giemsa cytospin or Papanicolaou cytology. Spinal fluid examinations in these cases were performed routinely, as is done for all leukemic patients at initial diagnosis, bone marrow relapse, or periodically in those with aneuploid DNA stemline in marrow. In five out of six samples, fewer than 5,000 cells were measured because of sparce cellularity. Between 3.6% and 32.3% of these cells had the same abnormal DNA stemline that had previously been found in the patient's bone marrow aspirate. In one patient with ALL, an abnormal DNA stemline was detected in CSF during remission, when the patient was on maintenance therapy. There was no evidence of leukemic involvement in marrow, and no cytologic abnormalities in CSF. In this case flow cytometry was the first indication of CNS involvement. The CSF sample contained 17 WBC/mm^3 of which 12.9% had a DNA index of 1.55, no blasts were present on cytospin, and cytology was negative. Six months later, the spinal fluid contained 97.2% cells with the same abnormal DNA stemline, and for the first time cytology was positive for leukemia (FIGURE 5,#F TABLE 5). This case enabled us to estimate cell cycle doubling time. The total leukemic cell number in the spinal fluid at the initial examination (assuming a CSF volume of 90 ml^{13} was 2 × 10^5 cells. At the diagnosis of CNS relapse 182 days later the CSF cell count was 1,230 WBC/mm^3 of which 97% had abnormal DNA content. To achieve a leukemic cell number of 1.1 × 10^8 cells in the spinal fluid at the time of CNS relapse, there would have to be nine doublings in 182 days indicating a doubling time for the leukemic cells in the CSF to be twenty days

DISCUSSION

The central nervous system can be a sanctuary for leukemic cells during therapy and therefore a site of possible extramedullary relapse. For cases with aneuploid leukemic cells, DNA ploidy can help identify relapse and distinguish the original leukemic clone from an emerging clone with another stemline. In a few cases we have seen two different leukemic stemlines simultaneously present in separate sites in one individual. One patient in this series (#17) had a diploid (DNA

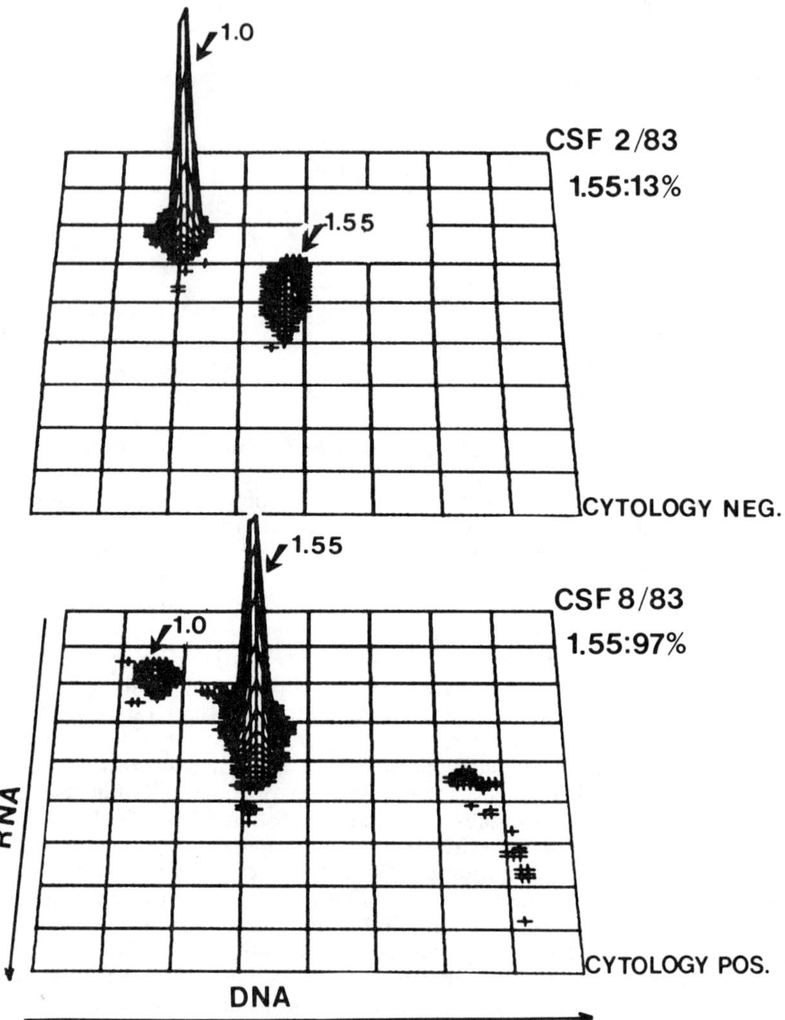

FIGURE 5. Computer drawn DNA-RNA histograms of CSF cells of a patient with ALL at two different time points six months apart. In the earlier sample (upper panel) 13% of the cells had a DI = 1.55 and cytology was negative. Six months later, (lower panel) 97% of cells in the CSF had DI = 1.55 and cytology was positive.

index = 1.0) leukemic line in the bone marrow at diagnosis, and bone marrow relapse one year later was characterized by triploid cells. Five months later central nervous system relapse occurred with the original diploid stemline while the bone marrow cells still contained the triploid leukemic cells. It was concluded that the triploid clone in the bone marrow developed by clonal evolution or *de novo*, whereas the cells with the initial diploid stemline persisted in the spinal fluid.

TABLE 5. Abnormal DI in Patients with No Clinical CNS Involvement

Case	DI[a]	Abnormal DI	Cell Count	Cytology
At Initial Diagnosis of Leukemia				
A	1.20	6.5% cells	0 RBC 1 WBC[e]	NEG[b]
B	1.30	10.8% cells	0 RBC 1 WBC[e]	NEG
Bone Marrow Relapse				
C	.70	32.3% cells	0 RBC 0 WBC	NEG
D	1.20	5.2% cells	400 RBC 5 WBC[e]	NEG
E	1.10	3.6% cells	0 RBC 0 WBC	ND[c]
In Remission				
F	1.55	12.9% cells	0 RBC 17 WBC[e]	NEG
	1.55[f]	97.2% cells[f]	0 RBC[f] 1230 WBC[f]	POS[df]

[a]DI = DNA index.
[b]NEG = negative.
[c]ND = not done.
[d]POS = positive.
[e]Differential counts showed no blasts.
[f]CNS relapse with positive cytology six months after abnormal DI was found by flow cytometry.

Flow cytometry can confirm the presence of CNS involvement by leukemias or lymphoma that have an abnormal DNA stemline (TABLE 1). The cells with abnormal DNA content (i.e. DNA index #1) correspond to leukemic blasts. Other markers are needed for leukemias with diploid DNA. In patients with ANLL who have high RNA content at diagnosis, for example, CNS relapse is characterized by high RNA content of the cells in the CSF. In cases of CNS leukemia with a diploid DNA stemline and low RNA, cell surface antigens may be useful markers, for example CALLA, as demonstrated in FIGURE 5. Increased CALLA immunofluorescence of cells in the spinal fluid identifies them as leukemic since CALLA-positive cells have not been reported in normal or "reactive" CSF.

We have measured 98 samples of CSF from patients without evidence of CNS disease and all had diploid DNA stemlines. These specimens had low S-phase fractions (1.68 ± 1.68) and low RNA indices of G_0/G_1 (9.44 ± .76), as do normal lymphocytes (TABLE 4). Therefore we consider high S-phase fractions, as well as high RNA indices, to indicate CNS involvement, even in the absence of DNA abnormality (see patient 14 and 19, TABLE 2 and FIGURE 2). Additional studies of other patients with encephalitides, demyelinating diseases, and other non-neoplastic central nervous system disease will be needed to confirm this data.

The Wright-Giemsa slide preparations were positive in 27 patients and not performed in three patients with a clinical diagnosis of CNS leukemia; they were negative in the six patients of TABLE 5 who had abnormal DNA stemlines as detected by flow cytometry without clinical or cytologic evidence of CNS leukemia. In these six cases, as demonstrated by the one patient not treated, flow cytometry may be the earliest indicator of CNS involvement.

Gondos and King[1] reported that Papanicolaou cytology was positive in only 40% of cases considered clinically to have central nervous system leukemia. Cytological examination of spinal fluid was positive in 10/22 cases (46%) of this present series, and was suspicious in three others (14%). Of the nine cases with clinical CNS leukemia and with negative cytology, eight were positive by flow cytometry (two had abnormal DNA stemlines; six had cells with high RNA content, including two with both high RNA and high proliferation).

Thus, flow cytometry of spinal fluid can be particularly valuable in diagnosing and monitoring CNS relapse, especially when leukemic cells are few in number, or uniform and resemble reactive lymphocytes. Abnormalities of DNA index, markedly increased proliferation, or a population of cells with higher than normal RNA content, were detected in 20/30 cases (67%) of CNS leukemia.

The significance of small numbers of aneuploid cells at diagnosis of leukemia in patients without clinical evidence of central nervous system involvement can not be determined by this present study. It is assumed that systemic relapse can evolve from minimal residual disease present in CSF or testis. Because of differences in the environment, leukemic cells may grow at a different rate in the CNS than in marrow, and may not be apparent at the time of marrow relapse. If so, this would increase the importance of early treatment of CNS relapse. The doubling time of the leukemic cells in the CSF was approximately 20 days in our one patient, who remained untreated. The doubling time of leukemic cells in the bone marrow has not been determined precisely, but a doubling time of five days has been reported[14] in one series.

We have shown that in some cases standard hematologic and cytologic examinations may not be able to detect leukemic cells in the cerebrospinal fluid. With better systemic therapy of leukemias, leukemic cell sanctuaries like cerebrospinal fluid have become increasingly important as sites of residual disease. Thus, improved monitoring of cerebrospinal fluid by flow cytometry may significantly contribute to the diagnosis and management of central nervous system involvement with leukemia or lymphoma.

SUMMARY

DNA/RNA flow cytometry studies were performed on the spinal fluid samples of thirty patients with acute leukemia or lymphoma at the time of clinical central nervous system relapse, and compared with similar studies of 56 patients (98 specimens) who had leukemia in remission with no evidence of CNS disease. Twelve of the 30 patients with CNS involvement had cells with abnormal DNA content in the spinal fluid (40%); the remaining eighteen had cells with diploid DNA content. In the group of 18 with diploid DNA, 10 had other abnormalities detected by flow cytometry. These included eight patients with acute leukemia who had cells with high RNA content, and two patients with non-Hodgkin's lymphoma who had markedly increased proliferation. Of the 22 patients studied by conventional cytology nine were negative for malignant cells, and in eight of these patients flow cytometry studies of DNA/RNA demonstrated abnormalities. Common ALL antigen was demonstrated by flow cytometry in three out of five cases studied.

Thus, abnormal DNA content, increased RNA content, increased proliferation and/or expression of the cell surface antigen CALLA identified CNS relapse by flow cytometry in 22 of 30 patients with acute leukemia or lymphoma. The technique appears to be at least as sensitive as conventional cytology and identifies CNS relapse in some patients with negative cytology.

ACKNOWLEDGMENTS

We wish to thank Ms. Edith Espiritu, Mr. Gordon Assing, and Ms. Jan Bressler for their excellent technical assistance.

REFERENCES

1. GONDOS, B. & E. B. KING. 1976. Cerebrospinal fluid cytology: diagnostic accuracy and comparison of different techniques. Acta Cytologica **20**:542–547.
2. SUAREZ, C., M. ANDREEFF, D. R. MILLER, P. G. STEINHERZ & M. R. MELAMED 1985. DNA and RNA determinations in 111 cases of childhood acute lymphoblastic (ALL) by flow cytometry: correlation of FAB classification with DNA stemline and proliferation. Brit. J. Hem. **60**:677–686.
3. ANDREEFF, M., Z. DARZYNKIEWICZ, T. SHARPLESS, B. CLARKSON & M. R. MELAMED. 1980. Discrimination of human leukemia subtypes by flow cytometric analysis of cellular DNA and RNA. Blood **56**:282–293.
4. REDNER, A., M. ANDREEFF, D. R. MILLER, P. STEINHERZ & M. R. MELAMED. 1984. Recognition of central nervous system leukemia by flow cytometry. Cytometry **5**:614–618.
5. TRAGANOS, F., Z. DARZYNKIEWICZ, T. SHARPLESS & M. R. MELAMED. 1977. Simultaneous staining of ribonucleic and dexoribonucleic acids in unfixed cells using acridine orange in a flow cytometric system. J. Histochem **25**:46–56.
6. DARZYNKIEWICZ, Z., F. TRAGANOS, T. SHARPLESS & M. R. MELAMED. 1975. Conformation of RNA in situ as studied by acridine orange staining and automated cytofluorometry. Exp. Cell Res. **95**:143–153.
7. KAPUSCINSKI, J., Z. DARZYNKIEWICZ & M. R. MELAMED. 1982. Luminescence of the solid complexes of acridine orange with RNA. Cytometry **2**:201–211.
8. SHARPLESS, T., F. TRAGANOS, Z. DARZYNKIEWICZ & M. R. MELAMED. 1975. Flow cytofluorimetry: discrimination between single cells and cell aggregates by direct size measurements. Acta Cytologica **19**:577–581.
9. SHARPLESS, T. 1979. Recording flow cytometric data. *In* Flow Cytometry and Sorting. M. R. Melamed, P. F. Mullany & M. L. Mendelsohn, Eds.: 359–379. Wiley. New York.
10. HIDDEMAN, W., J. SCHUMANN, M. ANDREEFF, B. BARLOGIE, C. HERMAN, R. C. LEIF, B. H. MAYALL, R. F. MURPHY & A. A. SANDBERG. 1984. Convention of nomenclature for DNA cytometry. Cytometry **5**:445–446.
11. KRUTH, H. S., R. C. BRAYLAN, N. A. BENSON & V. A. NOURSE. 1981. Simultaneous analysis of DNA and cell surface immunoglobulin in human B-cell lymphoma by flow cytometry. Cancer Res. **41**:4895–4899.
12. ANDREEFF, M., A. REDNER, S. THONGPRASERT, B. EAGLE, P. STEINHERZ, D. MILLER & M. R. MELAMED. 1986. Multiparameter flow cytometry for determination of ploidy, proliferation and differentiation in acute leukemia: treatment effects and prognostic value. *In* Tumor Aneuploidy. T. Buechner, Ed.:81–105. Springer Verlag. Berlin
13. ROSMAN, N. P. 1982. Elevated intracranial pressure. *In* The Practice of Pediatric Neurology. K. F. Swaiman & F. S. Wright, Ed 1:188–205. CV Mosby Co. St. Louis.
14. FREI, E. & E. J. FREIREICH. 1964. Progress and perspectives in the chemotherapy of acute leukemia. Adv. Chemotherapy **2**:269–298.

American Adult T Cell Leukemia/Lymphoma: A Flow Cytometric and Morphological Study

BERTRAM SCHNITZER, LARRY E. KAHN,
EDMUND J. LOVETT, III, AND JERRY L. HUDSON

Flow Cytometry Laboratory
Department of Pathology
University of Michigan
Ann Arbor, Michigan 48109

INTRODUCTION

T-cell lymphomas and leukemias with a mature or post-thymic immunophenotype are uncommon in the United States and Europe, in contrast to Japan where they are frequent and where a distinctive type of adult T-cell leukemia/lymphoma (ATLL) is endemic in the southwestern part of the country.[1] Occasional sporadic cases of ATLL have been reported in non-endemic areas of Japan, in the United States, and in several other countries.[2-9] Within a two-year period, we have seen four patients from Michigan with ATLL at the University of Michigan. The neoplastic cells of three of these patients were immunophenotyped by flow cytometry (FCM) and FCM DNA analysis was carried out. Although the neoplastic cells of these patients had a helper cell (T4+) phenotype, phenotypic heterogeneity was detected within this population by FCM. A fourth patient, residing in Michigan but originating from an endemic area in Georgia, was extensively studied and previously reported as part of a larger series of cases.[8,10,11] With the exception of the latter patient, all patients were leukemic and two of the four had hypercalcemia and antibodies to the C-type retrovirus, the human T-cell leukemia virus (HTLV).

MATERIALS AND METHODS

Light Microscopic Studies

Blood and bone marrow smears were prepared and stained with Wright's stain, acid phosphatase, periodic acid-Schiff (PAS), and non-specific esterase (alpha-naphthol butyrate) using standard techniques. Tissue was fixed in B5 or neutral buffered formalin and sections stained with hematoxylin and eosin (H&E).

Electron Microscopic Studies

Buffy coat preparations and tissue sections were fixed in a cacodylate-buffered solution containing 3% glutaraldehyde and 3% formaldehyde, processed according to previously reported techniques,[12] and examined and photographed in a Zeiss 9S electron microscope.

Immunologic Studies

Cell surface antigen phenotyping of Ficoll-Hypaque–separated peripheral blood cells or minced tissue (skin and lymph nodes) was carried out with a battery of monoclonal antibodies. These included: T11, T10, T9, T6, T8, T4, I2 (HLA/DR), J5 (CALLA), B1, B2, Mo1, Mo2, IgG, IgM, IgA, IgD, kappa, lambda (Coulter Immunology), Tac (kindly provided by Dr. T.A. Waldmann), Leu 7 (Becton-Dickinson), and TdT (immunofluorescence) (Bethesda Research Laboratories). Staining was performed by an indirect immunofluorescent procedure. Double staining in case 3 was carried out using a single fluorochrome (FITC) and a cocktail of primary antibodies. To determine whether the Tac antigen was present exclusively on T4+ or T8+ lymphocytes, the Ficoll-Hypaque–separated cells were stained with the following monoclonal antibodies: T4, T8, Tac, T4 and Tac, T8, and Tac (TABLE 1). Sorting of T4+, T8+, T4+ Tac+, and T8+ Tac negative populations was carried out and the sorted populations were examined morphologically. DNA staining was carried out on Ficoll-Hypaque–separated cells using a propidium iodide (PI) solution according to Vindelov.[13] Chicken erythrocytes were used as an internal DNA standard. An aliquot of the chicken erythrocytes was added to the suspension of leukocytes and stained with PI. DNA histograms were collected as a linear signal and the peak position of the chick cells was translocated to the same channel number. The number of cycling cells was calculated by integration of the S and G_2M regions of the histogram and expressed as percent of total nuclei counted. Assays for serum antibodies against disrupted HTLV were performed by previously published techniques.[14]

Case 1

A previously healthy 44-year-old white male from Michigan initially complained of constipation, abdominal discomfort, and distention. He had lost 8–10 pounds in the preceding three weeks. He noted generalized lymphadenopathy and a white blood cell count of over $100 \times 10^9/l$ was found. Pertinent physical findings included enlarged lymph nodes in the cervical and clavicular areas, a 2

TABLE 1. Relationship of T4+, T8+, and Tac+ Cells (Case 3)

Marker	% Positive
Control	0
T4	75
T8	17
Tac	40
T4 + Tac (cocktail)	75
SUM (T4, Tac)	115
T8 + Tac (cocktail)	56
SUM (T8, Tac)	57

cm right axillary lymph node and bilateral inguinal nodes 1–2 cm in diameter. Liver and spleen were enlarged. Laboratory values were: hemoglobin 14.0 g, hematocrit 43, white blood count $95 \times 10^9/l$ with 98% abnormal lymphocytes with hyperlobulated nuclei (FIGURE 1A), and a platelet count of $50 \times 10^9/l$. Serum calcium was 9.1 mg/dl. Serum antibodies to disrupted HTLV viral particles were not detected. A bone marrow biopsy revealed foci of normal hematopoiesis, but large areas of the marrow were overrun by hyperlobulated lymphocytes. The patient was treated with vincristine, prednisone, and intrathecal methotrexate. The white blood cell count initially fell to $50 \times 10^9/l$ but subsequently rose to $140 \times 10^9/l$. More aggressive therapy with vincristine, doxorubicin, and prednisone resulted in a transient disease in the WBC, but the patient's condition worsened, and he died 22 days after the diagnosis of his illness. At necropsy, widespread leukemic infiltrates were found.

Case 2

A 78-year-old white female was diagnosed as having chronic lymphocytic leukemia (CLL) at another hospital. Her white blood cell count was $98.9 \times 10^9/l$ with 91% lymphocytes. The patient refused therapy. No skin lesions were detected. The spleen was palpated 10 cm below the left costal margin, and the liver was enlarged below the right costal margin. Two months later she was admitted to the University of Michigan Hospital in a cachectic state. Bilateral pulmonary effusions were present. Sputum grew out acid-fast bacilli, and she was treated with INH, Rifampin, and pyridoxine. The white blood cell count was $174 \times 10^9/l$ with 91% lymphocytes. The hemoglobin was 12.7 mg/dl, hematocrit 34.5%, platelet count $124 \times 10^9/l$. The lymphocytes were small to medium-sized with multilobulated nuclei (FIGURE 1B). A bone marrow biopsy showed diffuse infiltration by lymphocytes with irregular nuclear contours. A morbiliform skin rash was biopsied and abnormal lymphoid cells were seen in the dermis as well as exocytosis of cells into the epidermis. The patient was treated with cytoxan, vincristine, and prednisone. As this therapy was not effective, she was given Deoxycoformycin, which was effective in reducing her white blood cell count but did not control her systemic disease. She continued to have pleural effusions containing neoplastic lymphocytes with the same phenotype as the cells in the peripheral blood. The patient died eight months after diagnosis. An autopsy was not performed.

Case 3

A 76-year-old male was admitted to the University of Michigan Hospital for "hepatitis" and "confusion." A serum calcium was found to be 17.1 mg/dl. Positive findings on physical examination included several 2×2 cm lymph nodes in the right axilla, shotty left axillary lymph nodes, and bilateral inguinal lymph nodes about 2×2 cm in size. The spleen was not palpable, and the liver was 3 cm below the costal margin with a 10 cm span. The hemoglobin at admission was 18.4 mg/dl, hematocrit 56.7%, white blood count $20.4 \times 10^9/l$, and the platelet count was $156 \times 10^9/l$. The differential white blood cell count contained 44% lymphocytes, most of which were hyperlobulated (FIGURE 1C). The patient was treated with saline, Lasix, a single dose (350 units) of calcitonin, and 100 mg Solu-medrol.

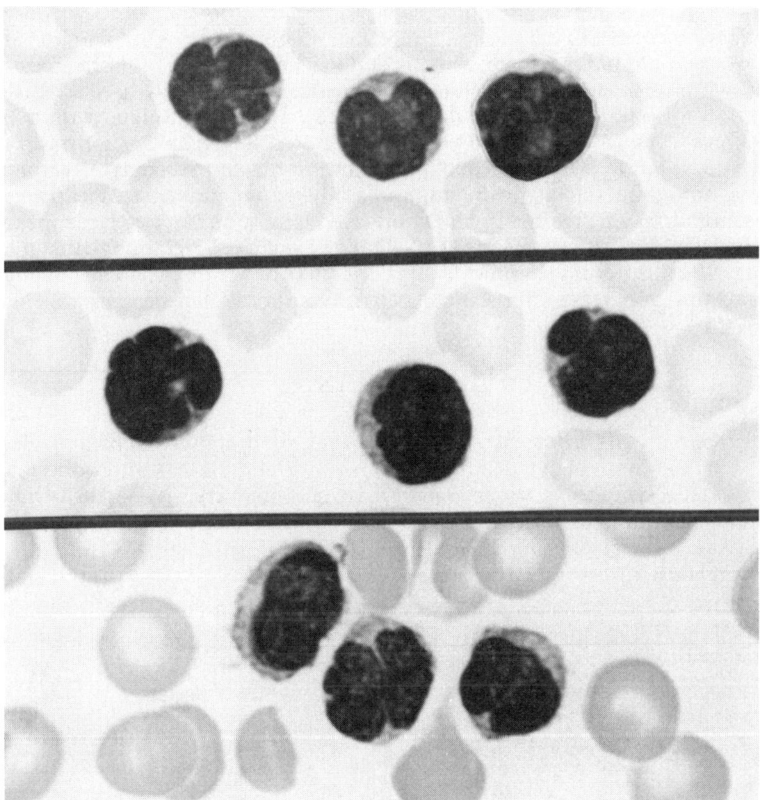

FIGURE 1. Multilobulated neoplastic lymphocytes from the peripheral blood of the three leukemic cases. (A) Case 1, (B) Case 2, (C) Case 3. Wright's stain × 1300.

During the next four days, the serum calcium dropped to 10.2 mg/dl. Over the next week, the white blood count rose to 32.8 × 10^9/l with 68% lymphocytes and then fell to 20 × 10^9/l with 80% lymphocytes. Examination of spinal fluid showed no abnormal lymphocytes. A bone marrow aspirate and biopsy contained less than 10% abnormal lymphocytes. During the course of illness, the patient developed pleural effusions and ascites. On numerous occasions, abnormal lobulated lymphocytes with the same phenotype as the blood lymphocytes were present in fluids from both cavities. Over the next four weeks, the enlarged lymph nodes became smaller, and the number of abnormal lymphocytes in the blood decreased progressively until several days before the patient died, when abnormal cells were no longer detected in the blood. Occasional hyperlobulated lymphocytes were still noted in the ascitic fluid several days before death. The patient died six weeks after onset of his illness. No evidence of lymphoma or leukemia was detected in autopsy tissue. Severe cytomegalovirus infection was detected in lungs, larynx, pancreas, adrenal glands, urethra, and kidneys.

Case 4

A 30-year-old black female, who resided in Michigan but was born in Georgia, initially presented with skin rash and hypercalcemia of 18 mg/dl. She refused an operation for parathyroidectomy and reappeared one month later with a serum calcium of 23 mg/dl. Lytic bone lesions were noted. Several skin biopsies (FIGURE 2) revealed a large cell lymphoma (T-cell immunoblastic sarcoma). A lymph node biopsy showed the same type of lymphoma. The patient was treated with a variety of chemotherapeutic regimens with short-lived remissions and with reappearance of skin lesions. Further treatment was carried out at the National Institutes of Health. She was never leukemic during the course of her disease. The patient died two and one-half years after a diagnosis of lymphoma had been established.

RESULTS

The leukemic cells from cases 1, 2, and 3 were similar by light and electron microscopy as well as cytochemically. Both in blood smears and bone marrow aspirates, the lymphocytes were medium sized with varying degrees of nuclear lobulations ranging from slight to pronounced, some having a mulberry

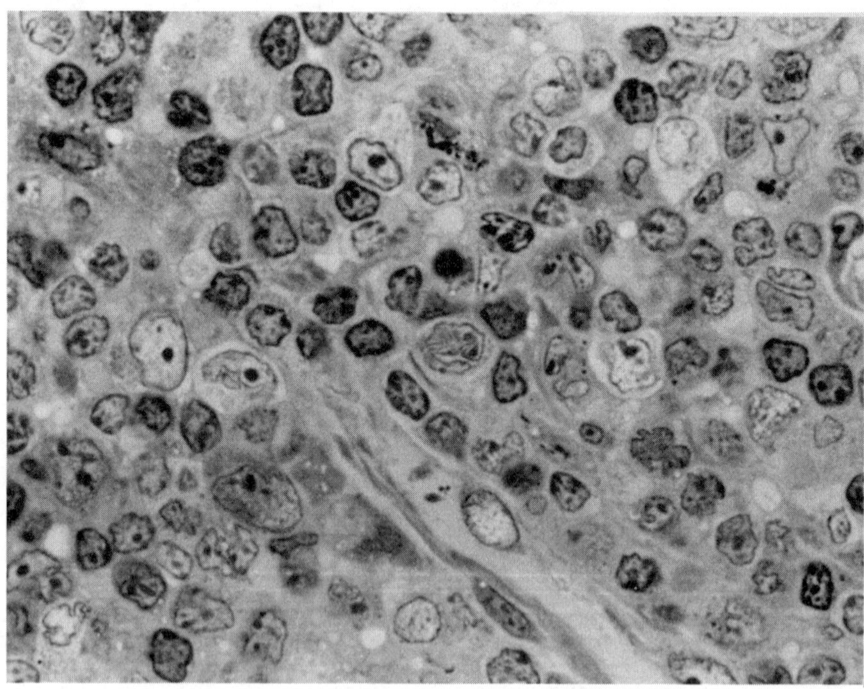

FIGURE 2. Lymph node biopsy from Case 4 showing a predominance of large neoplastic cells, some of which have markedly irregular nuclear contours. H&E × 500.

appearance (FIGURE 1). In most cells, in both peripheral blood as well as in smears of bone marrow aspirates, the nuclear chromatin was aggregated and nucleoli were not evident, while in somewhat larger cells, nuclear chromatin was less aggregated and a single nucleolus was noted (FIGURE 1). In all cases, the cytoplasm of the cells was diffusely punctate acid phosphatase positive, while NSE and PAS stains were negative. Ultrastructurally, the neoplastic cells showed varying degrees of nuclear lobulation and nucleoli were inconspicuous in most cells but prominent in a few cells (FIGURES 3–5). The neoplastic cells in the skin and lymph node biopsies of case 4 consisted predominantly of large cells, some of which had irregular nuclear contours, while others had round to oval nuclei with single nucleoli (FIGURE 2). Most of the cells had ample cytoplasm. Scattered smaller lymphoid cells with irregular nuclei and prominent nucleoli. The lymphocytes in case 3 had greater nuclear irregularity than in the other three cases and nuclear blebs and pocket formation were more often observed (FIGURE 5). In cases 2 and 4 small dense granules were noted in the cytoplasm of some cells. In all cases, the cytoplasm of the neoplastic cells contained moderate numbers of ribosomes, scattered mitochrondria, and rare segments of rough endoplasmic reticulum with collapsed cisternae. Serum antibodies to disrupted HTLV viral particles were detected in cases 3 and 4. Elevated levels of serum calcium were also noted in cases 3 and 4 (TABLE 2).

Cell surface antigen phenotyping by FCM varied slightly from patient to patient (TABLE 2). In case 1, over 90% of neoplastic cells were T11+, T3+, T4+,

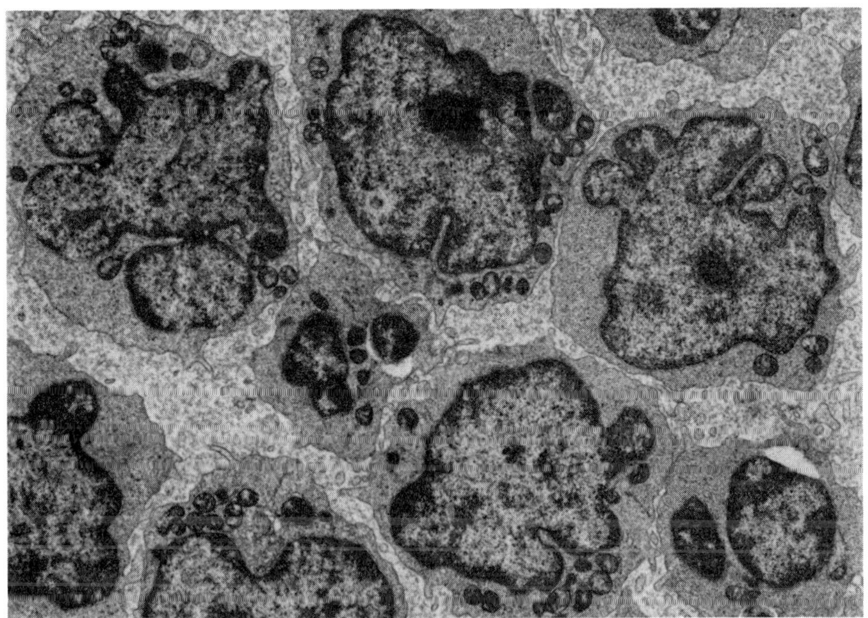

FIGURE 3. Electron micrograph of leukemic cells from the peripheral blood of Case 1. Multilobulated nuclei are evident. × 8,300. Figure reduction 70%.

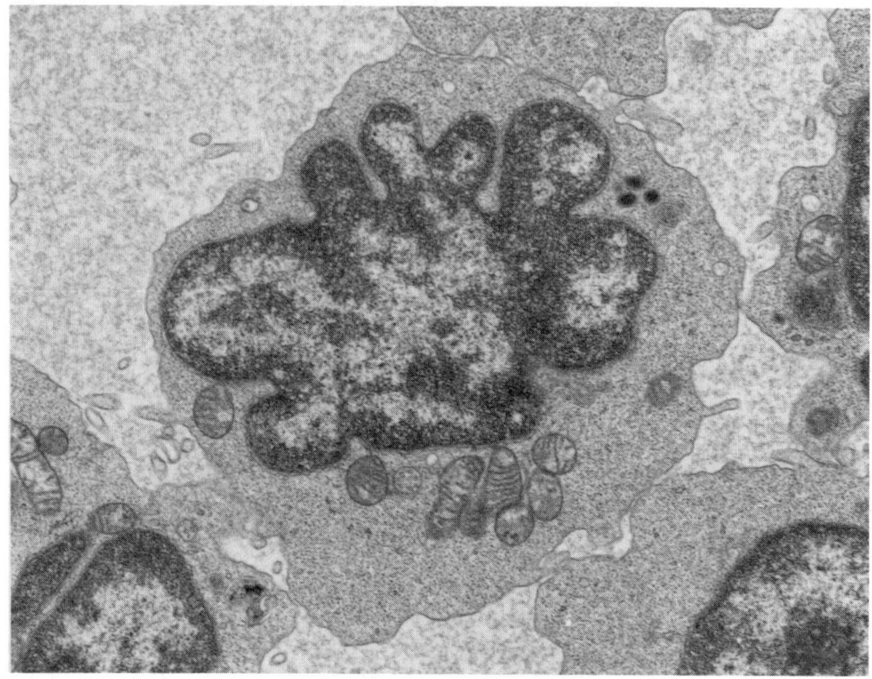

FIGURE 4. Electron micrograph of a leukemic lymphocyte of Case 2. The nucleus is multilobulated and the cytoplasm contains three dense granules and a number of mitochondria. × 10,000.

T8+. They did not express antigens detected by monoclonal antibodies T10, T9, T6, Ia, J5 (CALLA), Leu 7, B1, B2, and surface immunoglobulin (SIg) light and heavy chains were not detected. In case 2, the neoplastic cells had the following phenotype: T11+, T3+, T4+, T8−, T10−, T9−, Ia−, J5(CALLA)−, Leu 7−, B1−, B2−, and SIg light and heavy chain negative (TABLE 2), whereas the cells in case 3 had the phenotype: T11+, T10+, T3+, T4+, and Tac+ (TABLE 2). Forty percent of the neoplastic lymphocytes were Tac+, and all of these bore the T4 antigen, while none of T8+ cells were Tac+ (TABLE 1). The neoplastic cells did not express antigens detected by monoclonal antibodies T9, T8, T6, Ia, J5 (CALLA), Leu 7, B1, B2; and SIg light and heavy chains were absent. In case 4, the phenotype of the malignant cells in skin and lymph node was T11+, T3+, T4+, and T8−. In all cases, the neoplastic cells lacked TdT. DNA analysis failed to detect aneuploidy. In all of the cases, the number of cells in the S and G_2M phase of the cell cycle was not increased over those of normal peripheral blood lymphocytes.

DISCUSSION

It is important to differentiate T-cell from B-cell lymphoproliferative disorders, because the former often have a more fulminant clinical course and require more

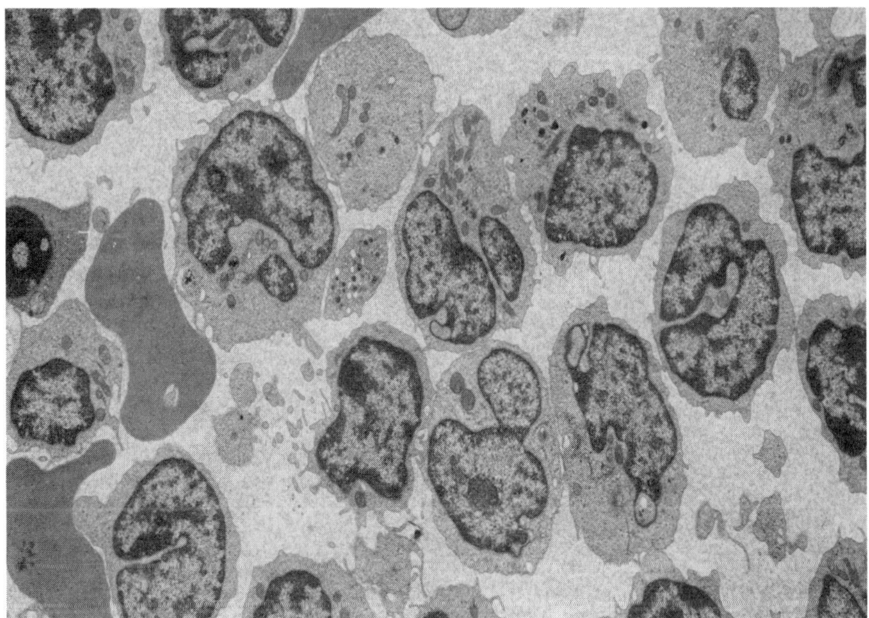

FIGURE 5. Electron micrograph of leukemic cells of Case 3. The nuclei are irregular and nuclear bleb formation is evident. X 4,300. Figure reduction 70%.

aggressive therapy. This is especially true of cases of ATLL, which in most instances are associated with a median survival of less than one year.[6-11]

In endemic as well as non-endemic regions, the most characteristic feature of ATLL is the presence of pleomorphic multilobulated lymphocytes in the peripheral blood.[6-11] Infiltrates of these abnormal lymphoid cells are the cause of lymphadenopathy, hepatosplenomegaly, skin lesions, and, less frequently, pulmonary lesions.[9-11] More common in endemic areas are hypercalcemia at the time of diagnosis or during the course of the disease, and antibodies to HTLV, which are present in 90 to 100% of cases from Japan and from the Caribbean.[7,9-11,15-19] The mediastinum, which is frequently involved in T-cell lymphoblastic lymphoma, is spared in ATLL. One of three of our non-endemic patients had serum antibodies to HTLV, and all patients were leukemic, while the patient born in the endemic region was not leukemic but presented with hypercalcemia and skin lesions and subsequent involvement of lymph nodes. As in almost all reported patients with ATLL, the lymphocytes of three of four of our patients had a helper cell phenotype.[7-11,20] The other patient had an anomalous phenotype, hitherto unreported in ATLL. The neoplastic cells had both a helper and suppressor cell phenotype (T11+, T3+, T4+, T8+, T6i, T10−),[6] which does not conform to the scheme of allegedly normal intrathymic T-cell differentiation.[21] In the three leukemic cases phenotyped by FCM, the neoplastic cells of case 3 did express the non-lineage specific T10 antigen. Thus, despite identical cytologic features of these T4+ leukemic cells, phenotypic heterogeneity was detected by FCM (TABLE 2).

TABLE 2. Features of Adult T Cell Leukemia/Lymphoma

Patient	Immunophenotype	DNA Analysis	White Blood Cell Count ($\times 10^9$/l)	Serum Calcium (mg/dl)	HTLV Antibody	Survival
1	T11+, T3+, T8+, T4+ T10−, T9−, T6−, Ia−, J5−, Leu 7−, B1−, B2− TdT−	Euploid Non cycling[a]	98−140	9.1	−	6 weeks
2	T11+, T3+, T4+ T10−, T9−, T8−, T6− Ia−, J5−, Leu 7−, B1− B2−, TdT−	Euploid Non cycling	62−402	10.2	+	6 weeks
3	T11+, T3+, T10+, T4+, Tac+ T9−, T6−, Ia−, J5−, Leu 7−, B1−, B2−, Tdt−	Euploid Non cycling	20−32.8	17.0	−	8 months
4	T11+, T3+, T4+ T8−, TdT−	ND[b]	5.1−28[c]	17.0−24.0	+	29 months

[a]Not greater than normal peripheral blood lymphocytes.
[b]Not done.
[c]Not leukemic.

Relatively little has been reported about HTLV seronegative cases of ATLL.[5,7,8] The two cases in this series in which no HTLV antibodies were demonstrated were, except for the absence of hypercalcemia, clinically and morphologically identical to our seropositive cases as well as to seropositive cases reported previously. In all three cases, the leukemic cells were multilobulated both by light and electron microscopy and had diffuse punctate acid phosphatase positivity. The HTLV's positive case is highly unusual in that the patient had a spontaneous remission of his disease prior to death, but died with a severe CMV infection similar to endemic cases of ATLL, in which patients not infrequently contract infections by opportunistic agents.[9] Another case of ATLL with spontaneous remission has been reported in Japan.[22]

The neoplastic cells of case 3 were analyzed with the anti-Tac monoclonal antibody. This antibody, which detects the receptor for the human T-cell growth factor or interleukin 2 (IL2) does not bind to normal resting T lymphocytes, B lymphocytes, or monocytes.[23-25] Tac monoclonal antibody does, however, identify HTLV-positive ATLL cells.[23,24] It is therefore useful not only in confirming a diagnosis of ATLL but also in differentiating other T-cell lymphoproliferative disorders, such as Sezary syndrome, in which the neoplastic T-helper cells may cytologically mimic those seen in ATLL.[26] Since it has been established that HTLV is T-cell tropic infecting primarily cells with a helper phenotype,[26,27] the Tac antigen should bind only to neoplastic T4+ cells and not to T8+ cells. In the HTLV+ case 3, in which the neoplastic cells were tested and found to react with anti-Tac antibody, multiparameter analysis was employed, revealing that the T4+, but not the T8+, cells were Tac+. We reached this conclusion on the basis of staining the cells with one fluorochrome and a cocktail of primary antibodies. It was assumed that if the Tac antigen is present on either the T4+ or T8+ lymphocytes, the sum of cells positive for the individual markers would be greater than that of their respective cocktails (T4 or T8+Tac). If the Tac antigen is not present on either T4+ or T8+ cells, then the sum of the percent of cells positive for the individual markers should be approximately equal to their respective cocktails. TABLE 1 summarizes the results of this double-staining procedure. Additional evidence that Tac was present only on T4+ was provided by sorting of T4+, Tac+ and T8+, Tac− populations. The sorted T4+ Tac+ cells were the neoplastic hyperlobulated acid phosphatase–positive lymphocytes, while the T8+, Tac− population were cytologically small, normal-appearing lymphocytes (FIGURE 6). The hypercalcemia in this patient may be explained by the presence of

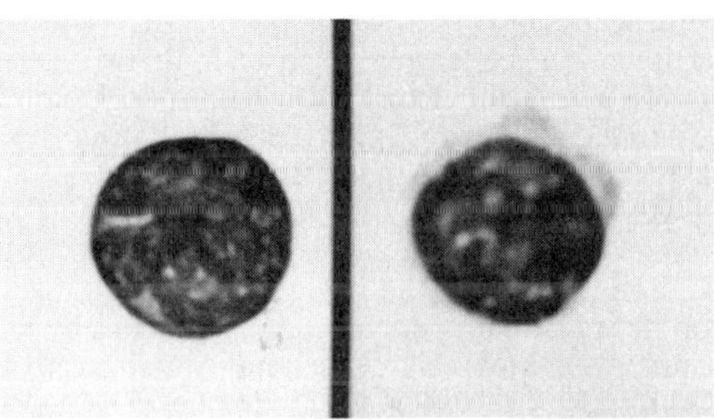

FIGURE 6. (Left) Sorted T4+ Tac+ leukemic cell with hyperlobulated nucleus. (Right) Sorted T8+ Tac− small lymphocyte. Wright's stain × 1,300.

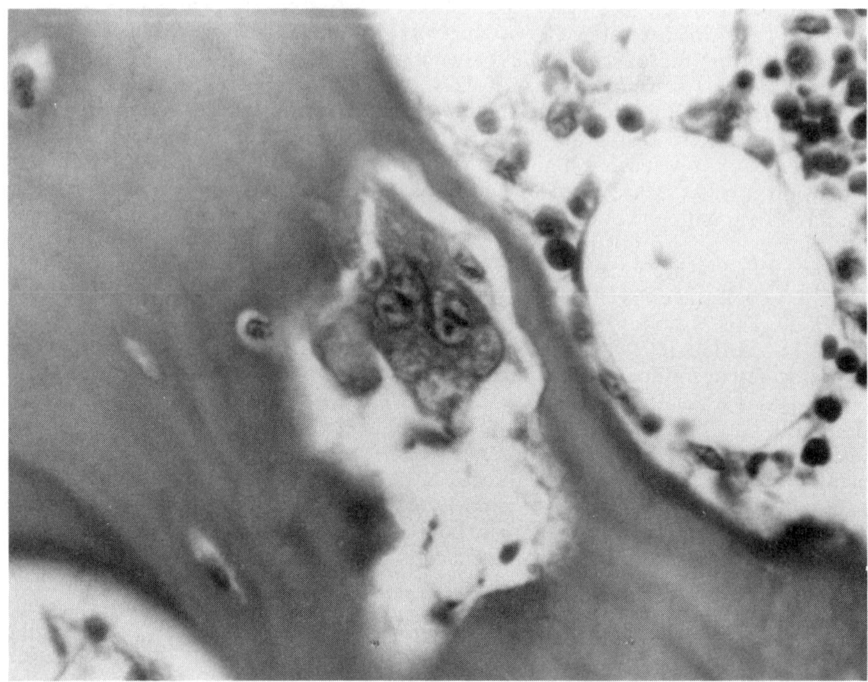

FIGURE 7. Bone and marrow from Case 3 showing scalloping of trabecular bone and an osteoclast. H&E × 200.

increased osteoclastic activity with scalloping of trabecular bone, which was evident in the bone of this patient despite the absence of associated leukemic marrow infiltrates (FIGURE 7). This feature has been reported in endemic cases from the United States and is thought to be due to a lymphokine produced by the neoplastic cells.[2]

In summary, three cases of non-endemic ATLL were analyzed by FCM. Cytologically, the leukemic cells were identical. In two cases, the cells had a helper phenotype, while in a third case both helper and suppressor cell markers were found in the same cells. A fourth patient, originating from an endemic area in Georgia, but residing in Michigan, also had a helper cell phenotype. Phenotypic heterogeneity among the neoplastic helper cells was detected by FCM. Mutliparameter analysis and cell sorting by FCM is useful in detecting more than one surface marker on the same cell and was essential in identifying the IL2 receptor on neoplastic helper cells. DNA analysis by FCM failed to detect an aneuploid population, and cell cycle analysis showed a low proliferative activity of ATLL cells.

REFERENCES

1. UCHIYAMA, T., J. YODOI, K. SAGAWA, L. TAKATSUKI & H. UCHINO. 1977. Blood **50**(3):481–491.
2. GROSSMAN, B., G. P. SCHECHTER, J. E. HORTON, L. PIERCE, E. S. JAFFE & L. WAHL. 1981. Am. J. Clin. Pathol. **75**(2):149–155.
3. KADIN, M. E. & M. KAMOUN. 1982. Lancet I(8279):1016–1017.
4. KADIN, M. E. & M. KAMOUN. 1982. Human Pathol. **13**(8):691–693.
5. TAKENAKA, T., H. NAKAMINE, I. OSHIRO, J. MAEDA, M. KOBORI, T. HAYASHI, H. KOMODA, Y. HINUMA & M. HANAOKA. 1983. Jpn. J. Clin. Oncol. **13**(Suppl. 2):257–268.
6. SCHNITZER, B., E. J. LOVETT, III, J. L. HUDSON, K. D. MCCLATCHEY, D. F. KEREN, L. DABICH & B. F. MITCHELL. 1982. Lancet II(8310):1273–1274.
7. SCHNITZER, B., E. J. LOVETT, III & L. E. KAHN. 1983. Lancet II(8357):1030.
8. KADIN, M. E., C. W. BERARD, K. NANBA & H. WAKASA. 1983. Human Pathol. **14**(9):745–772.
9. BUNN, P. A., G. P. SCHECHTER, E. S. JAFFE, D. W. BLAYNEY, R. C. YOUNG, M. J. MATHEWS, W. A. BLATTNER, S. BRODER, M. ROBERT-GUROFF & R. C. GALLO. 1983 N. Engl. J. Med. **309**(5):257–264.
10. JAFFE, E. S., W. A. BLATTNER, D. W. BLAYNEY, P. A. BUNN, JR., J. COSSMAN, M. ROBERT-GUROFF & R. C. GALLO. 1984. Am. J. Surg. Pathol. **8**(4):263–275.
11. BLAYNEY, D. W., E. S. JAFFE, W. A. BLATTNER, J. COSSMAN, M. ROBERT-GUROFF, D. L. LONGO, P. A. BUNN, JR. & R. C. GALLO. 1983. Blood **62**(2):401–405.
12. SCHNITZER, B. & L. KASS. 1974. Am. J. Clin. Pathol. **61**(2):176–187.
13. VINDELOV, L. L. 1977. Virchows Arch. (Cell Pathol.) **24**(3):227–242.
14. ROBERT GUROFF, M., V. S. KALYANARAMAN, W. A. BLATTNER, M. POPOVIC, M. G. SARNGADHARAN, M. MAEDA, D. BLAYNEY, D. CATOVSKY, P. A. BUNN, JR., A. SHIBATA, Y. NAKAO, Y. ITO, T. AOKI & R. C. GALLO. 1983. J. Exp. Med. **157**(1):248–258.
15. BLATTNER, W. A., V. S. KALYANARAMAN, M. ROBERT-GUROFF, T. A. LISTER, D. A. G. CALTON, P. SERIN, M. H. CRAWFORD, D. CATOVSKY, M. F. GREAVES & R. C. GALLO. 1982. Int. J. Cancer **30**(3):257–264.
16. ROBERT-GUROFF, M., Y. NAKAO, K. NOTAKE, A. SLISKI & R. C. GALLO. 1982. Science **215**(4535):975–978.
17. CATOVSKY, D., M. ROSE, A. W. G. GOOLDEN, J. M. WHITE, G. BOURIKAS, A. I. BROWNELL, W. A. BLATTNER, D. A. G. GALTON, D. R. MCCLUSKEY, I. A. LAMPERT, R. IRELAND, J. M. BRIDGES & R. C. GALLO. 1982. Lancet I(8273):639–643.
18. BLAYNEY, D. W., E. S. JAFFE, W. A. BLATTNER, J. COSSMAN, M. ROBERT-GUROFF, D. L. LONGO, P. A. BUNN & R. C. GALLO. 1983. Blood **62**(2):401–405.
19. BLATTNER, W. A., W. N. GIBBS, C. SAXINGER, M. ROBERT-GUROFF, J. CLARK, W. LOFTERS, B. HANCHARD, M. CAMPBELL & R. C. GALLO. 1983. Lancet II(8341):61–64.
20. HATFORD, T., T. UCHIYAMA, T. TOIBANA, K. TAKATSUKI & H. UCHINO. 1981. Blood **58**(3):546–547.
21. REINHERZ, E. L., P. C. KUNG, G. GOLDSTEIN, R. H. LEVEY & S. F. SCHLOSSMAN. 1980. Proc. Natl. Acad. Sci. USA **77**(3):1588–1592.
22. KIMURA, I., T. TSUBOTA, K. HAYASHI & T. OHNOSHI. 1983. Jpn. J. Clin. Oncol. **13**(Suppl 2):231–236.
23. UCHIYAMA, T., S. BRODER & T. A. WALDMANN. 1981. J. Immunol. **126**(4):1393–1397.
24. UCHIYAMA, T., D. L. NELSON, T. FLEISHER & T. A. WALDMANN. 1981. J. Immunol. **126**(4):1398–1403.
25. LEONARD, W. J., J. M. DEEPER, T. UCHIYAMA, K. A. SMITH, T. A. WALDMANN & W. C. GREENE. 1982. Nature **300**(5889):267–269.
26. BRODER, S., P. A. BUNN, E. S. JAFFE, W. BLATTNER, R. C. GALLO, F. WONG-STAAL, T. A. WALDMANN & V. T. DEVITA. 1984. Ann. Int. Med. **100**(4):543–557.
27. ROBERT-GUROFF, M. & R. C. GALLO. 1983. Blut **47**(1):1–12.

Detection of Minimal Residual Disease in Acute Leukemia by Flow Cytometry[a]

J. W. M. VISSER,[b,c] A. C. M. MARTENS,[c] AND
A. HAGENBEEK[c,d]

[c]*Radiobiological Institute TNO*
Rijswijk, The Netherlands

[d]*Dr. Daniël den Hoed Cancer Center*
Department of Hematology
Rotterdam, The Netherlands

INTRODUCTION

Treatment of acute leukemia is generally directed at induction of a state of remission. This state is characterized by the absence of recognizable leukemic cells in microscope preparations of bone marrow. Because of the limited possibilities of detecting rare abnormal cells by microscopy, remission is, by definition, already achieved if 5% or fewer of the bone marrow cells are leukemic. Since the body contains about 10^{12} bone marrow cells, a patient in remission may, therefore, still contain 10^{10} leukemic cells. It can be expected that 10^0 leukemic cell per body is sufficient to eventually give rise to a relapse. These figures indicate that today's treatment of acute leukemia should be monitored and quantified on a logarithmic scale. However, the techniques for enumerating and analyzing leukemic cells at low incidence are still poorly developed. The number of specific markers for leukemic cells is too low to give sufficient resolution in different leukemias. The automation of the detection of rare cells has in the past been mainly directed towards cervical abnormalities and blood disorders. Bone marrow provides its own specific problems and artifacts. At present, we are studying the detection of rare hematopoietic cell types in both normal and abnormal bone marrow from mice, rats, and man using a light-activated cell sorter. The transplantable Brown Norway Myeloid Leukemia (BNML),[1,2] which can be quantitated at various stages of disease in the rat by the lifespan of the animal or by means of a clonogenic assay, is used as a model system to evaluate the enumeration by flow cytometry.

[a]These investigations were supported in part by a program grant from the Netherlands Foundation for Medical Research (FUNGO), which is subsidized by the Netherlands Organization for the Advancement of Pure Research (ZWO) and by the Queen Wilhelmina Fund of the Dutch National Cancer League (grant RBI 80-1).

[b]Address correspondence to: Dr. J.W.M. Visser, Radiobiological Institute TNO, P.O. Box 5815, 2280 HV Rijswijk, The Netherlands.

MATERIALS AND METHODS

Animals

Male C57BL/Rij(H-2K^b) × C3H(H-2K^k)Fl hybrid (BC3) mice and female Brown Norway (BN) rats were bred and maintained under specific pathogen-free conditions in our institute.

Cell Suspensions

Bone marrow cells were obtained from 7-wk-old BC3 mice or from 10-wk-old BN rats by flushing the femoral shafts with Hanks' balanced salt solution buffered at pH 6.7 with HEPES (10 mM) (H.HBSS). Cell suspensions were filtered through a nylon sieve and kept on ice until use. Spleen and liver cells were obtained by mincing the organs and filtering through a sieve.

Rat Leukemia Model

The BN acute promyelocytic leukemia (BNML), which is transplantable in BN rats, shows great similarities with human AML.[1,2] Normal hematopoietic stem cells and leukemic clonogenic cells can be selectively discriminated by modified spleen colony assays.[3]

Lectin and Antibody Labeling

Cell suspensions were suspended in HASH (H.HBSS with 0.01% vol/vol NaN$_3$ and 5% vol/vol inactivated fetal calf serum) and incubated with antibody (45 min, 0–4°C). Mouse stem cell labeling was done by use of anti-H-2K^k-biotin (1 µg/10^6 cells). These cells were subsequently incubated with avidin-XRITC. In addition, the mouse bone marrow cells were labeled with fluoresceinated wheat germ agglutinin (WGA-FITC) (2 µg/10^7 cells, 15 min, 0–4°C). BNML cells were labeled with a monoclonal antibody RM124 which was produced and provided by Drs. H. Kaizer and R.J. Johnson (Johns Hopkins University, Baltimore, MD).[4] Various dilutions of this antibody were used. The second reagent was goat-anti-mouse IgM (Fc) conjugated to FITC (30 min; 0–4°C). After washing, the cells were resuspended in H.HBSS and analyzed by the flow cytometer.

Flow Cytometry

A modified FACSII light-activated cell sorter (Becton Dickinson, Sunnyvale, CA) with the laser tuned at 488 nm (0.5 W) was used. FITC fluorescence was measured by an S-20 type photomultiplier (PM) through a combination of a broad band multicavity interference filter (520–550 nm; Pomfret, Stamford, CT) and a 520 nm cut-off filter (Ditric, Hudson, MA). A logarithmic amplifier (T. Nozaki, Stanford, CA) was used for the fluorescence signals. Dual wavelength flow cytometry was performed using the RELACS-8 (Rijswijk Experimental Light-Activated Cell Sorter) with a R6G dye laser tuned at 580 nm (0.4 W) and an

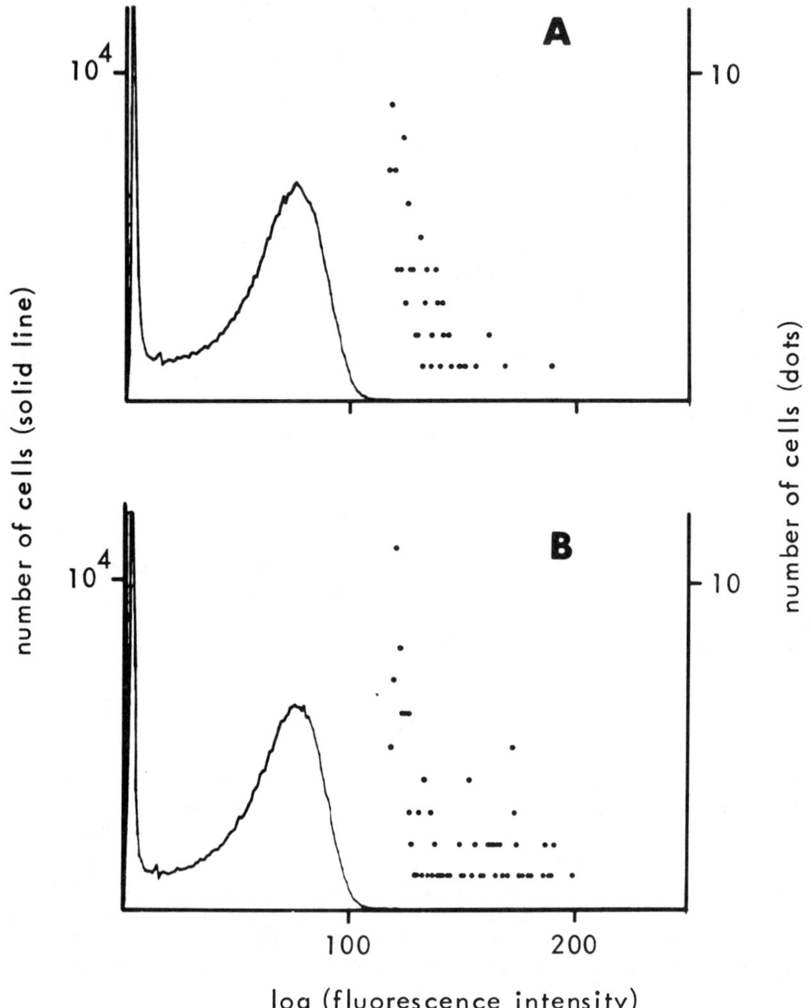

FIGURE 1. Histograms of the fluorescence intensities of mouse bone marrow blast cells, which were distinguished from the other cell types by using the simultaneously measured forward and perpendicular light scatter intensities (FLS-PLS windows; *cfr.* Ref. 7). The histograms have a different scaling along the ordinates at the right and at the left. (A) untreated cells. (B) treated cells containing one brightly fluorescent cell per 10^4 negative cells. 4×10^5 cells were recorded per histogram.

argon laser at 488 nm (0.5 W). XRITC fluorescence was measured by an S-20 type PM through a Schott RG 665 glass absorbance filter, FITC-fluorescence through the above-described filter combination by an S-11 type PM. Propidium iodide fluorescence was measured as described earlier.[5]

RESULTS AND DISCUSSION

The Detection Limit of Rare Cells by Flow Cytometry

Commercially available flow cytometers are capable of analyzing up to 10^4 cells per second. Therefore they could, theoretically, be used to accurately enumerate rare cell types that occur at frequences of 1 per 10^4 to 10^5 within two minutes measurement time. In practice, the technique is slightly less efficient. At 10^4 cells per second, the resolution is less than at slower cell rates, so that the distinction between the rare and the other cells cannot always be made. In addition, background and artificial signals contribute to the analysis; therefore, the measurement time has to be prolonged. Suspensions of unlabeled freshly isolated mouse bone marrow cells contain particles producing signals that the FACS interprets as fluorescent ones.

The incidence of these events is about 1 per 10^4 cells and they interfere with the detection of rare events. Most of these unwanted fluorescence signals arise from particles with a forward light scatter (FLS) intensity much higher than that of bone marrow cells. Therefore, a large proportion of the background signals can be excluded by using electronic windows that limit the analysis to events with FLS and PLS signals of bone marrow cells. FIGURE 1 shows that as a result of this gating labeled cells can be enumerated at 1 per 10^5 cells by analyzing 5×10^6 cells (30 minutes FACS time) from a mixture of labeled and unlabeled cells ($1:10^5$). This detection level must be considered as the lowest practically achievable using a commercially available flow cytometer.

The Detection of Rare Hematopoietic Cells

It can be calculated that between 0.2 and 1% of normal bone marrow cells consist of pluripotent hematopoietic stem cells (PHSC). The number of each of the earliest committed progenitor cells, the differentiated daughter cells of the PHSC, must be similar or probably somewhat lower, *viz.*, 0.03 to 0.1% of the normal bone marrow cells. Such incidences could be easily detected by flow cytometry if specific markers for each subpopulation existed. Up to now unique markers for the committed progenitor cells have not been reported. Mouse PHSC can be isolated using equilibrium density centrifugation, wheat-germ agglutinin-FITC, and anti-H-2K-FITC labeling using two sortings on the FACS.[6] Apparently, this PHSC can be uniquely marked. FIGURE 2 shows a dot plot of the red versus green fluorescence intensities of mouse bone marrow cells labeled with WGA-FITC and anti-H-2K-biotin-avidin-XRITC, which were electronically gated for PLS and FLS intensities of blast cells,[7] including CFU-S. A H-2K- and WGA-positive subpopulation comprising 0.9% of all nucleated cells can be distinguished, which, according to the previously studied properties, must consist primarily of PHSC. This illustrates that PHSC can be labeled specifically and

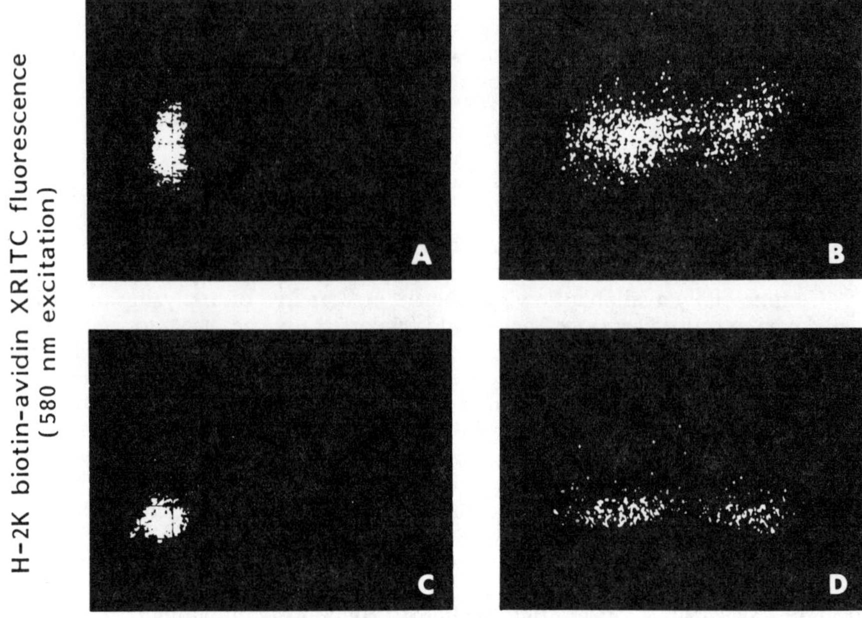

FIGURE 2. Dot plots of mouse bone marrow blast cells (distinguished by using FLS-PLS windows as for FIGURE 1) with: (A) anti-H-2K biotin-avidin-XRITC; (B) both anti-H-2K-biotin-avidin-XRITC and WGA-FITC; (C) avidin-XRITC only; and (D) WGA-FITC only.

that flow cytometers can be used for their detection and enumeration. It can be expected then that this can also be done with other species and that ultimately sorting, spleen colony assays, and *in vitro* culture methods will become superfluous for the examination of PHSC.

The Detection of Leukemic Cells during Remission in Bone Marrow and Other Organs by Flow Cytometry

The incidence of leukemic cells during remission is somewhere below five per 10^2 bone marrow cells. In man, the incidence may be as low as 1 per 10^{12} bone marrow cells. A maximum of about 10^8 cells can be taken for diagnosis; therefore, the sampling size limits the enumeration of leukemic cells. Residual leukemic cells at concentrations below 1 per 10^4 normal bone marrow cells cannot be detected due to this sampling limit. The conventional flow cytometers are too slow to efficiently process a sample of 10^8 cells. Analysis rates of up to 10^6 cells per second have been reported recently when newly designed high speed flow

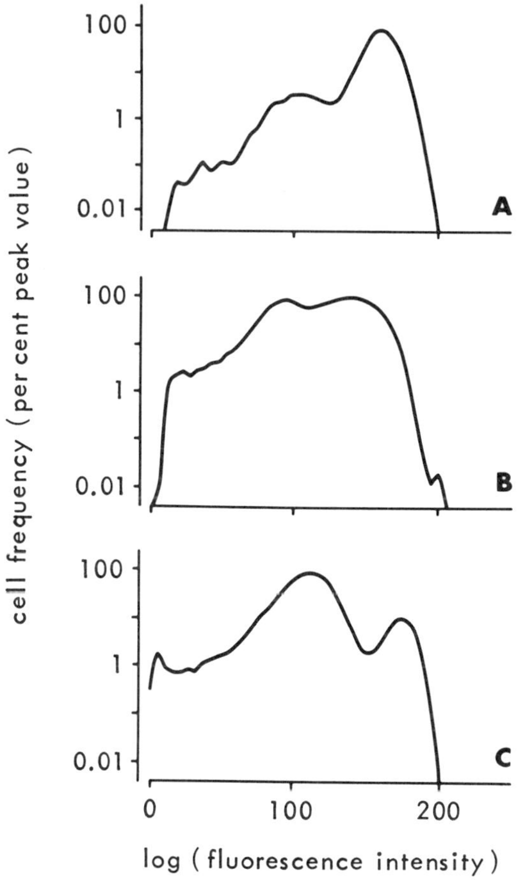

FIGURE 3. Histograms of the fluorescence intensities of rat bone marrow cells labeled with Rm124-FITC after high-dose cyclophosphamide treatment (100 mg/kg i.p.) of leukemic (BNML) rats. (A) bone marrow cells from rats taken shortly before cyclophosphamide treatment; (B) 9 days after treatment; and (C) 16 days after treatment. At 9 days after treatment, the frequency of BNML cells was 1 per 16,000 normal bone marrow cells as determined by the survival bioassay.[1,2,9]

cytometers are used.[8] These may be useful in the detection of minimal residual disease at the sampling limit, which is of importance for the evaluation of maintenance chemotherapeutic protocols.

At present, the nonavailability of specific labels for the detection of all kinds of leukemic cells with possibly varying membrane antigens is also limiting. In a rat model for human acute myelocytic leukemia (the BNML model) a specific monoclonal antibody against the leukemic cells is available.[4] In this system the detection of leukemic cells after remission-induction chemotherapy with cyclophosphamide in various organs can be followed by use of flow cytometry. The

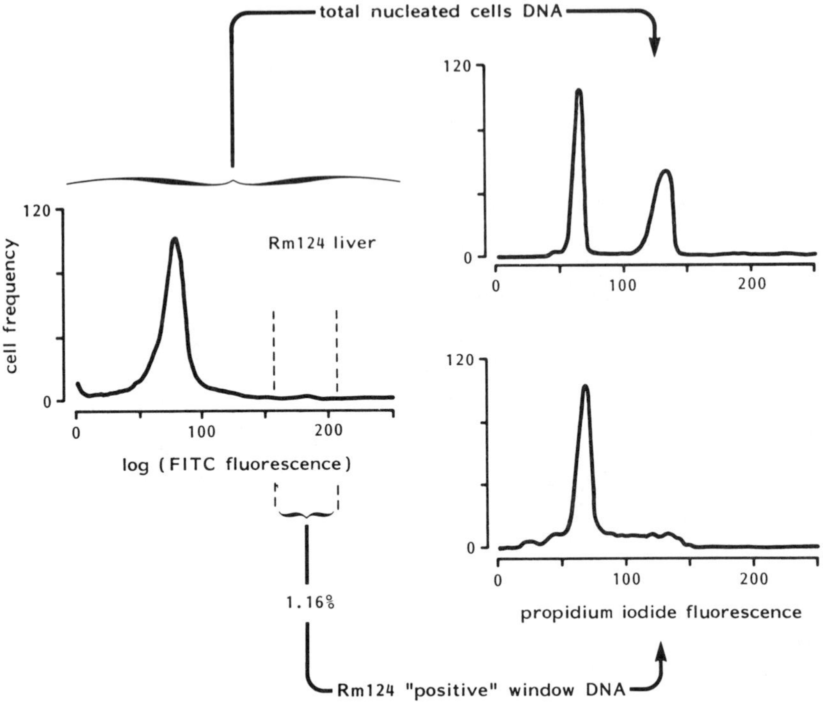

FIGURE 4. Histogram of the fluorescence intensities of cells from the liver of leukemic (BNML) rats. (Left panel) liver cells labeled with Rm124-FITC, dashed vertical lines indicate sorting window for BNML cells. (Right panels) DNA histograms of total cell suspension (top) and sorted BNML cells (bottom).

detection limit of leukemic cells was found to be 1 per 10^4 normal cells when this antibody was used.[9] FIGURE 3 shows the detection of residual leukemia with the RM124 antibody in bone marrow after high-dose cyclophosphamide treatment.

Cell Cycle Analysis of Minimal Residual Disease by Flow Cytometry

The choice of maintenance therapy may strongly depend on knowledge of the proliferative state of the residual leukemic cells and of the normal hematopoietic stem cells. To some extent, flow cytometry may be useful in providing such data. FIGURE 4 shows DNA histograms of all liver cells and of leukemic cells (BNML) that were sorted from a liver cell suspension containing one leukemic cell per 10^2 normal cells. The histogram of all liver cells (top panel) contains two peaks due to normal cells of 2n and 4n DNA content. The histogram of the sorted leukemic cells (lower panel) does not contain this 4n DNA peak. The cell cycle status of the leukemic cells can be monitored during remission by using this detection and sorting protocol.

SUMMARY

Commercial flow cytometers can detect and enumerate rare cells at the level of 1 per 10^5 other cells within a reasonable measuring time, provided that the rare cells can be uniquely labeled with a fluorescent marker. This detection level is sufficient for the enumeration of normal hematopoietic stem cells and committed progenitor cells. Detection at this level is useful for the quantitation of residual leukemic cells in remission bone marrow, for the analysis of the proliferative state of these cells as well as of normal stem cells, which are of importance in choosing the optimal chemotherapy regimen, and for monitoring the efficacy of maintenance chemotherapy. A further improvement in the speed of flow cytometers would be required, however, to make full use of the bone marrow samples.

REFERENCES

1. HAGENBEEK, A. & D. W. VAN BEKKUM, Eds. 1977. Proceedings of an international workshop on "Comparative evaluation of the L5222 and the BNML rat leukaemia models and their relevance for human acute leukaemia". Leukemia Res. 1:75–255.
2. BEKKUM, D. W. VAN & A. HAGENBEEK. 1977. Relevance of the BN leukemia as a model for human acute myeloid leukemia. Blood Cells 3:565–575.
3. HAGENBEEK, A. & A. C. M. MARTENS. 1981. Separation of normal hemopoietic stem cells from clonogenic leukemic cells in a rat model for human acute myelocytic leukemia. II. Velocity sedimentation in combination with density gradient separation. Exp. Hemat. 9:575–580.
4. MARTENS, A. C. M., R. J. JOHNSON, H. KAIZER & A. HAGENBEEK. 1984. Characteristics of a monoclonal antibody (RM124) against acute myelocytic leukemia cells. Exp. Hemat. 12:667–671.
5. MARTENS, A. C. M., G. J. VAN DEN ENGH & A. HAGENBEEK. 1981. The fluorescence intensity of propidium iodide bound to DNA depends on the concentration of sodium chloride. Cytometry 2:24–25.
6. VISSER, J. W. M., J. G. J. BAUMAN, A. H. MULDER, J. F. ELIASON & A. M. DE LEEUW. 1984. Isolation of murine pluripotent hemopoietic stem cells. J. Exp. Med. 159:1576–1590.
7. VISSER, J. W. M., G. J. VAN DEN ENGH & D. W. VAN BEKKUM. 1980. Light scattering properties of murine hemopoietic cells. Blood Cells 6:391–407.
8. PETERS, D., E. BRANSCOMB, P. DEAN, T. MERRILL, D. PINKEL, M. VAN DILLA & J. W. GRAY. 1985. The LLNL high speed sorter: design features, operational characteristics, and biological utility. Cytometry 6:290–301.
9. HAGENBEEK, A. & A. C. M. MARTENS. 1984. Detection of minimal residual leukemia utilizing monoclonal antibodies and fluorescence activated cell sorting (FACS). *In* Minimal Residual Disease in Acute Leukemia. B. Löwenberg & A. Hagenbeek, Eds.: 45–54. Martinus Nijhoff Publishers. Boston.

Prognostic Applications of DNA Flow Cytometry for Human Solid Tumors[a]

O. S. FRANKFURT,[b,c] S. G. ARBUCK,[c,d] J. L. CHIN,[c,f]
W. R. GRECO,[c,g] Z. P. PAVELIC,[c] H. K. SLOCUM,[c]
A. MITTELMAN,[d] S. M. PIVER,[c] E. J. PONTES,[b]
AND Y. M. RUSTUM[c]

[c]Grace Cancer Drug Center
[d]Department of Surgical Oncology
[e]Department of Gynecologic Oncology
[f]Department of Urologic Oncology
[g]Department of Biomathematics
New York State Department of Health
Roswell Park Memorial Institute
Buffalo, New York 14263

INTRODUCTION

The determination of biological characteristics of tumor cells that may predict the prognosis of malignant disease and that can be easily and precisely measured is an important goal. The search for such characteristics has been facilitated by the development of flow cytometry (FCM) with its application to clinical cancer research.[1] Recently many investigators have studied DNA content in the cells of human solid tumors and have attempted to relate DNA content abnormalities to clinical and pathological disease characteristics.[2-10] Some data suggest that for bladder, colon, and ovarian cancer, patients with diploid tumors have better prognoses than those with aneuploid tumors.[11-13] These observations require confirmation by other laboratories, since the number and selection of patients and the methods used for DNA measurements could affect conclusions.

To establish the usefulness of DNA ploidy as a potential prognostic tool, it is important to establish its relationship to accepted prognostic factors, such as tumor stage and grade. Staging and grading systems have provided useful prognostic information to guide the clinical management of patients with cancer. However, within particular stages there are often subgroups of patients with varying outcomes (e.g. patients with T_1 bladder tumors, stage C prostate tumors, and earlier stage colorectal tumors). Any biological characteristics that would permit early identification of those patients at greater risk would allow physicians to treat some patients more aggressively and to spare those less likely to benefit from additional or more toxic treatment.

[a]Supported by National Institutes of Health Grants CA-21071 and CA-28853 and Core Brant CA-24538. S. G. A. is recipient of an American Cancer Society Junior Faculty Clinical Fellowship. J. L. C. is recipient of Gordon E. Richards Fellowship Award of the Canadian Cancer Society.

[b]Address correspondence to: O. S. F., Grace Cancer Drug Center, Roswell Park Memorial Institute, 666 Elm St., Buffalo, NY 14263.

In this report, the results of FCM analysis of DNA distributions of tumors from 78 patients with bladder carcinomas, 45 patients with prostate carcinomas, 50 patients with renal carcinomas, 91 patients with colorectal carcinomas, and 50 patients with ovarian carcinomas are described. The relationship between DNA ploidy and the proportion of cells in S-phase and tumor stage and grade are presented. Although preliminary, early survival data for patients with colorectal and ovarian carcinoma are included.

MATERIALS AND METHODS

Methods for FCM analysis of human solid tumors have been described.[9,10,14] Unfixed cells were stained with the DNA-specific fluorochrome 4,6-diamidino-2-phenylindole (DAPI) in the presence of the detergent Triton X-100. DNA distributions were measured on an ICP-22 flow cytometer with mercury lamp excitation. Tumors were defined as aneuploid only if a separate G_0/G_1 aneuploid peak was observed. Staging of tumors was by standard criteria that have been described previously.[9]

Data were analyzed using the Fisher exact test, Chi Square analysis, the Mann-Whitney test, the Kruskal-Wallis test, linear logistic regression,[15] and Kaplan-Meier life table analysis[16] with the log rank test.[17] The last three techniques were performed with the software package, FREND,[18] which is an adjunct to SPSS.[19] Statistical tests of significance were performed at $\alpha = 0.05$.

RESULTS AND DISCUSSION

Bladder Tumors

DNA FCM of human bladder tumors has received much attention during recent years. Studies have established a relationship between DNA ploidy and both tumor characteristics and course of disease. DNA FCM has demonstrated significant diagnostic and prognostic potential for human bladder tumors.[6,10,11,20–22]

In this study, the DNA distributions of 99 bladder tumors (samples from biopsies and cystectomies) and 29 irrigation fluids from 78 patients with confirmed bladder tumors were evaluated. The frequency of DNA aneuploidy in bladder tumors increased with increasing stage and with decreasing cell differentiation (TABLE 1). All non-invasive tumors (T_a stage) were diploid. Twenty-seven percent of the tumors with invasion limited to the lamina propria (T_1 stage) were aneuploid. It is of interest that among T_1 tumors, grade appeared to be related to the frequency of aneuploidy. None of the well-differentiated (grade 1) tumors were aneuploid, whereas 30% of grade 2 tumors and 77% of poorly differentiated (grade 3) T_1 tumors contained aneuploid cells.

For bladder tumors with muscle invasion (stages T_2 and T_3) or pelvic extension (stage T_4), approximately three quarters were aneuploid. The presence of tumor cells in suspensions used for FCM measurements was confirmed by cytological analysis of smears prepared from these suspensions. Thus, the possibility of a sampling error for tumors with diploid DNA histograms was excluded.

TABLE 1. DNA-Ploidy Related to Stage and Grade of Human Bladder Tumors[a]

| Stage | Frequency of Aneuploidy[b] | | | | | | | | DNA Index of Aneuploid Tumors (Median) |
	All grades		Grade 1		Grade 2		Grade 3		
T_a (no invasive growth)	0/8	(0)	0/5	(0)	0/3	(0)	0		—
T_1 (invasion of lamina propria)	10/37	(27.0)	0/9	(0)	5/21	(23.8)	5/7	(71.4)	1.86
T_2 (superficial muscle invasion)	10/14	(71.4)	0		3/5	(60.0)	7/9	(77.8)	1.78
T_3 (deep muscle invasion)	21/28	(75.0)	0		2/4	(50.0)	19/24	(79.2)	1.69
T_4 (distant metastases)	9/12	(75.0)	0		0		9/12	(75.0)	1.76
Total	50/99	(50.5)	0/14	(0)	10/33	(30.0)	40/52	(76.9)	1.71

[a]99 specimens from 78 patients were analyzed by FCM. These included 67 biopsies from bladder tumors, 6 biopsies from metastatic lesions, and 26 cystectomies.

[b]Number of aneuploid tumors/the combined number of diploid and aneuploid tumors. The percent of aneuploid tumors is in parenthesis.

The relationship between tumor grade and the frequency of aneuploidy was also evident when tumors of all stages were considered together. None of the well differentiated, 30% of the moderately differentiated, and most of the poorly differentiated tumors were aneuploid (TABLE 1).

Thus, these data indicated a correlation between the frequency of aneuploidy and two important prognostic determinants in bladder tumors, i.e. extent of disease (stage) and degree of differentiation (grade): no aneuploid cells were detected in the least malignant tumors, a low frequency of aneuploidy was observed among tumors with intermediate characteristics, and aneuploid tumors predominated among the most malignant tumors.

Prognostic evaluation of T_1 bladder tumors is of significant clinical importance, since only 30% of these patients progress and need aggressive treatment.[23] It will be of interest to determine whether progression will be limited to the patients with aneuploid T_1 tumors (who represented 27% of the T_1 patients involved in this study) (TABLE 1). Indeed, FCM[24] and cytogenetic[25] studies have indicated a correlation between the presence of aneuploid cells and the development of subsequent invasive growth of superficial bladder tumors.

FCM analysis of bladder irrigation fluids is a convenient and clinically useful method for the detection of aneuploid cells in patients with bladder carcinoma.[20-22] The relationships between stage and grade of tumors and the frequency of aneuploid cells in irrigation fluids were in general the same as those observed in studies of biopsies: the frequency of aneuploidy increased with advancing of stage and with decreasing cell differentiation (TABLE 2). All patients with carcinoma *in situ* had aneuploid cells present in their irrigation fluid. This finding is consistent with the high frequency of aneuploidy in these lesions that has been described by others.[22,26]

In order to evaluate the representativeness of tumor cells in bladder irrigation fluids, FCM measurements were performed on the biopsy and irrigation specimens obtained at the same time from 22 patients. Good agreement was found between the measurements. Nine patients had aneuploid cells detected in biopsies, and eight of them had aneuploid cells with the same DNA index in irrigation fluids. No aneuploid cells were detected in the irrigation fluids from 12 patients with only diploid tumor cells in biopsies.

In one patient only diploid cells were detected in the cell suspension prepared from a biopsy, that on histologic examination revealed only inflammatory reaction. However, in the irrigation fluid obtained at the same time, aneuploid cells were found. A later biopsy at another site demonstrated an invasive tumor. FCM analysis of cells from the cystectomy specimen revealed aneuploid cells with the same DNA index as the cells in the irrigation fluid. Thus, FCM of irrigation fluids can supplement pathological analysis of bladder biopsies.

Prostate Tumors

For human prostate tumors, stage of disease and degree of glandular differentiation have established prognostic value.[27-29] To determine whether DNA ploidy has potential as an additional prognostic indicator, its relationship with pathological stage and degree of differentiation was determined. Gleason score[30] is widely used as a measure of glandular differentiation in prostate tumors. A correlation between cellular anaplasia and frequency of aneuploidy has been described by Tribukait *et al*[31]

Tumor tissue was obtained by radical prostatectomy from 34 patients with

TABLE 2. Frequency of Aneuploidy in Bladder Irrigation Fluids According to the Tumor Stage and Histological Grade[a]

Stage/Grade	Frequency of Aneuploidy[b]	
Stage T_a	0/2	(0)
T_1	2/7	(28.6)
T_2	3/4	(75.0)
T_3	9/12	(75.0)
Carcinoma *in situ*	4/4	(100.0)
Grade I	0/5	(0)
II	4/7	(57.1)
III	10/13	(76.9)

[a]Irrigation fluids were obtained from 29 patients with histologically confirmed bladder tumors. Cytology was positive or suspicious for these 29 irrigation fluids.

[b]Number of aneuploid tumors/the combined number of diploid and aneuploid tumors. The percent of aneuploid tumors is in parenthesis.

stages B, C, and D_1 disease and by biopsy from 11 patients with distant metastases (stage D_2). Staging of patients with prostatectomies was done by detailed histopathological analysis of the surgically excised prostate gland and lymph nodes.

The frequency of aneuploidy for all specimens was 44.4% (25 diploid and 20 aneuploid tumors). DNA indices for aneuploid tumors varied from 0.91 to 3.42. The median DNA index was 1.87. One tumor was hypodiploid and one tumor had two aneuploid stemlines. The frequency of aneuploidy increased with advancing stage (TABLE 3). All tumors confined to the prostate gland (stage B) were diploid. Both diploid and aneuploid tumors were encountered in specimens from stages C and D_1. Aneuploid tumors predominated in the patients with stage D_2 disease.

Thus, in our experience, the presence of aneuploid cells precluded disease

TABLE 3. DNA Ploidy in 45 Human Prostate Carcinomas Related to Pathological Stage[a]

Pathological Stage	Frequency of Aneuploidy[b]		DNA Index of Aneuploid Tumors (Median)
B (localized tumors)	0/11	(0)	—
C (invasion of prostate capsule)	6/15	(40.0)	1.72
D_1 (lymph node metastases)	5/8	(62.5)	1.68
D_2 (distant metastases)	8/11	(72.7)	1.75
Total	19/45	(42.2)	1.69

[a]Tumor tissue was obtained by prostatectomy from 34 patients with stages B, C, and D_1, and by biopsy from 11 patients with stage D_2 (8 primary and 3 metastatic tumors). Staging for prostatectomy specimens was done by histopathological analysis of prostate gland and lymph nodes.

[b]Number of aneuploid tumors/the combined number of diploid and aneuploid tumors. Percent of aneuploid tumors is in parenthesis.

confined to the prostate gland and also indicated a high probability of metastases formation. Diploid tumors, however, were encountered at all stages of disease.

The relationship among the degree of tumor spread, DNA ploidy, and glandular differentiation is summarized in TABLE 4. Frequency of metastases for relatively well differentiated tumors (Gleason score 5–6) was 18.2% and for less differentiated tumors (Gleason score 7–10) was 61.1%. Thus, a relatively low degree of glandular differentiation (Gleason score 7–10), as well as the presence of aneuploid cells, indicated a high probability of tumor spread outside the prostate gland.

To determine whether one could better predict tumor stage using both DNA ploidy and glandular differentiation, tumors were divided into four groups on the basis of the presence of aneuploid cells and the Gleason score (TABLE 4). This classification made it possible to identify a group of tumors of which only 7.1% formed metastases (diploid tumors with Gleason score 5–6) and a group of tumors of which 80% had metastasized at the time of presentation (aneuploid tumors with Gleason score 7 or above).

The relationship between Gleason score, DNA ploidy, and pathological stage was examined more formally with linear logistic regression. Gleason score was treated as a continuous predictive variable ranging from 5 to 10, ploidy was treated as a binary predictive variable with values of diploid or aneuploid, and pathological stage was treated as a binary outcome variable with values of no metastases (stages B and C) or metastases (stages D_1 and D_2). Significant relationships were found between Gleason score and stage, and between ploidy and stage ($\alpha = 0.05$). Moreover, using DNA ploidy and Gleason score as predictive variables simultaneously in the predictive equation was significantly better than using either of the variables alone ($\alpha = 0.05$).

From this small series, it appears that stage can be more accurately predicted

TABLE 4. Relationship Between DNA Ploidy, Glandular Differentiation, and Tumor Spread in Human Prostate Cancer

Tumor Characteristics		Number of Tumors (%)		
DNA Ploidy	Glandular Differentiation[a]	Localized (Stage B)	Invasive (Stage C)	Metastasized (Stages D_1 and D_2)
Diploid	5–10	10 (45.5)	8 (36.4)	4 (18.2)
Aneuploid	5–10	0 (0)	7 (38.2)	11 (61.1)
Diploid + Aneuploid	5–6	10 (45.5)	8 (36.4)	4 (18.2)
Diploid + Aneuploid	7–10	0	7 (38.9)	11 (61.1)
Diploid	5–6	10 (71.4)	3 (21.4)	1 (17.1)
Diploid	7–10	0 (0)	5 (62.5)	3 (37.5)
Aneuploid	5–6	0 (0)	5 (62.5)	3 (37.5)
Aneuploid	7–10	0 (0)	5 (20.0)	8 (80.0)

[a]Glandular differentiation in 40 primary prostate tumors was characterized by Gleason score,[25] which increases as tumors become less differentiated. In this study tumors were arbitrarily divided into two groups: relatively well differentiated (Gleason score 5–6) and less differentiated (Gleason score 7–10).

when DNA ploidy and glandular differentiation are considered together. Thus the classification of tumors based both on DNA ploidy and Gleason score could be useful for prognostic evaluation of patients with prostate tumors. This classification could have particular prognostic value for stage C patients, since prognosis varies[28] and 40% of tumors are aneuploid.

Renal Carcinomas

DNA distributions were obtained from 54 primary and metastatic renal tumors from 50 patients (TABLE 5). Most patients had advanced stages of disease, and over 70% of the tumors were aneuploid. Three of six renal carcinomas at stages 1 and 2 were aneuploid. Thus aneuploid cells may be present at any disease stage. There was no apparent difference in the frequency of aneuploidy or the DNA indices for primary and metastatic tumors, for tumors at stages III or IV or between tumors of different histologic subtypes (TABLE 5).

Colo-Rectal Carcinomas

DNA ploidy and S-phase index were determined for 100 colo-rectal carcinomas from 91 patients. The relationship between clinical and pathological characteristics of tumors, such as stage of the disease, tumor site, degree of differentiation, and FCM parameters, was analyzed and the survival of patients with diploid and aneuploid tumors or patients with tumors with high and low proliferative activity was compared.

The overall frequency of aneuploidy for all 100 specimens (59 primary and 41 metastatic tumors) was 68% (TABLE 6). There were no significant differences in frequency of aneuploidy or DNA indices for stages B, C, or D tumors or for primary and metastatic tumors, or for tumors with different degrees of differentiation (TABLE 6, Fisher exact test, Chi-square test, Mann-Whitney test, Kruskal-

TABLE 5. DNA Ploidy According to Tumor Characteristics in Human Renal Carcinomas

Tumor Characteristics	Frequency of Aneuploidy[a]	DNA Index (median)
Primary	31/43 (72.1)	1.54
Metastatic	9/11 (81.8)	1.63
Clear cell	15/24 (62.6)	1.40
Granular	9/12 (75.0)	1.61
Anaplastic	16/18 (88.9)	1.61
Stage I	2/4	—
Stage II	1/2	—
Stage III	13/17 (76.5)	1.50
Stage IV	24/31 (77.4)	1.64

[a]Number of aneuploid tumors/the combined number of diploid and aneuploid tumors. Percentage of aneuploid tumors is in parenthesis.

TABLE 6. DNA Ploicy in Relation to Tumor Characteristics in Human Colo-Rectal Cancer

Tumor Characteristics	Total Number of Tumors	No. of Aneuploid tumors		Frequency of Aneuploidy (%)	DNA Index (median)
		1 Stemline	2 Stemlines		
Dukes stage A	2	1	0	50	1.78
B	23	11	3	61	1.79
C	11	7	2	82	1.65
D	55	32	6	69	1.67
Total	91	51	11	68	1.72
Primary tumors	53	47	6	70	1.74
Metastases	46	40	6	70	1.69
Total	99	87	12	72	1.72
Ascending colon	13	5	3	62	1.72
Transverse	3	1	0	33	1.77
Descending colon	13	10	0	77	1.78
Rectum	14	9	2	79	1.78
Well differentiated	19	11	3	74	1.66
Moderately differentiated	36	22	1	64	1.75
Poorly differentiated	16	8	3	69	1.65

Specimens were obtained from primary tumors of 58 patients, from metastatic tumors of 33 patients, and from both primary and metastatic tumors of 7 patients. For characterization of Dukes stages, DNA ploidy of primary tumors was selected, if both primary tumor and metastases were analyzed.

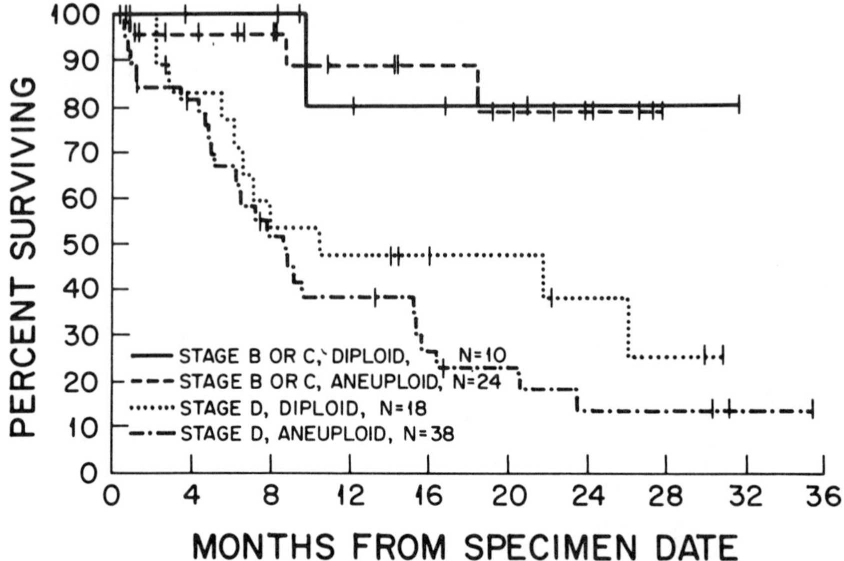

FIGURE 1. Survival of patients with diploid and aneuploid colo-rectal carcinomas. Survival of patients at stages B and C, and stage D shown separately. Vertical lines indicate patients on the study still alive. Estimated survival curves were made with the technique of Kaplan and Meier.[16] Tests of statistical significance were made with the log rank test.[7]

Wallis test, $\alpha = 0.05$). Tumors with two aneuploid stemlines (16% of aneuploid tumors) were observed with approximately the same frequency among different tumor subgroups.

Kaplan-Meier survival analyses with the log rank test were performed to determine the influence of DNA ploidy, the presence of a second aneuploid stemline, and the S-index on the survival time of patients from the date that the specimen was procured. Follow-up ranged from 0.5 to 35.3 months (median 17.3 months). There was no significant difference in survival between patients with diploid and aneuploid tumors for patients with stage D disease ($p = 0.25$, FIGURE 1). Follow-up has not been long enough to determine whether DNA ploidy has prognostic value for patients with earlier stage disease (B and C). However, relapses and deaths have occurred in patients both with diploid and aneuploid tumors.

The presence of a second aneuploid stemline had no significant effect on survival (FIGURE 2). Colo-rectal tumors have a higher frequency of two stemlines than other tumor types.[9,32] Such heterogeneity may complicate analysis of the effects of DNA ploidy on patient survival. Since measurement of ploidy in this study was based on the analysis of one specimen per tumor, the frequency of diploid tumors may be overestimated, and some tumors may contain regions with aneuploid cells in addition to the predominating diploid stemline. Such aneuploid cells, if they exist, could influence prognosis.

There was no significant difference in survival between patients with S index

TABLE 7. DNA Ploidy in Relation to Tumor Characteristics in Human Ovarian Carcinoma

Tumor Characteristics	Total Number of Tumors	No. of Aneuploid tumors		Frequency of Aneuploidy (%)	DNA Index (median)
		1 Stemline	2 Stemlines		
Stage at surgery 2	4	1	0	25.0	1.66
3	30	21	1	73.3	1.53
4	20	13	1	65.0	1.35
Total	54	35	2	64.8	1.42
Primary tumors	26	16	3	73.1	1.34
Metastases	37	23	1	64.9	1.58
Total	63	39	4	68.3	1.42
Residual disease < 2 cm	13	8	1	68.2	1.64
> 2 cm	23	15	1	69.6	1.35
Well differentiated	3	1	0	33.3	1.35
Moderately differentiated	19	12	1	68.4	1.32
Poorly differentiated	30	22	2	80.0	1.57

Specimens were obtained from primary tumors of 24 patients, from metastatic tumors of 30 patients, and from both primary and metastatic tumors of 3 patients.

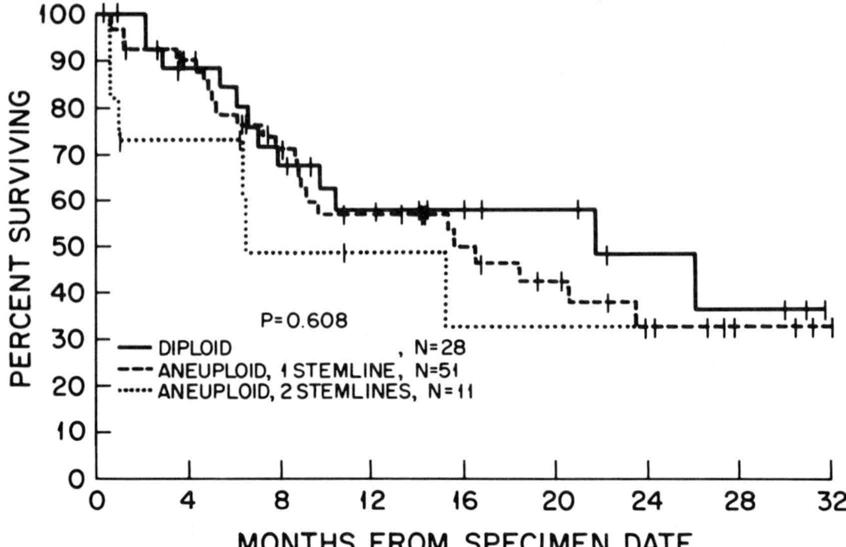

FIGURE 2. Survival of patients with colo-rectal carcinomas in relation to number of aneuploid DNA stemlines. Estimated survival curves were made with the technique of Kaplan and Meier.[16] Tests of statistical significance were made with the log rank test.[7]

above or below the median (19%) (FIGURE 3). Only S indices from aneuploid tumors were used for the analysis, since for these tumors the S index was calculated for aneuploid tumor cells only and reflects their proliferative activity.[14] The absence of an effect of S index on patient survival is in agreement with studies demonstrating that thymidine index did not affect survival time.[33]

There is only one published report indicating that patients with diploid colon tumors survive longer than patients with aneuploid tumors.[12] Most patients in that study had stage B or C disease and the frequency of aneuploidy (33%) was much lower than in this study. Our conclusions, which are based mainly on the survival analysis of patients at stage D, do not contradict data of Wolley *et al.*[12] It is possible that earlier stage aneuploid colo-rectal tumors have a higher potential for invasive growth and metastases formation and as a result have a worse prognosis than diploid tumors. For patients with already disseminated disease these characteristics may be less critical and thus may explain the similar poor survival of stage D patients regardless of tumor ploidy.

Ovarian Carcinomas

Frequencies of DNA-aneuploidy for various subgroups of ovarian carcinomas are presented in TABLE 7. Specimens were obtained from primary and metastatic solid tumors from 54 patients, 50 of whom had advanced disease (stages 3 and 4). Aneuploidy was detected in 35 advanced tumors (65%), which is similar to the frequency of aneuploidy described by others.[13] Four tumors had two

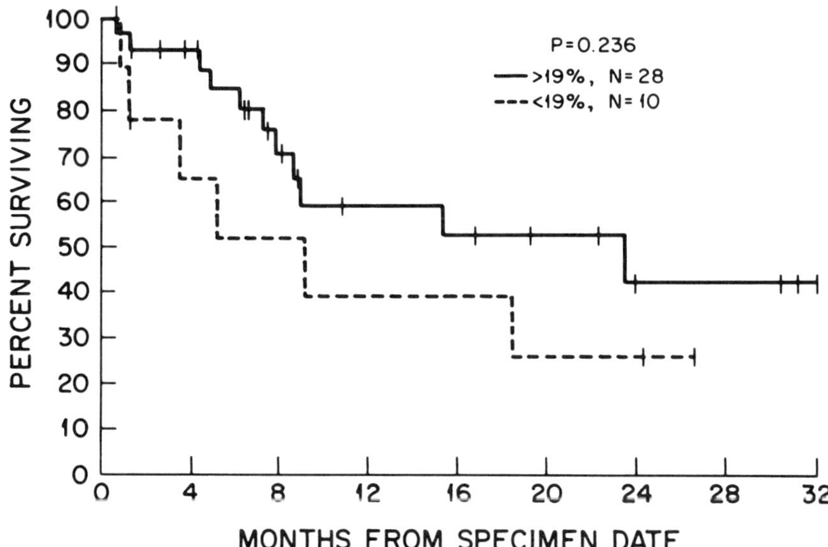

FIGURE 3. Survival of patients with aneuploid colo rectal carcinomas with S indices below or above median S index. Estimated survival curves were made with the technique of Kaplan and Meier.[16] Tests of statistical significance were made with the log rank test.[7]

aneuploid stemlines (9% of aneuploid tumors). There was no significant difference in the frequency of aneuploidy between primary and metastatic tumors, between tumors at stages 3 and 4, or between moderately and poorly differentiated tumors (Fisher exact test, $\alpha = 0.05$; TABLE 7). Median DNA index for ovarian carcinomas (1.42) was lower than for other tumors.[9] This reflected a relatively high frequency of hypodiploid tumors (four tumors or 9% of aneuploid tumors) and tumors with DNA indices in the range 1.05–1.3 (14 tumors or 33% of aneuploid tumors).

The median S index was significantly lower (Mann-Whitney test, $\alpha = 0.05$) for diploid, than for aneuploid tumors (11.4% and 17.6%, respectively). Similar results have been found for all other tumor types studied.[10] For diploid tumors analyzed separately and for aneuploid tumors analyzed separately, stage of disease and tumor site and grade did not influence the value of the S index (Mann-Whitney test, Kruskal-Wallis test, $\alpha = 0.05$).

A preliminary analysis demonstrated no significant difference in survival between patients with diploid and aneuploid tumors (FIGURE 4). However, this is a prospective study, the follow-up period has been relatively short (median 8 months, range 1–29 months) and many patients on the study are still alive.

Among patients with aneuploid tumors, S index had no significant influence on survival time (data not shown). In preliminary exploratory analyses, patients with diploid tumors that had DNA indices below 12% survived significantly longer than patients with tumors that had an S index of 12% or higher. The effect of DNA ploidy alone on the survival of patients with advanced ovarian carcinomas described by Friedlander et al.[13] was not apparent in our study.

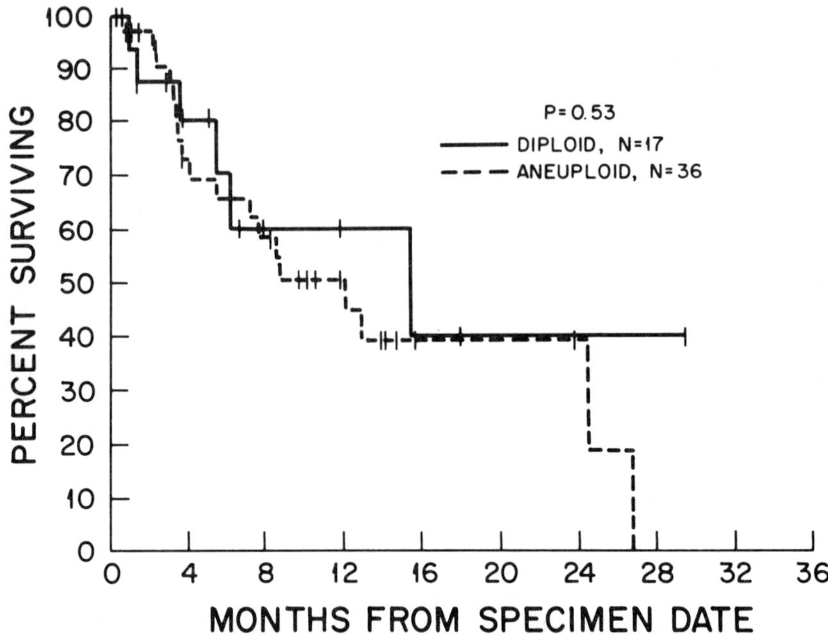

FIGURE 4. Survival of patients with diploid and aneuploid ovarian carcinomas. Estimated survival curves were made with the technique of Kaplan and Meier.[16] Tests of statistical significance were made with the log rank test.[7]

CONCLUSIONS

This study assessed the prognostic role of DNA ploidy and the proportion of cells in S phase for five solid tumor types: bladder, prostate, renal, colo-rectal, and ovarian carcinomas. The relationships between FCM parameters and well established prognostic factors, such as disease stage and tumor grade, were evaluated. The effects of DNA ploidy and S index on the survival of patients were also analyzed.

For bladder and prostate carcinomas, tumors of all stages were studied. For tumors of both diseases the frequency of aneuploidy increased with increased invasiveness, appearance of metastases, and less differentiation.

Approximately one third of bladder tumors with invasion limited to the lamina propria (T_1 stage) and 40% of locally invasive prostate tumors (stage C) were aneuploid. Only some of the patients with T_1 bladder tumors and stage C prostate tumors will develop progressive disease. For these patients measurement of DNA ploidy may have important clinical applications if progression and relapse occur predominantly or exclusively among patients with aneuploid tumors.

A classification of prostate tumors, which was based both on DNA ploidy and the degree of glandular differentiation, made it possible to select groups of tumors with low and high frequency of metastases. Among relatively well differentiated diploid tumors, only 7.1% metastasized and among poorly differentiated aneuploid tumors, 80% formed metastases.

For renal, colo-rectal, and ovarian carcinomas there was no relationship between DNA ploidy and disease stage or tumor grade. There was no difference in survival for advanced colo-rectal carcinoma patients with diploid and aneuploid tumors or for tumors with low or high S indices. For advanced ovarian carcinomas survival was similar for patients with diploid and aneuploid tumors. In an exploratory analysis it was found that patients with diploid ovarian tumors that had a low proportion of cells in S phase survived longer than patients with diploid tumors that had higher S indices.

REFERENCES

1. BARLOGIE, B., N. M. RABER, J. SCHUMANN, T. S. JOHNSON, B. DREWINKO, D. E. SWARTZENDRUBER, W. GOHDE, M. ANDREEFF & E. FREIREICH. 1983. Flow cytometry in clinical career research. Cancer Res. **43**:3982–3997.
2. BARLOGIE, B., B. DREWINKO, J. SCHUMANN, W. GOHDE, G. DOSIK, J. LATREILLE, D. A. JOHNSTON & E. J. FREIREICH. 1980. Cellular DNA content as a marker of neoplasia in man. Am. J. Med. **69**:195–203.
3. BEDROSSIAN, C. W. M., M. RABER & B. BARLOGIE. 1981. Flow cytometry and cytomorphology in primary resectable breast cancer. Anal. Quant. Cytol. **3**:112–116.
4. KUTE, T. E., H. H. MUSS, D. ANDERSON, K. CRUMB, B. MILLER, D. BURNS & L. A. DUBEN. 1981. Relationship of steroid receptor, cell kinetics and clinical status in patients with breast cancer. Cancer Res. **41**:3524–3529.
5. OLZEWSKI, W., Z. DARZYNKIEWICZ, P. P. ROSEN, M. K. SCHWARTZ & M. H. MELAMED. 1981. Flow cytometry of breast carcinoma. I. Relation of DNA ploidy level to histology and estrogen receptor. Cancer **48**:980–984.
6. TRIBUKAIT, B., H. GUSTAFSON & P. L. ESPOSTI. 1982. The significance of ploidy and proliferation in the clinical and biological evaluation of bladder tumors: A study of 100 untreated cases. Brit. J. Urol. **54**:130–135.
7. SONDERGAARD, K., J. LARSEN, U. MOLLER, I. Y. CHRISTENSEN & K. HOU-JENSEN. 1983. DNA-ploidy characteristics of human malignant melanoma analyzed by flow cytometry and compared with histology and clinical course. Virchows Arch. (Cell Pathol.) **42**:43–52.
8. VINDELOV, L. L., M. M. HANSEN, I. J. CHRISTENSEN, M. SPRANG-THOMSEN, F. R. MIRSCH, M. MANSEN & N. I. NISSEN. 1980. Clonal heterogeneity of small-cell anaplastic carcinoma of the lung as demonstrated by flow-cytometric DNA analysis. Cancer Res. **40**:4295–5300.
9. FRANKFURT, O. S., H. K. SLOCUM, Y. M. RUSTUM, S. G. ARBUCK, Z. P. PAVELIC, N. PETRELLI, R. P. HUBEN, E. J. PONTES & W. R. GRECO. 1984. Flow cytometric analysis of DNA ploidy in primary and metastatic human solid tumors. Cytometry **5**:10–20.
10. FRANKFURT, O. S. & R. P. HUBEN. 1984. Clinical applications of DNA flow cytometry for bladder tumors. Urology Suppl. **23**:29–34.
11. GUSTAFSON, H., B. TRIBUKAIT & P. L. ESPOSTI. 1982. DNA pattern, histologic grade and multiplicity related to recurrence rate in superficial bladder tumors. Scand. J. Urol. Nephrol. **16**:135–139.
12. WOLLEY, R. C., K. SCHREIBER, L. G. KOSS, M. KARAS & A. SHERMANN. 1982. DNA distribution in human colon carcinomas and its relationship to clinical behavior. J. Natl. Cancer Inst. **69**:15–22.
13. FRIEDLANDER, M. L., D. W. HEDLEY, I. W. TAYLOR, P. RUSSEL, A. S. COATES & M. M. N. TAHERSAL. 1984. Influence of cellular DNA content on survival in advanced ovarian cancer. Cancer Res. **44**:397–400.
14. FRANKFURT, O. S., W. R. GRECO, H. K. SLOCUM, S. G. ARBUCK, M. GAMARRA, Z. P.

PAVELIC & Y. M. RUSTUM. 1984. Proliferative characteristics of primary and metastatic human solid tumors by DNA flow cytometry. Cytometry. 5:629–635.

15. COX, D. R. 1970. The Analysis of Binary Data. Methuen.

16. KAPLAN, E. L. & P. MEIER. 1958. Nonparametric estimation from incomplete observations. J. Am. Stat. Assoc. 53:457–481.

17. PETO, R. & J. PETO. 1972. Asymptotically efficient rank invariant test procedures. J. R. Stat. Sco. A. 135:185–206.

18. EMRICH, L. J., P. A. REESE & J. D. KALBFLEISCH. 1981. COX MODEL: A proportional hazards model analyses for SPSS users. Proceedings of the Fifth Annual SPSS Users and Coordinators Conference, pp. 215–237.

19. NIE, N. J., C. H. HULL, J. G. JENKINGS, K. STEINBRENNER & D. H. BENT. 1975. Statistical Package for the Social Sciences. McGraw-Hill. New York.

20. COLLSTE, L. G., M. DEVONEC, Z. DARZYNKIEWICZ, F. TRAGANOS, T. K. SHARPLESS, W. F. WHITMORE JR. & M. R. MELAMED. 1980. Bladder cancer diagnosis by flow cytometry. Relation between cell samples from biopsy and bladder irrigation fluid. Cancer 45:2389–2394.

21. DEVONEC, M., Z. DARZYNKIEWICZ, W. F. WHITMORE & M. R. MELAMED. 1981. Flow cytometry for follow-up examinations of conservatively treated low stage bladder tumors. J. Urol. 126:166–170.

22. KLEIN, F. A., H. W. HERR, W. F. WHITMORE, JR. & M. R. MELAMED. 1982. An evaluation of automated flow cytometry in the detection of carcinoma-in situ of the urinary bladder. Cancer 50:1003.

23. HENEY, N. M., S. AHMED, M. J. FLANAGAN, J. FRABLE, M. F. CORDER, M. D. HAFERMAN & I. R. HAWKINS. 1983. For National Bladder Cancer Collaborative Group A. Superficial bladder cancer: progression and recurrence. J. Urol. 130:1083–1088.

24. GUSTAFSON, H., B. TRIBUKAIT & P. L. ESPOSTI. 1982. DNA profile and tumor progression in patients with superficial bladder tumors. Urol. Res. 10:13–18.

25. SUMMERS, J. L., J. S. COON, R. M. WARD, W. H. FALOR, A. W. MILLER & R. S. WEINSTEIN. 1983. Prognosis in carcinoma of the urinary bladder based upon tissue blood group ABH and Thomsen-Friedenreich antigen status and karyotype of the initial tumor. Cancer Res. 43:934–939.

26. GUSTAFSON, H., B. TRIBUKAIT & P. L. ESPOSTI. 1982. The prognostic value of DNA analysis in primary carcinoma in situ of the urinary bladder. Scand. J. Urol. Nephrol. 16:141–146.

27. GLEASON, D. F., G. T. MELLINGER & V. A. RESEARCH GROUP. 1974. Prediction of prognosis for prostatic adenocarcinoma by combined histological grading and clinical staging. J. Urol. 111:58–64.

28. GRAYBACK, J. T. & D. G. ASSIMOS. 1983. Prognostic significance of tumor grade and stage in the patient with carcinoma of the prostate. The Prostate 4:13–31.

29. GAETA, Y. F., Y. E. ASIZWATHAM, G. MILLER & G. P. MURPHY. 1980. Histologic grading of primary prostatic cancer. A new approach to an old problem. J. Urol. 123:684–693.

30. GLEASON, D. F. 1966. Classification of prostatic carcinomas. Cancer Chemotherapy Rep. 50:125–128.

31. TRIBUKAIT, B., L. RONSTROM & P. ESPOSTI. 1983. Quantitative and qualitative aspects of flow DNA measurements related to the cytologic grade in prostatic carcinoma. Anal. Quant. Cytol. 5:107–111.

32. PETERSON, S. E., M. LORENTZEN & P. BICHEL. 1981. A mosaic subpopulation structure of human colo-rectal carcinomas demonstrated by flow cytometry. Acta. Pathol. Microbiol. Scand. (A) 274:412–416.

33. MEYER, J. S. & P. G. PRIOLEAU. 1981. S-phase fractions of colo-rectal carcinomas related to pathological and clinical features. Cancer 48:1221–1228.

Potential Prognostic Significance of Cytometrically Determined DNA Abnormality in GI Tract Human Tumors[a]

L. TEODORI,[b] D. TIRINDELLI-DANESI,[c] E. CORDELLI,[b]
R. UCCELLI,[b] R. DE VITA,[b] M. SPANO,[b] F. MAURO,[b]
A. SCHILLACI,[c] A. MORALDI,[c] L. CAPURSO,[d]
AND S. STIPA[c]

[b]Laboratorio di Dosimetria e Biofisica
ENEA Casaccia
00060 Rome, Italy

[c]I Istituto di Chirurgia
Università di Roma
00161 Rome, Italy

[d]Servizio di Gastroenterologia
Ospedale San Filippo Neri
00135 Rome, Italy

INTRODUCTION

It is well known that human solid tumors are often characterized by an abnormal cellular DNA content. Recent studies after the advent of flow cytometry have confirmed that relative cellular DNA content can be a conclusive marker of malignancy. A recent review by Barlogie *et al.*[1] has reported that cytometric aneuploidy characterizes 75% human solid tumors. In our laboratory, where flow cytometry is routinely performed for tumors of certain sites and some "precancerous" conditions, we have obtained the comparable value of 79% for over 300 cases. Flow cytometric analysis of DNA content can also be used to study some aspects of tumor cell heterogeneity in terms of ploidy level(s), occurrence of multiclonality (that is, of more than one tumor stem-cell line), and proliferation pattern.[2-4] Furthermore, cytometric aneuploidy can be used in the early detection of neoplastic degeneration, especially in the instance of the GI tract.[5-7]

In the present investigation, DNA flow cytometric measurements were performed on samples from 64 patients affected by GI tract tumors in the attempt to evaluate the possible prognostic significance of cytometric parameters. In fact, the possibility of exploiting flow cytometric analysis to predict the natural history of the disease and/or eventually the response to treatment has already been suggested by some authors.[8-12]

[a]Partially supported by Progetto Finalizzato CNR "Controllo della crescita neoplastica," contract n° 80.01586.06. Part of the present data has been presented at the Engineering Foundation Conference on "Clinical Cytometry" held at The Cloister, Sea Island, Georgia, 1983.

MATERIALS AND METHODS

Patients

Flow cytometric measurements of cellular DNA content were performed on specimens from 64 patients affected by histologically confirmed tumors of the GI tract (24 colon-rectum, 34 stomach, and 7 esophagus). In TABLE 1, all patients are classified according to site, sex, age, histology, staging, and eventually grading.

Samples and Preparations

Cellular specimens were obtained by surgery and endoscopy or proctoscopy. In all cases, multiple site sampling (4 to 10 biopsy specimens per tumor) was carried out. Specimens were placed in RPMI 1640 culture medium (Grand Island, NY) supplemented with 5% fetal calf serum at 4°C. Parts of the same specimens were always used for double-blind histopathologic examinations. Monocellular suspension of biopsy material was achieved by mechanical and enzymatic treatment. The tissue was minced with scissors, washed in saline solution, and treated with 0.5% pepsin for 5–10 minutes (Serva, Heidelberg, FRG). Tris-buffer was then added, samples were centrifuged at 200 g for 5 minutes, and the pellet was resuspended in the same buffer.

Staining and Flow Cytometry

Twenty microliters of Nonidet P40 (Fluka, Buchs, Switzerland), 1 ml of 10 µg/ml ethidium bromide (Serva, Heidelberg, FRG), and 1 ml of 25 µg/ml mithramycin (courtesy of Pfizer, Italy) were added to 0.2 ml of cell suspension. The samples were then measured with an ICP arc-lamp pulse cytophotometer (Ortho, Westwood, MA) equipped with a glass flow chamber (kindly supplied by Dr. W. Göhde). The number of cells measured was generally about 10,000 per histogram. DNA content distributions were accumulated in a multichannel analyzer and data analysis performed on a Nuclear Data ND620 computer (Shaumberg, IL). In all cases, the diploid peak in the distribution was located by adding to the sample cells from a normal area from the same patients. The DNA index (DI) was calculated as the ratio of the $G_{1/0}$ aneuploid peak modal channel to the $G_{1/0}$ diploid peak modal channel.

RESULTS

Esophagus

The results for the esophagus samples are reported in TABLE 2 in terms of DI for all observed cell clones. In all cases, one or two aneuploid clones were present. In one of these cases (M.C.), samples from the mucosa at the resection margin were positive after flow cytometry, but negative after histopathology. A representative histogram of an esophagus epidermoid carcinoma is shown in FIGURE 1.

TABLE 1. Composition of the Groups of Patients under Investigation

Esophagus	
Total cases	7
Males	6
Females	1
Age (average and ranges)	63.6 (49–73)
Histology: epidermoid	6
adenocarcinoma	1
Stage: I	1
II	1
III	4
IV	1
Stomach (Surgery biopsies)	
Total cases	15
Males	12
Females	3
Age (average and ranges)	65.0 (57–83)
Histology: intestinal	8
spread	4
ND	3
Stage: I	—
II	1
III	4
IV	7
ND	3
Stomach (Endoscopy biopsies)	
Total cases	18
Males	12
Females	6
Age (average and ranges)	63.5 (42–83)
Histology: adenocarcinoma	18
Stage: see text	
Colon rectum	
Total cases	24
Males	16
Females	8
Age (average and ranges)	63.0 (25–78)
Histology: adenocarcinoma	24
Grade: I	—
II	15
III	2
ND	7
Duke's stage: A	3
B	7
C	7
D	7
Grand total number of cases	64

TABLE 2. Esophagus

Patient	Region	Stage	D	A1	A2	Histology
P.A.	medium 1/3	T1N0M0 I	1.00	1.63		Epidermoid
B.N.	lower 1/3	T3N1M0 III	1.00	1.86		Epidermoid
S.S.	lower 1/3	T2N0M1 IV	1.00	1.15	1.62	Epidermoid, poorly different.
P.D.	lower 1/3	T1N1M0 III	1.00	1.10	1.94	Epidermoid, giant cells
S.A.	medium 1/3	T2N1M0 III	1.00	1.61	([a])	Epidermoid, poorly different.
M.C.	lower 1/3	T3N0M0 III	1.00	1.61	1.67	Epidermoid
F.Q.	NR	T4N1M0 II	1.00	1.40	1.55	Adenocarcinoma

[a]Likely presence of a hypodiploid clone.

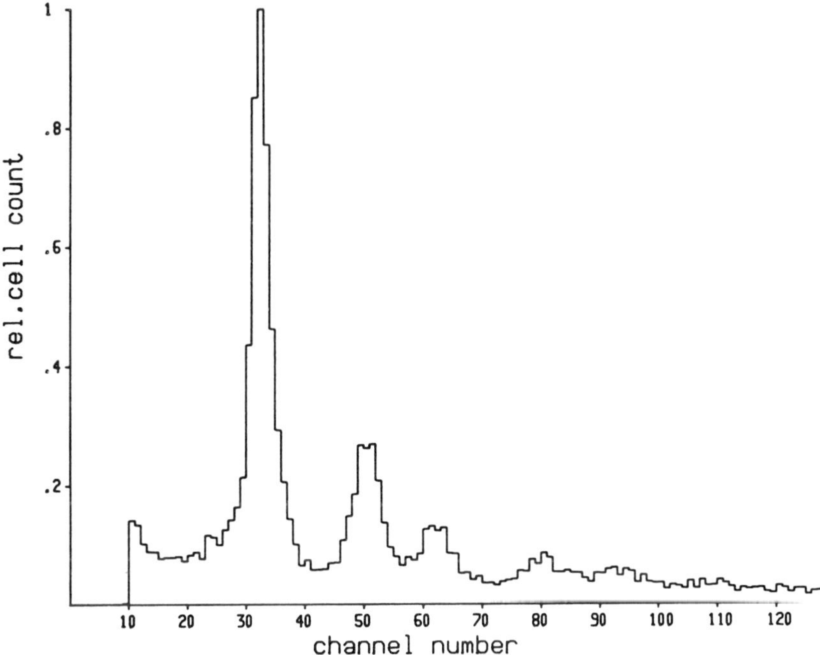

FIGURE 1. Flow cytometrically determined DNA content distribution of an esophagus epidermoid carcinoma.

Stomach

The results for the stomach samples are reported in TABLES 3 and 4 for the cases collected by endoscopy or surgery (in two different clinical outfits), respectively. It must be pointed out that these two groups of data differ also as far as tumor staging is concerned. In fact, while surgical cases could be carefully defined in terms of stage according to the Tumor Nodes Metastases (TNM) system, this was not possible for the endoscopic cases. However, all but two of the latter cases were classified as early stages. On the contrary, the large majority of the surgical cases were advanced tumors.

The two groups exhibited approximately the same frequency of aneuploidy (89% endoscopic and 87% surgical cases) but a different incidence of multiclonality (6% and 67%, respectively). A representative histogram of a multiclonal condition is shown in FIGURE 2. The data can also be analyzed in terms of DI. As shown in FIGURE 3, the endoscopic cases exhibit a modal DI of 1.3, an average DI of 1.57 ± 0.18, and a range from 1.0 to 1.8. Surgery cases exhibit a modal DI of 1.7, an average DI of 1.81 ± 0.10 SE (or 1.96 ± 0.15 SE if the clones with highest DIs are considered only), and a range from 1.4 to over 2. These data can be further compared to the aneuploid ("precancerous" carcinoma *in situ*) chronic atrophic gastritis we reported in a previous paper,[6] where the modal DI is 1.2, the average DI is 1.23 ± 0.04 SE, and the range is 1 to 1.6. This kind of analysis seems to

TABLE 3. Stomach (Endoscopy Biopsies)

Patient	Region	Stage	D	A1	A2	Histology
G01	Prepylorus		1.00	1.18		Adenocarcinoma
G07	Antrum		1.00	1.40		Adenocarcinoma
G32	Cardia		1.00	1.23		Adenocarcinoma
G48	Fundus		1.00	1.68		Adenocarcinoma
G55	Corpus		1.00	1.10		Adenocarcinoma
G54	Corpus		1.00	1.13		Adenocarcinoma
G58	Cardia		1.00	1.35		Adenocarcinoma
G62	Corpus	[a]	1.00	1.50	1.81	Adenocarcinoma
G64	Fundus		1.00	1.27		Adenocarcinoma
G65	Corpus		1.00	1.85		Adenocarcinoma
G67	Corpus		1.00	1.75		Adenocarcinoma
G67B	Corpus		1.00	1.54		Adenocarcinoma
G70	Fundus	[a]	1.00	3.13		Adenocarcinoma
G73	Angulus		1.00			Adenocarcinoma
G84	Fundus		1.00	1.37		Adenocarcinoma
GAS01	Corpus		1.00	1.76		Adenocarcinoma
GAS03	Fundus		1.00			Adenocarcinoma
GAS19	Fundus		1.00	1.60		Adenocarcinoma

[a]Advanced stages.

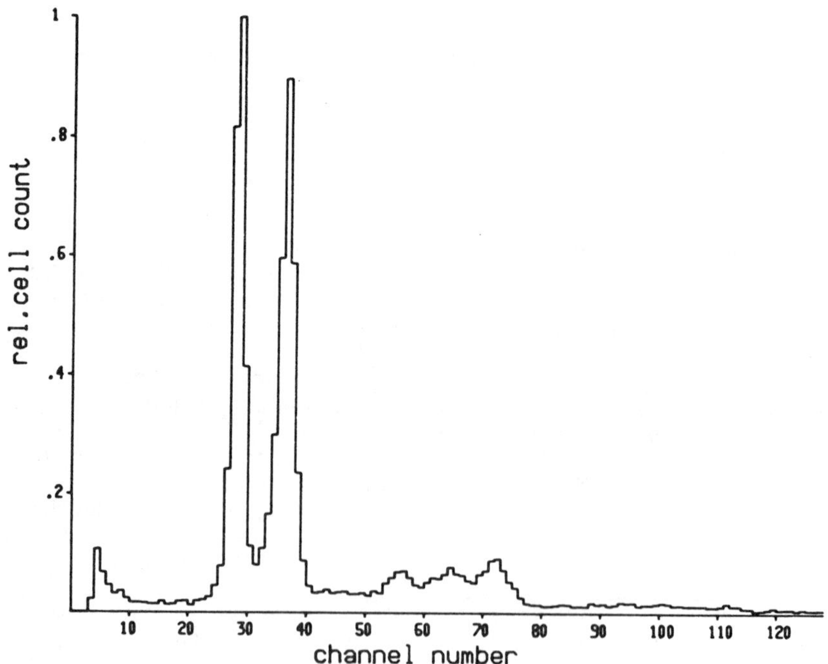

FIGURE 2. Flow cytometrically determined DNA content distribution of a gastric adenocarcinoma.

TABLE 4. Stomach (Surgery Biopsies)

Patient	Region	Stage	H1	D	A1	A2	A3	A4	Histology
T.S.	Cardia	T3N1M0 III		1.00	1.10	1.76			Adenocarcinoma, Intest., Poorly D.
A.A.	Ansa Parva	T3N0M0 III	0.90	1.00	1.84				Adenocarcinoma, Spread
D.M.	Ansa Parva	T4N1M0 IV	0.85	1.00	1.60	2.60			Adenocarcinoma, Intest.
C.F.	Cardia	T4N1M0 IV		1.00	1.42	1.55			Adenocarcinoma, Spread
G.A.	Antrum	T4N2M0 IV	0.76	1.00	1.24	2.24	2.44	2.64	Adenocarcinoma, Intest., Highly D.
M.F.	Cardia	T4N2M0 IV		1.00	1.48				Adenocarcinoma, Intest.
T.G.	Cardia	T4N2M0 IV		1.00	1.52	1.58			Adenocarcinoma, Colloidal Intest.
D.M.	Cardia	T4N3M0 III		1.00	2.06	2.96			Adenocarcinoma, Intest.
B.V.	Cardia–Fundus	T4N2M0 IV		1.00	1.57	1.79	2.59		Adenocarcinoma, Spread
P.L.	Antrum	T4N1M0 IV		1.00	1.62	1.66			Adenocarcinoma, Intest.
R.M.	Cardia	T4N1M0 IV		1.00	1.73				Adenocarcinoma, Spread
P.R.	Antrum	T2N0M0 II		1.00	1.37	1.63	1.80		Adenocarcinoma, Papill. Intest.
B.F.	Antrum	ND		1.00					NR
B.O.	Fundus	ND		1.00	1.24				NR
F.G.	Antrum	ND		1.00					NR

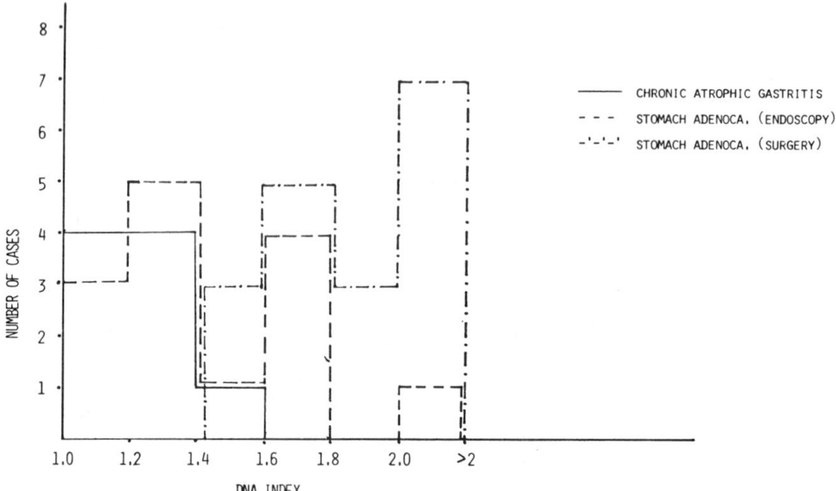

FIGURE 3. Frequency of DIs of the aneuploid samples of chronic atrophic gastritis (continuous line), endoscopic cases of gastric adenocarcinoma (dashed line), and surgery cases of gastric adenocarcinoma (dashed/dotted line).

indicate that ploidy level is approximately related to the stage of evolution of the disease.

Colon-rectum

The results for colon-rectum are reported in TABLE 5. In this instance, 63% cases were classed as aneuploid and 38% as multiclonal. It is remarkable that 8 out of 9 diploid tumors (microscopic examination of these samples has indicated that at least a fraction of the diploid cells can be recognized as malignant cells) belong to A or B Duke's stages, while 13 of 15 aneuploid tumors belong to C or D stages (TABLES 6). A representative histogram of a colon-rectum carcinoma is shown in FIGURE 4.

DISCUSSION

Flow cytometric analysis of relative DNA content of cellular samples from tumors of the GI tract has confirmed that aneuploidy and multiclonality characterize a certain fraction of esophagus, stomach, and colon-rectum malignancies.

The data collected for esophagus are too scanty to draw any indication, but have been presented because of the lack in the literature of cytometric reports on tumors of this site. Moreover, we can confirm that the use of multiple site sampling flow cytometry for esophagus tumors can be of help as a control of

TABLE 5. Colon Rectum

Patient	Region	Duke's Stage	Grade	H1	D	A1	A2	Histology
A.M.	Ascending colon	B	II		1.00			Adenocarcinoma
C.A.	Sigma-Rectum	C1	II		1.00			Adenocarcinoma
L.I.	Sigma	A	II		1.00			Adenocarcinoma (on tubulovillous adenoma)
F.G.	NR	B	NR		1.00			Adenocarcinoma
M.F.	NR	B	NR		1.00			Adenocarcinoma
F.V.	Caecum	B	II		1.00			Adenocarcinoma
S.G.	NR	B	NR		1.00			Adenocarcinoma
F.L.	NR	B	NR		1.00			Adenocarcinoma
V.M./BIS	Rectum	A	II		1.00			Adenocarcinoma
T.L.	Transversal colon	D	III		1.00	1.30	1.57	Adenocarcinoma
M.C.	Sigma	D	II		1.00	1.72		Adenocarcinoma
F.M.	Sigma	B	II		1.00	1.10		Adenocarcinoma
G.R.	Rectum	C1	II		1.00	1.71	2.48	Adenocarcinoma
V.D.	Descending colon	D	II		1.00	1.24	2.28	Adenocarcinoma
F.M.	Transversal colon	C2	NR		1.00	1.22		Adenocarcinoma
D.M.	Rectum	C1	II		1.00	1.45		Adenocarcinoma
M.V.	Ascending colon	C1	III		1.00	1.78	2.71	Adenocarcinoma
M.N.	Ascending colon	D	NR		1.00	1.70		Adenocarcinoma, signet ring cells
A.C.	Rectum	D	NR		1.00			Adenocarcinoma (on tubulovillous adenoma)
T.G.	Sigma	D	II		1.00	1.24	2.24	Adenocarcinoma
S.M.	Rectum	C1	II	0.86	1.00	1.76		Adenocarcinoma
V.M.	NR	C1	II		1.00	1.43	1.82	Adenocarcinoma
E.M.	Sigma	D	II		1.00	1.57	1.67	Adenocarcinoma
L.A.	Descending colon	A	II		1.00	1.16		Adenocarcinoma

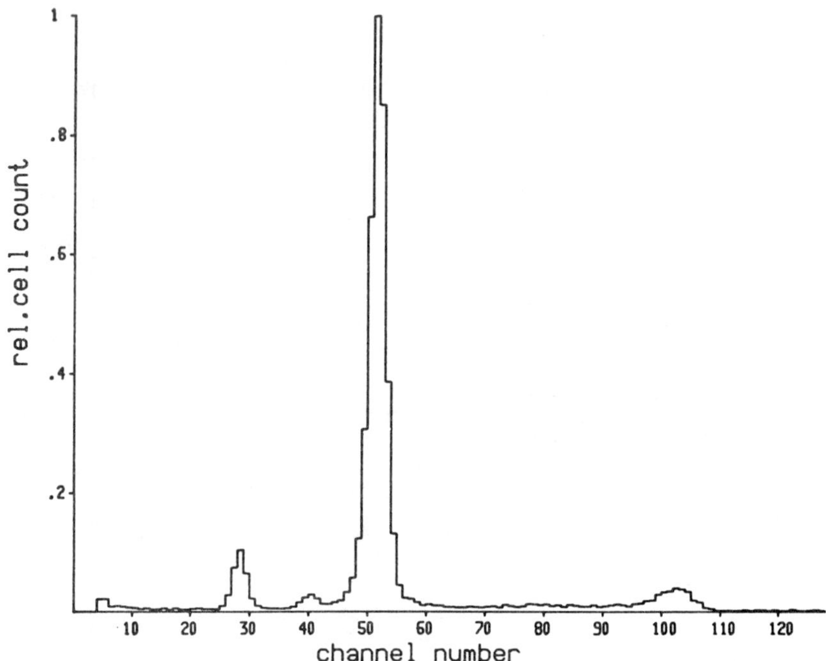

FIGURE 4. Flow cytometrically determined DNA content distribution of a colon adenocarcinoma.

disease dissemination (especially in relation to resection) or disease origin from other regions of the GI tract (e.g., case F.Q.).

In the instance of the stomach, both ploidy and multiclonality levels seem to be related to tumor stage. From this point of view, the results are more complex than we previously thought,[6] when our analysis was limited essentially to endoscopic samples only. Furthermore, the relation between DNA abnormalities and stage suggests that some kind of cellular evolution is taking place within a tumor, implying a tendency to increase in the ploidy level as a function of the tumor cell population life history. Such a phenomenon implies the existence of some kind of clonal evolution process based on readjustments of chromosome number and/or DNA content and eventual gradual appearance of various aneuploid stem-cell lines. Unfortunately, instances of sampling from the same untreated tumor at different times in its evolution are for the moment too rare and too difficult to obtain to allow the collection of the results necessary to clarify such a process. However, the case we have already reported[6] of an aneuploid chronic atrophic gastritis becoming a clear-cut early gastric cancer and exhibiting an increase in ploidy level seems to go in this direction.

In the instances of colon-rectum carcinoma, a relation between ploidy level and staging (according to Duke's classification) seems to be present. Thus, our data partially confirm those reported by Wolly *et al.*[8] These authors find a relation of ploidy level with disease outcome more than with Duke's staging. In our study, the results of patient follow-up is not yet available (at present, none of our diploid patients died and 4 survived at least 20 months; 4 aneuploid patients died, their average survival being 8 months). In our previous experience, the survival at 3 years of A and B stage patients is 90% to 67% while for C stage patients survival at 3 years is 43–15% according to substaging. Therefore, both staging and cytometric ploidy appear of prognostic significance.

More generally, at least in the instance of tumors of the GI tract, flow cytometry seems to be useful more from a prognostic (staging) than from a diagnostic (histological grading) point of view, the only exceptions being the early recognition of malignancies in "precancerous" conditions or of disease dissemination.

REFERENCES

1. BARLOGIE, B., M. N. RABER, J. SCHUMANN, T. S. JOHNSON, B. DREWINKO, D. E. SWARTZENDRUBER, W. GÖHDE, M. ANDREEFF & E. J. FREIREICH. 1983. Flow cytometry in clinical cancer research. Can. Res. **43**:3982–3997.
2. NERVI, C., G. BADARACCO, A. MAISTO, F. MAURO, D. TIRINDELLI-DANESI & G. STARACE. 1982. Cytometric evidence of cytogenetic and proliferative heterogeneity of human solid tumors. Cytometry **2**:303–307.
3. TEODORI, L., D. TIRINDELLI-DANESI, F. MAURO, R. DE VITA, R. UCCELLI, C. BOTTI, C. MODINI, C. NERVI & S. STIPA. 1983. Non-small-cell lung carcinoma: tumor characterization on the basis of flow cytometrically determined cellular heterogeneity. Cytometry **4**.174–183.
4. FRANKFURT, O. S., H. K. SLOCUM, Y. M. RUSTUM, S. G. ARBUCK, Z. P. PAVELIC, N. PETRELLI, R. P. HUBEN, E. J. PONTES & W. R. GRECO. 1984. Flow cytometric analysis of DNA aneuploidy in primary and metastatic human solid tumors. Cytometry **5**:71–80.
5. WEISS, H., G. P. WILDNER, H. J. GUTZ, K. EBELING, G. STEINHOFF & S. TANNEBERGER. 1981. DNA distribution patterns of preneoplastic cell and their interpretation. Oncology **38**:210–218.
6. TEODORI, L., L. CAPURSO, E. CORDELLI, R. DE VITA, M. KOCK, M. TARQUINI, F. PALLONE & F. MAURO. 1984. Cytometrically determined relative DNA content as an indicator of neoplasia in gastric lesions. Cytometry **5**:63–70.
7. CACCESE, W., D. BUDMAN, N. SCHEIDT, G. WEISSMAN & H. MCKINLEY. 1983. Flow-cytometry in the endoscopic diagnosis of malignant and premalignant conditions of the GI tract. Gastroint. Endoscopy **29**:184.
8. WOLLY, R. C., K. SCHREIBER, L. G. KOSS, M. KARAS & A. SHERMAN. 1982. DNA distribution in human colon carcinomas and its relationship to clinical behavior. J. Natl. Cancer Inst. **69**:15–22.
9. BARLOGIE, B., D. A. JOHNSTON, L. SMALLWOOD, M. N. RABER, A. M. MADDOX, J. LATREILLE, D. E. SWARTZENDRUBER & B. DREWINKO. 1982. Prognostic implications of ploidy and proliferative activity in human solid tumors. Cancer Genet. Cytogenet. **6**:17–28.
10. TRIBUKAIT, B. 1980. DNA content analysis in prostate cancer. Workshop, V International Symposium on Flow Cytometry. Rome, Italy.
11. JOHNSON, T. S., M. R. RAJU, R. K. GILTINAN & E. L. GILLETTE. 1981. Ploidy and DNA distribution analysis of spontaneous dog tumors by flow cytometry. Cancer Res. **41**:3005–3009.
12. ARCANGELI, G., F. MAURO, C. NERVI & G. STARACE. 1980. A critical appraisal of the usefulness of some biological parameters in predicting tumor radiation response of human head and neck cancer. Br. J. Cancer **41**(Suppl. IV):39–44.

Flow Cytometric Identification of Human Bladder Cells Using a Cytokeratin Monoclonal Antibody[a]

JEFFRY L. HUFFMAN,[b] PILAR GARIN-CHESA,
HELEN GAY, WILLET F. WHITMORE, JR.,
AND MYRON R. MELAMED

Memorial Sloan-Kettering Cancer Center
New York, New York 10021

The diagnosis of bladder cancer by flow cytometry (FCM) is based on identification of exfoliated epithelial cells that have aneuploid values of DNA. In patients with co-existing inflammation there is a sometimes large number of leukocytes and other cells present and a corresponding decrease in the proportion of epithelial cells. As a result, there may be difficulty in identifying some cases of carcinoma that shed only small numbers of tumor cells. To overcome this we have used a technique, described several years ago by Collste *et al.*,[1] to lyse polymorphonuclear leukocytes with Triton X-100. Positive identification of epithelial cells would be preferable, however, and we now report a mouse monoclonal antibody to epithelial cytokeratin that appears suitable for that purpose. The antibody, AE-1, is one of a series prepared against SDS-denatured human callus keratins by Sun *et al.*[2,3] to whom we are indebted for a gift of the antibody. In immunofluorescent and immunoperoxidase stained frozen and paraffin sections of urinary bladder, ureter, and more than 60 different bladder tumors studied by us, AE-1 has shown strong affinity for all the cells of the normal, reactive, and neoplastic urothelium with no binding to connective tissue, smooth muscle, endothelium, blood cells, or other non-epithelial elements.

MATERIALS AND METHODS

A series of 28 patients from the Urologic Service of Memorial Hospital, seen during a three-month period early in 1984, was studied simultaneously by flow cytometry with conventional acridine orange staining and flow cytometry using the AE-1 cytokeratin antibody with propidium iodide staining of DNA. All 28 patients underwent cystoscopic examination because of a history of bladder tumors, cytologic study of voided and/or catheterized urine, and biopsies when indicated by cystoscopic findings. In addition, benign and reactive urothelium and a spectrum of bladder papillomas and carcinomas were examined for AE-1 binding in histologic sections of frozen and paraffin-embedded tissue.

[a]Supported by National Cancer Institute Grant (Bladder Grant) CA-13434. Work by P.G.-C. performed during the tenure of a Fulbright Fellowship.
[b]Present address: University of California Medical Center, 225 Dickinson, H-897, San Diego, CA 92103.

BLADDER IRRIGATION SPECIMENS

Specimen Collection and Preparation

Specimens were collected for flow cytometry by vigorous irrigation at the time of cystoscopy, using an Ellik evacuator filled with 200–300 ml saline. The lavage fluid was refrigerated after collection and processed within 4–6 hours. After agitation to resuspend the cellular sediment, the specimen was sieved through a 53 μm nylon mesh to remove cell clumps, then centrifuged at 1500 rpm ($200 \times g$) for 10 minutes. The cell pellet was then resuspended in 2 ml $Ca^{2+} + Mg^{2+}$ free Hank's balanced salt solution (HBSS). The cell suspension was adjusted to approximately 10^6 cells/ml and aliquoted as indicated below for processing by flow cytometry.

Urine collected by catheterization prior to cystoscopy was fixed in an equal volume of 50% ethanol, and four Papanicolaou stained smears of the cellular sediment were prepared for routine cytologic examination.

Cystoscopic examination and biopsies obtained after irrigating for the flow cytometry specimen established the diagnosis of papillary carcinoma in five patients, flat carcinoma *in situ* with or without papilloma or papillary carcinoma *in situ* in six patients, chronic cystitis in seven patients, invasive carcinoma of bladder in three patients, papilloma in two patients, and carcinoma of prostate in one patient. Two patients had no suspicious findings, and no biopsies were taken.

Conventional Acridine Orange Staining

A 0.2 ml aliquot of the cell suspension containing approximately 4×10^5 cells was treated with 0.4 ml acid detergent (0.08 N HCl, 0.15 N NaCl, and 0.1% Triton X-100 [Sigma Chemical Co., St. Louis, MO]) at 4°C. The detergent treatment makes cells permeable to the dye, while at low pH the nucleic acids remain insoluble. After 30 sec in the detergent solution, chromatographically pure acridine orange (AO) (Polysciences, Warrington, PA) is added (1.2 ml AO 16 μg/ml; 20 μM) in 0.2 M phosphate-citric acid buffer at pH 6.0 with 1 mM EDTA, 0.146 M NaCl. Under these conditions AO intercalates into double-stranded DNA and fluoresces green (F_{530}) in blue light, while RNA is converted to its single-stranded form and interacts with the dye to fluoresce red.[4,5] The specificity of DNA and RNA staining is monitored by preincubation of aliquots of permeable cells with DNase I or RNase (Sigma).

Cytokeratin/DNA Staining

The remaining cell sample is divided into two aliquots, each is centrifuged, resuspended in 2 ml 70% ethanol at 4°C, vortexed to prevent cell clumping, and allowed to fix for 60 min. The cells are then washed twice in HBSS, and the cell pellets resuspended and incubated on ice for 60 min with either 0.1 ml AE-1 monoclonal antibody (Mab) to cytokeratin (Dr. Sun, N.Y.U. Medical Center) or NS-1 Mab (supernatant of the same Ig subtype from mouse NS/1 myeloma cells,[6] respectively. The AE-1 Mab recognizes low molecular weight acidic keratins.[3] The NS-1 Mab serves as a negative control. Following this incubation, both cell suspensions are again washed twice with HBSS, and the cell pellets resuspended and incubated on ice for 30 min with 0.1 ml goat anti-mouse immunoglobulin conjugated to FITC (Cappel Laboratories, Cochranville, PA), washed twice with HBSS, and treated with 0.2 ml RNase A (5000 units/ml) (Worthington Diagnostics, Freehold, NJ) for 30 min at 37°C. The samples are then stained for DNA by addition of propidium iodide (0.2 ml; 50 µg/ml, Calbiochem-Behring Corp., La Jolla, CA) for 30 min at room temperature, and measured by flow cytometry.

Flow Cytometry

All measurements were carried out on an Ortho FC 200 or System 50H Flow Cytometer (Ortho Diagnostics, Westwood, MA). A total of 5,000 cells per sample were measured. In the case of acridine orange–stained specimens, the DNA (green fluorescence), RNA (red fluorescence), and nuclear diameter (green fluorescence pulse width) of each cell was measured and recorded, and the distribution of measurements displayed and analyzed for aneuploid cell population(s), as previously described.[7] A specimen was considered negative for malignant cells if fewer than 10% of the cells were hyperdiploid and no aneuploid stemline could be detected. It was considered suspicious if no aneuploid stemline could be detected while 10–15% of the cells were hyperdiploid. If there was a definite aneuploid stemline or greater than 15% of the cells were hyperdiploid, the sample was considered positive.

For the samples stained with monoclonal antibody and propidium iodide, the following measurements were obtained and recorded for each cell: DNA [red (propidium) fluorescence], nuclear diameter (red fluorescence pulse width) and Mab binding [green (FITC) fluorescence]. The nuclear diameter measurements served to exclude cell doublets and larger aggregates.[8] The epithelial cells were identified as those single cells in the AE-1 stained specimen that had green (FITC) fluorescence greater than the fluorescence of cells in the control sample stained with NS-1. Cells that bound AE-1 by this definition were then replotted according to DNA content to determine the proportion of epithelial cells in the cell sample, the distribution of DNA content, and the percent hyperdiploid epithelial cells.

IMMUNOHISTOCHEMISTRY

Formalin-fixed, paraffin-embedded tissue was selected from patients who were operated on within the last year for: papilloma of the urinary bladder (seven cases), *in situ* carcinoma in flat epithelium or focally in papilloma (12 cases), non-invasive papillary carcinoma (five cases) and invasive carcinoma (10 cases) including two with lymph node metastases. Fresh tissue for frozen sections was obtained by transurethral biopsies of normal bladder mucosa (25 cases), bladder papilloma or low grade papillary carcinoma (15 cases), carcinoma *in situ* in flat epithelium or non-invasive papillary carcinoma (15 cases), from normal ureter of patients undergoing cystectomy for carcinoma (three cases), and from bladder mucosa showing squamous metaplasia (five cases) or cystitis glandularis (one case).

Paraffin blocks were cut into 5 µm sections, deparaffinized, and treated for 30 min at room temperature with 0.1% pepsin in 0.01 N HCl (pepsin 1:60,000, Sigma

Chemical Co., St. Louis, MO). Fresh tissue was collected in Hank's balanced salt solution and, within 4–6 hours, embedded in OCT solution in cryomolds and snap-frozen in liquid nitrogen. The following day, cryostat sections were obtained at 6–8 μm.

Paraffin and frozen sections were washed in phosphate-buffered saline (PBS) and stained by the following procedures (1) The unlabeled antibody enzyme procedure, the PAP method[9] and (2) The ABC-immunoperoxidase system.[10]

After blocking the endogenous peroxidase and the tissue background the sections were incubated overnight at 4°C with the primary antibody (AE-1).

As second antibodies, we used unconjugated goat anti-mouse at an approximate concentration of 50 μg/ml for the PAP procedure or biotinylated horse anti-mouse IgC at 1:100 dilution, given a final concentration of 0.015 mg/ml for the ABC system, followed by a mouse peroxidase anti-peroxidase complex from Ortho Diagnostic Systems (Carpinteria, CA) for the first procedure and the avidin-biotinylated horseradish peroxidase complex (Vector Labs, Burlingame, CA) for the ABC system.

The final reaction product in both procedures were visualized by incubating the sections in a solution 0.02% of 3,3′-diaminobenzidine (Sigma) in 0.01 M PBS pH 7.2 containing 0.002% of H_2O_2 for approximately 10 minutes. The sections were counterstained with hematoxylin, dehydrated, cleared, and mounted in Permont. Controls included the omission of the primary antibody and replacement with either a non-immune mouse serum or PBS.

RESULTS

Immunohistochemistry

There was intense staining of all urothelium by AE-1.[a] This was true of the normal bladder (FIGURE 1), orderly and atypical papillôma (FIGURE 2), carcinoma *in situ* in flat epithelium (FIGURE 3), papillary non-invasive carcinoma (FIGURE 4), and invasive (FIGURES 5–7) and metastatic (FIGURE 8) carcinoma. There was no staining of bladder wall—smooth muscle, connective tissue, endothelium, blood cells, basement membrane, etc.—all were entirely negative. The staining was uniform, not patchy, and slight variations in the generally intense reaction could not be related to cytologic differences or to invasiveness.

Flow Cytometry

In every bladder irrigation sample studied by flow cytometry there were cells present that bound AE-1 antibody, as judged by fluorescein fluorescence intensity greater than for the control aliquot treated with NS-1 antibody (FIGURES 9 and 10). The AE-1 positive cells (i.e. epithelial cells) ranged from 9% in one of the patients with chronic cystitis (FIGURE 11) to 83% in a patient with papillary carcinoma.

---

[a]The immunohistochemical preparations used for illustration in the following figures 1–12 are from formalin-fixed, paraffin-embedded tissue stained with AE-1 monoclonal antibody using the Avidin-Biotin peroxidase technique and counter-stained with hematoxylin.

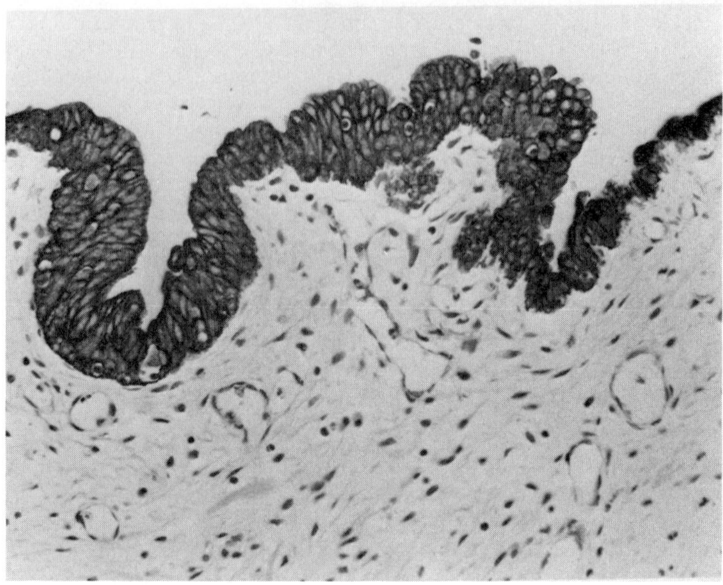

FIGURE 1. Normal bladder mucosa showing intense staining of all layers of the epithelium. There was no antibody binding to the underlying stroma. The section is counter-stained with hematoxylin.

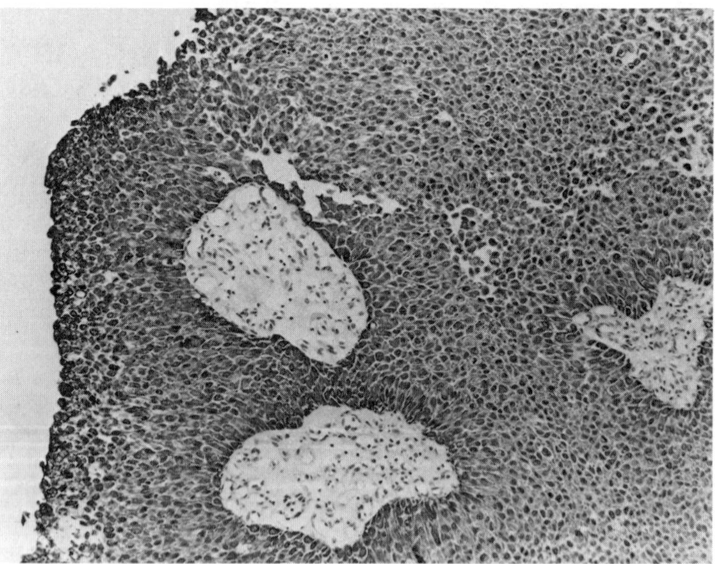

FIGURE 2. Papilloma. Note the uniform intense staining of epithelium.

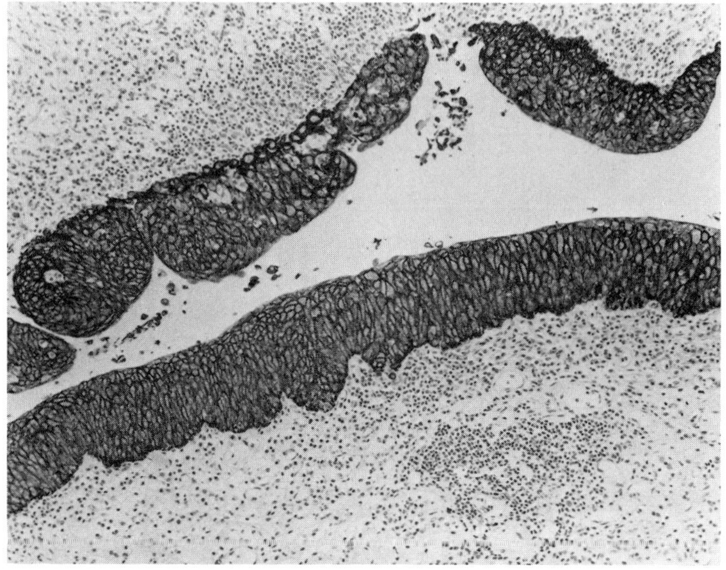

FIGURE 3. Carcinoma *in situ* in the upper strip of epithelium, atypia in the lower strip.

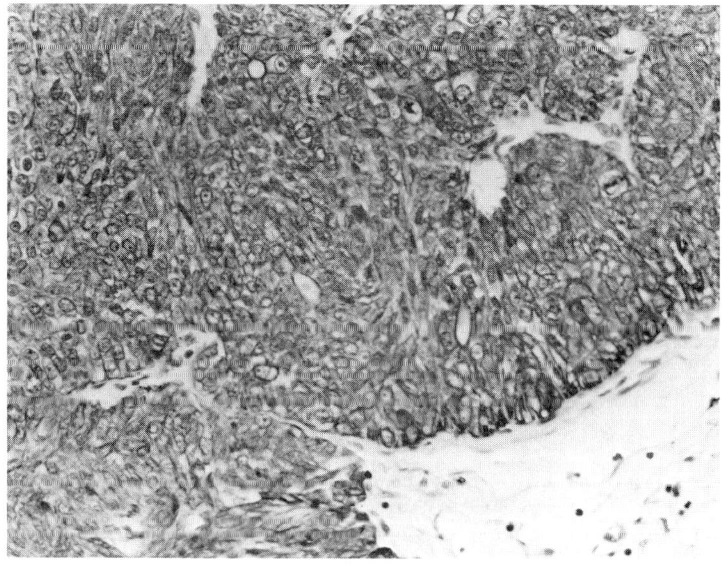

FIGURE 4. Papillary non-invasive carcinoma.

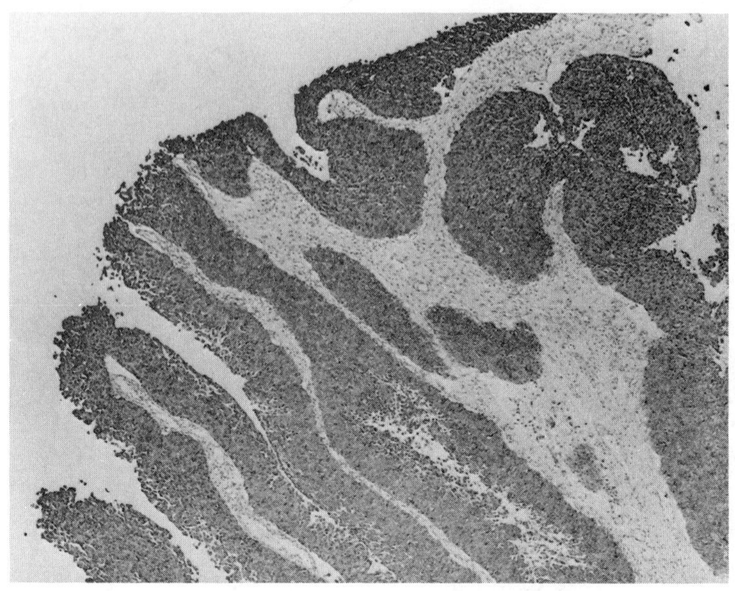

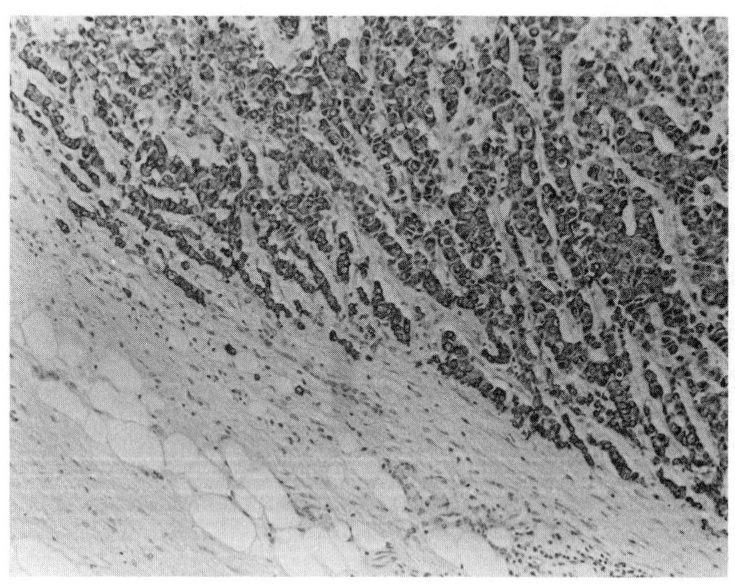

FIGURES 5 and 6. Invasive papillary carcinoma.

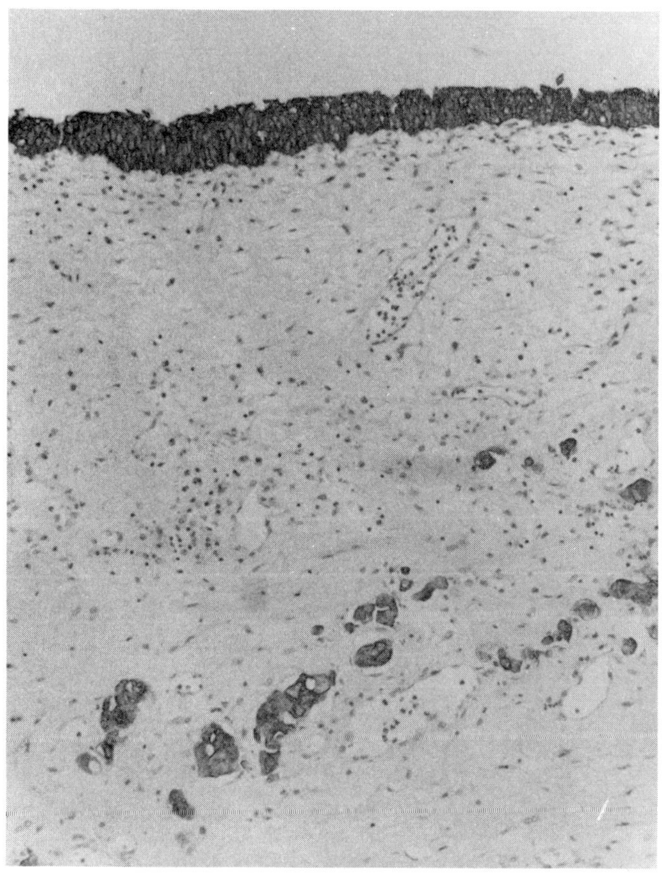

FIGURE 7. Nests of invasive carcinoma in bladder wall. Overlying epithelium is normal in this section.

About 40% of the patients had samples with more than 50% AE-1 positive cells; the fewest were in patients with chronic cystitis (9%, 15%, 20%), one with invasive carcinoma and inflammation (15%), one patient with normal cystoscopy (no biopsy) (10%), and one with prostate carcinoma (17%).

The percent of hyperdiploid cells was increased in all cases by excluding AE-1 negative cells (FIGURE 12 A and B). In the conventional AO-stained aliquots, 8 of the 14 patients with urothelial carcinoma had detectable aneuploid or tetraploid populations. In the P.I./AE-1 stained aliquot, aneuploid or tetraploid populations were detected in four additional cases. One patient with papilloma and one with atypical papilloma had tetraploid or hypertetraploid populations by P.I./AE-1 staining, not evident in the AO-stained aliquot, though both had abnormal RNA distributions by AO staining.[11] Aneuploid cells were also seen by P.I./AE-1 staining in the patient with prostate cancer, while the AO-stained aliquot

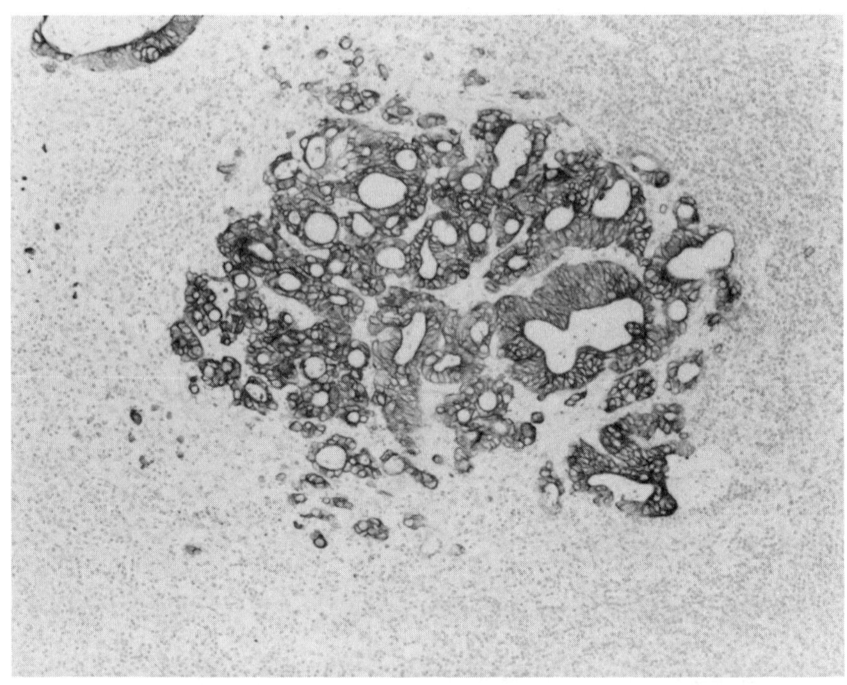

FIGURE 8. Metastatic carcinoma in lymph node.

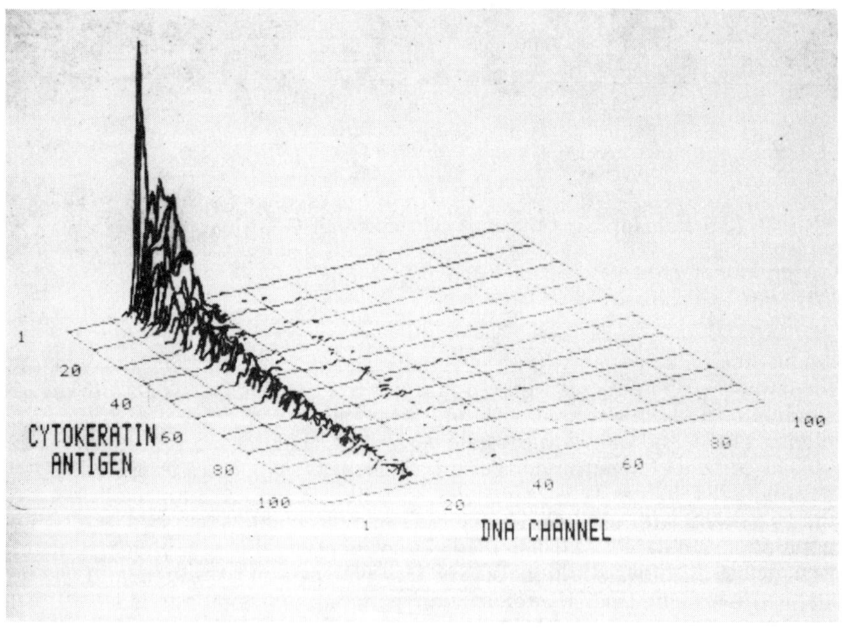

FIGURE 9. Flow cytometry histogram from a patient with papilloma showing a diploid population of cells that can be separated into two subpopulations; one with no cytokeratin antigen (the single sharp peak at channel 1 on the vertical axis) and a second subpopulation with variable content of cytokeratin antigen.

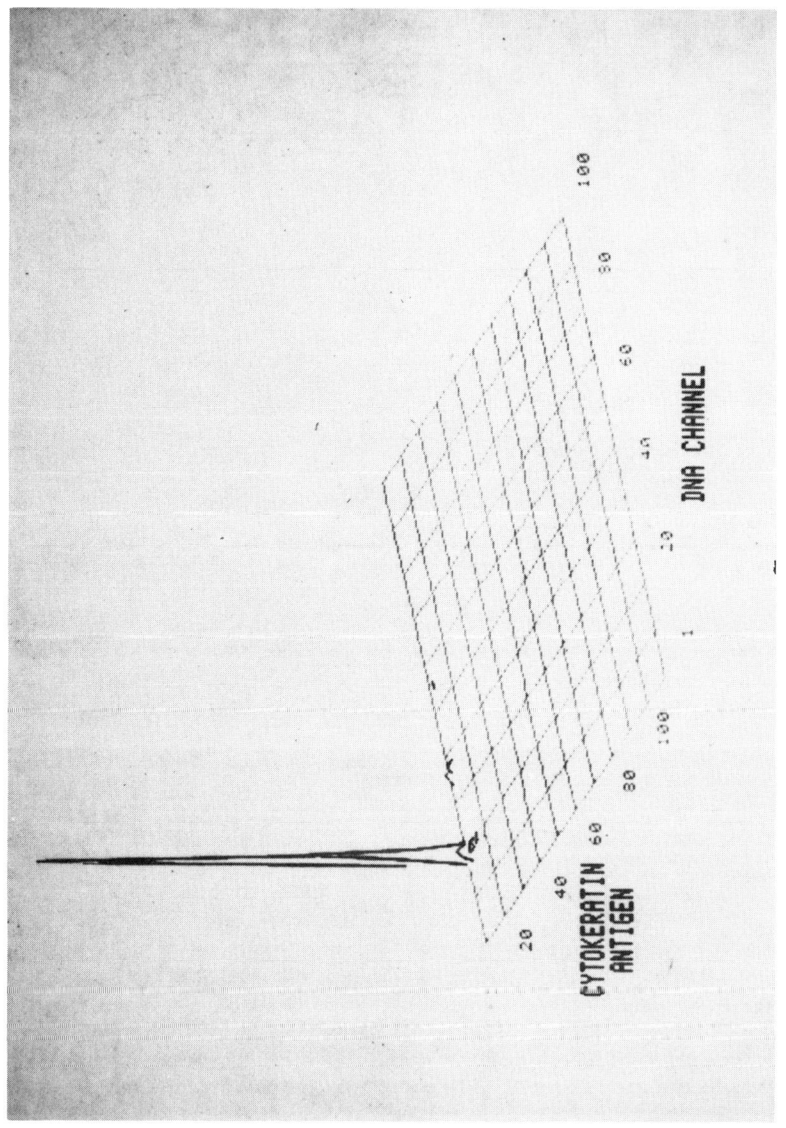

FIGURE 10. Same sample as in FIGURE 9, stained with the negative control monoclonal antibody NS-1.

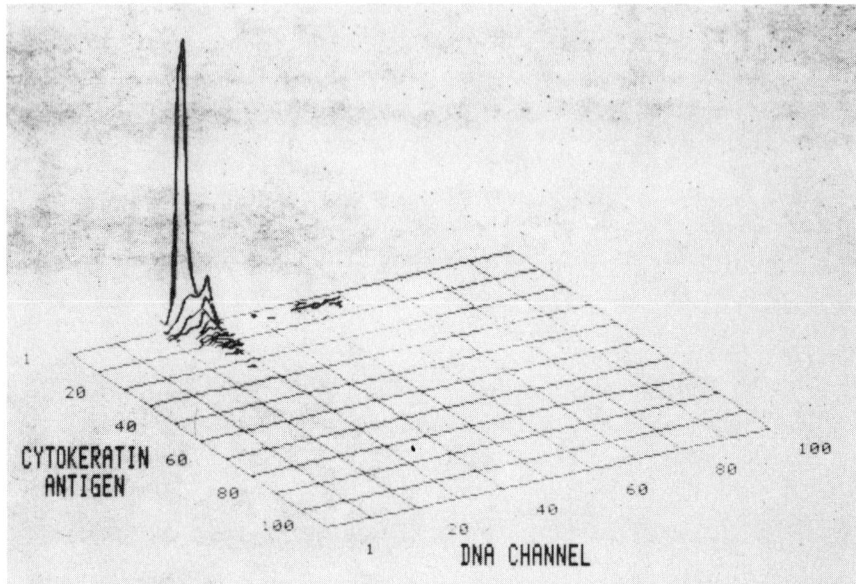

FIGURE 11. Chronic cystitis. In this sample there were many leukocytes and relatively few cells (9%) that had measurable staining with AE-1. The low intensity of staining probably was due to loss of cytoplasm of exfoliated cells as a result of the inflammation.

appeared normal. Of the seven patients with chronic cystitis, there were two who had tetraploid populations; the remaining five had a normal diploid distribution only.

DISCUSSION

Polyclonal and monoclonal antibodies are now used clinically with flow cytometry for sub-typing lymphocytes[12-14] and classification of leukemias and lymphomas,[15] and they have been used experimentally with tissue culture cell lines.[16] In principle, immunocytochemical reactions should be applicable to FCM of clinical specimens for identification and classification of epithelial cells, but so far the technique has been used only with histologic sections. Sun's monoclonal anti-cytokeratin antibody AE-1 is the first antibody that appears to be of real value for identifying epithelial cells in a clinical specimen. The antibody binds uniformly well to all of the cells of the bladder epithelium, normal and neoplastic. It is highly specific in that the fluorescence is bright and there is no discernable binding to other cellular components of the bladder wall, or to leukocytes or reactive cells present in an irrigation specimen. With the technique described here the cell membrane is permeabilized and the antibody can enter and bind to cytoplasmic cytokeratins of cells in suspension. When epithelial cells are specifically stained in this way, small numbers of aneuploid cells that might otherwise be overlooked can be identified even in the presence of many

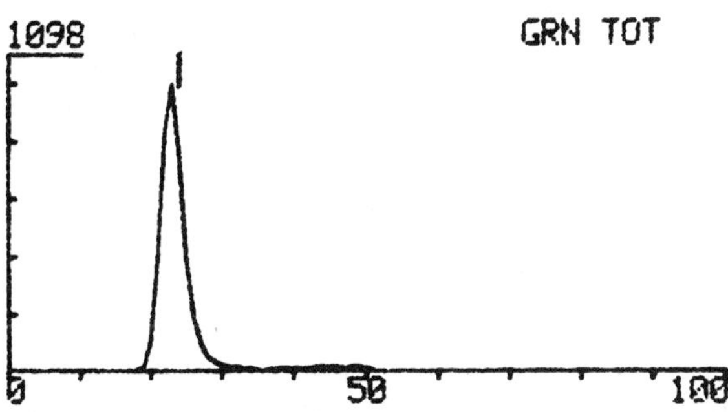

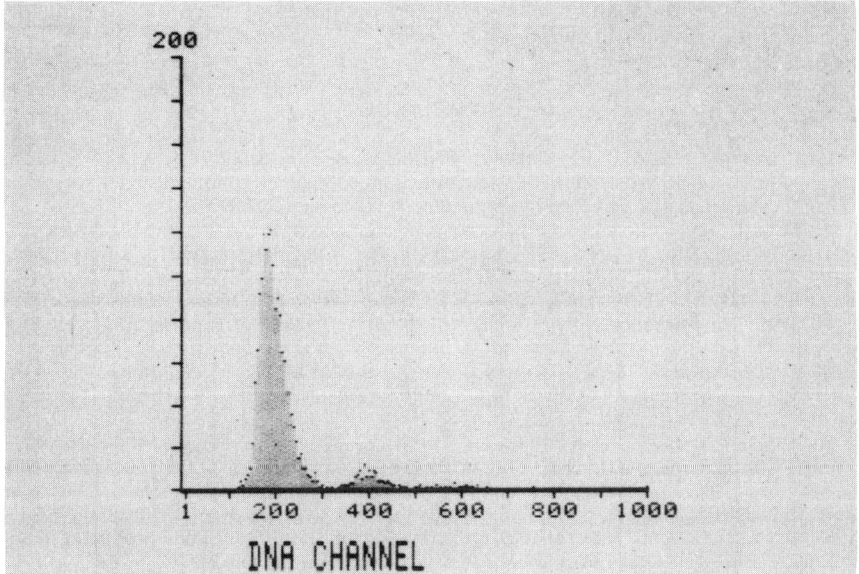

FIGURE 12. (Top) DNA histogram of the whole population of cells from a patient with a papillary tumor, showing a normal DNA distribution. (Bottom) DNA histogram of the AE-1 positive sub-population from the same sample as (top), showing a significant number of tetraploid cells.

inflammatory cells. In our study the number of cases of bladder cancer with identifiable aneuploid stemlines, and the percentage of aneuploid cells, was increased by excluding the non-staining (non-epithelial) cells. Estimates of the percentage of cycling cells also were increased, and differences in the significance attached to this parameter[17,18] may prove due to distortions of DNA distribution caused by inflammatory cells present.

The fluorescence signals of AE-1 stained cells are strong, and "positive" cells are easily distinguished from "negative." The few cells with relatively low fluorescence intensity we believe have lost cytoplasm during exfoliation, and it is likely that some stripped nuclei of epithelial cells were included among the negatively staining cells. We did not attach great significance to variations in the intensity of fluorescence because of probable variations in loss of cytoplasm during exfoliation. Normalizing these measurements will require a second cytoplasmic feature, perhaps total protein or RNA, and this now is feasible with dual laser flow cytometers.

This application of a monoclonal antibody to the identification and classification of epithelial cells in clinical specimens introduces a new era for flow cytometry. The monoclonal antibodies are powerful new tools. Over the next few years there are certain to be technical refinements in their use. New antibodies will be introduced and their specificities defined. Quantification of antibody binding, which is not practical and may not be feasible in histologic sections, will easily be accomplished with flow cytometry. Quantitative assays of multiple cellular constituents of individual cells in mixed populations, which we showed to be feasible more than 15 years ago,[19] now appear to be practical and clinically useful in the diagnostic cytology laboratory.

REFERENCES

1. COLLSTE, L. G., Z. DARZYNKIEWICZ, F. TRAGANOS, T. K. SHARPLESS, W. F. WHITMORE JR. & M. R. MELAMED. 1979. Identification of polymorphonuclear leukocytes in cytological samples for flow cytometry. J. Histochem. **27**:390–393.
2. SUN, T., R. EICHNER, W. G. NELSON, S. C. G. TSENG, R. A. WEISS, M. JARINEN & J. WOODCOCK-MITCHELL. 1983. Keratin classes: Molecular markers for different types of epithelial differentiation. J. Invest. Dermatol. **81**:109s–115s.
3. EICHNER, R., P. BONITZ & T. SUN. 1984. Classification of epidermal keratins according to their immunoreactivity, isoelectric point and mode of expression. J. Cell Biol. **98**:1388–1396.
4. DARZYNKIEWICZ, Z., F. TRAGANOS, T. K. SHARPLESS & M. R. MELAMED. 1976. Lymphocyte stimulation: a rapid, multiparameter analysis. Proc. Natl. Acad. Sci. USA **73**:2881–2884.
5. TRAGANOS, F., Z. DARZYNKIEWICZ, T. K. SHARPLESS & M. R. MELAMED. 1977. Simultaneous staining of ribonucleic and deoxyribonucleic acid in unfixed cells using acridine orange in a flow cytofluorometric system. J. Histochem. Cytochem. **25**:46–56.
6. KOHLER, G. & C. MILSTEIN. 1976. Derivation of specific antibody-producing tissue culture and tumor lines by cell fusion. Eur. J. Immun. **6**:511–519.
7. KLEIN, F. A., H. W. HERR, P. C. SOGANI, W. F. WHITMORE, JR. & M. R. MELAMED. 1982. Detection and follow-up of carcinoma of the urinary bladder by flow cytometry. Cancer **50**:389–395.
8. SHARPLESS, T. K., F. TRAGANOS, Z. DARZYNKIEWICZ & M. R. MELAMED. 1975. Flow cytofluorimetry: Discrimination between single cells and cell aggregates by direct size measurements. Acta Cytol. **19**:577–581.
9. STERNBERGER, L. A. 1979. Immunocytochemistry. pp. 82–159. 2nd edit. John Wiley & Sons. New York.
10. HSU, S., L. RAINE & H. FANGER. 1981. Use of Avidin-Biotin-Peroxidase complex (ABC) in immunoperoxidase techniques. J. Histochem. Cytochem. **29**:577–580.
11. KLEIN, F. A., M. R. MELAMED, W. F. WHITMORE JR., H. W. HERR, P. C. SOGANI & Z. DARZYNKIEWICZ. 1982. Characterization of bladder papilloma by two-parameter DNA-RNA flow cytometry. Cancer **42**:1094–1097.
12. REINHERZ, E. L. & S. F. SCHLOSSMAN. 1980. The differentiation and function of human T-lymphocytes. Cell **19**:821–827.
13. STASHENKO, P., L. M. NADLER, R. HARDY & S. F. SCHLOSSMAN. 1980. Characterization of a human B lymphocyte specific antigen. J. Immunol. **125**:1678–1685.
14. TODD, R., L. M. NADLER & S. F. SCHLOSSMAN. 1981. Antigens on human monocytes identified by monoclonal antibodies. J. Immunol. **126**:1435–1442.

15. ANDREEFF, M., Z. DARZYNKIEWICZ, T. K. SHARPLESS, B. D. CLARKSON & M. R. MELAMED. 1980. Discrimination of human leukemia subtypes by flow cytometric analysis of cellular DNA and RNA. Blood **55**:282–293.
16. RAMAEKERS, F. C. S., H. BECK, G. P. VOOIJS & C. J. HERMAN. 1984. Flow cytometric analysis of mixed cell populations using intermediate filament antibodies. Exp. Cell. Res. **153**:249–253.
17. COLLSTE, L. G., Z. DARZYNKIEWICZ, F. TRAGANOS, T. K. SHARPLESS, M. DEVONEC, M. L. CLAPS, W. F. WHITMORE & M. R. MELAMED. 1979. Cell-cycle distribution of urothelial tumor cells as measured by flow cytometry. Br. J. Cancer **40**:872–877.
18. TRIBUKAIT, B., H. GUSTAFSON & P. ESPOSTI. 1979. Ploidy and proliferation in human bladder tumors as measured by flow-cytometric DNA—analysis and its relations to histopathology and cytology. Cancer **43**:1742–1751.
19. KAMENTSKY, L. A. & M. R. MELAMED. 1969. Rapid multiple mass constituent analysis of biological cells. Ann. N.Y. Acad. Sci. **157**:310–323.

Determination of Cell Cycle DNA and 5′-Nucleotide Phosphodiesterase in Endometrial Cancer[a]

K. C. TSOU, D. H. HONG, M. VARELLO, J. E. WHEELER,
R. GIUNTOLI, C. MANGAN, AND J. MIKUTA

*Departments of Surgical Research, Pathology, Laboratory Medicine,
and Obstetrics-Gynecology
University of Pennsylvania School of Medicine
Philadelphia, Pennsylvania 19104*

INTRODUCTION

Endometrial cancer incidence has been reported to have risen in recent years not only in the United States[19,29] but also in Japan,[16] Norway,[18] England,[17] and other parts of the world.[6] Based on the recent epidemiological data, there were 15,432 cases of endometrial cancer diagnosed from 1973–1977,[31] out of a total of 31,540 cases of all female genital cancer. As a comparison, there were 7,386 cases of ovarian cancer and 6,672 cases of cervical cancer diagnosed in the same period. Therefore, the detection of this form of female cancer has received the attention of many research laboratories in recent years.[3,9] Indeed, it has been suggested that with frozen section diagnosis, the accuracy of an experienced surgical pathologist can be 95% or more.[11] In practice, many hospital clinics still favor the endometrial curettage as the preferred method[2,20] for detecting the presence of endometrial cancer. Since cellular samples can be made available,[21] the use of flow cytometry as a possible diagnostic aid for this form of cancer should be an encouraging prospect.

Because of the plausible association of estrogen levels and this form of cancer, a promising marker related to the progesterone or estrogen receptor[12,13] has occupied the interest of several laboratories[4,7] and, when available, such methods may be adaptable to flow cytometry. Bearing in mind that endometrium is a proliferative tissue, it is a difficult task to investigate the growth pattern of this tissue in an effort to search for direct evidence of neoplasia. However, recent research in our laboratory has suggested strongly that 5′-nucleotide phosphodiesterase (5′-NPD) is related to growth[15,23] and in fact, an isozyme-V of 5′-NPD has been found as a marker for human liver cancer.[24,25] Recent advances in our methodology have also enabled us to measure the cellular activity of this enzyme by fluorogenic or chromogenic methods.[28] It was therefore desirable for us to apply these methods with cell cycle kinetics to gather data in a series of endometrium tissues. The correlation of 5′-NPD activity and cell cycle kinetics with endometrial cancer forms the basis of this report. While preliminary in

[a]Supported by National Cancer Institute research grants CA 25376 and CA-28770, and by W.W. Smith Foundation. Presented in part at the Engineering Foundation Conference on "Clinical Cytometry," 1983.

nature, attempts have also been made to examine the relationship of this enzyme as a proliferative marker for patients in hormone replacement therapy and its possible significance to estradiol dehydrogenase, a receptor-regulatory enzyme.

MATERIALS AND METHODS

Sixty-eight endometrial specimens were provided from the Gynecologic Oncology Division of the Hospital of the University of Pennsylvania for this study. These specimens were divided into 28 normal, 25 cancer, 5 dysplasia, and 10 atrophy. The 10 atrophy patients had previous histories of hormone replacement therapy. Whenever possible the specimens were immediately processed for flow cytometric analysis on the same day, and sometimes within one hour of surgery. Otherwise, the specimens could be stored at $-70°C$ and used the next day. There was no apparent loss in enzyme activity based on the result. All tissue diagnosis was done at the surgical pathology laboratory and reviewed a second time. The pathological diagnosis was not initially available during the flow analysis to avoid any bias.

Flow Cytometric DNA Analysis

Flow cytometric DNA analysis (FCDA), was done as previously described[26] using 4',6-diamidino-2-phenylindole (DAPI) synthesized in our laboratory according to a modified procedure of Dann.[5] All specimens were syringed through an 18 gauge needle to a uniform suspension, then filtered through a 53 μm nylon mesh filter. Flow analysis was then carried out with an ICP-22A instrument (Ortho Instruments, Inc., Westwood, MA), which was equipped with a mercury lamp and a Zeiss 100 X objective. This instrument was calibrated weekly with a mouse liver specimen and had a coefficient of variation (cv) for a G_0/G_1 peak of 1.7%. The results were monitored on a Techtronics 2103 analyzer and plotted on a Hewlett-Packard chart recorder. The cell cycle data and distribution of the G_0/G_1, S, and G_2/M populations were then corrected by a program written by Dr. Walter Tsou for an Apple computer, based on the equation originally suggested by Wood and Todd.[30] This program is stored on a diskette and the data can be recalled or replotted with an Apple Plot program again for further analysis as illustrated in FIGURE 1.

Flow Cytometric 5'-Nucleotide Phosphodiesterase Analysis

Initially, flow cytometric analysis of this enzyme was tested with various fluorescent and chromogenic substrates. It was found that for the present study the reproducibility of the results favored the quenching method, based on the use of 5'-iodoindoxyl thymidine phosphodiester (IIpT) as a substrate.[28] Using the above DAPI-stained cell suspension and incubated with IIpT substrate solution (1.0 mg/ml in 0.1 M Tris-HCl buffer, pH 8.6) for 30 minutes, the flow analysis of DNA was then repeated again on the same instrument. Because of the quenching of the diiodoindigo dye formed on the cells as a result of the enzyme activity in the cells, this DNA analysis will yield a different distribution of the cell cycle kinetics from the above data (FIGURES 1 and 2). There will be a decrease of DAPI fluorescence in the cells where the enzyme activity is elevated. Therefore, qualitatively it is easy to assess which group of cells has more enzyme activity and

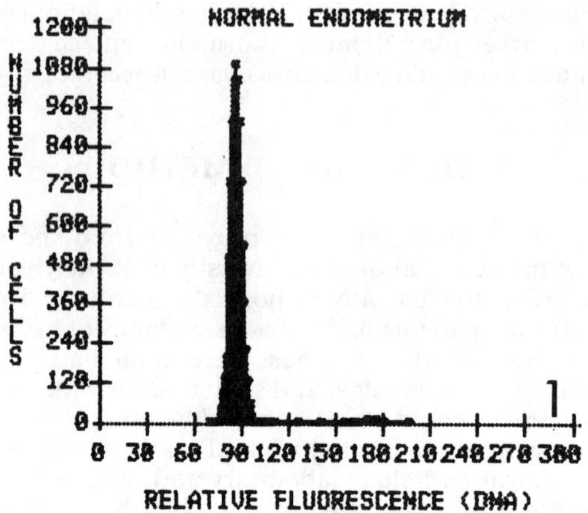

FIGURE 1. DNA histogram.

less fluorescence. Quantitative evaluation can also be made by a calculated quenching curve (omitted). Simultaneous fluorescence microscopy can be done with or without the enzyme substrate and a photograph can be taken for possible future comparison with morphology alteration (FIGURE 3, a and b). The results of this enzyme study are summarized in TABLES 1 and 2.

Quenching Curve Calibration

Standardization of the quenching effect of 5′-NPDase on DNA fluorescence was performed using constant concentrations of calf thymus DNA (2.0 µg/ml,

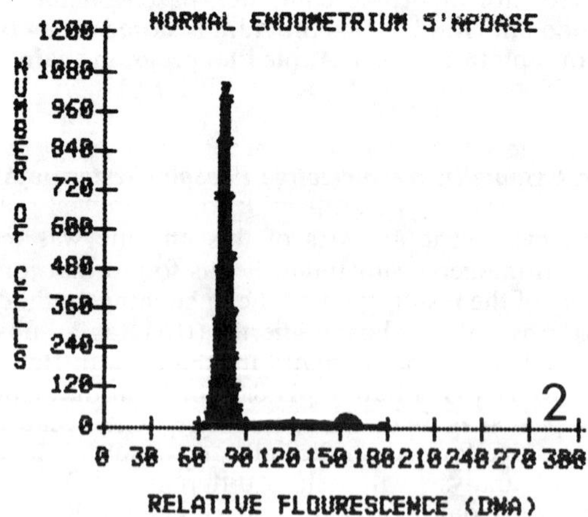

FIGURE 2. DNA histogram after IIpT treatment—note the decrease in G_0/G_1 peak due to quenching.

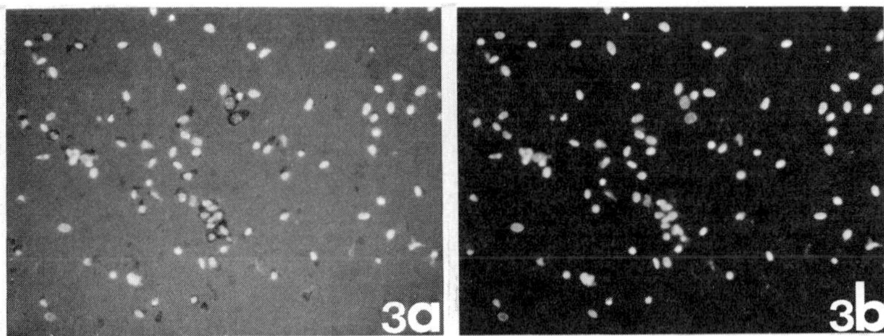

FIGURE 3. (A) Normal endometrial cells stained with IIpT and DAPI. Note not all cells have indigo dye 126X. (B) DAPI fluorescence for DNA only 126X.

Sigma Chemical Company, St. Louis, MO), 4',6-diamidino-2-phenylindole (5 µg/ml), and 5'-(5-iodo-3-indoxyl)-thymidine phosphodiester (1 mg/ml) in 3.0 ml of 0.1 M Tris-HCl buffer pH = 9.0. *Crotalus adameteus* venom phosphodiesterase (P.L. Biochemicals, Inc., Milwaukee, WI) was diluted to various concentrations, from a stock solution of 100 units/ml, and was added to the above mixture. This final solution was incubated for 30 minutes in a 37°C water bath. The relative percent of fluorescent decrease was measured using a Turner fluorometer model 111 (Amisco Instruments, Carpinteria, CA) equipped with excitation filters for peak transmittance at 360 nm and an emission filter for 450–490 nm. The results were recorded on semi-log graphing paper for which a linear relationship of the quenching effect of 5,5'-diiodoindigo on DNA fluorescence was obtained.

Calculation of 5'-NPD Units per Cell

The total cell concentration/ml (A) of each sample was determined using an American Optical counting chamber (Arthur Thomas, Inc., Philadelphia, PA). This number was then used to determine the number of cells/ml (b), using the DAPI histogram, of each corrected cell cycle population, i.e., A × (G_0/G_1%, S%, G_2 + M%) = B. The concentration of 5'-NPDase in units/ml was ascertained by comparing both the DAPI and DAPI plus IIpT histograms. The relative percent decrease in the quenched population was calculated and the concentration of 5'-NPDase of that population was determined using the standard quenching curve (X_i). Therefore, by dividing the units/ml of the quenched population by the cell number/ml of that quenched population the number of units/cell can be calculated (Y), i.e., $X_i/B = Y$. This number represents the average relative increase of 5'-NPDase in either G_0/G_1, S, or G_2 + M.

RESULTS

When the samples were decoded and the criteria for tissue diagnosis were included, these 68 specimens consisted of 25 cancer and 43 noncancer, based on histopathological diagnosis. As shown in TABLE 1 the noncancer group (A) could be separated into four subgroups: proliferative, secretory, inactive or atrophy, and dysplasia or hyperplasia. The cancer group (B) could be separated into three

TABLE 1. Results of Flow Cytometric DNA Analysis of Human Endometrium

	Pathologic Diagnosis		DNA Number of Cell Cycle				Total of Samples
			G_0/G_1	S	G_2/M	$S + G_2/M$	
A	Proliferative		85.11	10.8	4.5	15.4	24
	Secretory		86.8	9.8	3.4	13.2	4
	Inactive or Atrophy		86.6	9.8	3.5	13.3	10
	Dysplasia or Hyperplasia		89.1	7.2	3.8	11.2	5
B	Adenocarcinoma or Clear Cell Carcinoma	well differentiated	82.3	12.4	5.4	17.8	11
		moderately differentiated	80.4	14.8	5.0	19.8	9
		poorly differentiated	65.1	28.9	6.2	35.2	5
Total samples							68

TABLE 2. Localization of 5′-NPD in Cell Cycle of Human Endometrium

	No.	G_0/G_1%	S%	G_2/M%	S + G_2/M
Normal[a]	28	10^d (36%)	17 (61%)	7^d (25%)	5/28 (18%)[d]
Atrophy[b]	10	6 (60%)	9 (90%)	1 (10%)	0/10 (0)
Dysplasia	5	2 (40%)	3 (60%)	3 (60%)	2/5 (40%)
Cancer[c,e]	25	9 (35%)	16 (64%)	16 (64%)	9 (36%)
	68	27 (40%)	44 (65%)	27 (40%)	16 (23%)

[a]Includes 4 secretory, 4 proliferative.
[b]All 10 are in hormone replacement therapy.
[c]11 well-differentiated, 9 moderately differentiated, and 5 poorly differentiated.
[d]All are proliferative endometrium.

		G_0/G_1	S	G_2/M	S + G_2/M
[e]Well-differentiated	11	3 (27%)	9 (81%)	8 (73%)	6 (55%)
Moderately Differentiated	9	6 (66%)	4 (44%)	5 (56%)	2 (22%)
Poorly Differentiated	5	0 (0)	3 (60%)	3 (60%)	1 (20%)

groups: well-differentiated, moderately differentiated, and poorly differentiated. It was also noted that even the poorly differentiated cancer are not all polyploid. Only two (both poorly differentiated) samples were found to be polyploid in 25 cancer specimens. Therefore it was clear to us at the early stage of our investigation that heterogeneity of tumor is not the best index of endometrial cancer.

Flow Cytometric DNA Analysis

When the distribution of the normal endometrium specimens were examined (TABLE 1), there were apparent variations of G_0/G_1, S, and G_2/M from sample to sample, with an average value of G_0/G_1 = 86.17%; S = 10.13%; and G_2/M = 3.80%. There was no apparent relationship of the cell cycle data to the assignment of the proliferative or secretory phase of the endometrium. The difference between normal and hyperplasia seems to show a lower S + G_2/M in the latter group.

When the cancer group was examined the majority of the samples could be calculated for diploid and tetraploid populations, as in the normal specimens. There were only two samples having a polyploid pattern. These samples were both found to be poorly differentiated adenocarcinoma by pathology criteria. The use of flow analysis confirmed the surgical pathology diagnosis. The average DNA distribution in the cancer group was found to be 75.93% for G_0/G_1, 18.70% for S, and 5.53% for G_2/M. Thus, there was a significant difference in S and G_2/M between normal and cancer. A G_2/M higher than 5% was found in 80% of all cancers and less than 4% in all non-cancer ($p < 0.005$). If S + G_2/M may be considered as growth fractions, then the differences in TABLE 1 between normal cancer and group are significant—($p < 0.025$), even though it was not able to distinguish between dysplasia, proliferative, secretory, and atrophy without further quantitation. Such difference in DNA distribution is also presented in FIGURE 4 as a reference.

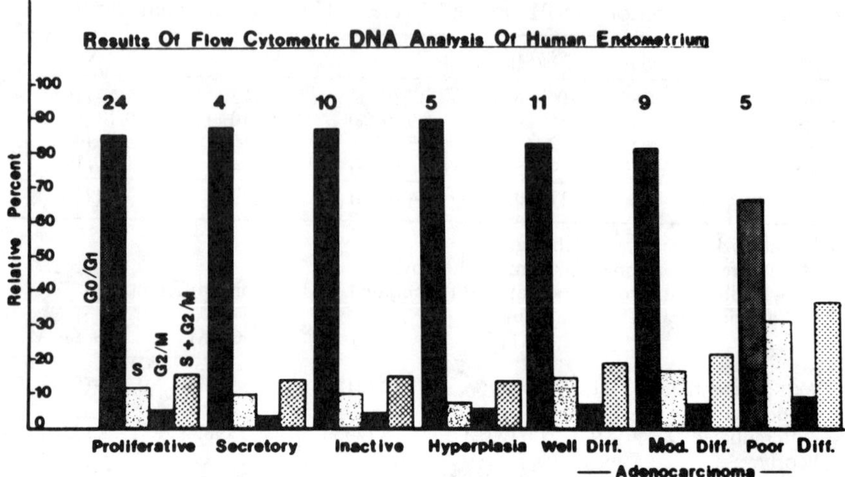

FIGURE 4. Results of flow cytometric DNA analysis of human endometrium. Note the gradual increase of S + G$_2$/M in the three classes of endometrial carcinoma (see also TABLE 1).

The flow cytograms as illustrated in FIGURE 1 form the basis of our DNA analysis. Thus, it shows DNA analysis based on the DAPI stain alone and FIGURE 2 shows DNA analysis after treatment with IIpT as a substrate for 5′-NPD activity. Usually similar numbers of cell were run through the flow analysis and therefore the qualitative difference becomes readily apparent. Similar differences were noted for the cancer group, or the polyploid cancer specimens. FIGURE 3 (a) demonstrates the cells stained with the 5,5′-diiodoindigo dye by IIpT and the fluorescent DNA stain with DAPI. FIGURE 3 (b) shows the same picture taken under the fluorescent microscope for DAPI only. Comparison can thus be made to see the cells with and without the exonuclease activity. An advantage of the DAPI stain lies also in the fact that it does not stain red cells, thus obviating the necessity of lysing the red cells in a limited number of hemorrhagic specimens, which are common in endometrial cancer. The distribution of 5′-NPD in different cell populations is summarized in TABLE 2 and the quantitative results are shown in TABLE 3. In the proliferative but not secretory normal endometrium group, IIpT quenched in G$_0$/G$_1$ (36%) and G$_2$/M (25%) with 18% in both S and G$_2$/M, whereas secretory normal endometrium had 5′-NPD only in S cells. Hyperplasia and the cancer group both have 40% and 36% of 5′-NPD in S + G$_2$/M, but hyperplasia never occurred in G$_2$/M alone. These data were interesting if one accepts the thesis that hyperplasia specimens may be preneoplastic. Quantitatively the concentration of 5′-NPD is 4 × higher in cancer specimens than that in normal specimens in G$_2$/M cells whereas in the hyperplasia specimens it is only about 2 × higher than that in normal specimens. Detailed comparison can also be seen in FIGURE 5.

The value of cell cycle data in providing quantitative information on the differentiation of these samples could be seen also in TABLE 3 and FIGURE 6. There were no significant differences between the proliferative and secretory samples in the normal specimens. Heterogeneity of the tumor population in the poorly

TABLE 3. Result of Flow 5'-NPD of Human Endometrium

Pathologic Diagnosis	G₀/G₁			S			G₂/M			S + G₂/M			
	DAPI	DAPI + IIpT	5'-NPD μU/cell	DAPI	DAPI + IIpT	5'-NPD μU/cell	DAPI	DAPI + IIpT	5'-NPD μU/cell	DAPI	DAPI + IIpT	5'-NPD μU/cell	Total
Proliferative	35.1	84.8	0.04	10.6	10.2	2.7	4.5	5.0	3.9	15.1	15.2	6.6	24
Secretory	36.8	84.9	0	9.8	11.9	7.2	3.4	3.2	2.1				4
Inactive or Atrophy	36.6	85.8	0.03	9.8	9.3	3.95	3.5	4.9	3.0				10
Hyperplasia or Dysplasia	39.1	89.2	0.1	7.2	6.4	3.55	3.8	4.3	5.3	11.2	10.7	8.85	5
Carcinoma	75.9	76.4	0.26	18.7	18.0	10.6	5.0	5.6	11.4	24.3	23.6	22.0	25
Total													68

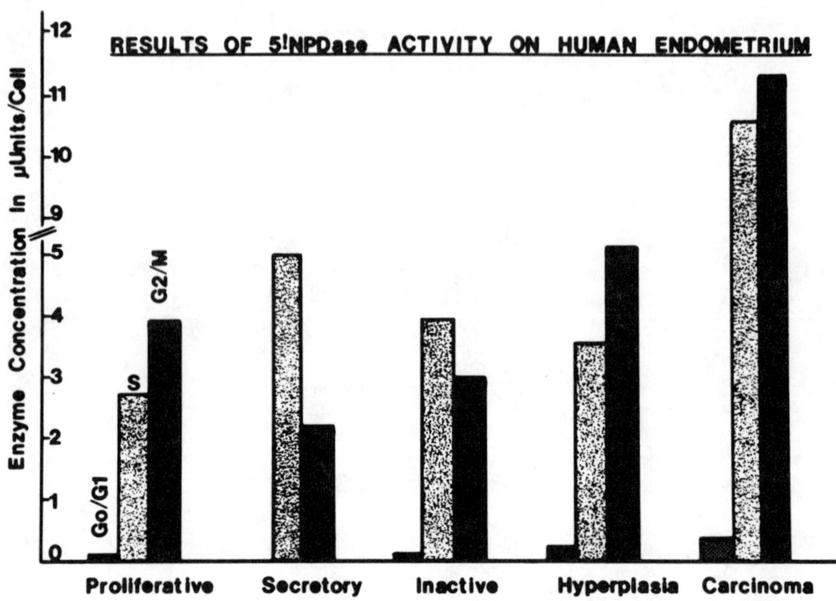

FIGURE 5. 5'-Nucleotide phosphodiesterase concentration in µU/cell. The enzyme is almost negative in G_0/G_1 in secretory phase and the higher concentration of 5'-NPD in G_2/M cells higher than that in hyperplasia.

differentiated specimen (not shown) is apparent both from the histopathology observation and flow analysis. Flow analysis may now provide another basis to subclassify poorly differentiated carcinoma for surgical pathology, as we noted that not all poorly differentiated endometrial carcinoma are aneuploidy. Details will be reported at a later date.[33] The 10 atrophy specimens provided a preliminary sampling for the value of using this enzyme marker as a proliferative index. As shown in TABLE 2, 90% of specimens in this group had elevation in S cell population.

DISCUSSION

In spite of the advances made in gynecological cancer diagnosis,[8] the detection of endometrial cancer remains a difficult task. With the advance of hormone maintenance therapy for postmenopausal women, there has been concern among physicians that the increase in endometrial cancer incidence as reported by some authors is related.[10] In this study, the average age of the cancer group was 60, whereas the average age of the non-cancer group was 43. Thus, even in this limited sampling, there was an indication that there may be a higher risk factor involved in the postmenopausal group.[8]

As seen in TABLE 1, the cell cycle analysis based on DNA in our normal group (A) did show a lower $S + G_2M$ than the cancer group (B). Using the ratio $(S + G_2/M)B/(S + G_2/M)A$ there was an increase in the growth fraction of $24.3\%/13.3\% = 1.83$, which is an 83% increase in the cancer group. All cases classified as cancer were diagnosed by histopathology. This significant difference is gratifying in spite of the fact that endometrial tissue is proliferative. The difference among the three histopathological grades for the G_2/M value may also be of real value for differentiation. With further study in a larger panel than ours the use of DNA S + G_2/M value may indeed be a useful diagnostic aid. We have, however, still sought other markers to augment this method.

In a previous study on the serum level of 5′-nucleotide phosphodiesterase[14,25] we noted with interest that there was a relatively high number of elevated enzyme samples in endometrial cancer. Subsequently, we studied the relationship of 5′-NPD to the cell cycle in mouse and rat liver and we were led to believe that this enzyme was intimately associated with the proliferation of liver in young rats.[15] It was therefore of interest for us to extend this marker study to endometrium in gynecological cancer, as we realized the similarity of the proliferative nature of both organs. Essentially, the problem was how to adapt our histochemical method to flow cytometry for this enzyme. For indeed, if the enzyme 5′-NPDase were a proliferative marker, then the variation of the distribution of this enzyme in G_0/G_1, S, and G_2/M or within a mixed population of cells undergoing proliferation may well be the key to malignancy. It is also necessary to emphasize that to study such differences of enzyme activity in a proliferative population, using histochemical methods, is virtually impossible. There is no method at present that yields a truly synchronous population of either of these three groups of cells. With the advance of automated cytology, such a study is now possible if a suitable marker can be found to label these populations. After several attempts in the development and use of fluorescent substrates, it occurred to us that a readily available substrate synthesized in our laboratory, 5′[5-iodo-3-indoxyl] thymidine phosphodiester (IIpT) was useful as a quenching substrate when used with DAPI as a DNA stain. Much of this new application of this indigogenic enzyme method has yet to be perfected and published, but a preliminary report on the use of this method for the identification of hairy cell leukemia has been presented.[27]

As shown in TABLE 3, when this method was applied to endometrial tissue, the flow cytometric DNA cytogram was markedly affected, and the alteration was directly related to the quenching effect of the indigo on the fluorescence of the cell nuclei. Careful examination of the cells under fluorescent microscope further confirms our suggestion that the enzyme distribution occurred in different populations of cells. Among the cancer patients in TABLE 2, there were two patients whose 5′-NPD were found in G_0/G_1 phase cells. These two patients had cancer only in the right corner endometrium, according to histopathology. Thus, there appears a shift of compartment from non-cancer to cancer in cell cycle of this enzyme. As shown in FIGURE 5, the enzyme per cell in the cancer group is also more than 5 × higher in G_2/M of cancer cells than in the normal proliferative cell. This is an important finding and with further study, it may be possible to assign 5′-NPDase as a proliferative marker related to neoplastic differentiation. A suggested scheme is shown in FIGURE 6.

Most other laboratories have reported the use of DNA aneuploidy as an indicator for malignancy.[1] Our study also confirmed the accuracy of these reports when such patterns were observed. However, based on histopathology, these samples in our study were all poorly differentiated types and therefore need no further confirmation by other methods than the diploid samples. In the polyploid endometrium samples, we also noted the presence of 5′-NPDase preponderantly in the higher ploid cells. Thus, in these cells, one may consider 5′-NPDase as a marker of endometrial cancer. It should be of interest, in the future, to further investigate whether there were isozymes present in malignant or proliferative endometrium.

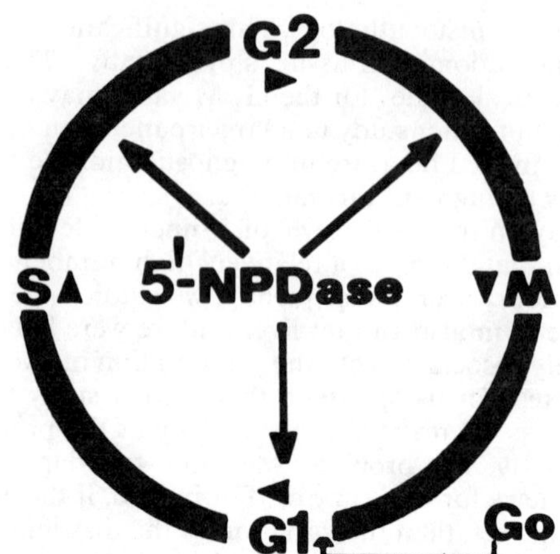

FIGURE 6. A schematic suggestion for the role of 5'-NPD in regulating proliferative population.

Although we have investigated the role of this enzyme as an exonuclease, it should be noted that 5'-NPDase may also be able to act as a high K_m phosphodiesterase on 3',5'-cyclic AMP and may therefore have a role in hormonal mediation. Our present investigation has not involved such a study, but should be considered in the future as an alternative mechanism to what was proposed in the present report. It was noted that in the atrophy specimens, there were four samples in which elevation of estradiol dehydrogenase was found.[33] Further improvement of cytochemical methods for the study of this steroid dehydrogenase may also be necessary for better correlation of 5'-NPD and this receptor regulatory enzyme. No conclusion could be drawn at present.

REFERENCES

1. BARLOGIE, D., B. DREWINKO, J. SCHUMAN, W. GOHDE, G. DOSIK, J. LATREILLE, D. A. JOHNSTON & E. I. FREIREICH. 1980. Cellular DNA content as a marker of neoplasia in man. Am. J. Med. 69:195–203.
2. BJERRE, B. A., S. A. GARDANSK & N. SJOBERG. 1971. Aspiration curettage—A new diagnostic method. J. Reprod. Med. 7:221–223.
3. CREASMAN, W. T. & J. C. WEED. 1973. Screening techniques in endometrial cancer. Cancer 23:96–105.
4. CREASMAN, W. T., K. S. MCCARTY, T. K. BARTON & K. S. MCCARTY, JR. 1980. Clinical correlates of estrogen and progesterone-binding proteins in human endocarcinoma. Obstet. Gynecol. 55:363.
5. DANN, O., G. BERGEN, E. DEMANT & G. VOLZ. 1971. Trypanocide diamidine des 2-phenylbenzofuran, 2-phenyl-inden und 2-phenyl-indols. Ann. Chem. Justus Liebigs 794:68–89.

6. DOLL, R., C. MUIR & J. WATERHOUSE. Eds. 1970. Cancer Incidence in Five Continents. Vol. 2. International Union Against Cancer. Springer-Verlag. Berlin.

7. DONALDSON, E. S., J. R. VAN NAGELL, JR., S. PURSELL, E. C. GAY, W. R. MECKER, R. KASHMIRI & J. C. VAN DE VORDE. 1980. Multiple biochemical markers in patients with gynecologic malignancies. Cancer **45**:948.

8. FOLEY, D. V. & T. MASUKAWA. 1980. Endometrial monitoring of high-risk women. Cancer **48**:511–514.

9. GUSBERG, S. B. 1973. An approach to the control of cancer of the endometrium. Cancer **23**:99–105.

10. GUSBERG, S. B. 1976. The individual at high risk for endometrial carcinoma. Am. J. Obstet. Gynecol. **126**:535.

11. HOFMEISTER, F. J. 1974. Endometrial biopsy: Another look. Am. J. Obstet. Gynecol. **118**:773–776.

12. JENSON, E. V., E. R. DESOMBRE & P. W. JUNGBLUTT. 1967. Estrogen receptors in hormone responsive tissues and tumors. *In* Endogenous Factors Influencing Host-tumor Balance. R. W. Wissler, T. C. Dao & S. Wood, Jr., Eds.:15. University of Chicago Press. Chicago, IL.

13. JENSON, E. V., P. I. BREDER, M. NOMATA, S. SMITH & E. R. DESOMBRE. 1975. Estrogen interaction with target tissues: Two step transfer of receptor for the nucleus. Methods Enzymol. **5**:46. Academic Press. New York.

14. LO, K. W., W. FERRAR, W. FINEMAN & K. C. TSOU. 1972. Fluoromedic assay of serum 5'-nucleotide phosphodiesterase in normal human and cancer patients. Anal. Biochem. **47**:609–613.

15. LO, K. W. & K. C. TSOU. 1982. Cell cycle and 5'nucleotide phosphodiesterase activity in rat liver. Cytometry **2**:414–420.

16. MASLUCHI, K. & H. NEMOTO. 1972. Epidemiologic studies on uterine cancer at Cancer Institute Hosptial, Tokyo, Japan. Cancer **30**:268.

17. MCBRIDE, J. M. 1959. Premenopausal cystic hyperplasia and endometrial carcinoma. J. Obstet. Gynecol. Brit. Commonw. **66**:288.

18. NAKHAS, W. A., C. J. LUND & J. H. RUDOLPH. 1971. Carcinoma of the corpus uteri: A 10 year review of 225 patients. Obstet. Gynecol. **38**:564.

19. SILVERBERG, E. Cancer Statistics. CA—A Cancer Journal for Clinicians. 1978, **28**:24–25. *Ibid.* 1979. **29**:6–7. *Ibid.* 1980. **30**:30–31. *Ibid.* 1981. **31**:20–21. *Ibid.* 1982. **32**:22–23.

20. SLAUGHTER, C. R. & E. J. SCHEWE, JR. 1962. Evaluation of biopsy of the endometrium by the Novak suction curette. Am. J. Obstet. Gynecol. **83**:1302–1305.

21. STUDD, J. W. W., M. THOM, F. DISHCE, M. DRIVRE, T. W. EVANS & D. WILLIAMS. 1979. Value of cytology for detecting endometrial abnormalities in climacteric women receiving hormone replacement therapy. Brit. Med. **1**:840–848.

22. TSOU, K. C., S. LEDIS & M. G. MCCOY. 1973. 5'-Nucleotide phosphodiesterase isoenzyme pattern in the serum of human hepatoma. Cancer Res. **33**:2215–2217.

23. TSOU, K. C., H. P. MORRIS, K. W. LO & J. J. MUSCATO. 1974. 5'-Nucleotide phosphodiesterase activity in rat hepatomas. Cancer Res. **34**:1295–1298.

24. TSOU, K. C. & K. W. LO. 1980. Serum 5'-nucleotide phosphodiesterase isozyme-V test for human liver cancer. Cancer **45**(2):209–213.

25. TSOU, K. C., K. W. LO, J. MULLEN, E. ROSATO, R. GUINTOLI, J. MIKUTA, C. MANGAN & J. MURPHY. 1980. Diagnosis of primary and secondary hepatoma with 5'-nucleotide phosphodiesterase isozyme test. *In* Prevention and Detection of Cancer. H. E. Nieburg, Ed. **2**.2199–2206. Marcel Dekker. New York.

26. TSOU, K. C., B. XU(HSU), E. E. MILLER & K. W. LO. 1981. Anti-tumor effect of (S)-10-hydroxycamptothecin on mouse hepatoma DW 7756 and its possible mode of action. Anticancer Res. **1**:115–119.

27. TSOU, K. C., K. W. LO, M. A. VARELLO, B. ATKINSON, E. ZISSELMAN & H. A. WURZEL. 1981. Enzyme and cytochemical study of hairy cell leukemia. Proc. for VIII Conf. Anal. Cytology and Cytometry. (Wentworth-by-the-Sea, New Hampshire). Cytometry **2**:132.

28. TSOU, K. C. & K. W. LO. 1982. 5'-Nucleotide phosphodiesterase and liver cancer. *In* Methods in Cancer Research. H. Busch, Ed. **19**:273–300. Academic Press. New York.

29. WEISS, N. S., D. R. SZEKELY & D. J. AUSTIN. 1976. Increasing incidence of endometrial cancer in the United States. N. Eng. J. Med. **294**:1259–1262.

30. WOOD, J. C. S. & P. TODD. 1979. Analysis of cellular DNA distributors. Cell Biophys. 1:211–218.
31. YOUNG, J. L., JR., C. L. PERCY & A. J. ASIRE, Eds. 1981. Surveillance, Epidemiology and End Results: Incidence and Mortality Data, 1973–1977. National Cancer Institute Monograph No. 57 (NIH Publication No. 81-2330). p. 59. U.S. Department of Health and Human Resources.
32. TSOU, K. C., D. H. HONG, M. VARELLO, J. E. WHEELER, R. GUINTOLI, C. MANGAN & J. MIKUTA. 1984. Flow cytometric DNA and 5'-nucleotide phosphodiesterase (5'-NPD) in endometrium. Cancer 56.
33. TSENG, L. & E. GURPIDE. 1974. Estradiol and 20αdihydroprogesterone dehydrogenase activities in human endometrium during the menstrual cycle. Endocrinology 94.

Flow Cytometry of Bacteria: Cell Cycle Kinetics and Effects of Antibiotics[a]

HARALD B. STEEN, KIRSTEN SKARSTAD,
AND ERIK BOYE

Department of Biophysics
Norsk Hydro's Institute for Cancer Research
Montebello, 0310 Oslo 3, Norway

INTRODUCTION

The kinetics of the cell cycle of bacteria are poorly understood. Several basic questions, such as the relative duration of the various phases of the cell cycle, have not been answered. Flow cytophotometry, which has been of crucial importance for revealing the cell cycle of mammalian cells, has so far been applied almost exclusively to eukaryotic cells. Only a few preliminary studies of prokaryotes, and of bacteria in particular, have been published. In most of the studies of bacteria, resolution was not sufficient to provide useful information on cell cycle kinetics. The primary reason for this is that the amount of DNA in bacteria is much less than that in mammalian cells. The chromosome of *Escherichia coli* (*E. coli*), for example, has about 1,400 times less DNA than that of a diploid human cell. Hence, investigators, using powerful flow cytophotometers (FCM) with argon laser excitation, have labeled DNA with ethidium bromide or propidium iodide to obtain as bright a fluorescence as possible.[1,2] However, these dyes have significant affinity for RNA as well. Because the relative amount of RNA in bacteria is usually much larger than in mammalian cells, a large portion of the fluorescence appears to have been associated with RNA. Thus, the significance of fluorescence histograms with regard to cellular DNA content and cell cycle kinetics was quite limited.

Using a new microscope-based FCM, which has a mercury arc lamp as the excitation light source, we have been able to obtain fluorescence and light scattering histograms of *E. coli*, stained with the highly DNA-specific dye mithramycin, which has a resolution approaching that normally measured for mammalian cells. It is thus feasible to do detailed flow cytometric studies of the cell cycle of bacteria. The present paper deals with some examples of such measurements and demonstrates how they can be used to evaluate the effects of antibiotics.

[a]Supported by the Norwegian Cancer Society.

METHODS

Cell Preparation

E. coli B/rA bacteria were grown as described elsewhere[3] in batch culture or in chemostat where the growth rate is controlled by the supply of glucose. Antibiotics were added to batch cultures in exponetial growth. Aliquots of cells were withdrawn at various times and immediately injected into 10 volumes of ice-cold 70% ethanol. In this fixative cells were stable for many weeks. Fixed cells were washed and resuspended in 50 µg/ml mithramycin (Pfizer) and 25 µg/ml ethidium bromide (Calbiochem) in Tris buffer, pH = 7.4, with 25 mM $MgCl_2$ and 100 mM NaCl. When excited near the absorption maximum of mithramycin, i.e., at 436 nm, this dye combination gave about five times more fluorescence than mithramycin alone, while DNA specificity appeared to be as good as that of mithramycin.

Flow Cytometry

The FCM has been described previously.[4,5] Briefly, it is a standard inverted Epi fluorescence microscope (Diavert, Leitz Wetzlar) fitted with a 100 W high pressure mercury arc lamp (Osram HBO 100 W) as excitation light source. The flow system comprises a nozzle with hydrodynamic focusing that produces a 80 µm diameter jet in air. The jet impinges at an oblique angle on the open surface of a microscope cover glass. The flat, laminar flow thereby formed on the glass is viewed by the oil immersion microscope objective (40 × /N.A. = 1.30). Light scattering is detected in dark field by a secondary microscope mounted opposite the primary microscope. The light scattering signal has been found to be proportional to the total cellular protein content.[6] Both microscopes have an adjustable rectangular slit in the image plane that can be set to eliminate background. The high numerical aperture of the primary microscope objective, which gives high excitation intensity and high fluorescence collection efficiency, coupled with the excellent imaging quality of the microscope optics, provide a high signal-to-noise ratio.

The cells were injected into the nozzle by a motor-driven syringe at a calibrated variable rate, so that the cell density of the sample could be calculated from the measuring rate, which was typically a few thousand cells per second.

RESULTS AND DISCUSSION

Cell Cycle Kinetics

The cell cycle kinetics of bacteria appear more complex than that of mammalian cells. The cell cycle time may vary widely, i.e. for *E. coli* from about 20 minutes to many hours, depending on growth conditions. The chromosome replication time is believed to have a lower limit of about 40 minutes.[7] Hence, chromosome replication time may exceed cell cycle time significantly. This is possible because the cells are able to initiate replication of a chromosome before the previous round of replication is completed; that is, there can be a replication fork on each branch of a replication fork.

Figure 1 depicts a fluorescence (DNA content) versus light scattering (protein content) dual parameter histogram of *E. coli* harvested during rapid exponential growth. The cellular DNA content can be seen to run from about three to about six chromosome equivalents, which is in quantitative accordance with the currently accepted model for rapid bacterial growth.[7] The absence of distinct peaks also supports the assumption that DNA replication time exceeds cell cycle time. Furthermore, it appears that cellular protein content is approximately proportional to the DNA content.

The various phases of the cell cycle of slowly growing bacteria are called B, C, and D, denoting, respectively, the pre-DNA replication phase where the cells

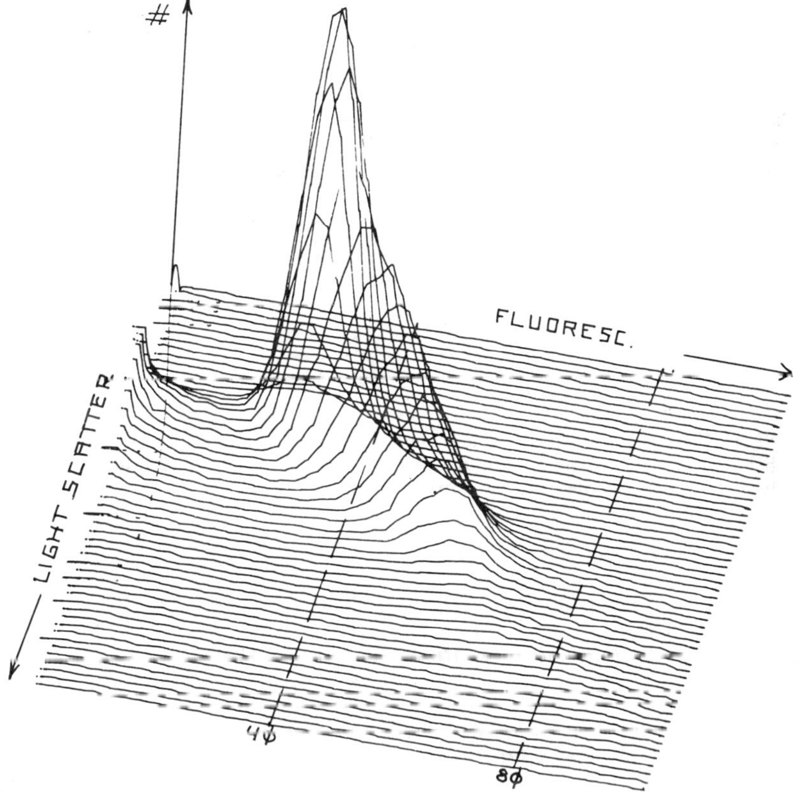

FIGURE 1. Fluorescence/light scattering histogram of *E. coli* harvested from a batch culture growing exponentially with a doubling time of 24 min. The cells were fixed in 70% ethanol and stained with a combination of mithramycin and ethidium bromide, which was excited around 436 nm. The light scattering signal is proportional to total cellular protein. φ denotes the DNA content of one chromosome.

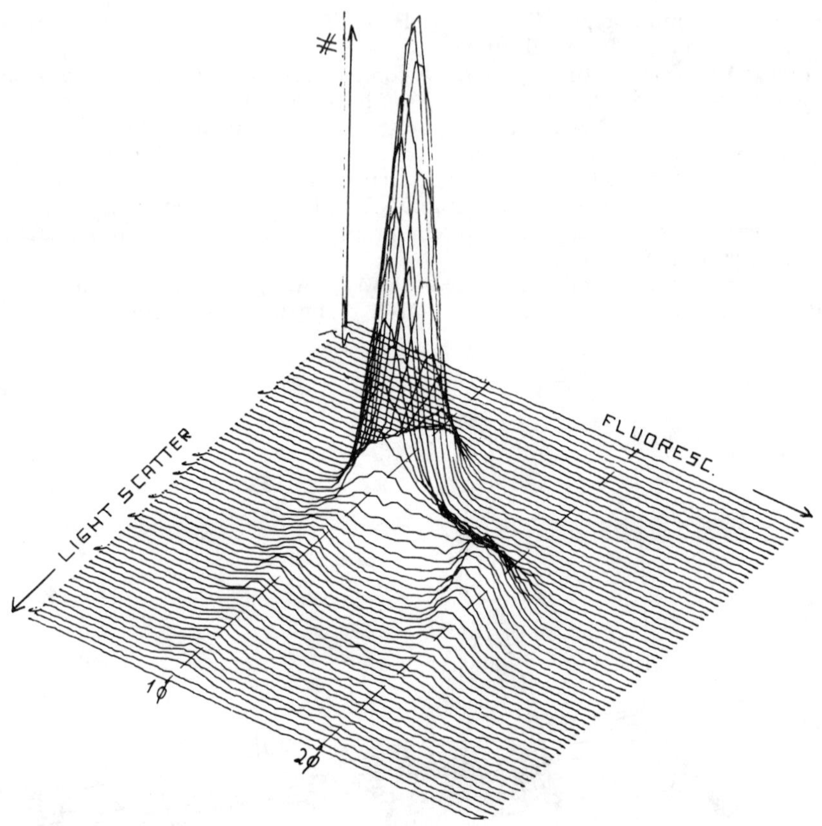

FIGURE 2. Fluorescence (DNA)/light scattering (protein) histogram of *E. coli* harvested from a chemostat culture with a doubling time of 17 hours. Measuring conditions as in Fig. 1. φ denotes the DNA content of one chromosome.

have a single chromosome, the DNA replication phase, and the post-DNA replication phase with two chromosomes per cell. The results of various studies differ widely as to the relative duration of the B and D phases.[8-10] There is also disagreement as to whether the cell must grow to a critical mass before it can enter C phase.

FIGURE 2 shows a histogram of a slowly growing chemostat culture with a cell cycle time much in excess of the chromosome replication time. As expected, the DNA content under these conditions runs between that of one and two chromosomes. An analysis of the histogram[3] shows that the duration of the B phase occupied about 80% of the cell cycle, or 13.6 hours, while the C and D phase each lasted for about 1.7 hours. Furthermore, the histogram seems to show that the cells in B phase must reach a certain size before they can enter C phase. It should be noted also that some cells grew to several times that size before entering C phase.

It may be intuitively expected that it should be easier to obtain reproducible results on relatively simple organisms, such as bacteria, than on the more complex mammalian cells. Using flow cytometry, which is able to reveal much more detail than conventional methods of cell cycle studies, we have found the opposite to be the case. In our experience *E. coli* is extremely sensitive to growth conditions. Even in the chemostat it has been difficult to maintain cultures perfectly stable with regard to cell cycle distribution. FIGURE 3 shows histograms of *E. coli* harvested at various times from a batch culture in a suboptimal medium while it was all the time under exponential growth as measured by the increase of cell number. It can be seen that the cell cycle distribution, i.e., the relative duration of the various cell cycle phases, changed significantly during this period. The most obvious explanation seems to be that under these conditions the duration of the C phase decreased significantly, probably because the cells were somehow conditioning the medium. These data throw doubt on the very basis of many studies of the bacterial cell cycle by conventional methods, namely that as long as cell doubling time remains constant the cell cycle distribution remains the same, and may thus point to an explanation for why it has proved so difficult to reach consensus on the cell cycle of slowly growing bacteria.

Effects of Antibiotics

Testing of antibiotic sensitivity is of vital importance in the treatment of some infectious diseases. Conventionally it is carried out by growing bacteria in the presence of the antibiotic(s) and assessing colony formation after one or two days. In FIGURE 4 are depicted histograms showing the effect of chloramphenicol on the DNA distribution of an *E. coli* culture. Obvious effects of the drug can be seen from DNA histograms after less than one hour in the presence of the drug.[11] After four hours the cells had accumulated in two distinct peaks representing cells with two and four chromosomes, respectively. Further exposure did not cause any further change. It appears that in the presence of the drug the cells were able to complete chromosomes under replication when the drug was given, whereas initiation of a new chromosomes was efficiently inhibited. Thus, cells having less than two chromosomes when the drug was given ended up in the two chromosome peak, while cells with more than two chromosomes ended up with four chromosomes. Cell division was also inhibited, as indicated by the small number of cells with one chromosome as well as by the total cell count. This interpretation is supported by the results depicted in FIGURE 5, which shows the result of chloramphenicol treatment of a culture growing at high rate, that is with DNA content running between three and six chromosomes. In accordance with this hypothesis the cells in this case end up with four or eight chromosomes. Chloramphenicol is known to inhibit protein synthesis at the ribosomal level.[12] Hence, the present results support reports claiming that *de novo* protein synthesis is required for initiation of chromosome replication[13] as well as for cell division.[14] The stagnation of protein synthesis is evident by comparing FIGURES 1 and 5. In fact, the cell mass of the drug-treated cells can be seen to be lower than when the drug was administered, demonstrating negative net protein synthesis.

Penicillin is known to inhibit cell wall synthesis and cell separation. FIGURE 6 shows histograms of *E. coli* harvested at different times from a culture grown in the presence of penicillin. It can be seen that cells grew to very large sizes and high DNA content, indicating filamentation. An increasing number of counts

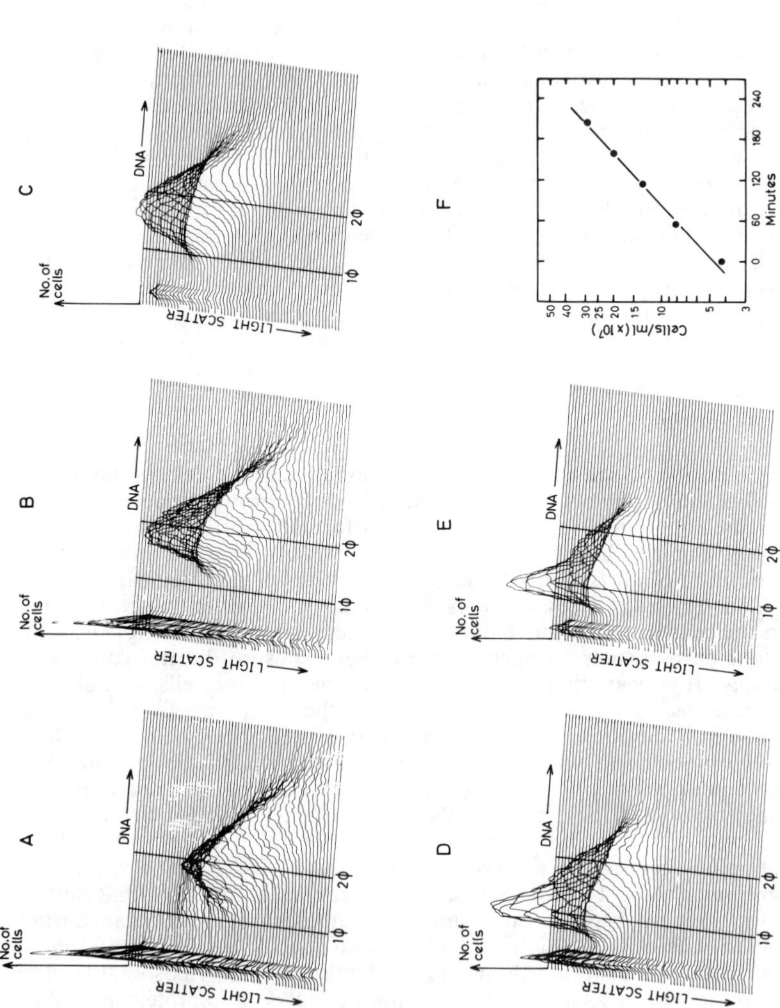

FIGURE 3. Fluorescence (DNA)/light scattering (protein) histograms of *E. coli* harvested from a batch culture growing in a proline medium[3] with a doubling time of 78 min. Samples B, C, D, and E were taken 55, 115, 160, and 250 min, respectively, after sample A. (F) growth curve covering the same period of time for the same culture. φ denotes the DNA content of one chromosome.

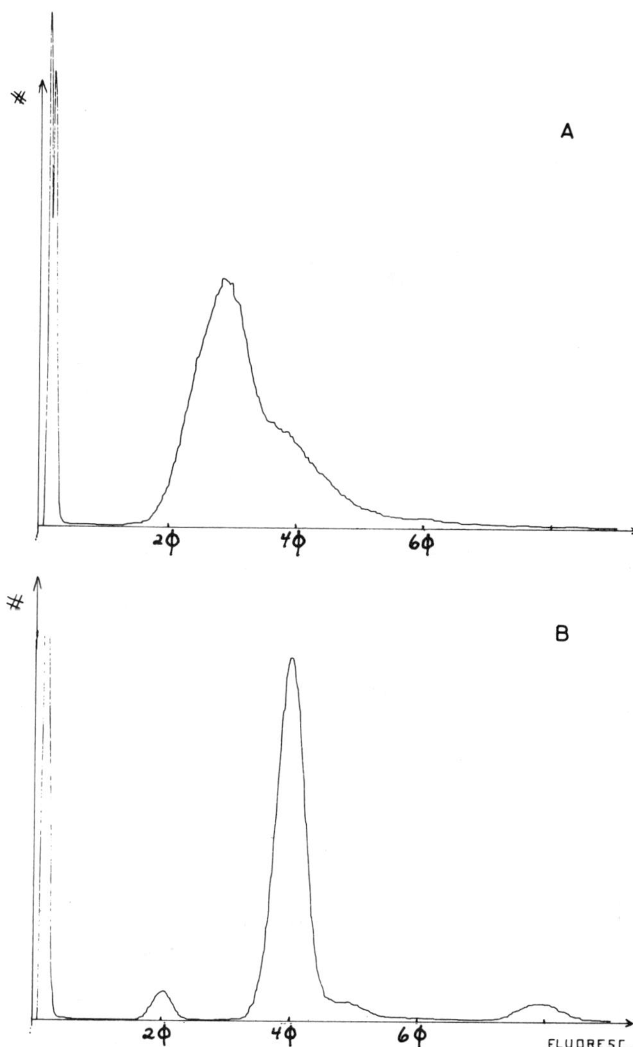

FIGURE 1. Fluorescence (DNA) histograms of *E. coli* harvested from batch culture (A) During exponential growth with doubling time of 28 min. (B) After 4 hours in the presence of 50 µg/ml chloramphenicol. φ denotes the DNA content of one chromosome. The width of the main peak in B corresponds to a cv of about 5%.

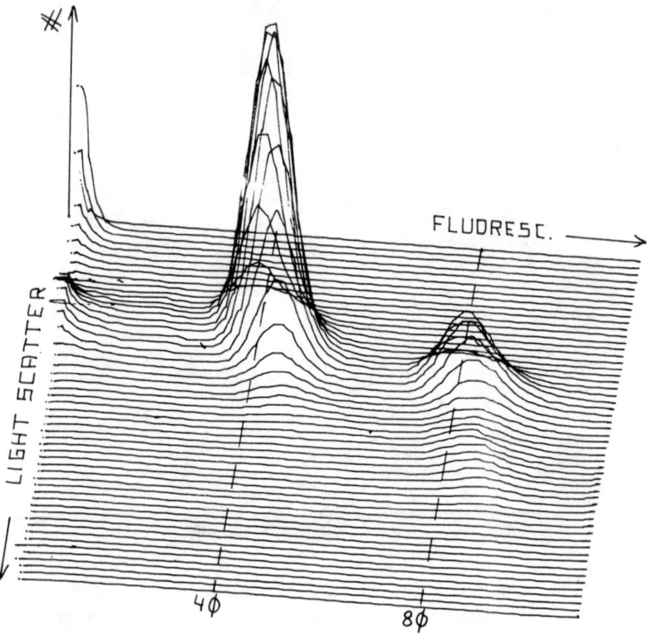

FIGURE 5. Fluorescence (DNA)/light scattering (protein) histogram of a rapidly growing *E. coli* culture (doubling time 24 min) 4 hours after it was given 50 µg/ml chloramphenicol. This histogram may be compared with that of FIGURE 1, which represents cells harvested at the time when the drug was given. φ denotes the DNA content of one chromosome.

close to zero represents dead and decaying cells. After four hours essentially all cells appear to be dead.

These data exemplify how effects of antibiotics can be assessed by means of flow cytometry in much more detail and in a much shorter time than is possible by conventional clinical methods.

Corresponding data for several other antibiotics have been published elsewhere.[6,15]

CONCLUSION

Flow cytophotometry of bacteria is technically feasible using a simple FCM with arc lamp excitation. DNA histograms with a resolution approaching that typically recorded for mammalian cells are readily obtained. Thus, it has become possible to do detailed studies of the cell cycle distribution of asynchronous cultures of bacteria and thereby answer several basic questions about the cell cycle kinetics of such cells.

By means of flow cytometry, effects of antibiotics on the cell cycle distribution and cell numbers of bacteria can be observed within a few hours of culture. Drug

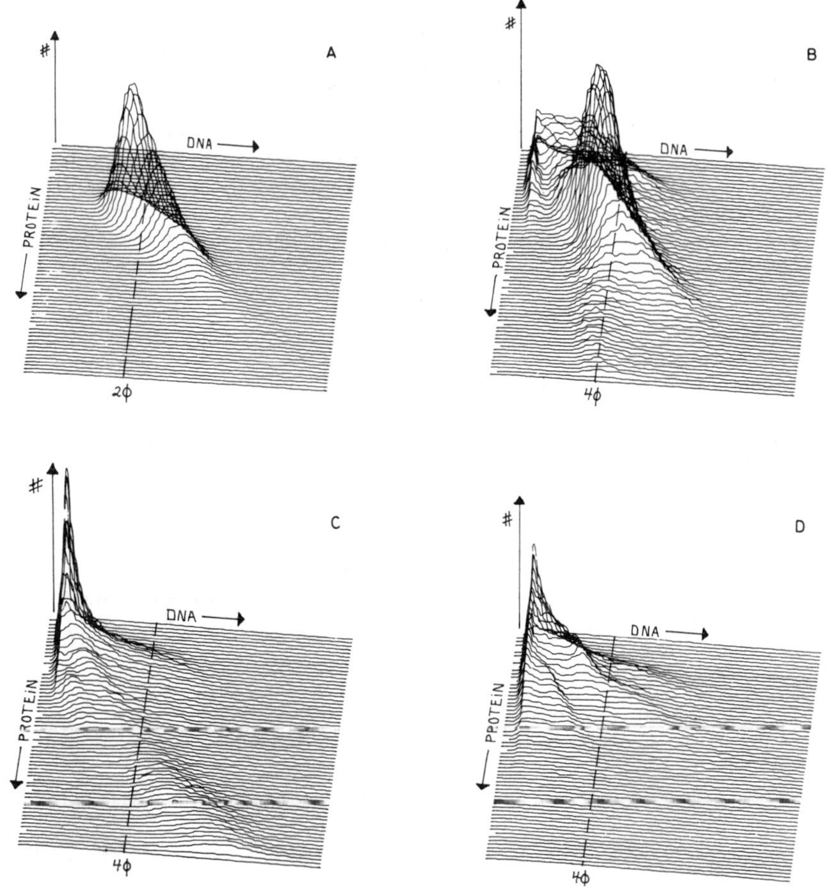

FIGURE 6. Fluorescence (DNA)/light scattering (protein) histogram of *E. coli* grown in optimal medium with 5 μg/ml benzylpenicillin for (A) 0 hours, (B) 1 hour, (C) 2 hours, and (D) 4 hours.

testing can thus be carried out much faster than by conventional clinical methods. In addition flow cytometry yields valuable information with regard to what processes are particularly affected by the drug.

SUMMARY

Flow cytometric determination of the DNA and protein content of *E. coli* has been carried out by means of a microscope-based flow cytophotometer with a high pressure arc lamp excitation light source. Fluorescence (DNA)/light scatter (total cell protein) dual parameter histograms with a resolution cv of 5% were obtained for cells labeled with a combination of mithramycin and ethidium bromide. Histograms of *E. coli* in rapid and slow exponential growth are presented to exemplify how the cell cycle kinetics of bacteria can be studied in much more detail than has been possible by other methods.

Significant effects of chloramphenicol and penicillin on the cell cycle distribution and cell numbers of *E. coli* cultures were evident after one hour of culture. The data provided information on which parts of the cell cycle and what types of processes were affected by the drug. It appears that flow cytometry may become a valuable tool in studies of the cell cycle of bacteria as well as in clinical drug testing.

REFERENCES

1. BAILEY, J. E., J. FAZEL-MADJLESSI, D. N. McQUITTY, L. Y. LEE, J. C. ALLRED & J. A. ORO. 1977. Science **198**:1175–1176.
2. PAAU, A. S., J. R. COWLES & J. A. ORO. 1977. Can J. Microbiol. **23**:1165–1169.
3. SKARSTAD, K., H. B. STEEN & E. BOYE. 1983. J. Bacteriol **154**:656–662.
4. STEEN, H. B. & T. LINDMO. 1979. Science **204**:403–404.
5. STEEN, H. B. 1980. Cytometry **1**:26–31.
6. BOYE, E., H. B. STEEN & K. SKARSTAD. 1983. J. Gen. Microbiol. **129**:973–980.
7. COOPER, S. & C. E. HELMSTETTER. 1968. J. Mol. Biol. **31**:519–540.
8. HELMSTETTER, C. E. & O. PIERUCCI. 1976. J. Mol. Biol. **102**:477–486.
9. KOPPES, L. J. H., C. L. WOLDRINGH & N. NANNINGA. 1978. J. Bacteriol. **134**:423–433.
10. KUBITSCHEK, H. E. & C. N. NEWMAN. 1978. J. Bacteriol. **136**:179–190.
11. STEEN, H. B. & E. BOYE. 1980. Cytometry **1**:32–36.
12. VAZQUEZ, D. 1966. Biochim. Biophys. Acta **114**:289–295.
13. JONES, N. C. & W. D. DONACHIE. 1973. Nature New Biol. **243**:100–103.
14. LARK, K. G. 1969. Ann. Rev. Biochem. **38**:569–604.
15. STEEN, H. B., E. BOYE, K. SKARSTAD, B. BLOOM, T. GODAL & S. MUSTAFA. 1982. Cytometry **2**:249–257.

Flow Karyology of Neoplastic Human Fibroblasts[a]

M. BARTHOLDI, G. TRAVIS, AND L. S. CRAM

Experimental Pathology Group
Los Alamos National Laboratory
Los Alamos, New Mexico 87545

P. PORRECA AND J. LEAVITT

Linus Pauling Institute of Science and Medicine
Palo Alto, California 94306

INTRODUCTION

Chromosomal changes play an important role in malignant transformation as demonstrated by the proximity of oncogenes to common translocations in Burkitt's lymphoma[1] and chronic myelocytic leukemia.[2] More generally, the process of karyotype instability with gross aneuploidy occurs in solid tumors. In 1914, Boveri suggested that chromosomal change causes cancer.[3] Testing this hypothesis has proven difficult, in part, due to technical problems in the analysis of chromosome rearrangements in tumor cells.

Chromosome banding methods, and more recently the technology of flow cytometry, can be applied in the analysis of the relationship between chromosomal changes and malignancy. Flow cytometry has both an analytical capability, in which the distribution of chromosome types in a cell population and their relative abundance can be rapidly determined, and a sorting capability in which single chromosome types can be purified for molecular analysis.[4]

Flow karyology has been effectively used to study karyotype instability in spontaneously transforming Chinese hamster cell strains and lines.[5] We have found that aneuploidy preceded the appearance of the tumorigenic phenotype as tested in nude mice as well as *in vitro* indicators of neoplasia. Additional chromosome changes continued to appear during neoplastic progression.[5]

Flow karyotype analysis, combined with Giemsa-banding (G-banding), provided the capability to detect chromosomal abnormalities early in the neoplastic process. The two techniques are complementary; flow karyology can rapidly characterize the range of chromosome types in a cell population and determine the extent of deviation from euploidy, whereas banding methods provide a detailed analysis of chromosome translocations. For example, in the study of Chinese hamster cell strains, a new peak was detected by flow karyology indicating that a chromosome aberration had arisen at a frequency of one in four cells.[6] By G-banding analysis, the chromosome giving rise to this peak was identified as chromosome 3 with an insertion of a single band in the long arm. However, considerable cell-to-cell variability in banding obscured the uniformity

[a]This work was performed under the auspices of the Los Alamos National Flow Cytometry and Sorting Research Resource funded by the Division of Research Resources of the National Institutes of Health (Grant P41-RR01315-02) and the Department of Energy.

of the insertion and its frequency of occurrence, both of which were readily determined from the flow histogram.

In the present study, the chromosomal changes accompanying incremental tumorigenicity in a series of neoplastic human fibroblast cell lines were analyzed by G-banding and high resolution flow cytometry. The parental human cell strain, termed KD cells, is diploid, of stable phenotype, and fails to transform or develop an unstable aneuploid karyotype spontaneously.[7] After treatment of KD cells with the chemical carcinogen, 4-nitro-quinolin-1-oxide, a number of neoplastic cell lines were obtained that showed high saturation density and the ability to grow in soft agar.[7] One of these neoplastic cell lines, termed HUT 14, produced solid tumors when injected subcutaneously into nude (athymic) mice. A substrain of HUT 14, termed HUT 14T, was derived from one of these tumors.[8]

A striking finding in the analysis of differences in gene expression between KD cells and HUT 14 cells was the expression of a mutant β-actin along with nearly an equal amount of normal β-actin in HUT 14 cells.[8,9] Furthermore, cells of the substrain HUT 14T expressed a more variant form of β-actin at twice the rate of normal β-actin. The second mutation in β-actin may have occurred during subcloning of HUT 14, during which a spontaneous 6-thioguanine–resistant subclone arose and ouabain resistance was induced by ultraviolet light. This selected subclone, termed HUT 14 uv cl 1 was actually used to produce the tumors from which HUT 14T was derived.

The tumorigenic potential of HUT 14, HUT 14 uv cl 1, and HUT 14T progressed through three distinct levels during the series of subcloning events.[9] Other phenotypic changes studied were: cytoplasmic structures that were found to decrease in order from linear actin cables in KD cells to diffuse actin in HUT 14T cells; and decreasing amounts of fibronectin in KD, HUT 14, and HUT 14T cells.

The HUT 14 and HUT 14T cells were near diploid but with marker chromosomes; the human origin of the tumor cells was confirmed by karyotype analysis.[7] In order to characterize the extent of chromosome change in more detail, and in particular to determine if an increase in number of a chromosome from two to three copies occurred that might be correlated with the increased rate of synthesis of mutant β-actin in HUT 14T cells, the chromosomes of each subclone were analyzed by high resolution flow cytometry.

We found that, although each subclone remained near diploid in DNA content and chromosome number, many chromosomal changes occurred. The chromosome changes detected by flow karyology were confirmed by banding analysis. Interestingly, an increase in the number of chromosome 7 from two to three copies from HUT 14 cells to HUT 14T cells was detected, but association of this particular chromosome change with the increased expression of mutant β-actin gene depends on chromosomal mapping of the active β-actin gene.

MATERIALS AND METHODS

Cell Culture

The origin of KD cells and the derivation of HUT 14T cells are described in references 7, 8, and 9. The fibroblast cell cultures were maintained in Dulbecco's medium with 10% fetal bovine serum at 37°C and 5% CO_2. The cultures were maintained in exponential growth and split 1:8 every 6 days.

Chromosome Preparation

Isolated metaphase chromosomes were prepared for flow karyology using hypotonic swelling of mitotic cells in the presence of propidium iodide.[5,10] Cells were blocked in mitosis by Colcemid, 0.1 µg/ml, for 16 hr. Cells released into the culture medium after a first shake-off were discarded. Following a second three-hour block and shake-off, the mitotic cells were resuspended in 1.0 ml of 75 mM KCl and 50 µg/ml propidium iodide. The cells were allowed to swell for 10 min at room temperature and 0.5 ml of the following solution was added: 75 mM KCl, 0.1% Triton X-100, 50 µg/ml propidium iodide, and 1.0 mg/ml RNase. After 3 min at room temperature the cells were syringed through a 1½ inch 22-gauge needle. The chromosome suspension was incubated for 30 min at 37°C and usually stored for less than 24 hr at 4°C before flow analysis.

Flow Cytometry

A flow cytometer[11,12] specially designed for analysis of chromosomes was used to measure HUT 14 flow karyotypes. The innovative features of this cytometer are: (a) a high power laser beam (2.0 watts at 488 nm) focused to a small illumination spot (2 µm), and (b) a triple nozzle flow chamber with a motor-driven syringe controlling the sample stream. This combination of features results in an instrument capable of very precise measurements of fluorescent intensities from individual chromosomes at rates up to 500/sec.

The high precision (1.0% coefficient of variation) resolves chromosomes differing by small amounts in fluorescent intensity. Propidium iodide intercalates in DNA without base specificity, and the fluorescent intensity from a particular chromosome type can be directly related to its relative DNA content.

Data Analysis

The flow karyotype is a histogram of the number of chromosomes in each peak resolved by a distinct fluorescent intensity. By this technique, the distribution of chromosome types within a cell population can be readily determined, the relative frequency of occurrence of each type can be measured, and the relative DNA content can be quantified.

The identification of the chromosome type(s) in each peak of the histogram is done by correlating the relative position of the peaks with chromosome size and DNA content measured by microscopy[13] or by banding sorted chromosomes.[4] Identification of aberrant chromosomes was done by correlating G-banding analysis with flow cytometric analysis. The relative peak positions of the normal human chromosomes may shift slightly due to homolog variations (polymorphisms) or variable heterochromatin content.[14,15]

Numerical analysis of the flow histograms is done by computer analysis in which the entire histogram is fit with a series of Gaussian distributions, one for each peak, which are free to vary in mode, width, and area, and an additional polynomial function fit to the underlying debris background.[16] The number of chromosomes counted in each peak is reduced to a relative frequency of occurrence as a percentage of the total number of chromosomes analyzed, and the mean peak values are reduced to relative DNA content values. An abnormal peak position indicates the presence of a marker or translocated chromosome, and an abnormal peak area indicates the loss or gain of a chromosome and the fraction of cells in the population in which the change occurred.

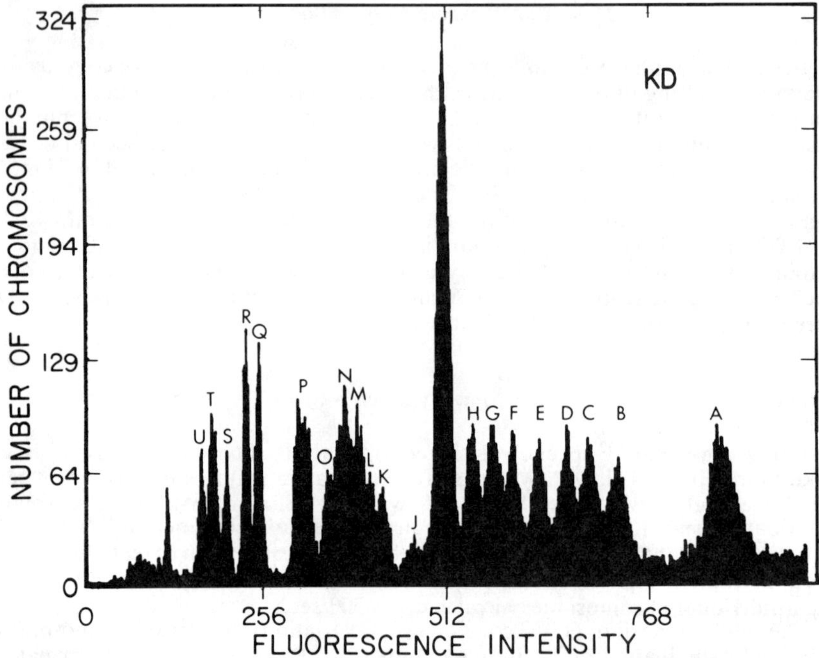

FIGURE 1. Flow histogram of chromosomes from KD cells. 50,000 isolated metaphase chromosomes were analyzed by flow cytometry at 100/sec. Chromosomes were stained with propidium iodide.

Three separate chromosome preparations were done for the KD cells and each subclone. At least two analyses were done on each preparation. The standard deviations used in the results to assess significant change were computed from six separate histograms. A difference larger than one standard deviation was required to determine if a significant chromosomal change between subclones had occurred.

Giemsa Banding

Metaphase cells were prepared for Giemsa banding analysis by fixation in methanol, acetic acid (3:1), spread on cold, wet slides, and air dried. The slides were treated with 0.05% trypsin for varying durations, up to 5 min, rinsed successively in 70, 70, 95, and 100% ethanol, air dried, and stained for 2 to 5 min in 2% Giemsa (G. T. Gurr, London, England). Banding patterns were identified according to the Paris Conference (1972).

RESULTS

KD Cells

The KD cells are diploid, 46 XX, with no chromosome abnormalities detected by G-banding analysis. The flow histogram of chromosomes from the KD cells

(FIGURE 1) contains 21 peaks. (The leftmost peak is from microspheres.) The peaks, lettered from right to left, A to U, are listed in TABLE 1 with the relative frequency of occurrence of chromosomes in each peak reduced to the number of chromosomes per cell (multiplied by 46), and the relative peak position reduced to a percentage of the autosomal cellular DNA content (determined by adding each of the peak means for the autosomes, multiplied by the number of autosomes in each peak). The relative DNA content of each chromosome type agrees well with those determined by absorption microscopy[13] and are more precise then those determined by previous flow analyses.[4,14] The agreement of the relative DNA content values with microscopy confirms the identity of the chromosomes in each peak.

Chromosomes that were not resolved as single peaks, but overlapped each other are: chromosomes 1 and 2; 9, 10, 11, and 12; 15 and 16; and 17 and 18. Homolog variation (polymorphism) was detected in chromosomes 9, 13, 16, and 21 by flow karyology. In the G-banding analysis, a difference between the chromosome 9 homologs was seen, but homolog differences in chromosomes 13, 16, and 21 were difficult to confirm. In this study, the pattern of flow karyotype from the KD cells was used as a baseline against which chromosomal changes in the derived subclones could be detected.

HUT 14 Cells

The HUT 14 and HUT 14T cells were each near diploid, with modal chromosome numbers of 44 and 45, respectively, and with several marker

TABLE 1. Chromosome Frequency and Relative DNA Content in KD Cells

Peak[a]	Chromosomes Per Cell[b]	DNA Content[c]	Chromosome
A	4.1 ± .1	4.22 ± .02	1,2
B	2.1 ± .1	3.52 ± .01	3
C	2.0 ± .1	3.34 ± .02	4
D	2.1 ± .2	3.18 ± .02	5
E	1.9 ± .1	3.00 ± .02	6
F	1.9 ± .2	2.82 ± .02	7
G	2.1 ± .1	2.70 ± .02	X
H	1.9 ± .2	2.56 ± .01	8
I	7.9 ± .9	2.36 ± .01	9a,10,11,12
J	0.7 ± .1	2.15 ± .01	9b
K	1.0 ± .1	1.96 ± .01	13a
L	0.9 ± .2	1.87 + .01	13b
M	2.0 ± .1	1.79 ± .01	14
N	3.0 ± .2	1.68 ± .01	15,16a
O	1.2 ± .2	1.55 ± .02	16b
P	3.2 ± .4	1.42 ± .01	17,18
Q	2.1 ± .2	1.14 ± .01	20
R	2.0 ± .2	1.04 ± .01	19
S	1.0 ± .2	0.92 ± .02	21a
T	2.0 ± .5	0.82 ± .01	22
U	1.0 ± .2	0.75 + .01	21b

[a]Peaks from FIGURE 1 lettered from right to left.
[b]Relative frequency of chromosomes in each peak multiplied by 46.
[c]Percentage of total autosomal DNA content.

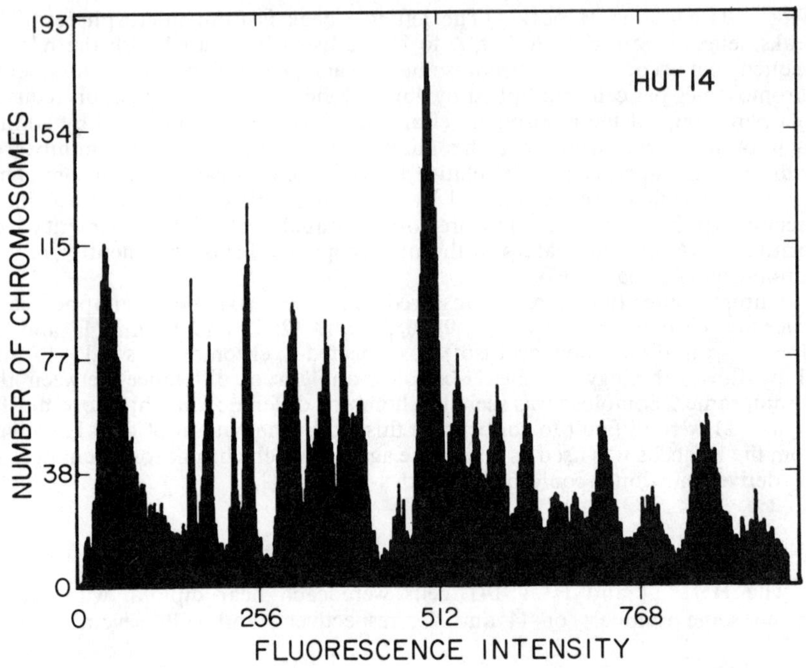

FIGURE 2. Flow histogram of chromosomes from HUT 14 cells. Details as in FIGURE 1.

chromosomes detected by banding analysis. The total cellular DNA content of KD, HUT 14, and HUT 14T cells were equal as measured by flow cytometry (data not shown).

The flow karyotype of HUT 14 cells (FIGURE 2) exhibits a significantly altered pattern from that of the KD cells. Analysis of the histogram was done as for the KD cells. Peaks containing unchanged chromosomes were identified by determining the relative DNA content as a fraction of the normal (KD) autosomal cellular DNA content. For example, the second peak from the right in the HUT 14 karyotype contains chromosomes with 4.21% of the autosomal DNA content. This value agrees with the rightmost peak in KD flow karyotype (4.22%) and identifies this peak in HUT 14 as chromosomes 1 and 2. This peak also represents chromosomes occurring at a relative frequency of 4 (3.8 ± .4) chromosomes per cell indicating that there was no loss or gain of chromosomes 1 and 2.

The peaks to the right and left of the chromosome 1 and 2 peak in HUT 14 have relative DNA content values of 4.51% and 3.85% and contain 1 (1.1 ± .1 and 1.2 ± .2) chromosome per cell. These values do not correspond to any peaks in the KD flow karyotype.

On this basis, and on the basis of length measurements of G-banded marker chromosomes, the chromosomes in these peaks were identified as marker chromosomes M4 and M6. Also, marker chromosomes M2, M3, and M5 were placed in peaks containing the normal chromosomes 13b, 9–12, and 17–18. The assignment of marker M3 to the 9–12 peak is ambiguous, given the standard deviation of nearly one chromosome copy with a mean value near 8 (see peak I,

TABLE 2. Summary of Chromosome Changes

Chromosome	Relative Frequency of Chromosomes Per Cell			
	KD	HUT 14	HUT 14 uv cl 1	HUT 14T
M4	0	1.1[a]	1.1	1.0
1,2	4.1	3.8	4.0	4.2
M6,4p$^+$	0	1.2[a]	1.4	2.3[a]
3	2.1	2.2	2.4	2.5
4	2.0	1.0[a]	1.0	0.0[a]
5	2.1	1.0[a]	1.2	0.9
6	1.9	2.3	2.3	2.4
7	1.9	2.4	2.4	3.2[a]
X	2.1	1.5[a]	1.2	1.1
8	1.9	2.6	2.6	2.7
9a,10,11,12,M3	7.9	8.3	7.5	8.0
9b	0.7	0.8	1.1	0.9
13a	1.0	0.0[a]	0.0	0.0
13b,M2	0.9	2.0[a]	1.9	1.8
14	2.0	2.3	2.1	1.7
15,16a	3.0	2.5	3.1	2.4
16b	1.2	1.2	1.0	1.1
17,18,M5	3.2	3.9[a]	3.9	4.5[a]
20	2.0	2.3	1.9	1.4[a]
19	2.0	0.8[a]	1.0	0.8
21a	1.0	0.0[a]	0.0	0.0
22	2.0	1.8	1.9	1.4[a]
21b	1.0	1.1	0.9	0.7
M1	0.0	0.0	0.0	0.7[a]

[a]Significant changes detected between KD and HUT 14 and between HUT 14T and HUT 14.

TABLE 1). A similar degree of counting error occurs in the peak containing chromosomes 17–18 and M5, in which only three chromosomes are counted in the KD cells and only four chromosomes in the HUT 14 cells, indicating loss of a copy of either chromosome 17 or 18 that was not confirmed by banding analysis. However, five instances of loss of a single copy of normal chromosome homologs were reliably detected by flow cytometry that included chromosomes 4, 5, X, 13a, and 19 (summarized in TABLE 2). Thus, when chromosome types are not clearly resolved by flow cytometry but fall in overlapping peaks, the error in counting may be as large as one chromosome per cell. But when single chromosome types are well resolved, the error is reduced to 0.1 chromosomes per cell (TABLE 1).

The flow karyotype of HUT 14 uv cl 1 cells (FIGURE 3) shows no significant chromosomal change from the HUT 14 cells. Confirming that the karyotypes of the two cells strains were identical was difficult by G-banding given the cell-to-cell variability in the appearance of the marker chromosomes.

The flow karyotype of HUT 14T cells is shown in FIGURE 4. Seven chromosome changes were detected as significantly different from the HUT 14 cells (identified in TABLE 2). Another marker chromosome, M1, appeared as the smallest chromosome. The remaining normal chromosome 4 homolog was lost. A 4p$^+$ chromosome, identified by G-banding, overlapped marker chromosome M6 in the flow histogram. On the basis of the flow karyotype alone, the assumption that the copy number of M6 increased from 1 to 2, would have been

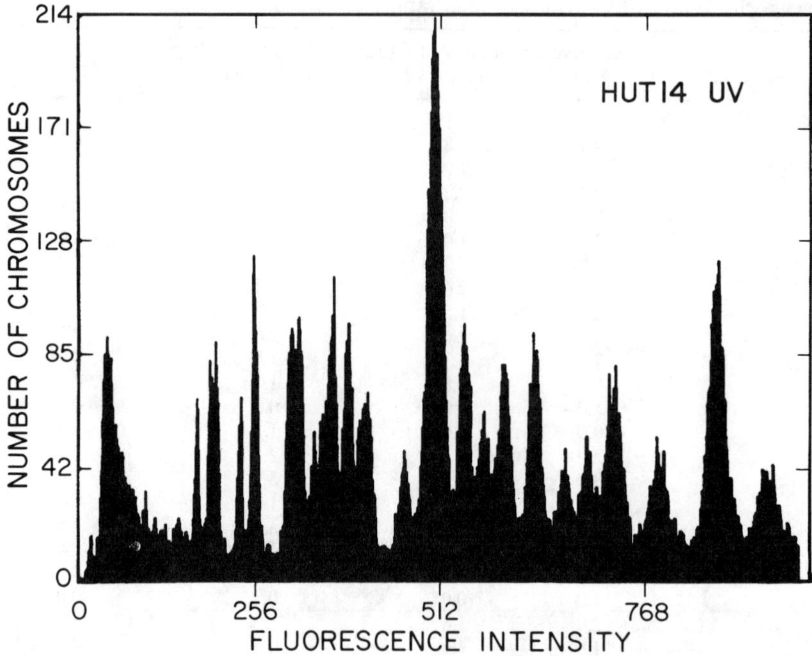

FIGURE 3. Flow histogram of chromosomes from HUT 14 uv cl 1 cells. Details as in FIGURE 1.

considered. Three chromosomes, 22, 20, and one in the 17–18, M5 peak, appeared to alter their relative frequency of occurrence. Chromosome 7 was measured to increase from a copy number of 2 to 3 (TABLE 2). Decreases in the relative frequency of chromosomes 20 and 22, and an increase in the frequency of chromosome 7, were also seen in the banding analysis.

The constitution of the marker chromosomes, as identified by G-banding analysis was: M1 contained material from chromosomes 21 and 13; M2 contained material from chromosomes 4 and 5; M3 contained material from chromosome 19; M4 contained material from chromosome 5; M5 contained material from chromosome 21; and M6 contained material from chromosome 9.

DISCUSSION

The role of chromosomal change in neoplasia and malignancy can be studied by establishing correlations between chromosome aberration and gene expression.[1,2] The HUT 14 cells expressed a mutant β-actin, and the rate of synthesis of mutant β-actin doubled in the HUT 14T cells. The tumorigenic potential of the HUT 14T cells was also increased. Possibly, trisomy of chromosome 7 detected in the HUT 14T cells doubled the gene number of the mutant β-actin allele and accounts for the increased synthesis of β-actin. Chromosomal gene mapping of the actin gene

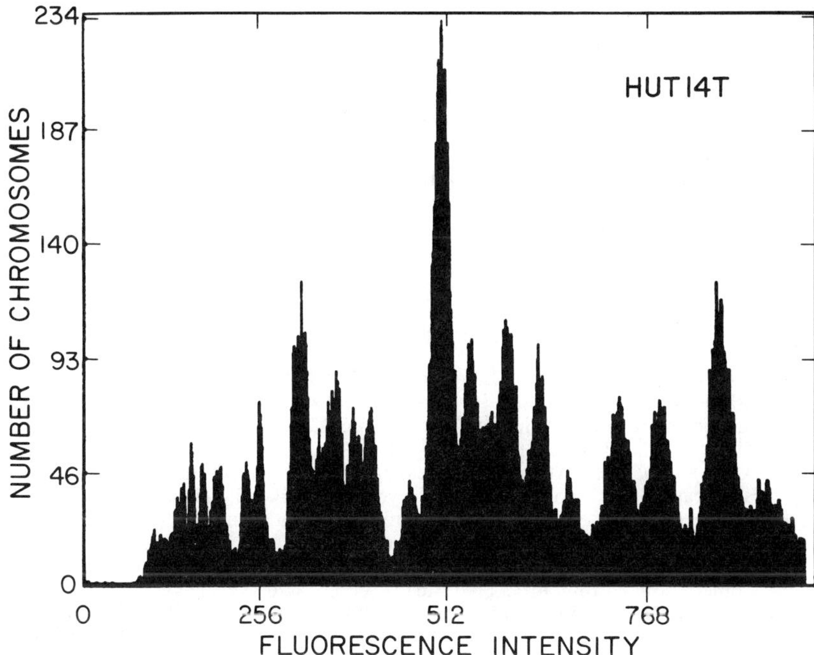

FIGURE 4. Flow histogram of chromosomes from HUT 14T cells. Details as in FIGURE 1.

is needed to firmly establish the correlation. Other chromosome changes also occurred that may influence the synthesis of β-actin, or the defect may lie elsewhere.

The utility of flow karyology to characterize cell populations with aneuploid chromosome complements is that deviations from euploidy can be readily detected. Comparison of the patterns of flow histograms obtained at different stages of progression complements banding analysis in that chromosome changes within cell populations can be monitored without detailed microscopic observation. Flow karyotypes obtained from highly heteroploid Chinese hamster ovary (CHO) cell lines indicate that cell-to-cell karyotype diversity can be due to variable copy number of individual chromosomes in each cell, and not to the occurrence of structurally unique chromosomes.[17] Perhaps the chaotic chromosome complement often found in tumor cells can be analyzed by flow cytometry for outstanding recurrent chromosome abnormalities. Also, the process of karyotype instability can be followed by comparison of flow histograms at different stages of tumor progression.[5]

Currently, flow analysis of chromosomes requires at least 5,000 mitotic cells for preparation and sample handling. However, analysis of as few as 500 mitotic cells should be feasible (G. van den Engh, personal communication) and would be a minimum for significant histograms. This number would require only short term culture of tumor specimens. Recent developments in both resolution of human chromosomes by flow cytometry[18] and in the purification of human chromosomes for the construction of chromosome-specific DNA libraries[19] highlight the current utility and future potential of flow cytometry in the study of chromosomes and cancer.

ACKNOWLEDGMENT

Dr. Brian Crawford, Genetics Group, Los Alamos National Laboratory provided valuable assistance and advice.

[NOTE ADDED IN PROOF: The actin gene has now been mapped to human chromosome 7.]

REFERENCES

1. DALLA-FAVERA, R., M. BREGNI, J. ERIKSON, D. PATTERSON, R. C. GALLO & C. M. CROCE. 1982. Human c-*myc onc*gene is located on the region of chromosome 8 that is translocated in Burkitt lymphoma cells. Proc. Natl. Acad. Sci. USA **79**:7824–7827.
2. HEISTERKAMP, N., J. R. STEPHENSON, J. GROFFEN, P. F. HANSEN A. deKLEIN, C. R. BARTRAM & G. GROSVELD. 1983. Localization of the c-*abl* oncogene adjacent to a translocation break point in chronic myelocytic leukemia. Nature **306**:239–242.
3. BOVERI, T. 1914. Zur Frage der Entwichlung malignen Tumoren. Gustav. Jena.
4. CARRANO, A. V., J. W. GRAY, R. G. LANGLOIS, K. J. BURKHART-SCHULTZ & M. A. VAN DILLA. 1979. Measurement and purification of human chromosomes by flow cytometry and sorting. Proc. Natl. Acad. Sci. USA **76**:1382–1384.
5. CRAM, L. S., M. F. BARTHOLDI, F. A. RAY, G. L. TRAVIS & P. M. KRAEMER. 1983. Spontaneous neoplastic evolution of Chinese hamster cells in culture: multistep progression of karyotype. Cancer Res. **43**:4828–4837.
6. CRAM, L. S., M. F. BARTHOLDI, F. A. RAY, G. L. TRAVIS, J. H. JETT & P. M. KRAEMER. 1983. Quantitation of one aspect of karyotype instability associated with neoplastic transformation in Chinese hamster cells. Prog. Nucleic Acid Res. **29**:39–42.
7. KAKUNAGA, T. 1978. Neoplastic transformation of human diploid fibroblast cells by chemical carcinogens. Proc. Natl. Acad. Sci. USA **75**:1334–1337.
8. LEAVITT, J., D. GOLDMAN, C. MERRILL & T. KAKUNAGA. 1982. Changes in gene expression accompanying chemically-induced malignant transformation of human fibroblasts. Carcinogenesis **3**:61–70.
9. LEAVITT, J., G. BUSHAR, T. KAKUNAGA, H. HAMADA, T. HIROKAWA, D. GOLDMAN & C. MERRILL. 1982. Variations in expression of mutant β-actin accompanying incremental increases in human fibroblast tumorigenicity. Cell **28**:259–268.
10. ATEN, J. A., J. B. A. KIPP & G. W. BARENDSEN. 1980. Flow-cytometric determination of damage to chromosomes from x-irradiated Chinese hamster cells. *In* Flow Cytometry IV. O. D. Laerum, T. Lindmo & E. Thorud, Eds.:287–292. Universitetsforlaget. Oslo.
11. CRAM, L. S., D. J. ARNDT-JOVIN, B. G. GRIMWADE & T. M. JOVIN. 1980. One-dimensional image analysis of microspheres and Chinese hamster cells in a flow cytometer/flow sorter. *In* Flow Cytometry IV. O. D. Laerum, T. Lindmo & E. Thorud, Eds.:251–259. Universitetsforlaget. Oslo.
12. BARTHOLDI, M. F., D. C. SINCLAIR & L. S. CRAM. 1983. Chromosome analysis by high illumination flow cytometry. Cytometry **3**:395–401.
13. MENDELSOHN, M. L., B. H. MAYALL, E. BOGART, D. H. MOORE & B. H. PERRY. 1973. DNA content and DNA-based centromeric index of the 24 human chromosomes. Science **179**:1126–1129.
14. YOUNG, B. D., M. A. FERGUSON-SMITH, R. SILLAR & E. BOYD. 1981. High resolution analysis of human peripheral lymphocyte chromosomes by flow cytometry. Proc. Natl. Acad. Sci. USA **78**:7727–7731.
15. LAGNLOIS, R. L., L. C. YU, J. W. GRAY & A. V. CARRANO. 1982. Quantitative karyotyping of human chromosomes by dual beam flow cytometry. Proc. Natl. Acad. Sci. USA **79**:7876–7880.

16. BARTHOLDI, M. F., F. A. RAY, L. S. CRAM & P. M. KRAEMER. 1984. Flow karyology of serially cultured Chinese hamster cell lineages. Cytometry **5:**534–538.
17. DEAVEN, L. L., E. W. CAMPBELL & M. F. BARTHOLDI. 1983. On the nature of stem lines in tumorigenic cell populations. *In* Cancer: Etiology and Prevention. R. G. Grispen, Ed.:61–70 Elsevier. New York.
18. MEYNE, J., M. F. BARTHOLDI, G. L. TRAVIS & L. S. CRAM. 1985. Counterstaining human chromosomes for flow karyology. Cytometry **5:**580–583.
19. LALANDE, M., L. M. KUNKEL, A. FLINT & S. A. LATT. 1984. Development and use of metaphase chromosome flow sorting methodology to obtain recombinant phage libraries enriched for parts of the human X chromosome. Cytometry **5:**101–107.

Male Germ Cell Analysis by Flow Cytometry: Effects of Cancer, Chemotherapy, and Other Factors on Testicular Function and Sperm Chromatin Structure[a]

D. P. EVENSON

Chemistry Department
South Dakota State University
Brookings, South Dakota 57007

INTRODUCTION

Human male fertility is possible only if the complex and multifaceted process of morphological and biochemical germ cell differentiation is successfully completed leading to mature sperm. This process is quite sensitive to environmental factors; testicular exposure to a broad array of agents will lead to reduced sperm production often with a concomitant reduction of sperm quality.

Current clinical evaluation of semen quality is primarily dependent on light microscope measurements to measure cell concentration, motility index, and morphology. These indicators are very useful in a routine evaluation of male fertility potential. However, fertility is a complex phenomenon and is dependent on many additional factors. In this context, flow cytometry methods are being developed to measure more rapidly some of the same parameters currently obtained by light microscopy. In addition, other assays are being developed for parameters that may be uniquely determined by flow cytometry. This chapter is focused on our approach to the subject and does not review the work of other investigators.

FLOW CYTOMETRY MEASUREMENTS ON MALE GERM CELLS

Testicular Cells

Flow cytometry measurements of cellular suspensions prepared from testicular biopsies provide a rapid evaluation of cell types present in testes including altered ratios of normal cell types and the presence of tumor cells. This methodology has the distinct advantage of very rapid analysis (e.g. 5–10 min per testis biopsy) of large numbers of cell types in contrast to very slow analysis by histological scoring.

[a]Supported by National Institutes of Health grants No. ES03035, ESO 339, and EPA grant No. R810986.

The distribution of acridine orange (AO) stained testicular biopsy cells obtained from a normal testis is seen in FIGURE 1(A). Computer analysis of the green fluorescence frequency histogram showed that the ratios of cell types here were 10% tetraploids, 15% diploids, 61% round and elongating spermatids, and 14% elongated spermatids. Testicular function is sensitive to environmental factors that will cause an alteration of these cell type ratios. FIGURE 1(B) shows the AO cellular staining pattern of biopsy material obtained from a "grossly normal" region of a testis that contained a tumor. Disturbances of testicular function shut down normal spermatogenic functions; as a consequence, the somatic and germ diploid cells predominated and the relative numbers of differentiating germ cells were very low. Thus, the presence of a tumor in this testis apparently caused a reduction of normal sperm production. FIGURE 1(C) shows that the tumor cells had a diploid DNA content as evidenced by the stainability being equivalent to the marker lymphocytes.

Prior to puberty, the testis consists primarily of non-germ, supporting cells. FIGURE 2 shows the AO staining pattern of cells isolated from a cryptorchid testis of a sexually immature juvenile patient as compared to the staining pattern of somatic mononuclear blood cells and normal semen cells. Mature sperm cells lack RNA[3] and have minimal red fluorescence. Even though the DNA content of sperm is one half that of diploid cells, due to the high level of chromatin condensation, they have only about one tenth the AO stainability.[2,4] The juvenile testicular cells have the same DNA stainability and a slightly lesser RNA stainability than peripheral blood lymphocytes. These data and that from light microscope evaluations are consistent with the cells being primarily Sertoli cells. Examination of 30 cases revealed no differences among the samples with one questionable exception of a higher ploidy level, which is of interest due to the higher incidence of testicular carcinoma among patients with cryptorchid testes.

Semen

Standard clinical assessment of semen quality includes light microscope examination of several hundred sperm cells or less to determine morphology, motility, and cell concentrations, providing a first indication of fertility potential. The low number of cells evaluated makes it difficult to obtain highly accurate data. In contrast, flow cytometry methods described here allow for a very rapid determination of cell concentration, ratios of cell types present including somatic cells (e.g., leukocytes), chromatin structure, viability, and mitochondrial function.

Chromatin Structure of Sperm

AO Staining of Semen Cells. Two-step AO staining of fresh or frozen semen: An aliquot of either fresh semen, a dilution of semen in HBSS, or a frozen ($\leqslant -20°C$) for up to several months) sample (1:1 in glycerol) is first mixed with a detergent-acid solution for 30 sec and then stained with AO as described in the legend for FIGURE 1; the acid-detergent solution greatly reduces the viscosity of semen. During spermiogenesis somatic histones are exchanged for sperm-specific basic proteins, S-S crosslinks are formed between these basic proteins, and the chromatin undergoes a high degree of condensation; all of these events apparently contribute to the fivefold reduction of DNA accessibility to AO stain relative to somatic cells as seen in FIGURE 2(B). Since cells at both staining levels and also intermediate levels of stainability may be present in the sample, it is useful to

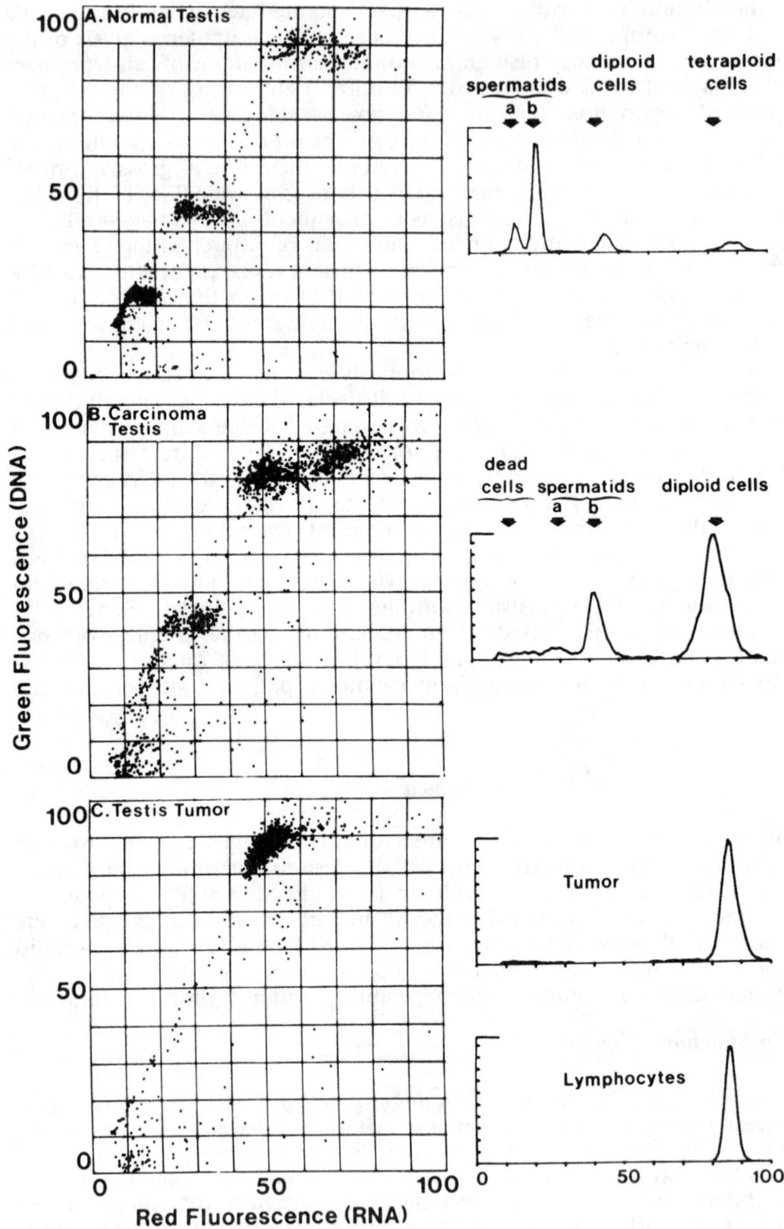

FIGURE 1. Computer-drawn scattergrams of the distribution of testicular biopsy cells from a normal (A) and tumor-bearing testis (B and C) according to their green (DNA) and red (RNA) fluorescence intensities after staining with acridine orange (AO) by the two-step method as previously described[1,2] and outlined below. In the scattergrams, each dot represents a single cell; the distance from the abscissa and ordinate corresponds to the

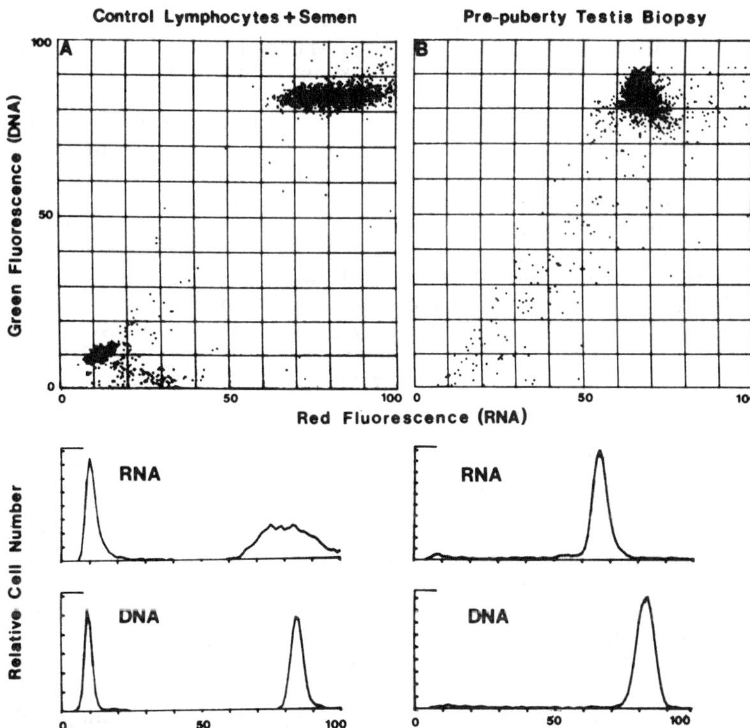

FIGURE 2. Computer-drawn scattergrams of the distribution of (A) mixture of normal human fresh semen cells and Ficoll-Hypaque isolated peripheral blood lymphocytes and (B) pre-puberty testis biopsy cells according to their green and red fluorescence intensities after staining with AO and measuring in a flow cytometer as described in the legend to FIGURE 1. The RNA and DNA staining frequency histograms are below each scattergram.

green and red fluorescence of the cell, respectively. Note that the photomultipliers (PMTs) were set so as to include tetraploid cells in (A) but were raised to a level to exclude this population in (B) and (C). To the right of each scattergram are the green fluorescence frequency histograms. The bottom frequency histogram represents the green fluorescence of AO-stained peripheral blood lymphocytes, which were used as a marker for normal diploid DNA stainability. The biopsies were minced with scissors in Hanks balance salt solution (HBSS) at 4°C and the liberated cells filtered through 53 μm nylon mesh. A 0.2 ml aliquot of the filtrate containing about 2×10^5 cells was mixed with 0.4 ml of a detergent solution consisting of 0.1% Triton X-100 (Sigma Chemical Co.) in 0.08 N HCl and 0.15 M NaCl. Thirty seconds later, 1.2 ml of a solution containing 0.2 M Na_2HPO_4 − 0.1 M citric acid buffer (pH 6.0), 1 mM ethylenediaminetetraacetic acid (EDTA), 0.15 M NaCl, and 6 μg AO/ml (chromatographically purified, Polysciences, Inc.) was admixed, and the sample was measured within 1–3 min in a FC-200 Cytofluorograf (Ortho Diagnostic Instruments, Westwood, MA) interfaced to a Nova 1220 minicomputer. The cells suspended in the dye solution were transported through the flow cell at a rate of about 200 per second. Red and green fluorescence signals generated by the 488 nm argon-ion laser beam correspond under these staining conditions to RNA and native DNA, respectively. The data were based on 5,000 cells per sample.

initially set the photomultipliers (PMT) at values that will include both sperm and somatic cells as shown in FIGURE 2. The PMT levels should then be raised in order to assess smaller alterations in DNA stainability (chromatin structure) as described below.

A great variety of patterns representing different ratios of cell types in semen have been seen in clinical samples. For example, a sample obtained from a chemotherapy-treated patient (FIGURE 3A) had a concentration of 5×10^4 morphologically identifiable sperm per ml semen. The majority of total cells were morphologically rounded cells with staining characteristics of round spermatids. Thus, the chemotherapy apparently caused an interruption in spermiogenesis such that the great majority of semen cells were undifferentiated. Patient B (FIGURE 3) had stage I testicular cancer. No normal mature sperm were seen; immature sperm were present as verified by electron microscopy. Note the great heterogeneity of staining ranging from mature sperm to diploid cell level.

Clinical samples often have a less dramatic alteration of staining patterns than seen in FIGURE 3(B). For example, FIGURE 4 compares the green fluorescence versus red/total fluorescence of a control donor sample with a sample obtained from a newly diagnosed Hodgkin's disease patient prior to chemotherapy. The increased green fluorescence is most likely due to a lack of a normal exchange of histones for protamines and/or a lack of appropriate condensation. The disease may be causing these alterations of testicular function. Note in the green fluorescence histograms that 93% of the control cells are within three standard deviations (SD) of the control mean whereas only 60% of the patient cells are within 3 SD of the control mean. In contrast to the increased green fluorescence, there was no significant increase in red fluorescence indicating no increase in single-stranded DNA or RNA. Also, there was no significant increase in green pulse width, i.e., the time taken for the green fluorescence signal to pass through the laser beam indicating no significant increase in the size/shape of the sperm.[6]

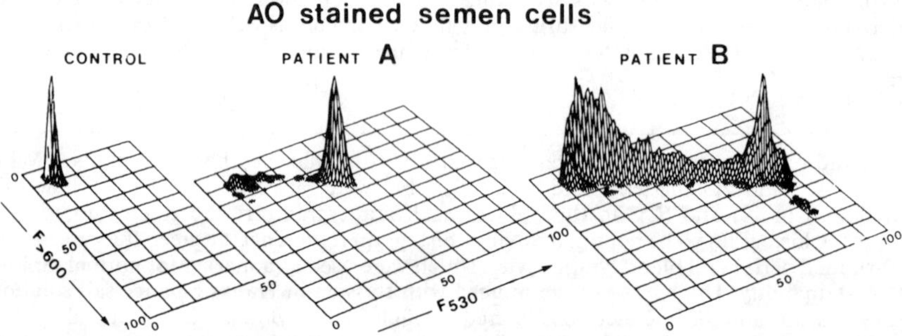

FIGURE 3. Computer-drawn two-parameter (F_{530} versus $F_{>600}$) histogram representing distribution of AO-stained human semen cells. The control semen sample was obtained from a healthy fertile donor. Patient A, age 18, had stage III embryonal cell testicular carcinoma. At 3 months postorchiectomy the semen had a sperm count of 5×10^6/ml, 45% sperm motility 1 hr after emission, and 55% of the cells were morphologically normal. The sample represented here was obtained 3 months later (6 months post-orchiectomy and 3 months post-VAB 6 induction chemotherapy). The sperm concentration was 5×10^4/ml with 0% motility. Patient B, age 28, had stage I testicular cancer with teratoma; this sample was obtained 11 months post-orchiectomy. (From Evenson *et al.*[5] With permission from *Journal of Histochemistry and Cytochemistry*.)

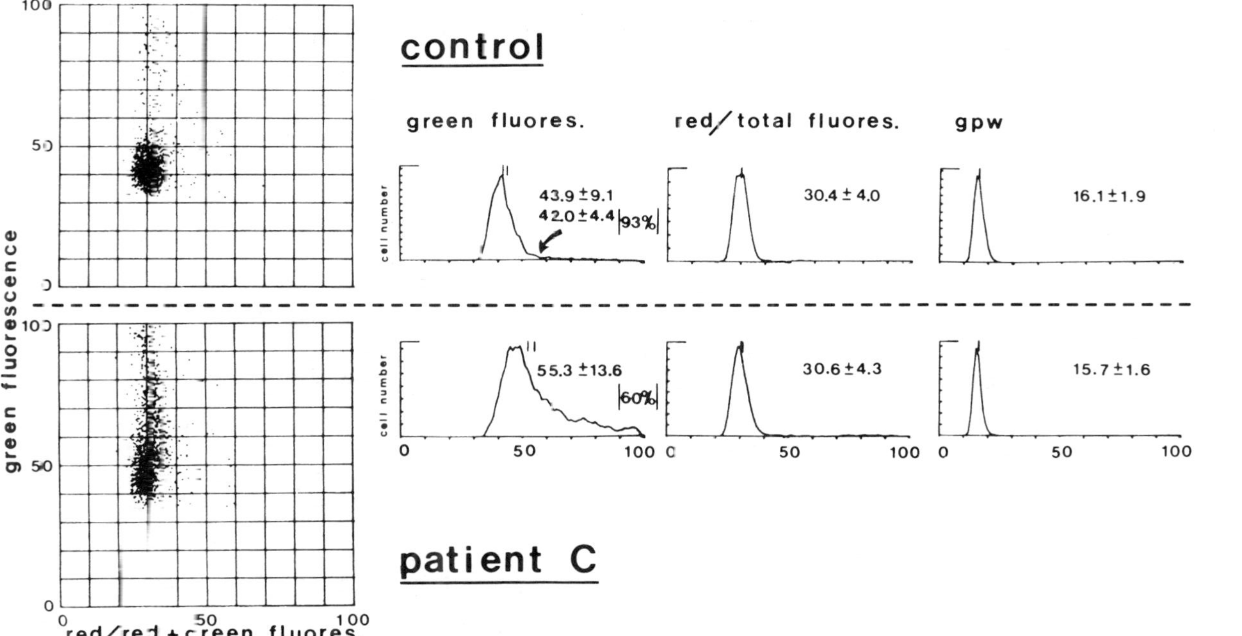

FIGURE 4. Computer-drawn scattergrams of the distribution of individual semen cells from control and patient C according to their green and red/(red + green) (αt) fluorescence intensities after staining with AO; to the right of each scattergram are the green fluorescence, red/total fluorescence, and green pulse-width (gpw) frequency distribution histograms. The numbers correspond to the mean value ± standard deviation (SD) (upper number), and the mean green fluorescence ± SD of a selected population within the normal range (less than channel 58; see arrow). The % number is that percentage of cells within 3 SD of the mean of the selected population. Patient C, age 18, was diagnosed as having Hodgkin's disease and the sample was obtained prior to therapy. The sperm concentration was 30×10^6 per ml and 70% of sperm were motile 1 hr after emission. Sperm morphology appeared normal when examined by light microscopy. (From Evenson *et al.*[5] With permission from *Journal of Histochemistry and Cytochemistry*.)

AO staining of heat-denatured sperm nuclei: Often a more sensitive way to study alterations of sperm chromatin structure is to subject isolated, fixed sperm cell nuclei to heat, partially denaturing the DNA *in situ*.[7] The extent of denaturation can be determined by FCM measurements of these nuclei after staining with AO as green fluorescence corresponds to native double-stranded DNA and red fluorescence to single-stranded DNA. The degree of this metachromatic shift from green to red fluorescence is quantitated by the expression alpha t (αt) defined as red/total (red + green) fluorescence.[8] Thus, a cell with only green fluorescence (native DNA) would have a value of 0 and a cell with only red fluorescence (denatured DNA) would have a value of 1. It should be noted that the αt index is also used to quantitate the metachromatic shift sometimes observed with the two-step AO technique (FIGURE 4) as well as with thermal denaturation.

Examples of the use of the heat denaturation technique are shown from studies on testicular cancer patients and infertility clinic patients.

(i) *Testicular carcinoma*. Testicular carcinoma is known to cause abnormal sperm production in the contralateral testis after unilateral orchiectomy. A recent FCM study[9] analyzed semen obtained from 3 controls and 14 patients with testicular cancer using both the two-step AO and the heat denaturation techniques.

Flow cytometry of semen aliquots showed, with only one exception, an increased AO staining of the sperm nuclear chromatin attributable to abnormal chromatin composition and/or condensation. FIGURE 3 (patient B) demonstrates a dramatic alteration of two-step AO stainability of semen cells. The chromatin was also abnormally sensitive to thermal stress, indicated by elevated αt values. These changes were present up to 15 months post unilateral orchiectomy, which in most cases ruled out surgical trauma as a cause and suggests continued abnormal spermiogenesis in the contralateral testis. FIGURE 5 illustrates raw data on unheated and heated sperm nuclei obtained from a control donor and a patient with testicular carcinoma previously treated by unilateral orchiectomy. Note the homogeneity of staining for the control sample in contrast to the heterogeneous staining of the patient sample. TABLE 1 illustrates how data are currently being analyzed. After rehydration, the fixed, unheated nuclei are first stained with AO and measured for red and green fluorescence. The αt values for the unheated nuclei are in column 2. Another aliquot of the same sample is then heated at 100°C for 5 min, stained with AO and measured (αt values in column 3). The ratio of αt heated patient sample/αt control unheated sample is then determined. This ratio is considered to provide a better determination of the level of alterations than that obtained by the ratio of heated/unheated values of patient samples since abnormally high values in column 2 would cancel out the effect of a high value in column 3.

The αt index (column 5) is then determined by the ratio of the value in column 4 to the control values in column 4. Thus, the αt index becomes normalized to 1.0 for controls. Patient αt index values shown here range from 1.1 (patient #7) to 2.2 (patient #3). It is difficult to define the αt index level that may be consistent with fertility; however, from previous work[7,9,10] we have chosen an αt index of 1.5 as that upper level. The αt index is the mean of the sample and is generally used to classify the sample. However, the same mean could be arrived at by two very different situations: all of the nuclei having a moderate level of DNA denaturation or half of the nuclei having a low level of denaturation such as seen for control nuclei and the other half extensively denatured. Columns 6–8, TABLE 1, show the percentage of nuclei that are or are not within a cluster prior to or after heat denaturation. Thus, for control #1, 97% of the nuclei in a cluster had an

TABLE 1. Effect of Testicular Carcinoma on Numerical Index (αt) of Sperm Chromatin Quality

1.	2.	3.	4.	5.	6.	7.	8.
			αt Patient (H)	αt Index	Mean αt S.D.		Mean αt ±S.D.
	Mean αt Total Pop.	Mean αt Total Pop.			of Least Denatured	%	of Remaining
Patient	±S.D. (UH)	±S.D. (H)	αt Control (UH)	αt Patient Control	Group UH/H	Cells UH/H	Cells UH/H
C #1	0.15 ± 0.03	0.29 ± 0.09	1.93	1.0	0.15 ± 0.01	97	0.24 ± 0.07
					0.25 ± 0.02	83	0.46 ± 0.10
C #2	0.15 ± 0.02	0.28 ± 0.08	1.87	1.0	0.15 ± 0.01	96	0.25 ± 0.07
					0.25 ± 0.02	81	0.45 ± 0.09
C #3	0.15 ± 0.02	0.28 ± 0.07	1.89	1.0	0.15 ± 0.02	98	0.27 ± 0.08
					0.26 ± 0.02	85	0.48 ± 0.09
#1	0.25 ± 0.04	0.43 ± 0.19	2.85	1.4	0.24 ± 0.02	89	0.36 ± 0.07
					0.29 ± 0.04	55	0.63 ± 0.16
#2	0.29 ± 0.14	0.58 ± 0.20	3.87	2.0	0.21 ± 0.02	63	0.44 ± 0.14
					0.36 ± 0.05	34	0.72 ± 0.13
#3	0.53 ± 0.23	0.64 ± 0.30	4.27	2.2	0.27 ± .04	24	0.59 ± 0.18
					0.27 ± 0.03	9	0.67 ± 0.16
#4	0.27 ± 0.14	0.56 ± 0.22	3.73	1.9	0.18 ± 0.02	55	0.38 ± 0.15
					0.33 ± 0.05	37	0.70 ± 0.16
#5	0.32 ± 0.16	0.59 ± 0.16	3.93	2.0	0.18 ± 0.02	55	0.43 ± 0.12
					0.48 ± 0.09	45	0.74 ± 0.13
#6	0.18 ± 0.07	0.56 ± 0.14	3.73	1.9	0.15 ± 0.02	82	0.30 ± 0.11
					0.48 ± 0.07	69	0.73 ± 0.10
#7	0.15 ± 0.03	0.32 ± 0.16	2.13	1.1	0.15 ± 0.01	97	0.28 ± 0.10
					0.24 ± 0.04	69	0.51 ± 0.17
#8	0.20 ± 0.04	0.42 ± 0.16	2.80	1.4	0.18 ± 0.02	88	0.31 ± 0.08
					0.33 ± 0.06	70	0.66 ± 0.11
#9	0.13 ± 0.01	0.45 ± 0.13	3.27	1.6	0.13 ± 0.01	93	0.26 ± 0.09
					0.45 ± 0.13	82	0.69 ± 0.09
#10	0.23 ± 0.09	0.58 ± 0.18	3.87	2.0	0.16 ± 0.02	56	0.31 ± 0.08
					0.44 ± 0.09	54	0.76 ± 0.08
#11	0.22 ± 0.09	0.60 ± 0.17	4.00	2.1	0.18 ± 0.01	77	0.35 ± 0.09
					0.50 ± 0.08	67	0.82 ± 0.09
#12	0.16 ± 0.06	0.38 ± 0.16	2.53	1.3	0.15 ± 0.01	90	0.30 ± 0.11
					0.29 ± 0.04	67	0.56 ± 0.15

C = control; SD = standard deviation; UH = unheated; H = heated. After Evenson *et al*.[9]

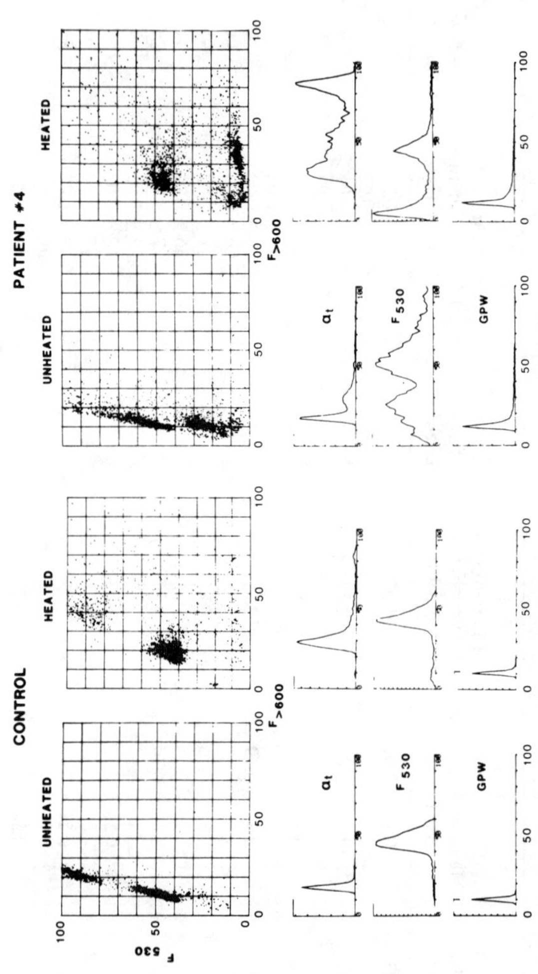
CONTROL
PATIENT #4
UNHEATED
HEATED
UNHEATED
HEATED
F>600
F>600
F 530
F 530
F 530
αt
αt
F 530
F 530
GPW
GPW
0 50 100
0 50 100
0 50 100
0 50 100
0 50 100

FIGURE 5. Comparison between red (denatured DNA) and green (native DNA) fluorescence values, computed αt values, and green pulsewidths of nuclei isolated from sperm of control donor and from a testicular carcinoma patient (same as #4 in TABLE 1). Below the scattergrams are frequency histograms representing the αt, F_{530} and gpw values of unheated and heated samples. The preparation of sperm cell nuclei, previously described.[7] was recently modified[9] as follows. Frozen aliquots of semen:glycerol mixtures were thawed, diluted to 12 ml in TNE buffer (0.01 M Tris, 0.15 M NaCl, and 0.001 M EDTA, pH 7.4) then washed twice by centrifuging (2500 g) through TNE buffer. The semen cells were resuspended in 3.6 ml TNE buffer in a Falcon plastic test tube (#3033 Falcon, Oxnard, CA) that was immersed in an ice water slurry, then sonicated for 30 sec with a model 185 Branson Sonifier set at a power scale of 3.5. After 30 sec pause to cool in the same ice bath, the cells were sonicated for an additional 30 sec. The sonicate was then mixed with 1.2 ml of 60% sucrose (w/w) in 0.01 M Tris-HCl (pH 7.4) and 0.002 M EDTA. The cell suspension was layered over 9 ml of the same sucrose solution in a 16 ml Sepcor polycarbonate tube with cap (Separation Science Corp., Stratford, CT) and centifuged at 25,000 g in a Sorvall HB4 rotor for 60 min. The supernatant, including the sperm tails located at the sucrose gradient interface, was removed by aspiration using a Pasteur pipette. The nuclear pellet was resuspended in 1.0 ml of 0.15 M NaCl, 0.005 M $MgCl_2$, and 0.02 M Tris-HCl (pH 7.4) and then forcefully pipetted into 9.0 ml of a 1:1 mixture of 70% ethanol and acetone in a glass test tube. All of the above procedures were done at 4°C. The sperm nuclei were stored in the fixative for several days to several months at -20°C. Prior to staining and measurement the nuclei were pelleted in the same glass test tube (1600 g × 10 min) and following careful aspiration of the fixative the nuclei were resuspended in 15 ml of "heating buffer" consisting of 0.02 M sodium cacodylate, 10^{-4} M EDTA, and 40% (vol/vol) ethanol (pH 6.6). After 10 to 15 min, the nuclei were again pelleted and resuspended in a volume of heating buffer such that the cell concentration was about 10^6 cells per ml. After approximately 30 min to allow for exchange of the heating buffer for fixative, 0.20 ml aliquots of suspended nuclei were admixed with 2.0 ml staining solution [0.15 M NaCl, 0.005 M $MgCl_2$, 0.02 M Tris-HCl (pH 7.4)] and 8 µg AO per ml. After 2 min to allow the dye to equilibrate, the sample was measured by flow cytometry. For thermal denaturation, a 0.40 ml aliquot (increased volume to compensate for loss of nuclei that adhere to the plastic tube during the heating process) was transferred to a #3033 Falcon test tube that was then placed in a boiling water bath for 5.0 min. transferred to an ice water slurry for 15 sec, and then admixed with 2.0 ml of the same staining solution as above. After 2 min the flow cytometry measurements were made. (From Evenson *et al.*[9] With permission from *Journal of Urology*.)

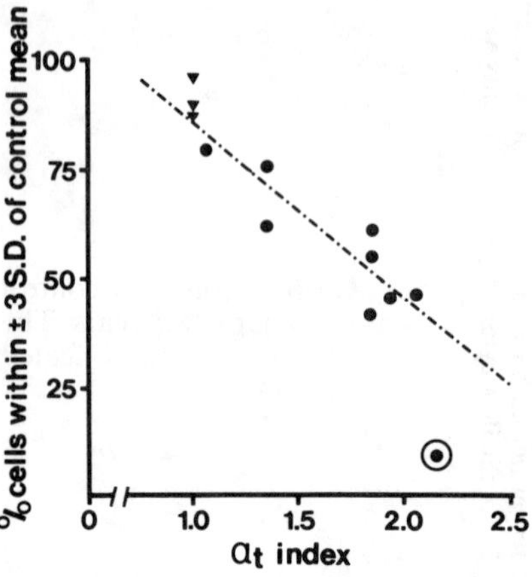

FIGURE 6. Correlation between percentage of AO-stained whole sperm cells with green fluorescence values within 3 SD of mean and αt index of heated and AO-stained nuclei isolated from the same sample. The green fluorescence values were analyzed as shown in FIGURE 5. (▲) Control samples, (●) patient samples, and (○) patient sample point excluded from regression analysis used to draw line. $r = 0.94$. (From Evenson *et al.*[9] With permission from *Journal of Urology*.)

unheated (UH) αt value of 0.15 ± 0.01; the other 3% of the nuclei had an αt value of 0.24 ± 0.07. For the heated (H) aliquot, 83% of the nuclei remained in a cluster with an αt of 0.25 ± 0.02 while the remaining 17% had an αt value of 0.46 ± 0.10. Some patient samples were very heterogeneous, e.g., for patient #4 (also shown in FIGURE 5) 55% of the unheated nuclei were in a near-normal cluster. After heating, 37% of the nuclei had a relatively normal staining distribution. The unheated nuclei that had a low level of AO staining (which are intact sperm nuclei but with staining alterations that are not understood) demonstrated a high level of denaturation. In general, the mean values of αt should be adequate for most clinical studies and perhaps only in special cases would the split values be of particular interest.

In this limited study on the testicular carcinoma patients there was a high correlation $(r = 0.94)$ between the percentage of sperm cells that had green fluorescence staining values within 3 SD of the control mean and the αt index, shown in FIGURE 6. Likewise there was a correlation $(r = 0.78)$ between the

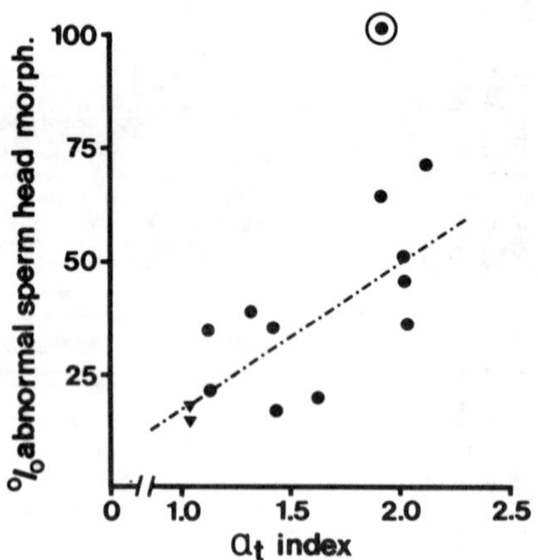

FIGURE 7. Correlation between percentage of sperm with abnormal sperm head morphology and αt index of AO-stained nuclei. Sperm head morphology was assayed using semen that had been smeared on glass microscope slides, air-dried, fixed in 95% ethanol, air-dried again, and stained with Giemsa dye. Morphology was categorized as either normal, tapered, micro-, macrocephalic, amorphous, or immature. (▲) Control samples, (●) patient samples, (○) patient sample point excluded from regression analysis used to draw line. $r = 0.78$. (From Evenson *et al.*[9] With permission from *Journal of Urology*.)

percentage of abnormal sperm head morphology and the αt (FIGURE 7). Current research has shown a positive correlation between chemical-induced sperm head abnormalities in mice and the αt index of unheated and heated sperm nuclei.

(ii) *Infertility clinic.* The correlations seen in FIGURES 6 and 7 have not held for all samples studied however. Samples obtained from infertility clinics have demonstrated cases classified as excellent with regard to sperm concentration, morphology, and motility but which have exhibited high αt values after heat denaturation.

FIGURE 8 shows the raw data from semen samples obtained from four control donors in good health who had fathered a child within the past two years. The relatively homogeneous AO staining profile seen for both unheated and heated nuclei shows that the sperm nuclear DNA was highly resistant to *in situ* thermal denaturation. In sharp contrast, FIGURE 9 shows that five semen samples obtained from patients attending an infertility clinic with various symptoms demonstrated a wide range of AO staining heterogeneity both prior to and following heat denaturation. Some samples were considered to be of excellent quality by classical criteria indicating that this technique brings out a defect not detectable by light microscope analysis.

Residual Effects of Drug Treatment. Cancer chemotherapy: Recovery from vigorous chemical therapy such as cancer chemotherapy is possible even when potent alkylating drugs are used. A prospective study of patients who were successfully treated with the L-10 protocol for acute lymphocytic leukemia (ALL) showed that after 17–52 months all of the patients had recovered to a near normal or normal status of fertility, determined by classical criteria and FCM measurements of AO stained sperm cells.[10] Patients who undergo such therapy successfully and are in the category of wanting to father a child are obviously very concerned that the harsh chemical treatment may have caused a detrimental effect on their fertility status. Our chromatin assay will detect abnormalities in the chromatin structure of sperm but would most likely not detect mutations unless they would affect a gene product involved in spermiogenesis. Although previous and current studies show a correlation between fertility status and profiles of AO stained sperm, it is premature to conclude that a certain pattern rules out fertility.

Antibiotic therapy: A study on a limited number of patients with urogenital infection was set up with the intent of correlating the AO staining distribution of pre- and post-treatment semen samples. Unfortunately only three of twelve patients returned to donate a post-antibiotic treatment sample. Of interest, as shown in FIGURE 10, all three had similar abnormal AO staining of sperm nuclei before and after heat denaturation. First, there was a reduced level of fluorescence, and secondly, the nuclear DNA was highly resistant to thermal denaturation, even to a greater extent than seen for normal control samples. Thus, it seems likely that the antibiotic was associated with the sperm DNA such that it reduced the accessibility of the DNA to AO and also increased resistance to *in situ* DNA denaturation.

FCM Analysis of Sperm Cell Viability and Mitochondrial Function Related to Motility

FCM also offers a rapid means to determine cell viability as determined by dye exclusion and mitochondrial membrane potential, which has been related to motility.[11] An aliquot of semen is first diluted with a physiological buffer and stained with rhodamine 123 (R123) followed by ethidium bromide (EB). R123 specifically stains mitochondria;[12] R123 green fluorescence is localized in the sperm midpiece containing the mitochondria. Previous[11] and current work

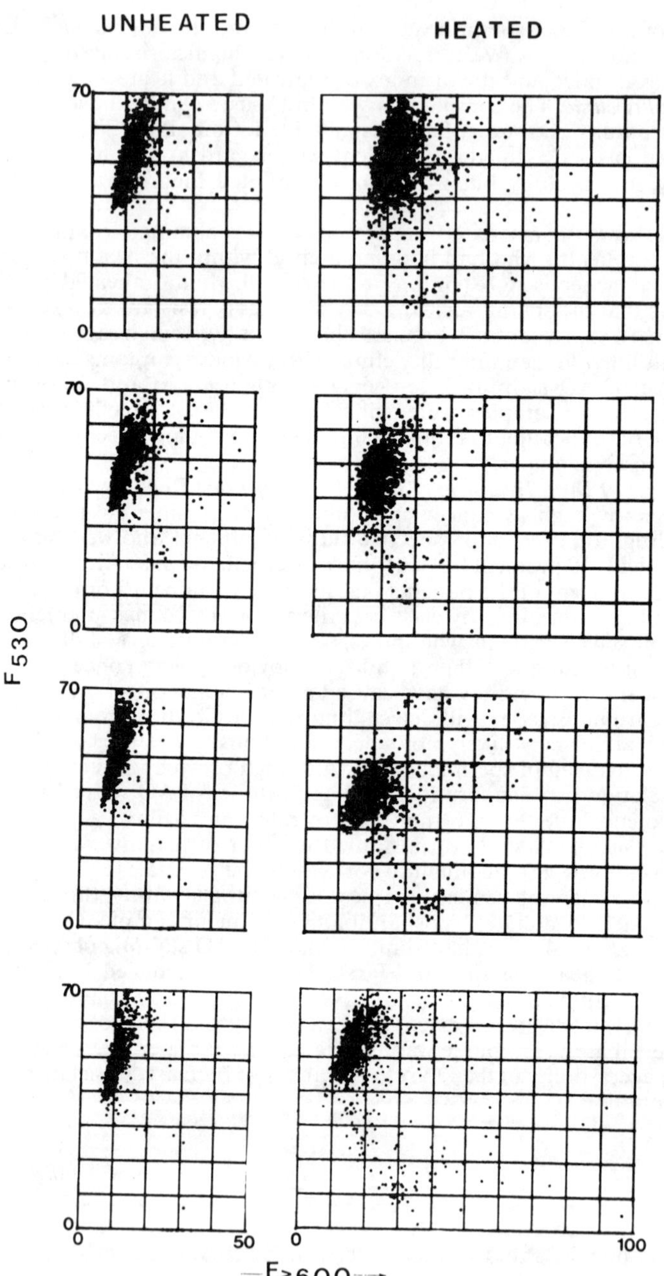

FIGURE 8. Computer-drawn scattergrams of the distribution of measurements on individual sperm nuclei from control donor semen samples according to their green (F_{530}) and red ($F_{>600}$) fluorescence intensities after heating (or not heating) and staining with AO as described in the legend to FIGURE 5.

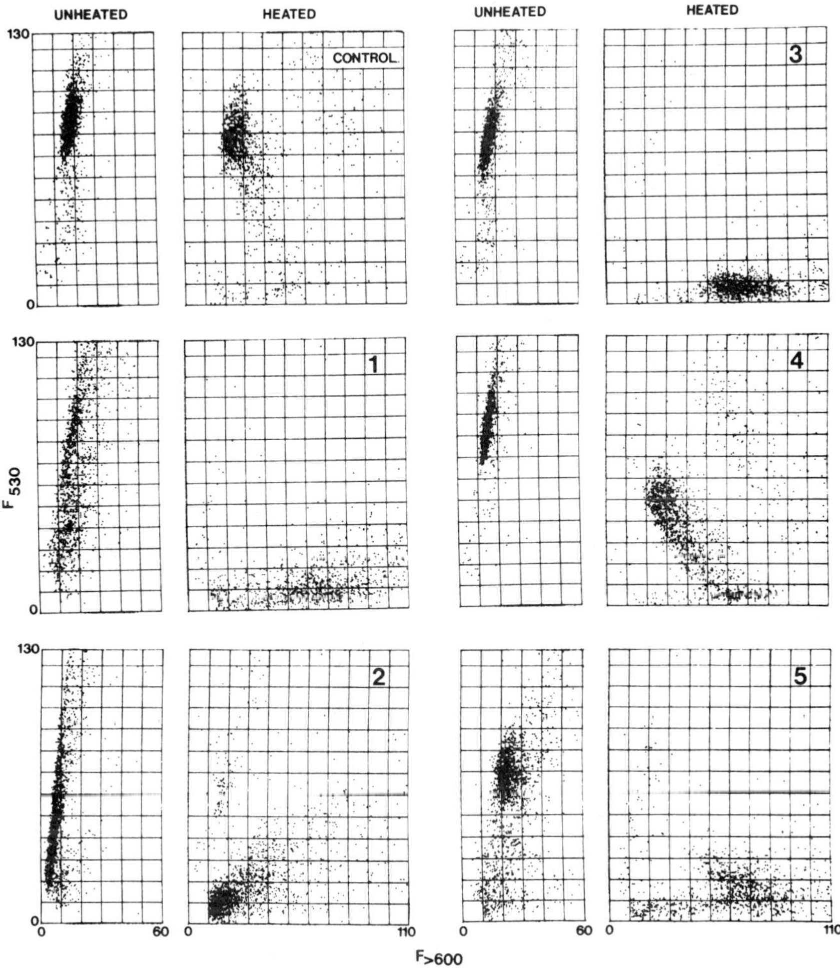

FIGURE 9. Computer-drawn scattergrams of the distribution of measurements on individual sperm nuclei from one control donor and five patients attending an infertilty clinic according to their green (F_{530}) and red ($F_{>600}$) fluorescence intensities after heating (or not heating) and staining with AO as described in the legend to FIGURE 5. Patients were diagnosed as follows: (1) Morphology poor to borderline. Infection $(I)^+$; (2) Prostatitis/non-specific urethritis (NSU), morphology minimal, I^-; (3) Prostatitis/NSU; good specimen, I^+, (4) Morphology and motility good, 2 miscarriages, I^-; (5) Excellent specimen in all respects, I^-.

suggest that the level of dye uptake is related to the level of motility. EB serves as a viability stain; sperm with intact plasmalemma exclude the dye while those with broken membranes take up the dye, which stains the nuclear DNA and fluoresces red. FIGURE 11 shows the raw data of measurements of R123 and EB stained semen 4 and 30 hr after collection. Note the loss of green fluorescence and gain of red fluorescence with time. Fluorescent light microscope observations showed

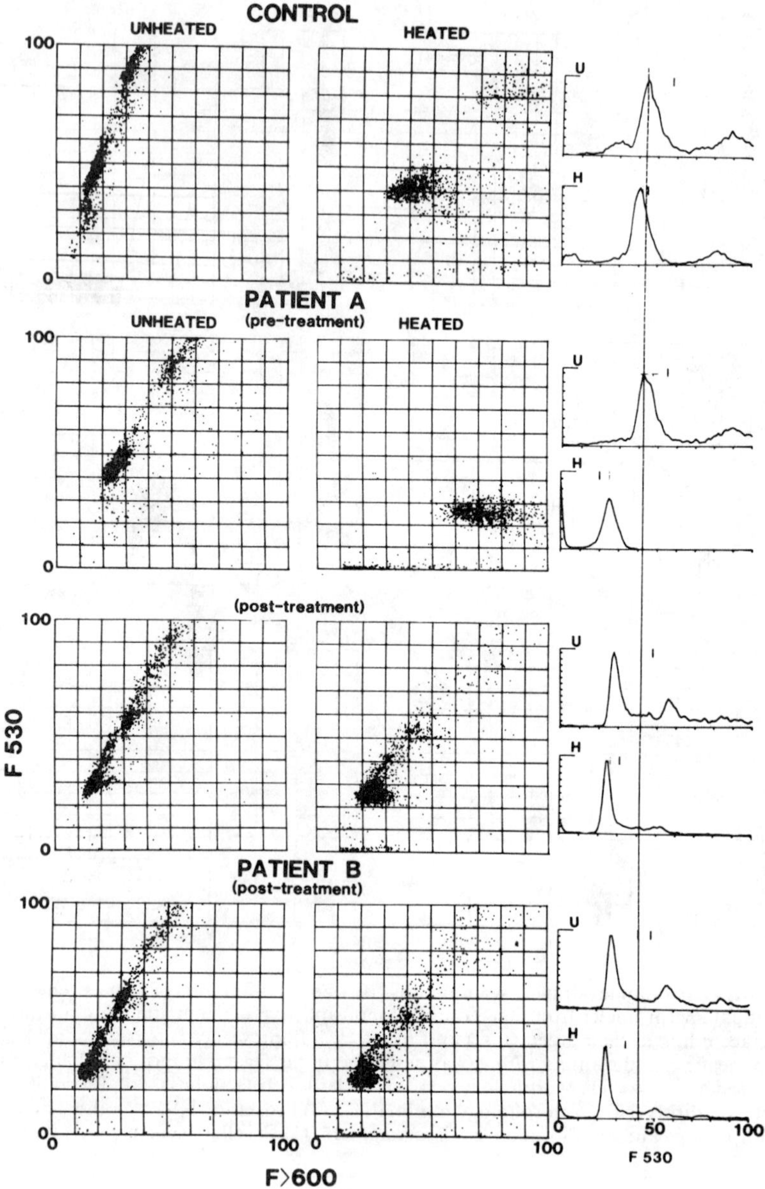

FIGURE 10. Computer-drawn scatterplots of the distribution of individual sperm nuclei obtained from a control donor and two patients with urogenital infections. On the right are the green fluorescence frequency histograms of the unheated (U) and heated (H) samples. Patient A: beta Streptococcus treated with 500 mg erythromycin four times daily for 2 wk followed by 100 mg vibramycin twice daily for 2 weeks; sample collected prior to and 2 wk post therapy. Patient B: Chlamidia plus nonhemolytic Streptococcus treated with above dosage of vibramycin for 2 wk followed by 2 wk of erythromycin followed by 4 wk of vibramycin; sample collected prior to (data not shown) and 2 wk post therapy.

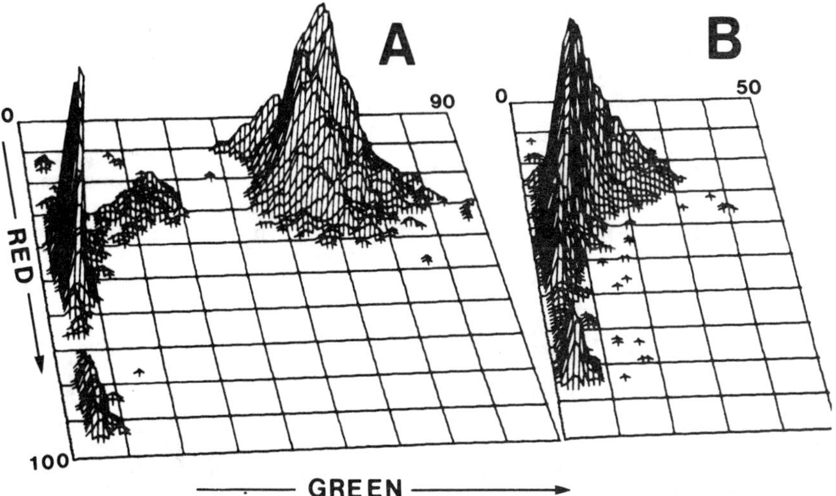

FIGURE 11. Two-parameter frequency histograms demonstrating stainability of human sperm cells with rhodamine 123 and ethidium bromide. A single semen sample obtained from a normal individual of proven fertility was stained and measured at (A) 4 hr after collection and (B) 30 hr after collection. The sample was kept at room temperature in a flat-bottomed plastic jar. Prior to measurement, the semen was diluted to approximately 1–2×10^6 cells per ml in RPMI 1640 media (GIBCO, Grand Island, NY) containing 10% fetal calf serum (GIBCO). Alternatively, samples have been suspended and stained in phosphate-buffered salt solution (PBS) containing 1% (w/vol) sucrose with similar results. The samples were stained with R123 (Eastman Organic Chemicals, Rochester, NY, laser dye purity) at 10 µg/ml final concentration for 10 min; then the sperm were pelleted, resuspended in the same medium, and counterstained with EB (10 µg/ml, Polysciences, Warrington, PA). A 1 mg/ml stock solution of R123 was made in distilled water and diluted into the sample solution. All procedures were done at room temperature. (From Evenson *et al.*[11] With permission from *Journal of Histochemistry and Cytochemistry*.)

that bright green fluorescence in the sperm midpiece, which contains the mitochondria, correlated with vigorous sperm motility and dull green fluorescence correlated with slow motility. Bright red and green fluorescence were mutually exclusive; however, some cells exhibited pale nuclear red and green midpiece fluorescence suggesting that some cells lose green fluorescence as they begin to gain red fluorescence. Recently, known numbers of green fluorescent beads have been added to the R123 and EB stained cell suspensions enabling the measurement, in a single sample, of the concentration of sperm, live/dead ratios, and mitochondrial function (which is related to motility).

CONCLUSIONS

Flow cytometry can provide valuable, rapidly obtained data concerning male fertility. Two-step AO staining of testicular biopsies allows rapid measurements of ratios of cell types present, which indicate the functional status of the testis. Of more practical value for fertility analysis is the protocol to stain semen aliquots by the two-step AO technique. This technique measures the levels of red and green fluorescence providing an index (αt) of chromatin structure. Measurements on heat-treated isolated nuclei stained with AO will give, in some cases, additional information on sperm chromatin structure that may be related to fertility. These simple measurements on chromatin stainability plus the measurements of R123 and EB stained sperm samples admixed with known numbers of fluorescent beads provide data on ratios of cell types in semen, chromatin structure, cell count, viability, and mitochondrial function related to motility. Since many medical centers have flow cytometers in a core facility, it would seem efficient and practical to utilize flow cytometry as a tool to aid in the diagnosis of male fertility.

ACKNOWLEDGMENTS

Financial assistance for attending the conference was provided by the National Engineering Foundation and the Pathology Department, Memorial Sloan-Kettering Cancer Center (MSKCC). Testicular biopsies were provided by Dr. M. Melamed of the Pathology Department of MSKCC and Dr. D. T. Mininberg, the New York Hospital, Cornell Medical College. Semen samples were provided by Dr. A. Toth, the New York Hospital, Cornell Medical College and Drs. J. Redman, D. Bajorunas, F. Klein, and Z. Arlin of the outpatient clinic at MSKCC. All FCM measurements shown here were made in the Investigative Cytology Laboratory at the MSKCC, New York. I thank Drs. M. Melamed and Z. Darzynkiewicz for their interest and valuable discussions. E. Wood typed the manuscript.

[NOTE ADDED IN PROOF: Recent studies (Evenson *et al.*, Cytometry **6**:238–253, 1985 and Evenson *et al.*, Toxicol. Appl. Pharmacol. **82**:151–163, 1986) have shown that acid treatment is as effective as heat to induce partial denaturation of DNA in sperm nuclei. This procedure is easier and more rapidly accomplished. Using this recent work in our laboratory has demonstrated a correlation between αt values and bull fertility (Ballachey *et al.*, unpublished).]

REFERENCES

1. DARZYNKIEWICZ, Z., F. TRAGANOS, T. SHARPLESS & M. R. MELAMED. 1976. Lymphocyte stimulation: a rapid multiparameter analysis. Proc. Natl. Acad. Sci. USA **73**:2881–2884.
2. EVENSON, D. P., Z. DARZYNKIEWICZ & M. R. MELAMED. 1980. Comparison of human and mouse sperm chromatin structure by flow cytometry. Chromosoma **78**:225–238.
3. MONESI, V. 1971. Chromosome activities during meiosis and spermiogenesis. J. Reprod. Fertil. (Suppl.) **13**:1–14.
4. GLEDHILL, B. L., M. P. GLEDHILL, R. RIGLER, JR. & N. R. RINGERTZ. 1966. Changes in deoxyribonucleoprotein during spermiogenesis in the bull. Exp. Cell Res. **41**:652–665.
5. EVENSON, D. P. & M. R. MELAMED. 1983. Rapid analysis of normal and abnormal cell types in human semen and testicular biopsies by flow cytometry. J. Histochem. Cytochem. **31**:248–253.
6. SHARPLESS, T., F. TRAGANOS, Z. DARZYNKIEWICZ & M. R. MELAMED. 1975. Flow cytometry: discrimination between single cells and cell aggregates by direct size measurements. Acta Cytol. **19**:577–581.
7. EVENSON, D. P., Z. DARZYNKIEWICZ & M. R. MELAMED. 1980. Relation of mammalian sperm chromatin heterogeneity to fertility. Science **240**:1131–1133.

8. DARZYNKIEWICZ, Z., F. TRAGANOS, T. SHARPLESS & M. R. MELAMED. 1975. Thermal denaturation of DNA in situ as studied by acridine orange staining and automated cytofluorometry. Exp. Cell Res. **90**:411–428.
9. EVENSON, D. P., F. A. KLEIN, W. F. WHITMORE & M. R. MELAMED. 1984. Flow cytometric evaluation of sperm from patients with testicular carcinoma. J. Urol. **132**:1220–1225.
10. EVENSON, D. P., Z. ARLIN, S. WELT, M. L. CLAPS & M. R. MELAMED. 1984. Male reproductive capacity may recover following drug treatment with the L-10 protocol for acute lymphocytic leukemia (ALL). Cancer **53**:30–36.
11. EVENSON, D. P., Z. DARZYNKIEWICZ & M. R. MELAMED. 1982. Simultaneous measurement by flow cytometry of sperm cell viability and mitochondrial membrane potential related to cell motility. J. Histochem. Cytochem. **30**:279–280.
12. JOHNSON, L. V., M. L. WALSH & L. B. CHEN. 1980. Localization of mitochondria in living cells with rhodamine 123. Proc. Natl. Acad. Sci. USA **77**:990–992.

Prognostic Value of DNA/RNA Flow Cytometry of B-Cell Non-Hodgkin's Lymphoma: Development of Laboratory Model and Correlation with Four Taxonomic Systems[a]

MICHAEL ANDREEFF,[b,d] HERBERT HANSEN,[b]
CONNIE CIRRINCIONE,[c] DANIEL FILIPPA,[d]
AND HOWARD THALER[c]

[b]Department of Medicine
[c]Biostatistics Laboratory and
[d]Department of Pathology
Memorial Sloan-Kettering Cancer Center
New York, New York 10021

Non-Hodgkin's lymphomas are conventionally diagnosed and classified by morphological criteria, and much effort has been put into the development of taxonomic systems based on these criteria. Four different systems are presently being used, i.e. Rappaport's, Lukes' and Collins', the Kiel classification, and the "Working Formulation."[1] The reproducibility of the subtypes identified, comparability of different systems with each other, and prognostic value of these classifications, however, have been questioned.[2] In addition, inter- and intra-observer variability add to the uncertainties associated with morphological approaches to a group of diseases characterized by lymphoid cell populations of not always dissimilar morphological appearance. Immunological studies have greatly expanded our understanding of lymphoma biology but are of limited value in classifying this disease.

In this study, diagnostic lymph node biopsies from 96 patients with non-Hodgkin's lymphomas of B-cell type have been analyzed by several quantitative cytometric techniques in an attempt to characterize disease based on biological cell features, to correlate these with morphological classification systems, and to investigate their prognostic value.

These techniques include multi-parameter acridine orange flow cytometry for the simultaneous determination of cellular DNA and RNA content, which allows us to analyze cell cycle distribution, DNA stemline, and RNA content; autoradiographic[^{3}H]thymidine labeling index for the determination of the number of cells synthesizing DNA; and cell volume measurements using the Coulter counter.

Several papers have investigated the importance of pretreatment proliferative parameters on survival of patients. The most widely used parameter was the S phase labeling index,[3–9] but more recently, cytometric[10] and flow cytometric

[a]Supported by grants from the National Cancer Institute CA-20194, CA-29564, and CA-05826.

investigations of DNA content have been carried out.[11-28] We attempt here to expand these parameters and to analyze them in a multi-variate context in a group of defined B-cell lymphomas.

MATERIALS AND METHODS

Lymph nodes were surgically removed and dissected. One part was used for morphological classification, while the other part was mechanically minced thoroughly with scalpels in saline solution until homogeneity was achieved. The sample was then filtered through 75 μm nylon mesh, and aliquots were taken for [³H]thymidine pulse labeling, as previously described,[9] volume measurement in a Coulter counter as previously described,[9] and DNA/RNA flow cytometry.[29]

[³H]Thymidine-labeling Index

After incubation with thymidine, cells were spun down using a cytocentrifuge and coated with autoradiographic emulsion. After exposure and development, labeled cells were counted and expressed as labeling index (percent labeled cells) as previously described.[9]

DNA/RNA Flow Cytometry

The method is based on the ability of acridine orange to stain DNA and RNA selectively, under certain conditions. Acridine orange (AO) intercalates into double-stranded base pairs, which results in green luminescence, and is bound to single-stranded nucleic acids, i.e. RNA, by sandwich intercalation, as described by Kapuscinski.[30] Aliquots (0.2 ml) of cell suspensions containing $0.2-0.4 \times 10^6$ cells in HBSS are mixed with 0.4 ml of 0.05 N NaCl and 0.1% Triton X-100 (vol/vol, Sigma Chemical Co., St. Louis, MO). After 30 seconds, 1.2 ml of a solution containing Na_2HPO_4(0.2 M)-citric acid (0.1 M)buffer (pH6.0), 1 mM EDTA-Na, 0.15 N NaCl, and 6 μg/ml of acridine orange (AO) are added. Chromatographically purified AO obtained from Polysciences Inc. (Warrington, PA) is used. The cells are pretreated with a solution containing Triton-X to permit entry of the metachromatic dye into the cells. The solution makes nucleic acids insoluble at low pH. Double-stranded RNA is denatured by the presence of EDTA, which binds Mg required to maintain the double-stranded configuration of RNA. Thus, acridine orange stains DNA and RNA selectively, resulting in green and red fluorescence.

Flow Cytometric Analysis

The cells were measured in a FC-201 flow cytometer (Ortho Instruments, Westwood, MA). The instrument is connected to a dedicated Nova 1220 minicomputer. Green fluorescence (F530, from 515–575 nm), red fluorescence (F600, from 600–650 nm), and green pulse width (GPW) are measured. This last parameter serves to discriminate single cells from doublets. Data are stored in list mode for 5,000 cells in each sample. Data analysis was performed, excluding dead cells and cell aggregates. Cell cycle analysis was performed according to our "Peak symmetry model"[31] and different cell cycle phases ($G_{0/1}$, S, and G_2M) were distinguished. RNA content was determined for $G_{0/1}$ cells as described previously. The "RNA index $G_{0/1}$" is calculated from the mean RNA content of cells in $G_{0/1}$, determined by their DNA content, times 10, divided by the median RNA content of control lymphocytes.[32]

DNA index (DI) was calculated by measuring the sample at standard instrument settings, then measuring control lymphocytes, followed by measurement of a mixture of control cells and lymph node cells at standard and high instrument settings for DNA. Only double peaks in the mixture experiment were considered proof of DNA aneuploidy, as requested by the "Convention on Nomenclature of DNA Aneuploidy."[32] The method allows the discrimination of differences in DNA content of 5%. The mean coefficient of variation for measurement of control cells was 2.9 + 0.9%.

Treatment of Non-Hodgkin's Lymphomas

Patients with low-grade malignant lymphoma were treated with (1) cytoxan, vincristine, and prednisone (CVP), (2) the Memorial Hospital NHL-4 protocol, or single-agent chemotherapy, containing no adriamycin or low-dose adriamycin.

High-grade malignant lymphoma, identified by standard morphological criteria, were treated with an adriamycin-containing regimen including CHOP, or the L-17M protocol, originally developed for acute lymphoblastic leukemia.[34]

This paper includes only B-cell non-Hodgkin's lymphoma. The B-cell nature was determined by the presence of a single light chain in 50% or more of cells, identified by fluorescence microscopy of kappa- or lambda-stained cell suspensions, with FITC-conjugated polyclonal antibodies. If the other light chain was present, the ratio of the predominant to the secondary light chain was to be greater than 3:1.

Statistical Analysis

Each variable was analyzed independently, different variables were correlated with each other, factor score analysis combining two variables, and Cox regression analysis were performed. Survival curves were generated by the Kaplan-Meier method.

RESULTS

DNA/RNA Flow Cytometry

DNA/RNA flow cytometric analysis of lymph node suspensions revealed "typical" patterns of proliferation, RNA content, and DNA aneuploidy. In general, low-grade non-Hodgkin's lymphomas were characterized by low proliferation, low RNA index, and mostly diploid DNA stemlines. A typical example is given in FIGURE 1. High-grade lymphomas, by contrast, were characterized by high proliferation, high RNA index, and frequently by the presence of DNA aneuploid cell populations. An example for a diffuse histiocytic lymphoma is also given in FIGURE 1. As in other series, large variability of the individual variables was observed. Some high-grade lymphomas had low RNA content, low proliferation, and no DNA aneuploidy. FIGURE 2a shows the large variability of cells in S phase in the different groups of the Rappaport system. Clearly, the well-differentiated and poorly differentiated lymphocytic lymphomas have a lower S phase than histiocytic and unclassified lymphomas. A similar pattern can be seen for the Lukes-Collins system (FIGURE 2b), for the Kiel classification (FIGURE 2c), and for the Working Formulation (FIGURE 2d). Of note is the non-linearity of the S phase scale in these figures.

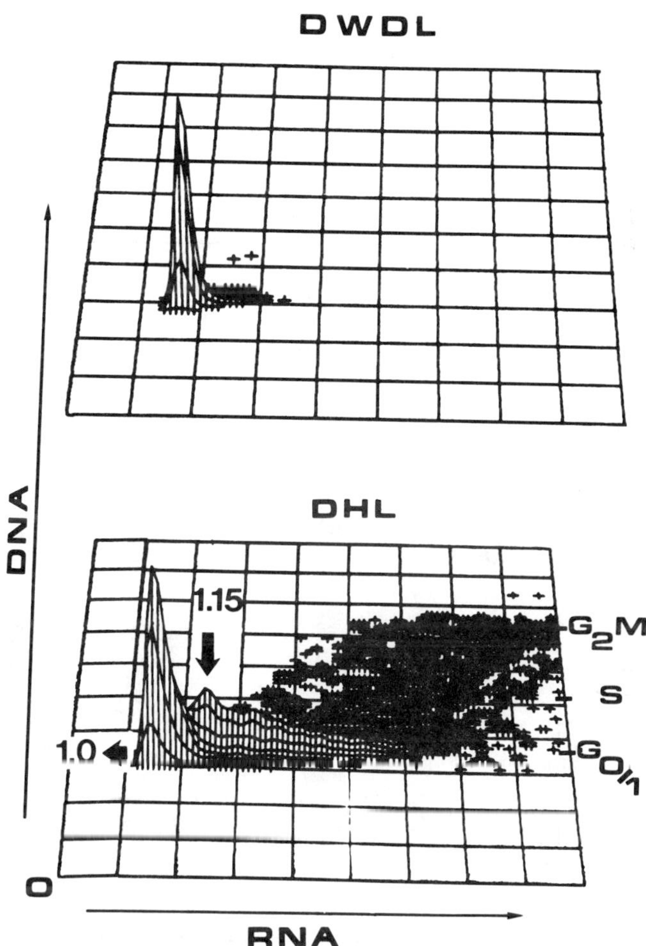

FIGURE 1. Correlated DNA/RNA histograms of non-Hodgkin's lymphoma cells from lymph nodes. x - axis = RNA, y - axis = DNA, z - axis = number of cells. Top: typical low-grade NHL (DWDL) is characterized by low RNA content, almost complete absence of proliferating cells (S, G_2 M), and diploid DNA stemline. Bottom: typical example of high-grade NHL (DHL) is characterized by the presence of a hyperdiploid DNA stemline (DNA index = 1.15), increased RNA content and increased proliferation.

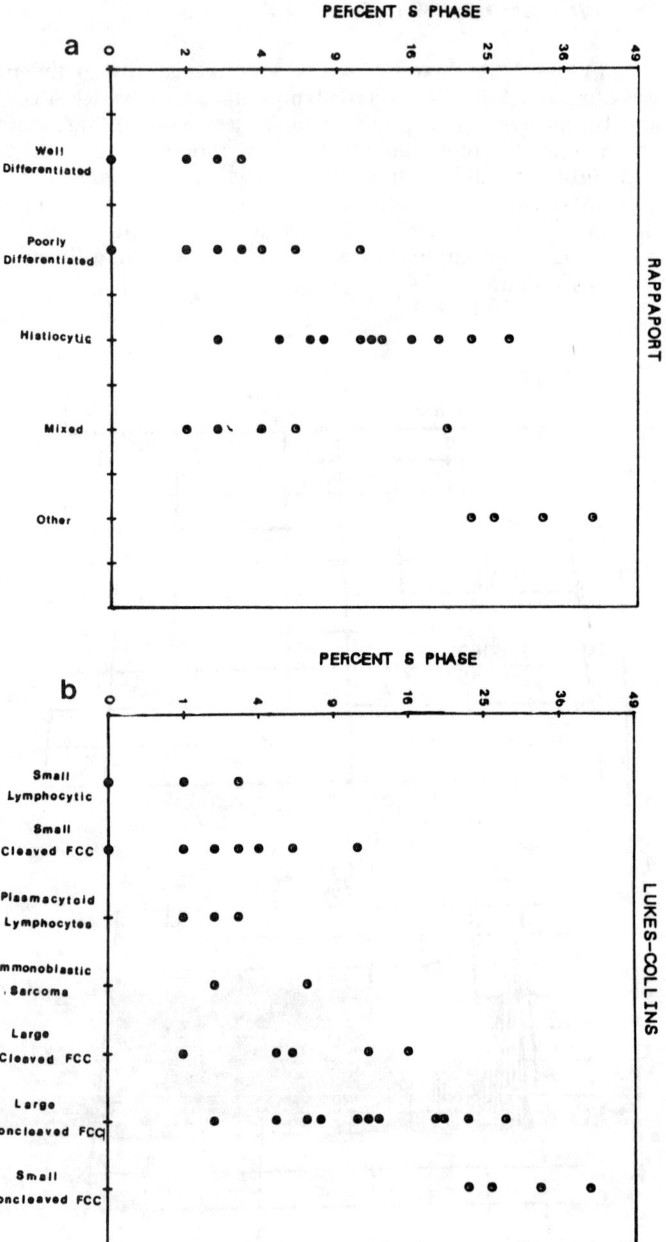

FIGURE 2. Correlation of proliferation and morphological subgroups in non-Hodgkin's lymphoma. Lymphoma were classified according to Rappaport (a) Lukes Collins (b) Kiel (c), and Working Formulation (d). S phase is plotted on the *y*-axis (note: non-linearity of scale). S phase is low in most low-grade malignant categories, regardless of the classification system used. S phase increases with increasing degree of malignancy. However, some cases of high-grade malignant NHL have low S phase.

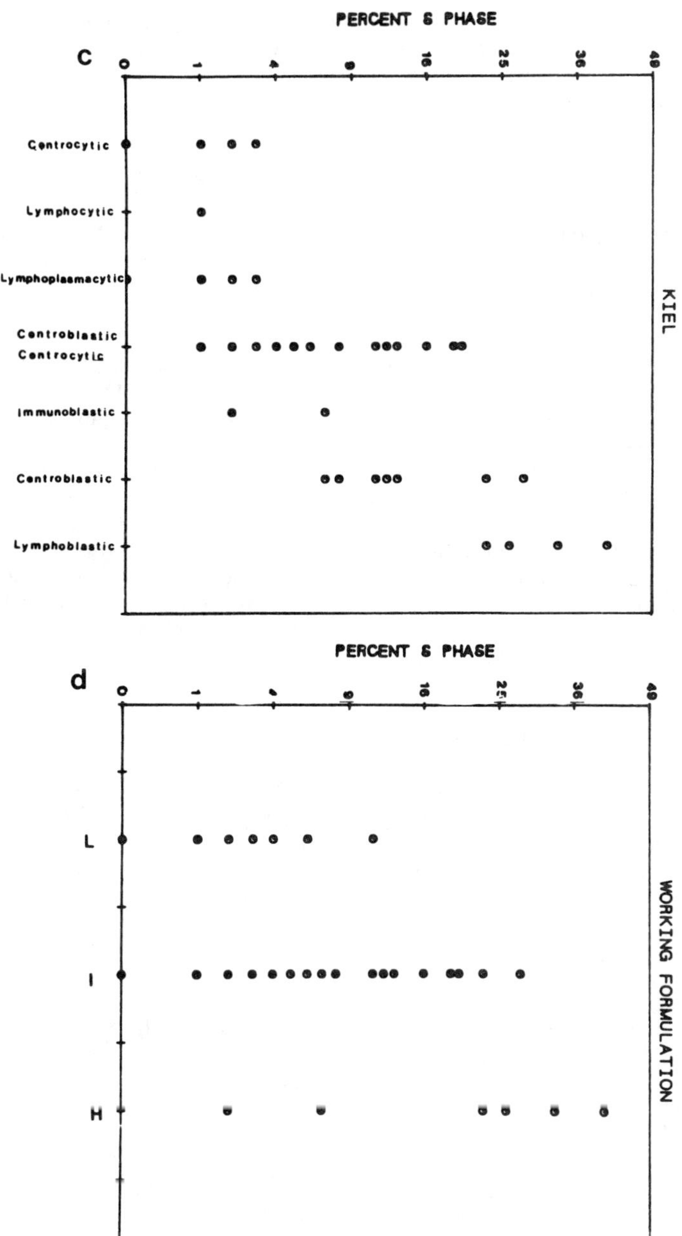

PERCENT S PHASE
c
0 1 4 8 16 25 35 40
Centrocytic
Lymphocytic
Lymphoplasmacytic
Centroblastic
Centrocytic
Immunoblastic
Centroblastic
Lymphoblastic
KIEL
PERCENT S PHASE
d
0 1 4 8 16 25 35 40
L
I
H
WORKING FORMULATION

RNA content of $G_{0/1}$ cells (RNA index) varied widely, being low in most low-grade and high in most high-grade lymphomas (data not shown).

DNA aneuploidy was rare in low-grade and more frequent in high-grade lymphomas (data not shown).

Factor Score Analysis of DNA and RNA

Factor score analysis, which combines two variables, S phase and RNA index, and summarizes them to form one factor, was different for the different Rappaport, Lukes-Collins, Kiel, and Working Formulation subgroups (FIGURE 3).

A significant difference was seen between nodular and diffuse lymphomas: the percentage of cells in S phase and their RNA index were both lower in nodular, as compared to diffuse lymphomas. FIGURE 4a shows differences in S phase (note non-linearity of scale) for nodular and diffuse lymphomas in the Rappaport system. FIGURE 4b shows both RNA index and S phase for nodular and diffuse lymphomas. While some diffuse lymphomas have both low S phase and RNA index, only two cases of nodular lymphoma have more than 4% S phase or log RNA index greater than 1.125. It appears that these two parameters are better correlated to the nodular or diffuse pattern than to low- or high-grade malignancy. While this is true for the Rappaport, Lukes-Collins, and Kiel classifications, RNA index and S phase appear to separate clearly low and intermediate, from high grades of malignant non-Hodgkin's lymphomas in the Working Formulation (FIGURE 5.)

Correlation of Laboratory Variables

A number of variables were correlated with each other in order to determine whether or not they measured similar cellular characteristics. The number of proliferating cells, determined by DNA flow cytometry (S phase) and thymidine pulse labeling (LI), was almost equally well-estimated by both methods. FIGURE 6a shows the excellent correlation of both variables (correlation coefficient $r=0.84$). Of interest is the higher labeling index as compared to flow cytometric S phase, especially at low values: a number of cases was read to have 1% S phase by FCM, but 2, 3, and 4% S phase cells by labeling index. One would expect that the large number of cells analyzed by flow cytometry would better identify low numbers of S phase cells. It may be, however, that these cells are "hidden" in the right shoulder of the G_1 peak. Labeling index was also higher for cases with very high proliferation. This phenomenon is known and was systematically investigated.[31] Our present method of analyzing DNA histograms underestimates S phase cells in situations with very high proliferation. S phase by FCM was also correlated with serum levels of lactate dehydrogenase (LDH). The correlation coefficient was $r = 0.496$ ($p = 0.001$), indicating a very weak positive correlation. The correlation between cell volume measured by the Coulter method (MCV) and S phase ($r = 0.54$), correlation between RNA index and labeling index ($r = 0.57$), and between RNA index and cell volume ($r = 0.53$) were equally weak. These correlations were significant with p values of 0.001 for each r. The correlations indicate that the variables investigated are largely independent from each other, except for labeling index and S phase (TABLE 1). FIGURE 6b shows how RNA index and cell volume are correlated: at low RNA levels, there is considerable

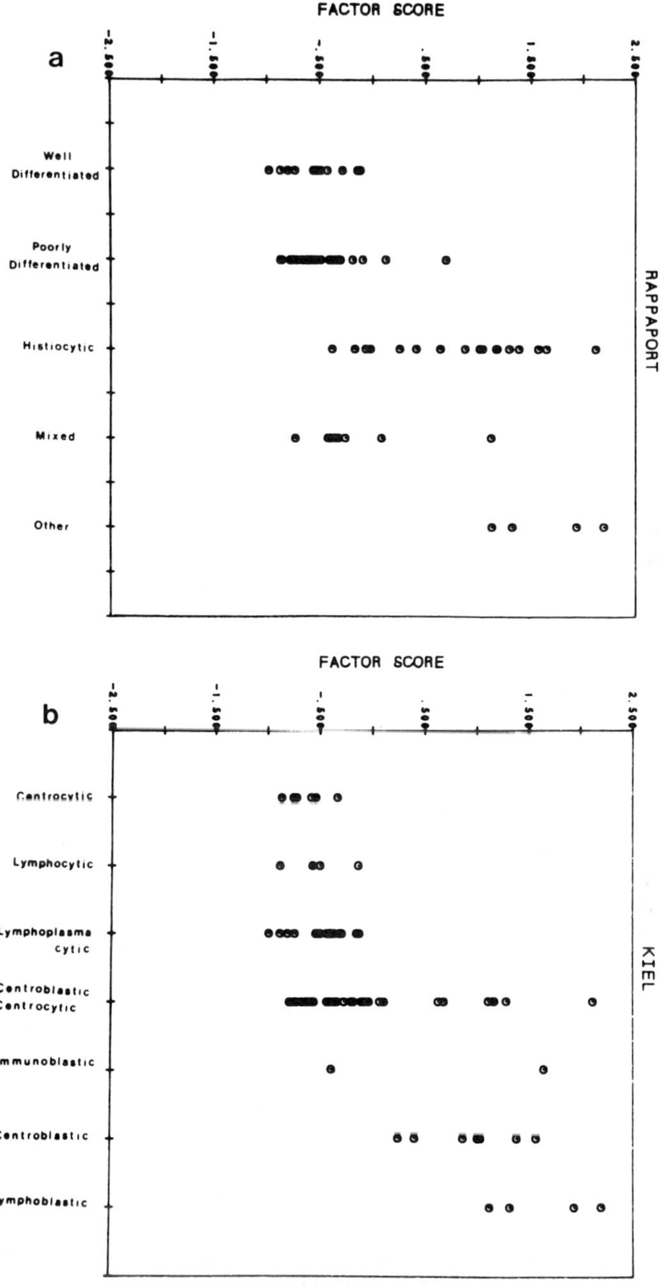

FIGURE 3. Factor score analysis, combining S phase and RNA index, was obtained for the Rappaport (a) and Kiel (b) classifications. Factor score increased with increasing grade of malignancy in both systems.

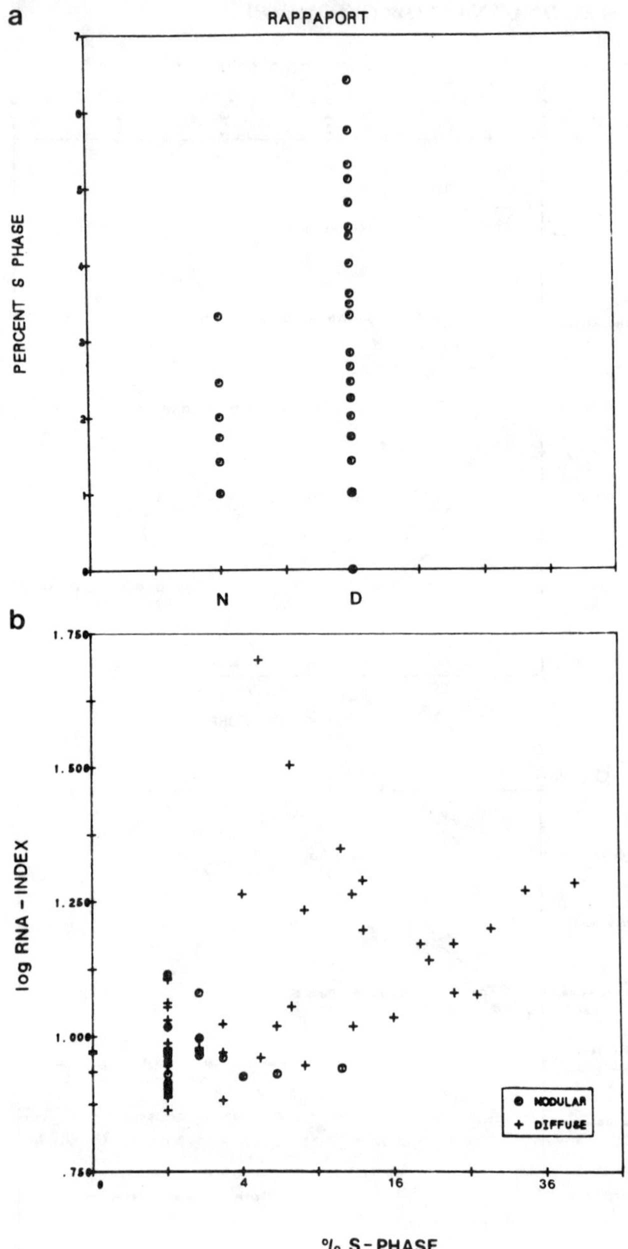

FIGURE 4. Correlation of S phase, RNA index, and growth pattern in NHL. (a) In the Rappaport system, nodular lymphoma has lower S phase than diffuse lymphoma (N = nodular, D = diffuse). The plotting program cannot discriminate between different events falling in the same range. Some dots therefore represent multiple data points. (b) S phase and RNA index in nodular and diffuse lymphoma. Log RNA index is plotted on the y-axis, percent S phase on the x-axis. This plot shows the low correlation between S phase and RNA index for the entire group and the characterization of diffuse lymphomas by both, increased S phase and increased RNA index.

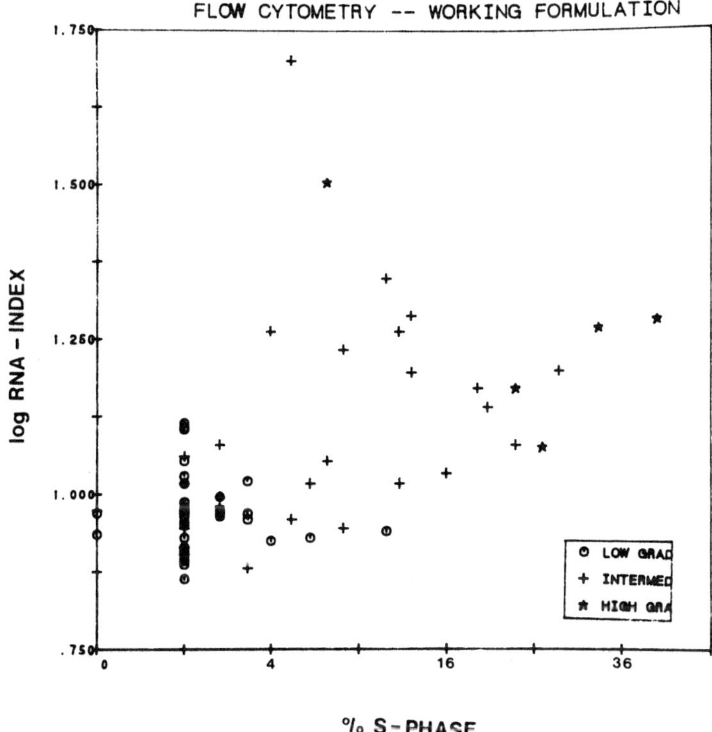

FIGURE 5. Correlation of S phase and RNA index, identified as low-, intermediate-, or high-grade lymphoma by Working Formulation criteria. Comparison with FIGURE 4B shows that the majority of cases with diffuse lymphoma and high S phase and RNA index (S phase over 9%, log RNA index over 1.125) are classified as intermediate- or high-grade malignant NHL.

heterogeneity of modal cell volume; while cells with high RNA content appear to have a rather homogeneous volume. The two variables therefore do not measure the same property.

Prognostic Relevance of Clinical and Laboratory Variables

Survival was determined from time of diagnosis and was analyzed in a variety of ways. FIGURE 7 shows survival based on different subgroups of the Rappaport classification. As expected, low-grade lymphomas including diffuse well-differentiated lymphocytic and nodular lymphomas had 90% and 80% survival at four years, respectively. Patients with diffuse poorly differentiated and in particular with diffuse histiocytic lymphomas had lower survival rates. For DHL, only 40% of patients were alive at four years. These results are consistent with most therapeutic trials reported to date with similar therapeutic regimens. The prognostic importance of the flow cytometry-derived variables for survival was as follows: S phase ($p = 0.05$), DNA stemline ($p = 0.02$), RNA index ($p = 0.002$), and factor analysis (combination of S phase and RNA index, $p = 0.001$).

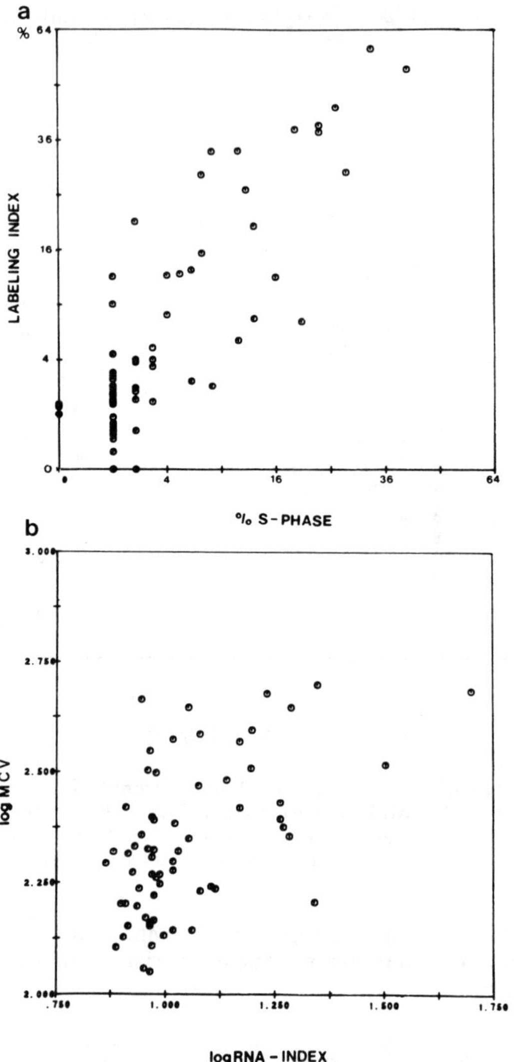

FIGURE 6. Correlation between S phase (*x*-axis) and labeling index (*y*-axis). (a) S phase was determined by flow cytometric measurement of acridine orange green fluorescence, labeling index by incorporation of [³H]thymidine into proliferating cells and auto-radiography. Correlation coefficient $r = 0.845$, $p = 0.00001$. Of note is a generally higher labeling index, as compared to S phase. For explanation see text. (b) Correlation of cell volume and RNA content. MCV was measured in a Coulter counter, the same sample was also measured by acridine orange flow cytometry and the RNA index was calculated as described in the text. Correlation coefficient $r = 0.5334$, $p = 0.00001$. The two variables are correlated, but not significantly.

TABLE 1. Correlation of Laboratory Variables in Patients with B-Cell NHL

Variables	Coefficient of Correlation	p Value
L. I. − S (FCM)	0.845	0.00001
MCV − S (FCM)	0.548	0.00001
L. I. − RNA-$G_{0/1}$	0.574	0.00001
MCV − RNA-$G_{0/1}$	0.5334	0.00001
L. I. − RCG_1	0.5798	0.00001
MCV − RCG_1	0.5848	0.00001

Abbreviations: L. I. = $[^3H]$TdR-pulse labeling index, S(FCM) = S phase determined by acridine-orange flow cytometry, MCV = modal cell volume determined by Coulter counter, RNA-I. $G_{0/1}$ = RNA index of $G_{0/1}$ cells (see Methods), RCG_1 = product of % cells in G_1 and RNA index of G_1 cells. G_1 cells were discriminated from G_0 cells by their RNA content. G_1 cells are defined as cells with unit DNA content and RNA content higher than that of normal control lymphocytes.

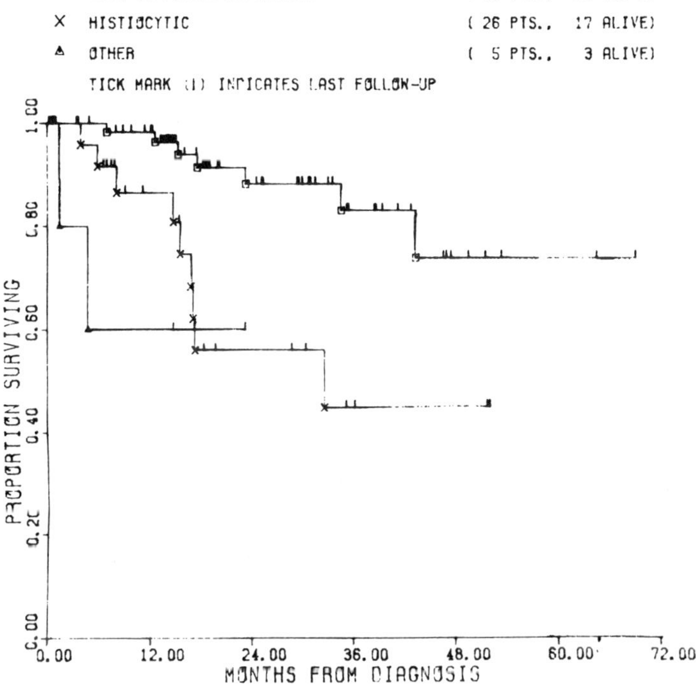

FIGURE 7. Kaplan-Meier plot of survival in 96 patients with B-cell NHL, classified according to the Rappaport system. Patients with well-differentiated, poorly differentiated, and mixed cell lymphomas have significantly better survival as compared to patients with histiocytic lymphoma.

Survival was predicted differently by the different taxonomic systems. In the Rappaport system, nodular lymphomas had a borderline better survival than did diffuse lymphomas ($p < 0.06$). In the Lukes-Collins system, prediction of survival was different for small lymphocytic, small-cleaved, plasmocytic and immunoblastic, compared to large-cleaved, small non-cleaved, and large non-cleaved lymphomas ($p = 0.017$). The Working Formulation, subdivided into low-, intermediate-, and high-grade lymphomas predicted survival with $p = 0.004$. The best prediction of survival was achieved with the Kiel system, which was subdivided into lymphocytic, lymphoplasmocytic, centrocytic, centrocytic-centroblastic versus centroblastic, lymphoblastic, and immunoblastic lymphoma with $p < 0.0003$.

FIGURE 8a shows the prognostic value of RNA index. Patients with an RNA index less than 9.5 had an almost flat survival curve at 95%, while patients with RNA index greater than 9.5 had a median survival of 43.5 months ($p = 0.002$). If the RNA index was subdivided into three groups (< 9.5, 9.5–11.0, and > 11.0), survival was predicted for these three groups with $p = 0.001$. Within the group of high-grade lymphomas, patients with low RNA index had a borderline better survival than patients with high RNA index ($p = 0.068$) (FIGURE 9).

Cox Regression Analysis of Prognostic Variables

Analysis of individual variables seemed to indicate that the variables investigated, i.e. RNA index, S phase, and DNA ploidy, had prognostic significance. This notion was confirmed by the prognostic significance of a factor score that encompasses RNA and S phase. Cox regression analysis was subsequently performed because of its ability to evaluate variables individually and in different combinations. As shown in TABLE 2, S phase alone was not significant in predicting survival. This appears to be at variance with the analysis of individual data presented above. One has to consider, however, that the Cox regression analysis is the more appropriate method. DNA stemline (aneuploid versus diploid) was of prognostic value in the Cox analysis, but RNA alone was much more significant than each of the other variables ($p = 0.001$). The addition of ploidy and S phase to RNA did not add significance to the ability of RNA index to predict survival. RNA differed significantly between the diploid and the aneuploid groups ($p = 0.0056$). A 50% increase in RNA index corresponds to a 2.3-fold increase in risk, i.e. decrease in survival (exponent 2.65).

A "laboratory model" consisting of S phase and RNA index was developed and Cox regression analysis was performed comparing this model with the "Working Formulation." As shown in TABLE 2, the laboratory model predicts survival with a p value of 0.001. The Working Formulation was divided into low-, intermediate-, and high-grade malignant lymphoma. It predicts survival with a p value of 0.0025. The combination of intermediate- and high-grade malignant lymphomas resulted in a coefficient of 5.2, i.e. it is 5.2 times more likely for this group to be a failure compared to the low-grade group. And indeed, the 11 deaths observed were all in the high or intermediate group. Comparison of the laboratory model to the morphological model clearly demonstrates superiority of the laboratory model.

Statistical analysis was then done to evaluate whether laboratory variables could improve the prediction generated by the morphological models. Again, the "Working Formulation" was stratified into low- and intermediate-/high-grade lymphoma. The RNA index was still predictive and added information regarding

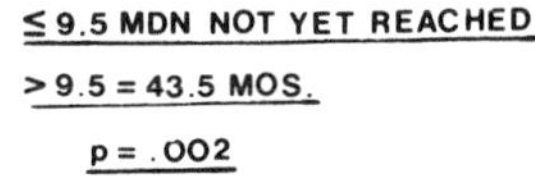

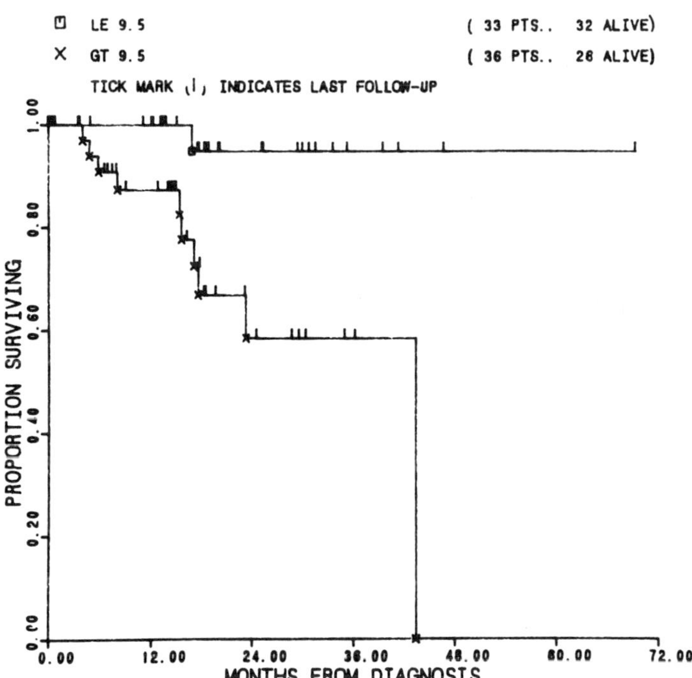

FIGURE 8. Flow cytometry in B-cell lymphomas. Survival duration by RNA index. Kaplan-Meier plot of sixty-nine patients with B-cell NHL. Patients with RNA index $G_{0/1}$ less than 9.5 have significantly better survival as compared to patients with a RNA index greater than 9.5 ($p = 0.002$). Median survival has not been reached for the group with low RNA index. Median survival was 43.5 months for the group with high RNA index.

survival with $p < 0.0155$. A similar approach was used for the Kiel classification, which was divided into low and high risk groups. RNA added prognostic value with $p < 0.0003$.

DISCUSSION

Several authors have investigated the importance of cell size or volume[9,11,17,18,25] of labeling index[7,8,9,15,19] and of S phase, determined by DNA flow cytometry[11,15-17,19,25,28] in non Hodgkin's lymphoma. Because these reports have been generated in different centers, their comparability is limited, as different treatment strategies and classification schedules have been used. Also, as indicated above, reproducibility of morphological diagnosis in non-Hodgkin's lymphoma is poor.

TABLE 2. Cox Regression Analysis of NHL

Cox Regression Analysis of B-Cell NHL
  S phase alone $p = .1835$
  Ploidy alone exp. 3.72 $p = .0264$
  RNA alone exp. 2.65 $p = .001$
  Ploidy and S do not add significance to RNA

Comparison of Two Models to Predict Survival in NHL
  Laboratory model:

{ S phase
{ RNA index
  test of model: $\chi^2 = 17.92$, D.F. = 2, $p = .0001$
    coefficient RNA = 2.8

Working formulation model:

  low-intermediate-high
  test of model: $\chi^2 = 11.94$, D.F. = 2, $p = .0025$
    coefficient I + H − 5.2

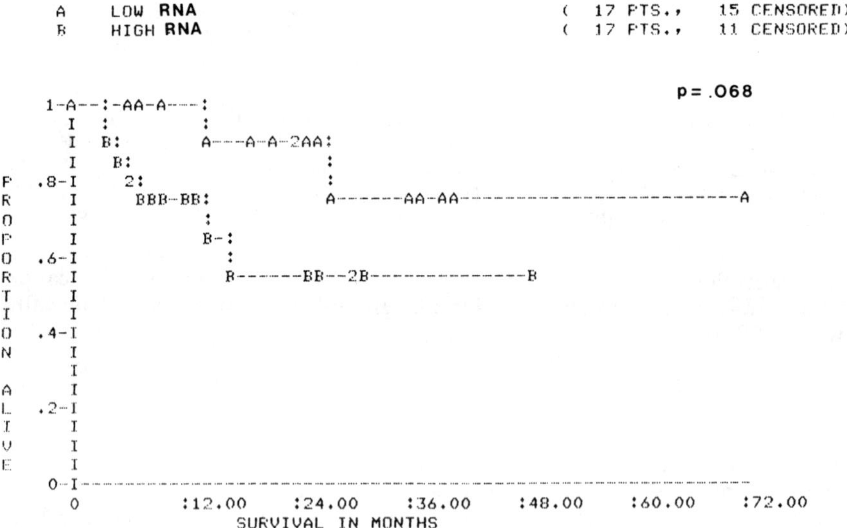

FIGURE 9. Kaplan-Meier plot for 34 patients with high-grade non-Hodgkin's lymphoma, according to the Rappaport classification. Survival is marginally better for patients with low RNA, as compared to those with high RNA ($p = 0.068$).

In this study we have attempted multi-variate analysis of a complex set of data to determine their interrelations and prognostic value. We provide evidence that S phase determined by DNA flow cytometry using acridine orange is highly correlated with labeling index determined by autoradiography of [³H]thymidine labeled cells. Braylan *et al.*[15] found a similar correlation for labeling index and S phase determined by propidium iodide flow cytometry. A potential disadvantage of the flow cytometric method lies in its inability to identify cells morphologically. An elegant solution to this problem has been developed by Braylan *et al.*[27,35] They devised a method for double staining of cells for kappa or lambda light chain and DNA. The method allows the identification of the monoclonal B-cell population and the estimation of its proliferative potential and DNA stemline (FIGURE 10). The contamination with non-malignant lymphocytes, frequency found in non-Hodgkin's lymphomas, can therefore be discriminated from the malignant B-cell process. This technique may also help in detecting very small abnormalities in DNA stemline, which would otherwise not be resolved by standard DNA flow cytometry. The coefficient of variation of our DNA measurements using acridine orange as a fluorescent dye (2.9 + 0.9%) is at least comparable to those of other reported series. Once an aneuploid cell population has been identified by flow cytometry, it can be used to detect disease elsewhere, e.g. in the peripheral blood and bone marrow. We have investigated 51 bone marrow samples from patients with aneuploid non-Hodgkin's lymphomas. Marrow involvement was found in 35% of cases by morphology and/or flow cytometry and in 8% of cases by flow cytometry only. Results will be reported in detail elsewhere.

The large variability of the number of cells involved in DNA synthesis confirms previous observations.[25] Generally speaking, low-grade malignant non-Hodgkin's lymphomas have uniformly low S phase, with the group of inter-mediate- and high-grade lymphomas showing higher S-phase with larger variability. S phase alone is not useful in discriminating low- and high-grade non-Hodgkin's lymphomas. The same appears to be true for labeling index,[9] cellular volume,[20] and RNA index. Because the RNA index, i.e. the cell cycle corrected and normalized RNA content, is calculated only for cells in $G_{0/1}$, it is independent of cells in S, G_2, and mitosis. Cells progressing through the cell cycle increase their RNA content and a non-corrected RNA measurement would rather reflect cell proliferation, which could be better measured by DNA cytometry. However, Darzynkiewicz and co-workers[36] have clearly demonstrated the correlation of RNA content and cell cycle stage. Cells in kinetic quiescence are characterized by low RNA content, which increases when cell progress to G_{1A}, the non-committed part of the G_1 phase, and subsequently to G_{1B}, when cells are invariably committed to enter S phase. These different stages of cell kinetic quiescence and proliferation can clearly be distinguished by RNA content. The highly significant correlation of RNA index and survival suggests that cells with higher RNA content have an increased potential to recur after initial chemo-therapy. The same was demonstrated for adult acute lymphoblastic leukemia, where the RNA index $G_{0.1}$ was identified as an independent prognostic variable.[37] The weak correlation between RNA index and S phase underlines its importance as an independent parameter of cell proliferation, not directly correlated with S phase.

The analysis of prognostic importance of laboratory and clinical variables was performed using a variety of statistical methods, including Kaplan-Meier plots, univariate analysis, and Cox regression analysis. Of these methods, Cox regression analysis is the most appropriate way to analyze these data. Inter-estingly, RNA index turned out to be more important in predicting survival than

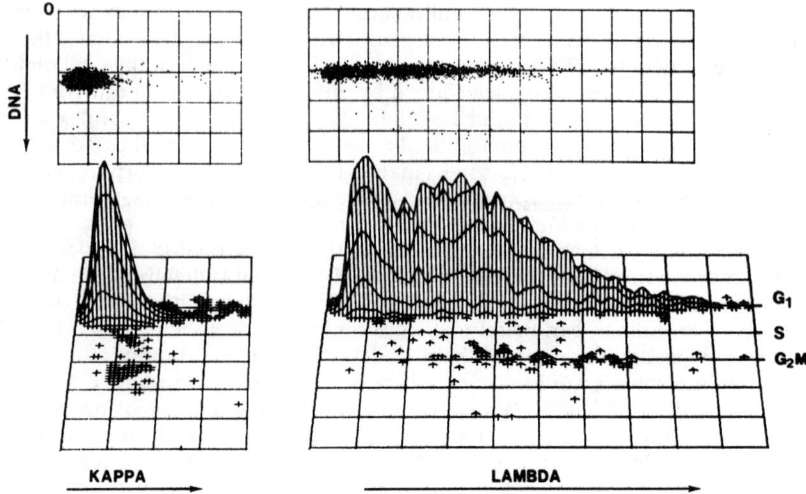

FIGURE 10. Flow cytometric determination of DNA content and immunoglobulin light chain in non Hodgkin's lymphoma. Cells were first stained with FITC-labeled antibodies to kappa or lambda light chains, then fixed, treated with RNase and counterstained with propidium iodide for DNA.[35] Red and green fluorescence was measured, to discriminate DNA content from immunoglobulin light chain expression. Three-dimensional plots (lower panel) of the same data shown as scatter plots in the upper panels. x-axis = kappa or lambda light chain expression, y-axis = DNA content (G_1, S, G_2 M). This sample is characterized by the predominance of cells expressing lambda light chains with low proliferation.

any other single method. A laboratory model consisting of RNA index and S phase was superior in predicting survival compared to all standard morphological methods, if analyzed with identical statistical techniques.

It should be emphasized that the acridine orange technique used in our studies is highly reproducible, requires only a few minutes from sample preparation to statistical analysis, can easily be performed, and therefore should be included in all ongoing studies evaluating prognostic factors in non-Hodgkin's lymphoma. A combination of RNA content, immunoglobulin light chain expression, and DNA content analysis by flow cytometric techniques should be sufficient in diagnosing non-Hodgkin's lymphoma of B-cell type and in the identification of low and high risk groups. It is of interest, that the flow cytometric analysis of clonal excess, as described by Ault and co-workers,[38] was recently correlated with immunoglobin gene rearrangement studies. The excellent correlation found between the two methods further enhances the value of flow cytometric techniques in diagnosing non-Hodgkin's lymphoma, especially in cases with questionable morphology. RNA index could then be used as a predictive tool alone or in conjunction with a standard taxonomic system.

REFERENCES

1. The Non-Hodgkin's Lymphoma Pathologic Classification Project: National Cancer Institute sponsored study of classifications of Non-Hodgkin's lymphomas. 1982. Cancer **49**:2112–2135.
2. ERSBOLL, J., H. B. SCHULTZ, P. HOUGAARD, N. I. NISSEN & K. HOU-JENSEN. 1985. Comparison of the working formulation of non-Hodgkin's lymphoma with the Rappaport, Kiel, and Lukes & Collins classifications: Translational value and prognostic significance based on review of 658 patients treated at a single institution. Cancer **55**:2442–2458.
3. COOPER, E. H., M. J. PECKHAM, R. E. MILLARD, I. M. E. HAMLIN & R. GERARD-MARCHANT. 1968. Cell proliferation in human malignant lymphomas. Analysis of labelling index and DNA content in cell populations obtained by biopsy. Eur. J. Cancer **4**:287.
4. PECKHAM, M. J. & E. H. COOPER. 1970. The pattern of cell growth in reticulum cell sarcoma and lymphosarcoma. Eur. J. Cancer **6**:453.
5. SILVESTRINI, R., R. PIAZZA, A. RICCARDI & F. RILKE. 1977. Correlation of cell kinetic findings with morphology of non-Hodgkin's malignant lymphomas. J. Natl. Cancer Inst. **58**:499.
6. MEYER, J. S. & E. HIGA. 1979. S-phase fractions of cells in lymph nodes and malignant lymphomas. Arch. Pathol. Lab. Med. **103**:93.
7. KVALOY, S., T. GODAL, P. F. MARTON, H. STEEN, I. O. BRENNHOVD & A. F. ABRAHAMSEN. 1981. Spontaneous (^{3}H)-thymidine uptake in histological subgroups of human B-cell lymphomas. Scand. J. Haematol. **26**:221–234.
8. COSTA, A., G. BONADONNA, F. VILLA, P. VALAGUSSA & R. SILVESTRINI. 1981. Labeling index as a prognostic marker in non-Hodgkin's lymphomas. J. Natl. Cancer Inst. **66**:1–5.
9. HANSEN, H., B. KOZINER & B. CLARKSON. 1981. Marker and kinetic studies in the non-Hodgkin's lymphomas. Am J. Med. **71**:107–123.
10. SANDRITTER, W. & H. GRIMM. 1977. DNA in non-Hodgkin's lymphoma, a cytophotometric study. Beitr. Path. **160**:213–230.
11. BRAYLAN, R. C., B. J. FOWLKES, E. S. JAFFE, S. K. SANDERS C. W. BERARD & C. J. HERMAN. 1978. Volumes and DNA distributions of normal and neoplastic human lymphoid cells. Cancer **41**:201–209.
12. ANDREEFF, M., D. FILIPPA, J. STEINMETZ, R. MERTELSMANN, M. R. MELAMED, R. A. GOOD & B. CLARKSON. 1979. Malignant lymphomas: Classification and staging by multiparameter analysis. Engineering Foundation Conference. Automated Cytology VII. (Asilomar, California, November 25–30).
13. ANDREEFF, M., D. FILIPPA, M. R. MELAMED & B. D. CLARKSON. 1980. Flow cytometry in the evaluation of non-Hodgkin lymphomas: DNA/RNA measurements in 107 cases. Proc. 18th Congress Intl. Soc. Hematol. Abstract #839.
14. ANDREEFF, M., D. FILIPPA, T. ROSEN, J. STEINMETZ, T. GEE, R. A. GOOD, B. D. CLARKSON & M. R. MELAMED. 1980. Non-Hodgkin lymphomas: Classification by DNA/RNA flow cytometry and surface maker analysis. Proc. Amer. Assoc. Clin. Oncol. **21**:585.
15. BRAYLAN, R. C., L. W. POWELL & B. MARTY-GOLDER. 1980. Percentage of cells in the S phase of the cell cycle in human lymphoma determined by flow cytometry: Correlation with labeling index and patient survival. Cytometry **1**:171–174.
16. DIAMOND, L. W., R. C. BRAYLAN, R. M. BEARMAN, C. D. WINBERG & H. RAPPAPORT. 1980. The determination of cellular DNA content in neoplastic and non-neoplastic lymphoid populations by flow cytofluorometry. Flow Cytometry **4**:478–482.
17. DIAMOND, L. W. & R. C. BRAYLAN. Flow analysis of DNA content and cell size in non-Hodgkin's lymphoma. Cancer Res. **40**:703–712.
18. SHACKNEY, S. E., K. S. SKRAMSTAD, R. E. CUNNINGHAM, D. J. DUGAS, T. L. LINCOLN & R. J. LUKES. 1980. Dual parameter flow cytometry studies in human lymphomas. J. Clin. Invest. **66**:1281–1294.

19. COSTA, A., G. MAZZINI, G. DEL BINO & R. SILVESTRINI. 1981. DNA content and kinetic characteristics of non-Hodgkin's lymphoma: Determined by flow cytometry and autoradiography. Cytometry 2:185–188.

20. DIAMOND L. W., B. N. NATHWANI & H. RAPPAPORT. 1982. Flow cytometry in the diagnosis and classification of malignant lymphoma and leukemia. Cancer 50:1122–1135.

21. BRAYLAN, R. C., N. A. BENSON, V. NOURSE & H. S. KRUTH. 1982. Correlated analysis of cellular DNA, membrane antigens and light scatter of human lymphoid cells. Cytometry 2:337–343.

22. BRAYLAN, R. C., N. A. BENSON & V. A. NOURSE. 1982. Flow cytometry: A new approach toward characterizing lymphomas. In B and T Cell Tumors: Biological and Clinical Aspects. E. Vitetta & C. F. Fox, Eds.: 1–5. Academy Press. New York.

23. ANDREEFF, M., H. HANSEN, C. CIRRINCIONE, G. ASSING & M. R. MELAMED. 1983. Prognostic value of DNA-RNA flow cytometry of B-cell non-Hodgkin lymphoma (NHL) in conjunction with four taxonomic systems. Proc. Amer. Soc. Clin. Onc. 2:13.

24. PORWIT-KSIAZEK, A., B. CHRISTENSSON, C. LINDEMALM, H. MELLSTEDT, B. TRIBUKAIT, G. BIBERFELD & P. BIBERFELD. 1983. Characterization of malignant and non-neoplastic cell phenotypes in highly malignant non-Hodgkin lymphomas. Br. J. Cancer 32:667–674.

25. SHACKNEY, S. E., A. M. LEVINE, R. I. FISHER, P. NICHOLS, E. JAFFE, W. H. SCHUETTE, R. SIMON & C. A. SMITH. 1984. The biology of tumor growth in the non-Hodgkin's lymphomas: A dual parameter flow cytometry study of 220 cases. J. Clin. Invest.: 1201–1214.

26. ANDREEFF, M., H. HANSEN, C CIRRINCIONE, H. THALER, S. GROSHEN, G. ASSING, B. D. CLARKSON & M. R. MELAMED. 1984. Cellular RNA content: a major prognostic factor in adult acute leukemia and non-Hodgkin lymphoma. Proc. Intl. Conf. Analyt. Cytol. 10:25.

27. BRAYLAN, R. C., N. A. BENSON & V. A. NOURSE. 1984. Cellular DNA of human neoplastic B-cells measured by flow cytometry. Cancer Res. 44:8010–8016.

28. SRIGLEY, J., B. BARLOGIE, J. J. BUTLER, B. OSBORNE, M. BLICK, D. JOHNSTON, H. KANTARJIAN & J. REUBEN. 1985. Heterogeneity of non-Hodgkin's lymphoma probed by nucleic acid cytometry. Blood 65:1090–1096.

29. ANDREEFF, M., J. BECK, Z. D. DARZYNKIEWICZ, F. TRAGANOS, S. GUPTA, M. R. MELAMED & R. A. GOOD. 1978. RNA content of human lymphocyte subpopulations. Proc. Natl. Acad. Sci. USA 75:1938–1942.

30. KAPUSCINSKI, J., Z. DARZYNKIEWICZ & M. R. MELAMED. 1982. Luminescence of the solid complexes of acridine orange with RNA. Cytometry 2:202–211.

31. ANDREEFF, M., A. REDNER, S. THONGPRASERT, B. EAGLE, P. STEINHERZ, D. MILLER & M. R. MELAMED. 1985. Multiparameter flow cytometry for determination of ploidy, proliferation and differentiation in acute leukemia: treatment effects and prognostic value. In Tumor Aneuploidy. Th. Buechner, Ed.: 81–105. Springer-Verlag. Berlin.

32. ANDREEFF, M., Z. DARZYNKIEWICZ, T. K. SHARPLESS, B. D. CLARKSON & M. R. MELAMED. 1980. Discrimination of human leukemia subtypes by flow cytometric analysis of cellular DNA and RNA. Blood 55:282–293.

33. HIDDEMANN, W., J. SCHUMANN, M. ANDREEFF, B. BARLOGIE, C. HERMAN, R. C. LEIF, B. H. MAYALL, R. F. MURPHY & A. A. SANDBERG. 1984. Convention on nomenclature for DNA cytometry. Cytometry 5:445–446.

34. SLATER, D., E., R. MERTELSMANN, B. KOZINER, C. HIGGINS, S. MCKENZIE, P. SCHAUER, T. GEE, D. STRAUS, S. KEMPIN, Z. ARLIN & B. D. CLARKSON. 1986. Lymphoblastic lymphoma in adults. J. Clin. Oncol. 4:57–67.

35. KRUTH, H. S., R. C. BRAYLAN, N. A. BENSON & V. A. NOURSE. 1981. Simultaneous analysis of DNA and cell surface immunoglobulin in human B-cell lymphomas by flow cytometry. Cancer Res. 41:4895–4899.

36. DARZYNKIEWICZ, Z. & M. ANDREEFF. 1981. Multiparameter flow cytometry. Part 1: Application in analysis of the cell cycle. MSKCC Clin. Bull. 11:47–57.

37. CLARKSON, B., S. ELLIS, C. LITTLE, T. GEE, Z. ARLIN, R. MERTELSMANN, M. ANDREEFF, S. KEMPIN, B. KOZINER, R. CHAGANTI, S. JHANWAR, S. MCKENZIE, C. CIRRINCIONE & J. GAYNOR. 1985. Acute lymphoblastic leukemia in adults. Sem. Oncol. 12:160–179.

Prognostic Value of DNA/RNA Flow Cytometry in Myeloblastic and Lymphoblastic Leukemia in Adults: RNA Content and S-Phase Predict Remission Duration and Survival in Multi-Variate Analysis[a]

MICHAEL ANDREEFF,[b,c] GORDON ASSING,[b] AND
CONNIE CIRRINCIONE[d]

bLeukemia Cell Biology Laboratory
Departments of cMedicine and dBiostatistics
Memorial Sloan Kettering Cancer Center
New York, New York 10021

In the pre-chemotherapeutic era, the prognosis for patients diagnosed as having acute lymphocytic or acute non-lymphocytic leukemia was uniform: all patients died within a rather limited period of time. The exception to this rule were very few patients, who had prolonged survival or even remissions that occurred spontaneously. With the development of increasingly effective chemotherapeutic regimens, it became more important to predict which patients would do well on a given treatment schedule and which would not.

The most impressive gains in remission incidence and survival have been obtained in pediatric acute lymphoblastic leukemia. In adults, significant progress has been achieved in the treatment of ALL with approximately 44% of patients being long-term survivors.[1,2] In non-lymphoblastic leukemia (ANLL), remission rates have been steady between 60 and 80%, but relapse rate is high: less than 20% of all patients survive five years and the majority of relapses occur during the first and second year after the achievement of remission.[1] In a number of studies, prognostic factors have been investigated and age and white blood cell count were identified as significant prognostic parameters in ALL, while age, presence of Auer rods, and absence of terminal deoxynucleotidyl transferase (TdT) were found to be good prognostic signs in ANLL.[3]

In this study, we present data from a multi-variate analysis of a large number of patients with adult ANLL and a smaller number of patients with ALL. The special emphasis of this study was the evaluation of the possible prognostic importance of flow cytometry–derived variables, such as S-phase, RNA content, and DNA index.

[a]Supported by grants from the National Cancer Institute CA-20194, CA-29564, and CA-05826.

MATERIALS AND METHODS

Patients

One hundred twenty patients with acute non-lymphoblastic leukemia, diagnosed between March 1977 and January 1983, are included in this study. They were newly diagnosed and untreated at the time when studies were performed. Treatment consisted of Memorial Hospital protocols L-14M, L-16, or L-16M.

Forty-four patients with newly diagnosed and untreated acute lymphoblastic leukemia (ALL) were studied. These patients were treated according to the L-17 and L-17M protocols.[2]

Preparation of Cells

Bone marrow aspirates were taken from the posterior iliac crest under local anesthesia. Bone marrow (0.5 ml) was aspirated into heparinized syringes. Whenever possible, bone marrow biopsies were obtained from an immediately adjacent site. After complete mechanical dispersion, the bone marrow biopsy suspension was drained through a nylon filter to remove residual bone chips. Bone marrow aspirates were subjected to Ficoll-Hypaque gradient separation and washed twice.

Simultaneous Staining for DNA and RNA

Staining for cellular DNA and RNA content was done using the acridine orange (AO) two-step technique. Cells were pretreated with Triton X-100, which makes cells permeable to the dye at low pH when nucleic acids remain insoluble. Subsequent staining with AO in the presence of EDTA results in denaturation of cellular RNA, which then stains metachromatically red, while native double-stranded DNA intercalates AO and stains orthochromatically green.[6] Aliquots (0.2 ml) of cell suspensions containing 0.2–0.4×10^6 cells in HBSS were mixed with 0.4 ml of 0.05 N HCl, 0.15 N NaCl, and 0.1% Triton X-100 (vol/vol, Sigma Chemical Co., St. Louis, MO). After 30 seconds, 1.2 ml of a solution containing Na_2HPO_4 (0.2 M), citric acid (0.1 M) buffer (pH 6.0), 1 mM EDTA-Na, 0.15 N NaCl, and 6 µg/ml of acridine orange (AO) were added. Chromatographically purified AO obtained from Polysciences Inc. (Warrington, PA) was used.

Flow Cytometric Instrumentation and Multi-Parameter Data Analysis

A computer-interfaced research cytofluorograph, Model FC-201 (Ortho Instruments, Westwood, MA) was used to obtain simultaneous measurements of cells in two separate wave length bands (F 530 in the band from 515 to 575 nm; and F 600 in a band from 600 to 650 nm). Red fluorescence and green fluorescence emission for each cell was optically separated and measured by separate photomultipliers. 5,000 cells were counted in each sample and background fluorescence of the dye solution was automatically subtracted. Red pulse width, i.e. the time the cell nucleus takes to pass through the illuminating beam, was recorded and used to distinguish single cells from cell aggregates. Cells were illuminated with a 488 nm argon ion laser (Lexel, 30 mW).

All measurements were stored in a computer memory (1220 Nova mini-computer) in list mode. Software developed in our laboratory by Sharpless was used to analyze each sample. Statistical analysis was performed on the total and on sub-populations and includes total number of cells, mean, median, peak values, standard deviations, co-efficient of variation, and skewness. DNA histograms, scattergrams, and three-dimensional plots were all computer generated. The details of the method have been described elsewhere.[4,7]

Cell cycle analysis was performed using our "Peak Symmetry Model." RNA index (RI) was calculated as described.[5] A refinement of the analysis technique was applied, in that G_0 and G_1 cells were separately determined, based on differences in their RNA content. The upper limit of G_0 was determined by the RNA content of normal control peripheral blood lymphocytes. Cells with a higher RNA and G_1 DNA content were termed "G_1". This discrimination is based on differences in RNA content observed in quiescent and stimulated peripheral blood lymphocytes. They may not apply to cells of non-lymphoid origin, e.g. myeloid cells. We will, therefore, use the terms "G_0" and "G_1" in an operational, and not strictly in a functional sense.

Definition of G_0, G_1, $G_{0/1}$ Cells

G_0 cells are cells with unit DNA content and RNA content of control lymphocytes (RNA index = 10) or less. G_1 cells are cells with unit DNA content and RNA content higher than that of control lymphocytes (RNA index > 10). $G_{0/1}$ cells are all cells with unit DNA content regardless of their RNA content.

Definition of RNA Indices

$$RIG_0 = \frac{\text{Mean RNA content of } G_0 \text{ cells (sample)} \times 10}{\text{Median RNA content of } G_0 \text{ control lymphocytes}}$$

$$RIG_1 = \frac{\text{Mean RNA content of } G_1 \text{ cells (sample)} \times 10}{\text{Median RNA content of } G_0 \text{ control lymphocytes}}$$

$$RIG_{0/1} = \frac{\text{Mean RNA content of } G_{0/1} \text{ cells (sample)} \times 10}{\text{Median RNA content of } G_0 \text{ control lymphocytes}}$$

Statistical Analysis

A large number of variables was collected for each patient, including diagnosis, FAB classification for ALL and ANLL, terminal transferase, LDH, WBC, marrow blast number, age, sex, G_0, G_1, $G_{0/1}$, RNA index G_0, RNA index G_1, RNA $G_{0/1}$, S, G_2M, DNA index, and karyotype.

Uni-variate and multi-variate statistical analysis (Step-wise Cox regression analysis) was also performed correlating these data with remission incidence, time to remission, remission duration, and survival.

RESULTS

Pretreatment Variables in ALL

A typical DNA/RNA histogram for ALL cells is shown in FIGURE 1. DNA aneuploid cells can be distinguished from DNA diploid cells. DNA aneuploidy was previously found in 30% of adult ALL and 40% of pediatric ALL.[8] DNA aneuploidy is an excellent tumor marker. Some of the diploid cells, however, may also be of leukemic origin. This can be shown in double-staining experiments where cells are stained for DNA and CALLA.[9] RNA content of these leukemias is generally lower than that of ANLLs. The median RNA index $G_{0/1}$ was 11.4 for this patient group (TABLE 1). Median S-phase was 4%, WBC 8,950/µl, LDH 741 units/ml, and the median age for the group was 24.79 years. There was a predominance of male ($N = 25$) to female ($N = 19$) patients. TdT was studied in 30 patients with a biochemical assay and was found to be positive in 28 out of 30 patients. In 4 patients, immunofluorescence of TdT was positive: in these samples, 42% of cells were positive. DNA aneuploidy was found in 8 out of 39 patients studied (20%). Cytogenetic analysis in 30 patients revealed no analyzable metaphases in 9 patients, diploid karyotype in 11 patients, pseudodiploid karyotype in 4 patients, hypodiploidy in 3 patients, and hyperdiploidy in 3 patients.

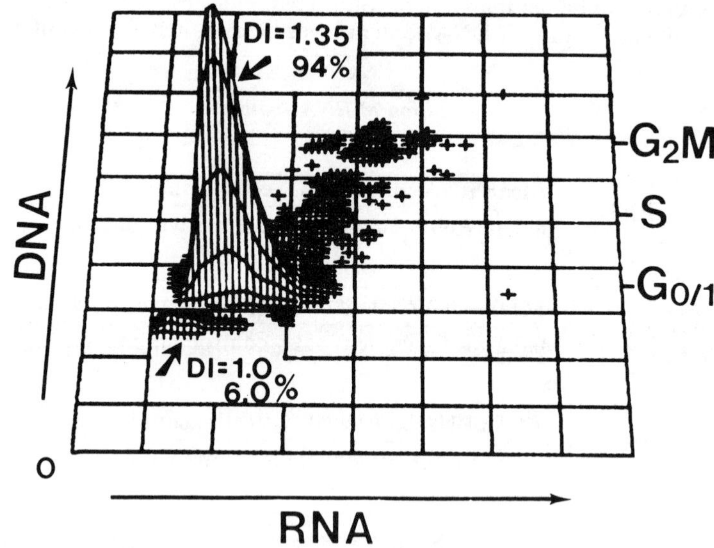

FIGURE 1. DNA/RNA histogram of bone marrow cells from patient with acute lymphoblastic leukemia. Diploid (DNA index = 1.0) and hyperdiploid (DNA index = 1.35) cells can be clearly distinguished. The hyperdiploid cells are proliferating. RNA index of the hyperdiploid cells is low (RNA index $G_{0/1}$ = 14.3).

TABLE 1. ALL

	x	SD	Median	Patients Studied
S-phase	5.526	4.298	4.00	38
RNA-I.$G_{0/1}$	11.997	2.763	11.40	39
RNA-I.G_0	10.938	2.825	10.30	29
RNA-I.G_1	16.790	4.031	16.10	29
$G_{0/1}\%$	91.737	6.070	93.50	38
$G_0\%$	73.448	20.060	82.00	29
$G_1\%$	17.517	17.788	11.00	29
RC $G_{0/1}$	1090.995	258.967	1067.50	38
RC G_0	808.100	328.70	817.80	29
RC G_1	294.134	316.986	180.40	29
G_2M	3.00	2.528	2.00	37
WBC($\times 10^3/\mu l$)	21.603	25.692	8.95	30
LDH (U/ml)	1245.654	1589.641	741.00	26
Survival (months)	21.893	15.289	18.73	44
Remission duration (months)	18.455	15.871	14.45	38
Days to CR	41.789	36.342	35.00	38
Age (Total)	30.482	15.173	24.79	44
Age (Male)	30.630	14.704	24.52	25
Age (Female)	30.286	16.174	25.00	19
Sex	M = 25	F = 19		44
Response	Fail = 4	CR = 38	Died/Rx = 2	44
TdT (biochemistry)	Pos = 28	Neg = 2		30
IF (TdT by immunofluorescence)	Pos = 4	Neg = 2		6
IF% Pos (TdT)	49.250	32.469	42.00	4

Ploidy:DNA Index		Karyotype:diploid	(11)
1.00	(31)	hypodiploid	(3)
1.05	(4)	hyperdiploid	(3)
1.15	(1)	no analyzable	
1.20	(1)	metaphases	(9)
1.40	(1)	pseudodiploid	(4)
1.60	(1)		
Total = 39			Total = 30

Pretreatment characteristics of 44 patients with adult, newly diagnosed, untreated lymphoblastic leukemia. S-phase and RNA index are derived from acridine orange flow cytometric measurements. RC is the product of RNA index and percent of cells. Survival is measured from time of diagnosis, remission duration from time of achievement of remission.

Pretreatment Variables in ANLL

One hundred twenty patients with ANLL were included in this study. FIGURE 2 shows a "typical" DNA/RNA histogram: cells in G_0 and G_1 can be discriminated from cells in S and G_2M. This particular patient has a monocytic leukemia, subtype M5A by the FAB classification. His marrow contains 87% blasts and leukemic monocytes. Twenty-four percent of cells are in G_0 with an RNA index $G_0 = 10.4$, 65% of cells are in G_1 with RNA index $G_1 = 29.0$. The figure illustrates the discrimination of G_0 and G_1. The over-all RNA index $G_{0/1}$ is 23.9. This is in contrast to the considerably lower RNA index seen in the ALL shown in FIGURE 1.

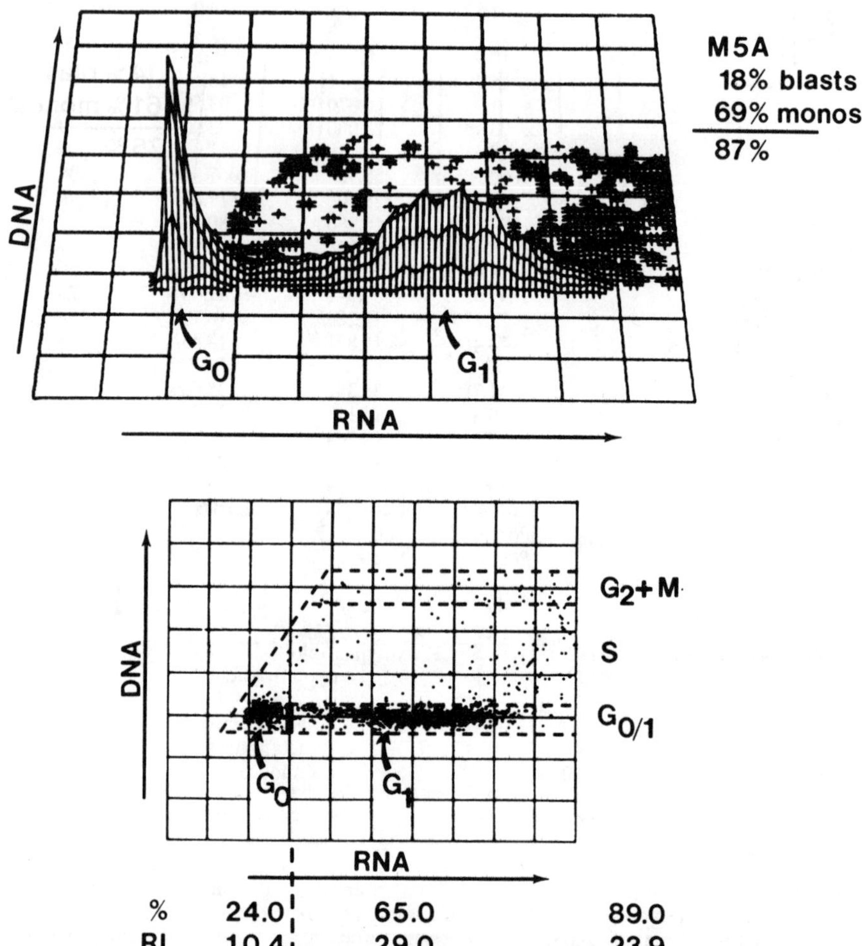

FIGURE 2. DNA/RNA histogram of bone marrow cells from patient with ANLL (M5A). Marrow contains 87% leukemic cells. The DNA/RNA histogram shows 24% G_0 cells with RNA index 10.4; 65% of cells have RNA index 29.0; 89% of cells are in $G_{0/1}$ with RNA index 23.9. The top panel represents a three-dimensional display of the data, the z axis represents the number of cells. The bottom panel shows the same data, displayed as scatter plot.

However, a number of patients with ANLL are characterized by low RNA index as shown in FIGURE 3. This monoblastic leukemia (FAB M5B) has 75% blasts and monocytes in the bone marrow. Though the composition of cell types is similar to that of the patient shown in FIGURE 2, the RNA content is significantly lower: 75%

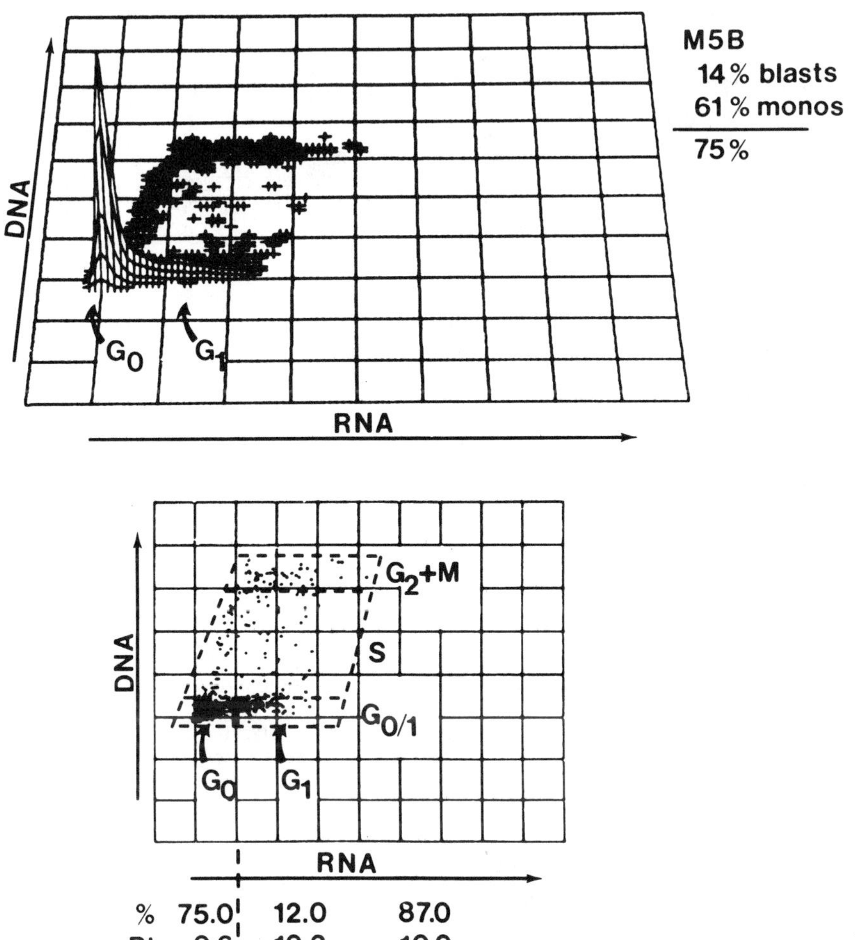

FIGURE 3. DNA/RNA histogram of ANLL, characterized by low RNA content (M5B) This bone marrow sample contains 75% leukemic cells. The majority of cells is in G_0 (75%) with RNA index 9.6; 12.0% of cells are in G_1 with RNA index 19.3; 87% of cells are in $G_{0/1}$ with RNA index 10.9. In addition, cells in S and G_2M can be identified.

of cells are in G_0 with RNA index $G_0 = 9.6$, 12.0% of cells are in G_1 with RNA index $G_1 = 19.3$. The over-all RNA index $G_{0/1}$ is 10.9.

TABLE 2 gives a summary of kinetic and clinical data in ANLL. The median S phase was 4% and is not different in ALL. RNA index $G_{0/1}$ is 13.9 with 41.5% cells

TABLE 2. ANLL

	x	SD	Median	Patients Studied
S-phase	5.061	3.796	4.00	115
RNA-I.$G_{0/1}$	14.678	4.308	13.90	120
RNA-I.G_0	9.754	2.411	9.90	110
RNA-I.G_1	19.222	3.928	19.05	110
$G_{0/1}\%$	92.733	4.636	94.00	120
$G_0\%$	50.718	25.090	48.00	110
$G_1\%$	42.182	24.680	41.50	110
RC $G_{0/1}$	1360.694	407.685	1272.40	120
RC G_0	493.676	285.344	454.20	110
RC G_1	852.475	593.397	788.80	110
G_2M	2.351	1.805	2.00	114
WBC ($\times 10^3/\mu l$)	38.533	59.851	9.70	69
LDH (U/ml)	782.082	820.690	473.00	61
Survival (months)	13.303	15.206	7.92	121
Remission duration (months)	13.008	13.824	7.98	65
Days to CR	29.723	13.648	26.00	65
Age (Total)	48.255	18.171	49.75	122
Age (Male)	48.455	18.445	51.02	69
Age (Female)	48.008	17.981	48.48	53
Sex	M = 69	F = 53		122
Response	Fail = 26	CR = 65	Died/Rx = 25	116
TdT (biochemistry)	Pos = 19	Neg = 39		58
IF (TdT by Immunofluorescence)	Pos = 4	Neg = 11		15
IF% Pos (TdT)	18.750	20.775	12.00	4

Ploidy:DNA Index		Karyotype		
	0.70 (1)	diploid		(20)
	1.00 (95)	hypodiploid		(7)
	1.05 (7)	hyperdiploid		(9)
	1.10 (11)	no analyzable		
	1.15 (2)	metaphases		(13)
	1.20 (2)	pseudodiploid		(10)
	1.30 (1)			
	2.15 (1)			
	Total = 120			Total = 59

Laboratory and clinical variables in acute non-lymphoblastic leukemia. For details see text and legend for TABLE 1.

having an RNA index G_1 of 19.05. This is in contrast to ALL, where only 11% have an RNA index G_1 of 16.1. In some ALL, this population with high RNA content probably represents residual non-leukemic myeloid cells. WBC in ANLL is lower than in ALL (9,700/μl), LDH is also significantly lower (473 units/ml). The age of ANLL patients is approximately twice as high (49.75 years), as compared to ALL patients. Sixty-nine patients were males and 53 patients were females. TdT was positive in 19/58 patients studied by biochemical methods and was positive in 4/15 patients studied by immunofluorescence.

DNA aneuploidy was found in 25 of 120 patients studied. Ploidy ranged from hypodiploid (DNA index 0.70) to hypertetraploid (DNA index 2.15) with a

TABLE 3. Correlation of FAB, Proliferation, and RNA Index in ANLL

| FAB | S ± SD | RNA Index | |
		G_1 ± SD	$G_{0/1}$ ± SD
M0	10.3 ± 9.05	18.9 ± 5.34	15.1 ± 7.15
M1	3.8 ± 1.67	19.0 ± 2.09	12.3 ± 3.04
M2	5.4 ± 3.37	18.2 ± 3.52	13.8 ± 2.68
M3	3.1 ± 1.39	20.9 ± 3.25	17.3 ± 4.01
M4	5.0 ± 3.56	18.5 ± 2.91	13.7 ± 2.67
M5	5.0 ± 3.11	20.0 ± 3.95	14.9 ± 4.29
Total	5.0 ± 3.96	19.3 ± 3.58	14.6 ± 4.15

Correlation of FAB subgroups in ANLL with proliferation (S-phase) and RNA index (G_1 and $G_{0/1}$).

maximum in the hyperdiploid range between DNA index 1.05 and 1.10 (diploid DNA index = 1.0).

Cytogenetic analysis was done in 59 patients. Thirteen patients had no analyzable metaphases, 20 patients were diploid, 7 patients were hypodiploid, 9 hyperdiploid, and 10 were pseudodiploid.

Proliferation and RNA index were correlated with FAB-subgroups. TABLE 3 shows that only unclassified acute leukemias had a higher S-phase (10.3%). Other subgroups are not significantly different from each other. Similarly, RNA index G_1 and $G_{0/1}$ were not significantly different for the different subgroups.

In order to better understand the correlation between myeloid and lymphoid phenotype and high and low RNA content, a group of patients was selected that had diagnoses of ALL and ANLL established by a variety of criteria. ALL were all TdT positive, ANLL were TdT negative. Cytochemistry and morphology of peripheral blood and bone marrow cells were all interpreted anonymously by at least two investigators as consistent with ALL or ANLL. FIGURE 4 shows the correlation between percent cells in G_0 and the percentage of lymphoblasts in patients with ALL. The correlation coefficient was 0.8339 (top). The plot on the bottom of FIGURE 4 correlates the percentage of cells with G_1 RNA content and the percent of blasts and myeloid cells in patients with ANLL. The correlation coefficient again is 0.84. These figures demonstrate the close association of low RNA index with lymphoblasts in ALL and of high RNA index with myeloid cells in ANLL.

Reproducibility of DNA and RNA Measurements by Flow Cytometry

In order to determine the reproducibility of DNA/RNA measurements, three different individuals, all trained extensively in flow cytometry in our laboratory, measured the same sample repeatedly and analyzed the individual measurements separately. A hyperdiploid leukemia was selected and TABLE 4 gives an account of the five measurements done by each flow cytometrist. Measurements 1–5, 6–10, and 11–15 were done by the same operator. Samples were analyzed blindly, without comparison between the different individuals. TABLE 5 gives the mean values, median values, and standard deviations for each measurement, i.e. ploidy, % aneuploid cells, cell cycle, and RNA analysis. The measurements demonstrate the excellent reproducibility of the acridine orange technique, if performed by well-trained flow cytometrists.

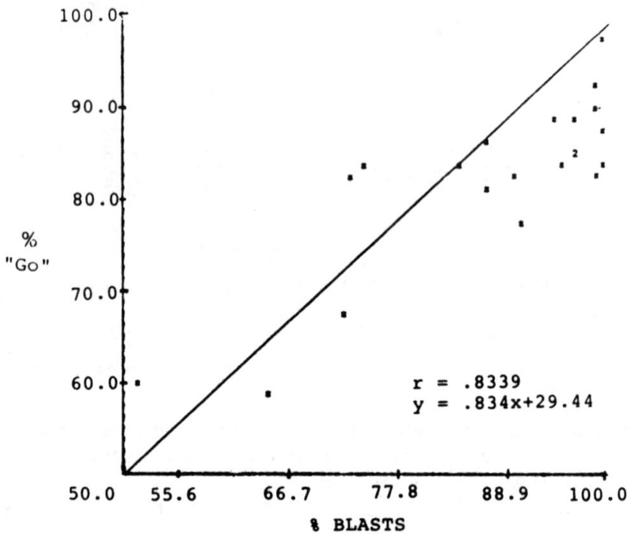

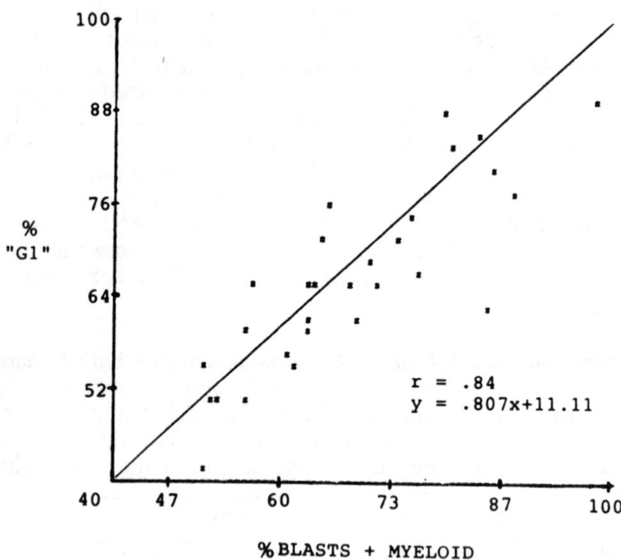

FIGURE 4. Correlation of flow cytometric data and morphology in ALL (top) and ANLL (bottom). Percent blasts is plotted on x axis for ALL and percent blasts and myeloid cells for ANLL. y axis gives percent of cells in G_0 for ALL (low RNA), and percent of cells in G_1 for ANLL (high RNA). The correlation coefficient is $r = 0.8339$ in ALL and $r = 0.84$ in ANLL. For details see text.

TABLE 4. Flow Cytometry Assay for DNA and RNA: Quality Control

	DNA Index	Aneuploid %	G_0 (%)	RNA-I	G_1 (%)	RNA-I	$G_{0/1}$ (%)	RNA-I	S (%)	G_2 M (%)	$S + G_2 + M$ (%)
1.	1.2	97.3	82.0	11.1	10	13.4	92.4	11.4	6.7	0.9	7.6
2.	1.2	98.3	83.0	11.1	11	13.3	94.1	11.4	5.6	0.3	5.9
3.	1.25	98.3	82.0	11.1	11	13.5	92.9	11.4	6.2	0.9	7.1
4.	1.2	97.3	83.0	11.1	10	13.5	93.0	11.4	6.4	0.6	7.0
5.	1.2	97.5	80.0	11.3	14	13.4	94.3	11.5	5.3	0.4	5.7
6.	1.2	97.7	89.9	12.1	3	15.8	92.9	12.2	6.7	0.4	7.1
7.	1.2	97.5	79.9	13.0	13.5	15.6	93.4	13.4	6.4	0.2	6.6
8.	1.2	97.8	79.9	12.9	13.3	15.6	93.2	13.4	6.7	0.1	6.8
9.	1.2	98.0	78.3	13.0	15	15.7	93.3	13.4	6.5	0.2	6.7
10.	1.2	97.1	66.4	13.3	26.9	15.5	93.3	13.9	6.6	0.1	6.7
11.	1.2	97.5	81.0	12.1	13	15.6	94.0	12.6	5.8	0.2	6.0
12.	1.2	97.1	81.0	11.7	13	14.8	93.9	12.1	5.5	0.6	6.1
13.	1.2	97.7	80.0	11.8	14	14.7	93.6	12.3	6.2	0.2	6.4
14.	1.2	97.4	80.0	12.4	13	16.6	92.8	12.9	6.8	0.4	7.2
15.	1.2	97.6	80.0	11.8	14	14.7	94.2	12.3	5.4	0.4	5.8

Quality control of flow cytometric assay for DNA and RNA in acute leukemia. Three flow cytometrists measured the same sample five times and analyzed the results independently. The sample was characterized by hyperdiploid DNA stemline. Readings 1–5, 6–10, and 11–15 relate to the three investigators.

TABLE 5. Flow Cytometry Assay for DNA and RNA Quality Control

	$\bar{X}$	SD	Median
Ploidy	1.203	0.013	1.20
% Aneuploid	97.680	0.353	97.70
% G_0	80.427	4.714	80
RI G_0	11.987	0.783	11.80
% G_1	12.980	4.830	13
RI G_1	14.780	1.103	14.80
% $G_{0/1}$	93.427	0.586	93.30
RI $G_{0/1}$	12.373	0.866	12.30
% S	6.187	0.528	6.40
% $G_2 + M$	0.393	0.258	0.40
% $S + G_2 + M$	6.580	0.577	6.70

Quality control of flow cytometric assay for DNA and RNA: this table summarizes the results of TABLE 4 and provides mean values, medians, and standard deviations for each variable measured.

Response to Treatment in ALL

Complete remission was achieved in 38/44 patients, 2 patients died during induction therapy, and 4 patients failed. Remission duration for the responding patients was 14.45 months, survival was 18.73 months. The median time to achievement of complete remission was 35 days (TABLE 1). The achievement of complete remission was invariably associated with disappearance of aneuploid cells determined by flow cytometry. In some cases, clinical remission was achieved but DNA aneuploid cells persisted. Results from this study have been reported elsewhere.[7]

Response to Treatment in ANLL

Complete remissions were achieved in 65/116 patients. Twenty-five patients died during induction therapy, and 26 patients failed therapy. The median remission duration was 7.9 months and the median survival was also 7.9 months. Mean remission duration was 13.0 months and mean survival was 13.3 months. The median time to complete remission was 26 days.

An association was seen between pretreatment RNA index and remission rate: patients with RNA index $G_{0/1}$ over 19 achieved CR in 8/9 cased = 89%; patients with RNA index $G_{0/1} \geqslant 16$ achieved CR in 31/48 = 65% of cases, and patients with RNA index $G_{0/1} < 16$ achieved CR in 32/65 = 49% of cases. This association between response or failure is shown in FIGURE 5. TABLE 6 shows the difference in response between patients with low or high RNA index. Patients with RNA index above the mean have a 64.9% CR rate, while patients with low RNA index have a 47% CR rate. The difference is significant with $p = 0.059$. There was no significant correlation between remission incidence and S-phase or absence or presence of DNA aneuploidy. Time to remission was not correlated with remission duration in this series, and was weakly correlated with survival ($p = 0.453$ and 0.732, respectively in the Yates corrected chi-square test).

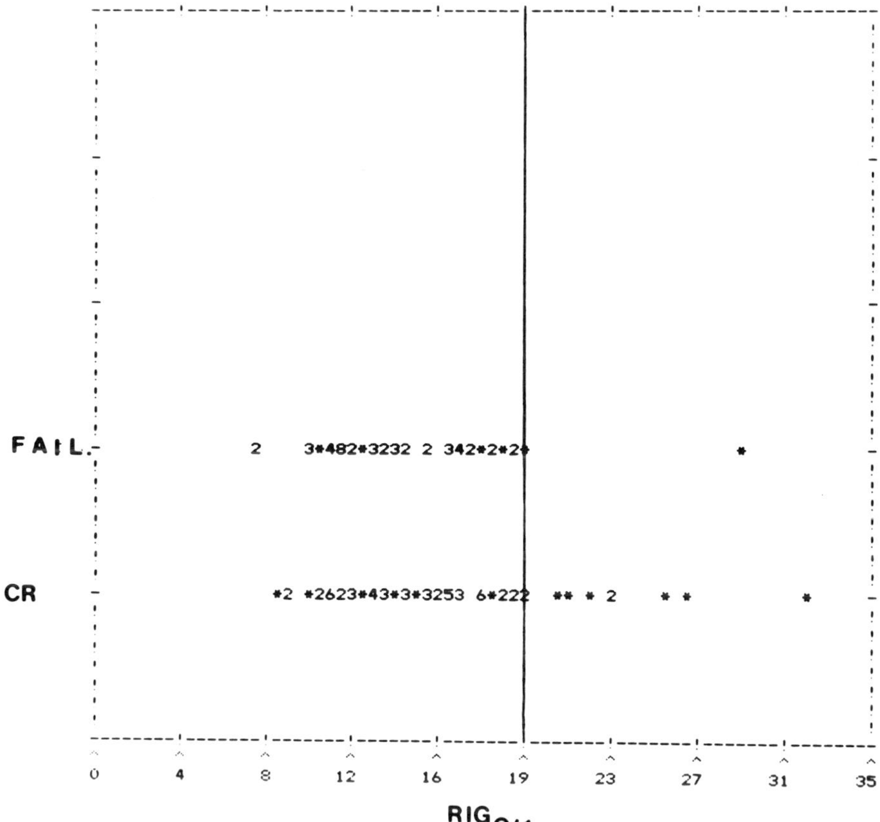

FIGURE 5. Correlation of RNA index $G_{0/1}$ and response to induction therapy in ANLL. Patients with RNA index $G_{0/1}$ over 19 achieved complete remission with one exception only. Numbers indicate patients with identical RNA-index.

There was also no discernible correlation of FAB morphology and achievement of complete remission.

Remission Duration in ALL

Remission in adult ALL was highly correlated with age and WBC at diagnosis. No difference was seen between DNA aneuploid and diploid ALL. S-phase was found to be of borderline prognostic significance: patients with low S-phase (less than 7%), had longer remission durations than patients with higher S-phase ($p = 0.06$). RNA index was found to be highly significant: patients with RNA index less than 12.7 had significantly longer remission durations than patients with higher RNA index ($p = 0.017$) (FIGURE 6). A detailed analysis of prognostic factors affecting remission duration in ALL is in preparation.

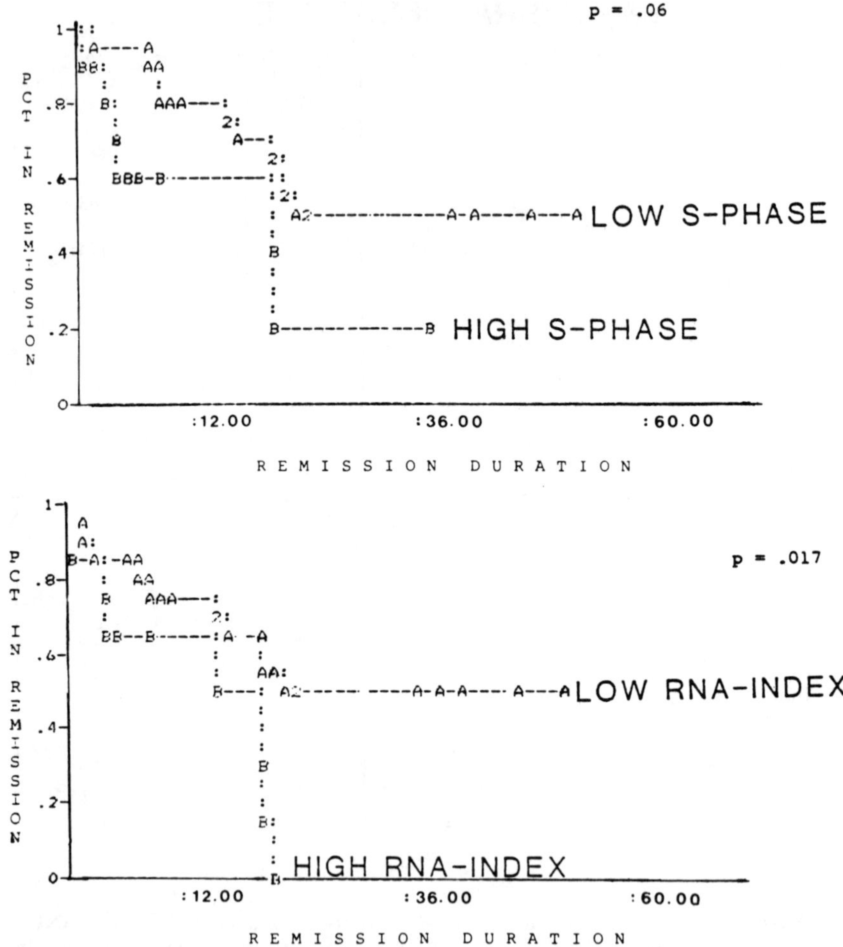

FIGURE 6. (Top) Kaplan-Meier plot of remission duration in ALL. Patients were separated into those with low and high S-phase (above or below 7%). Patients with low S-phase have borderline improved remission duration ($p = 0.06$). (Bottom) Kaplan-Meier plot of remission duration with adult ALL. Patients were separated into those with high RNA index (above 12.7) and low RNA index. Patients with low RNA index have significantly longer remission duration than those with high RNA index ($p = 0.017$).

Remission Duration in ANLL

Step-wise Cox regression analysis was performed for prognostic factors related to remission duration in ANLL. Age and RNA index were found to be significant (TABLE 7). The most significant p value was found for log RNA index G_1 ($p = 0.026$). DNA stemline, S-phase, and sex were not significant. The log-rank test was significant for age (above and below 50 years). FIGURE 7 shows Kaplan-Meier plots of patients with ANLL by remission duration. The four groups

TABLE 6. RNA Index and Remission Rate in ANLL

RNA-I	$G_{0/1}$	Low	High	
CR	N	27	37	64
	%	47.3	64.9	56.1
	%	42.1	57.8	
NR	N	30	20	50
	%	52.6	35.0	43.8
	%	60.0	40.0	
Total	N	57	57	114
	%	50	50	

$$p = 0.059$$

Correlation between achievement of complete remission (CR) or no response (NR) in ANLL. One hundred fourteen patients were investigated, 64 achieved CR, 50 failed. CR rate 56.1%, NR rate 43.8%. RNA index $G_{0/1}$ was subdivided into low and high and patients were distributed accordingly. Of patients achieving CR, 27/64 had low RNA index. Of patients failing, 30/50 had low RNA content. The difference is significant, $p = 0.059$.

TABLE 7. Remission Duration in ANLL

RNA Index G_1	$p = .048$
LG Index G_1	$p = .026$
LG Index $G_{0/1}$	$p = .044$
DNA Stemline	N.S.
S-phase	N.S.
Sex	N.S.
Age	$p = .037$
Log-Rank Test	$p = .001$ ($\gtrless$ 50 yr)
	$p = .36$ ($\gtrless$ 50 yr)

Step-wise Cox regression analysis of prognostic factors in ANLL. The parameters listed were related to remission duration. Log RNA index G_1 and age were most significant.

distinguish patients in remission with young age (below the median of 49.75 years) and low RNA index (RNA index $G_{0/1} < 13.90$); young patients with high RNA index; old patients with low RNA index; and old patients with high RNA index. The differences are significant ($p = 0.059$); young patients with high RNA index do significantly better than the other groups.

Survival in ANLL

The importance of the RNA index of G_1 cells was also apparent for survival of patients with ANLL. FIGURE 8 shows the correlation between high RNA index and survival. Patients with RNA index $G_1 < 16$ did not survive 12 months. All patients surviving longer than 12 months were in the group with higher RNA index. Only 1 patient with low RNA index survived 15 months. His diagnosis was AUL (arrow in FIGURE 8).

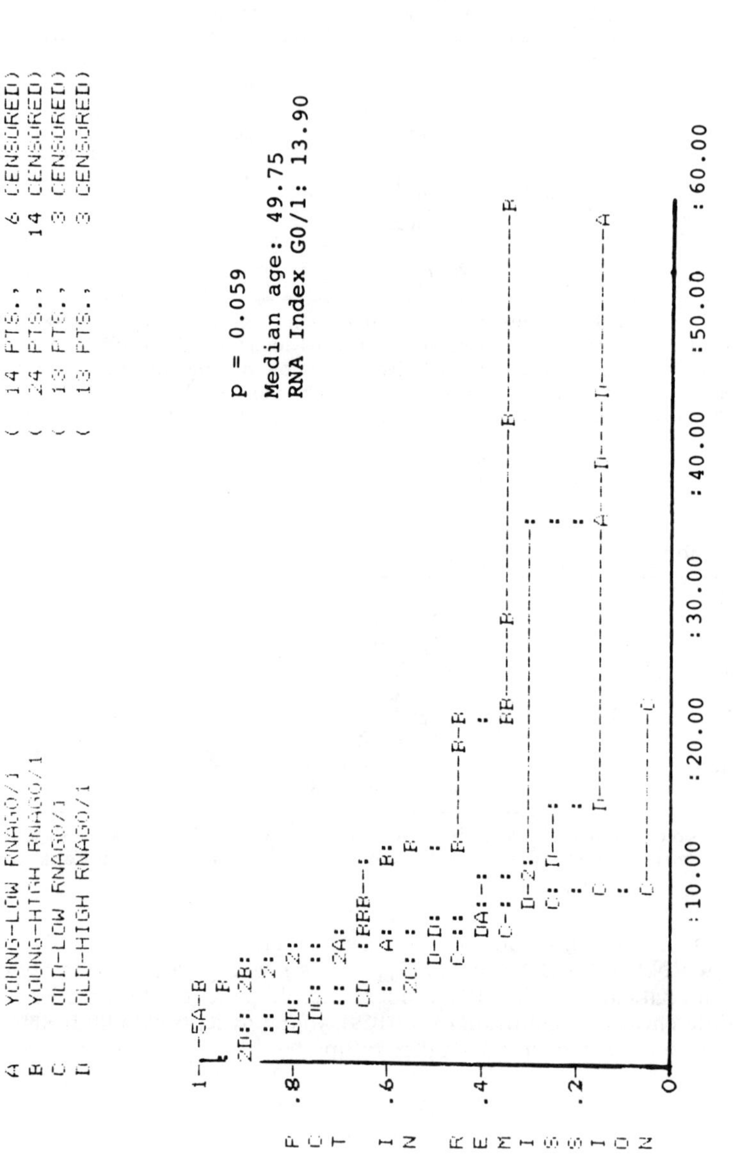

FIGURE 7. Kaplan-Meier plot of remission duration in ANLL. Four groups are discriminated, based on age and RNA index $G_{0/1}$. Median age: 49.75 years, RNA index $G_{0/1}$: 13.90. The different remission duration curves are significant at $p = 0.059$.

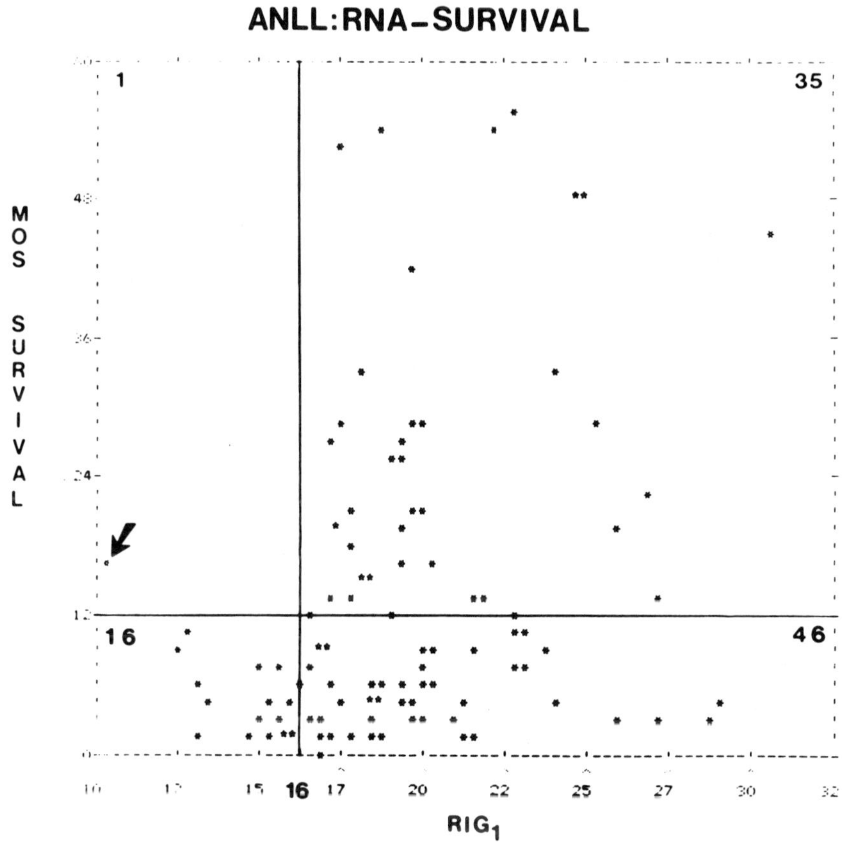

FIGURE 8. Correlation between RNA index G_1 and survival in patients with ANLL: patients with RNA index G_1 less than 16 did not live longer than 12 months. The one patient indicated by arrow was diagnosed as AUL.

Step-wise Cox regression analysis again showed RNA index to be significant ($p = 0.026$). Sex and age were not significant in this analysis, but age was highly significant in the log-rank test (TABLE 8).

DISCUSSION

Results presented here attempt to identify new prognostic parameters, mainly based on clinical and on cell biological characteristics determined by DNA/RNA flow cytometry. DNA measurements provide information on DNA ploidy and on cell cycle distribution, while RNA measurements allow us to further subdivide the cell cycle, in particular to separate quiescent from proliferating cells.

TABLE 8. Survival in ANLL

lg RNA index G_1	$p = 0.026$
lg RNA index $G_{0/1}$	$p = 0.044$
Sex	NS
Age	NS
Log-Rank Test	$p = 0.000$ ($\gtrless 50$ yr)
	$p = 0.04$ ($\gtrless 60$ yr)

Step-wise Cox regression analysis of prognostic factors in ANLL. Each parameter listed was investigated for its importance regarding survival. Log RNA index G_1 was most significant, age was not significant in Cox regression analysis regarding survival, but log-rank test was significant for both 50 and 60 years.

DNA aneuploidy was found in only 20% of cases. This is a significantly lower rate, as compared to pediatric ALL, which demonstrated aneuploidy in 40% of cases.[8] The rate of DNA aneuploidy in ANLL was also 20%. DNA aneuploidy is a very reliable marker of malignancy and its application in monitoring and diagnosing disease has been described elsewhere.[7,9]

Proliferation, as measured by the percentage of cells in S-phase, was identical in ALL and ANLL. As previously described, the RNA index was higher in ANLL, as compared to ALL.[5] However, there was a large variability in RNA content in ANLL. Some patients were found to have low RNA index, otherwise noted only in ALL. Two alternative explanations are possible: leukemic cells identified as non-lymphocytic could be of lymphoid origin, or low RNA content in ANLL could reflect the morphological criteria of micromyeloblasts.

Cell separation studies in suspected biphenotypic leukemias provided evidence for the presence of two distinct populations, i.e. lymphoid and myeloid.[10] In cases with uncontested diagnosis of ALL or ANLL, blast percentage was highly correlated with the percentage of cells in G_0 and G_1, respectively (FIGURE 4). This could serve as evidence that lymphoid and myeloid leukemic cell populations are indeed characterized by differences in RNA content.

Our previous studies on separated subpopulations of normal bone marrow have demonstrated that myeloid cells lose RNA during differentiation. When the RNA content of different subgroups of ANLL was correlated with degree of differentiation by FAB criteria, no significant differences were found between the different groups. Obviously, this represents a significant difference from normal myeloid cells. In terms of proliferation, only acute unclassified leukemias (MO) had a significantly increased S-phase, as compared to M1 to M6 ANLL.

Remission duration in ALL was largely determined by age and white blood cell count at diagnosis. These have been prognostic variables of well established importance. In addition, RNA index and S-phase appeared to be significant. The patients with low RNA index and low S-phase had significantly longer remission durations than patients with high proliferation and high RNA content. The significance of RNA index is higher than that of S-phase. High values of these two measures of proliferative activity appear to confer a higher potential for relapse. A subsequent multi-variate study in adult ALL confirmed the prognostic importance of RNA index,[2] and patients with high RNA index at diagnosis are now treated differently, after they achieve complete remission (L-20 protocol).

In acute non-lymphocytic leukemia, patients with high RNA index at diagnosis had a significantly higher remission incidence. In contrast to the ALL study, patients with ANLL and high RNA content had longer remission rates and longer survival. RNA index was found in Cox regression analysis to be a significant, independent, new variable. The biological basis for this observation remains to be fully understood, but low RNA index may reflect the presence of a second, non-myeloid leukemic population in myeloid leukemia, or a lower sensitivity to drugs. In sequential studies in individual patients, cells with high RNA are invariably the ones that are first eliminated by chemotherapy. In cell kinetic terms, cells with low RNA are believed to be quiescent and should therefore be less sensitive to chemotherapy that is mostly aimed at proliferating cells.

In conclusion, this study identifies RNA index as a new and independent prognostic variable in adult acute lymphoblastic and non-lymphoblastic leukemia. In our experience, in ANLL, young age, the presence of Auer rods, absence of TdT, and high RNA index confer improved remission duration and survival. In acute lymphoblastic leukemia, young age, low white count, and low RNA index are prognosticators of long survival. As a clinical consequence, patients who do not have the criteria predicting good survival should be treated with experimental protocols.

ACKNOWLEDGMENT

The authors would like to thank Gail Guillet for her excellent assistance in the preparation of this manuscript.

REFERENCES

1. CLARKSON, B., T. GEE, Z. A. ARLIN, R. MERTELSMANN, S. KEMPIN, R. DINSMORE, R. O'REILLY, M. ANDREEFF, E. BERMAN, C. HIGGINS, C. LITTLE, C. CIRRINCIONE & S. ELLIS, 1984. Current status of treatment of acute leukemia in adults: An overview. *In* Therapie der akuten Leukaemien. Th. Buechner, D. Urbanitz & J. Van de Loo, Eds.:1–31. Berlin.

2. CLARKSON, B., S. ELLIS, C. LITTLE, T. GEE, Z. ARLIN, R. MERTELSMANN, M. ANDREEFF, S. KEMPIN, B. KOZINER, S. CHAGANTI, S. JHANWAR, S. MCKENZIE, C. CIRRINCIONE & J. GAYNOR. 1985. Acute lymphoblastic leukemia in adults. Sem. Oncol. **12**:160–179.

3. MERTELSMANN, R., M. A. S. MOORE & B. CLARKSON. 1982. Leukemic cell phenotype and prognosis: An analysis of 519 adults with acute leukemia. Blood Cells **8**:561–583.

4. ANDREEFF, M., J. BECK, Z. D. DARZYNKIEWICZ, F. TRAGANOS, S. GUPTA, M. R. MELAMED & R. A. GOOD. 1978. RNA content of human lymphocyte subpopulations. Proc. Natl. Acad. Sci. USA **75**:1938–1942.

5. ANDREEFF, M., Z. DARZYNKIEWICZ, T. K. SHARPLESS, B. D. CLARKSON & M. R. MELAMED. 1980. Discrimination of human leukemia subtypes by flow cytometric analysis of cellular DNA and RNA. Blood **55**:282–293.

6. KAPUSCINSKI, J., Z. D. DARZYNKIEWICZ & M. R. MELAMED. 1982. Luminescence of the solid complexes of acridine orange with RNA. Cytometry **2**:202–211.

7. ANDREEFF, M., A. REDNER, S. THONGPRASERT, B. EAGLE, P. STEINHERZ, D. MILLER & M. A. MELAMED. 1985. Multiparameter flow cytometry for determination of ploidy, proliferation and differentiation in acute leukemia: Treatment effects and prognostic value. *In* Tumor Aneuploidy. Th. Buechner, Ed.:81–105. Springer-Verlag. Berlin.

8. SUAREZ, C., D. R. MILLER, P. G. STEINHERZ, M. R. MELAMED & M. ANDREEFF. 1985. DNA and RNA determinations in 111 cases of childhood acute lymphoblastic leukemia (ALL) by flow cytometry: Correlation of FAB classification with DNA stemline and proliferation. Br. J. Haematol. **60**:677–686.

9. REDNER, A., E. GRABOWSKI, A. AL-KATIB & M. ANDREEFF. 1986. Lymphoblastic leukemia with four DNA stemlines: Heterogeneity of RNA and differentiation antigen expression. Leukemia Res. (In press.)

10. ANDREEFF, M., J. D. BECK, N. KAPOOR, J. STEINMETZ, M. R. MELAMED, T. GEE, D. MILLER & B. CLARKSON. 1979. Clonal evolution in acute leukemia: Analysis of subpopulations by velocity sedimentation and multiple markers. Proc. Am. Assoc. Cancer Res. 20:112.

Index of Contributors